12 40202
B4

Engineering
and Society

Engineering and Society
Challenges of Professional Practice

Stephen F. Johnston
University of Technology, Sydney, Australia

J. Paul Gostelow
The University of Leicester, United Kingdom

and

W. Joseph King
University of the Pacific, Stockton, CA

Prentice Hall
Upper Saddle River NJ 07458
http:/www.prenhall.com

Library of Congress Cataloging-in-Publication Data

Johnston, Stephen, 1940-
  Engineering and Society/ Stephen F. Johnston, J. Paul Gostelow, and W. Joseph King.
    p. cm.
  Includes bibliographical references and index.
  ISBN 0-201-36141-8
    1. Engineering--Social aspects. I. Gostelow, J. P. II. King, W. Joseph III. Title.

TA157 .J58 1999
303.48'3--dc21   99-050120

Editor-in-Chief: *Marcia Horton*
Acquisitions editor: *Eric Svendsen*
Editorial/production supervision: *Scott Disanno*
Executive managing editor: *Vince O'Brien*
Managing editor: *David A. George*
Cover design director: *Jayne Conte*
Cover design: *Bruce Kenselaar*
Manufacturing manager: *Trudy Pisciotti*
Manufacturing buyer: *Pat Brown*
Assistant vice-president of production and manufacturing: *David W. Riccardi*

© 2000 by Prentice Hall, Inc.
Pearson Higher Education
Upper Saddle River, New Jersey 07458

Printed in the United States of America

10 9 8 7 6 5 4 3 2 1

ISBN 0-201-36141-8

Prentice-Hall International (UK) Limited, *London*
Prentice-Hall of Australia Pty. Limited, *Sydney*
Prentice-Hall Canada Inc., *Toronto*
Prentice-Hall Hispanoamericana, S.A., *Mexico*
Prentice-Hall of India Private Limited, *New Delhi*
Prentice-Hall of Japan, Inc., *Tokyo*
Simon & Schuster Asia Pte. Ltd., *Singapore*
Editora Prentice-Hall do Brasil, Ltda., *Rio de Janeiro*

Acknowledgments

We would like to thank everyone who has contributed to the making of this book.

Evan Jones, a Senior Lecturer in Economics at the University of Sydney, Australia, who since 1975 has taught in their unique 'Political Economy' program; Evan wrote Chapter 5, several case studies and made valuable comments and suggestions in other areas.

Faculty and staff in the Science, Technology and Society Program at Stanford University, California, hosted and assisted Stephen Johnston as a Visiting Scholar for the Fall Quarter of 1997-98, and provided stimulating and valued criticism and encouragement. Robert McGinn, the Director of the Program, and Naushad Forbes, a Consulting Professor, were particularly helpful and supportive. So were their Teaching Assistants, Mehul Patel and Katherine Kao Cushing, and their students in STS 101 'Science, Technology and Contemporary Society' and IE/STS 279 'Technology, Policy and Management in Newly Industrializing Countries'. As noted in the Preface, the present text includes a number of case studies which were either substantially written or based on ideas contributed by Stanford students, usually (but not always) as part of the requirements for these courses. They were Richard Bjorn, Amy Chang, Stephanie Di Chiara, Glenn Davis, Monika Ellis, Randall Fish, Jennifer Gadda, Steve Gu, Ruth Kim, Daniel Kramer, Beatrice Lee, Angela Lim, Wendy Marinaccio, Seema Patel, Shannon Quek, Phillip Riekert, Daniel Skarbek, Sean Smith, Katherine Steel, Shelley Toich, Lan Tran, John Wayland and Jaime Wong.

Colleagues in Australia at the University of Technology, Sydney (UTS), who have taught in the subject to which the original Australian book was particularly directed, and who contributed to earlier drafts, include Robyn Fourikis, Rod Belcher, Helen McGregor, Garry Marks, Lance Reece and Frank Swinkels. Successive cohorts of our UTS students who worked through these issues helped us to deepen our own understanding, and their enthusiasm encouraged us to persevere.

Robert Hudspith, Director of the Engineering and Society Program at McMaster University in Canada, provided real encouragement for the project, including a case study. Norman Ball from Waterloo University and Peter Lissaman of the University of Southern California each contributed a case study. Adam Lucas offered useful suggestions on Chapter 1. Alan Wightley, General Manager, Group Development, and Doug Jones, Manager Intellectual Property for Warman Group Technology, Warman International, provided material on Warman's experience with protection of intellectual property that was used in Chapter 6. Input by Dr Chris Coats at the University of Leicester significantly improved Chapter 9.

Margaret Johnston made a major contribution, including reviewing the whole text and writing several case studies. The improvements she suggested have increased both the clarity and coherence of the whole book, and will be appreciated by our readers as much as by us. Indeed, the production of this book has touched and drawn upon all those close to the authors. Family and friends have borne with us loyally for what at times seemed to be an eternity. The book now bears a dedication to Paul's late wife Val, who maintained a keen interest in the project. Joe's own acknowledgement is to his wife Patty and their children Jim, Dare and Allison, who have suffered through nine books now.

We also gratefully acknowledge the efforts of numerous others who helped turn our early drafts into their present more comprehensive, accurate and readable form. They include the editorial staff at HarperEducational (which produced the first Australian edition), and the reviewers whose very constructive comments helped to clarify and fill out the work. Finally, this edition has been made possible by the strong support of the editorial and production staff of Prentice Hall in the USA, the United Kingdom, and Australia. Such cooperation is increasingly important in our global marketplace.

Preface

The authors began with the question: What preparation for professional practice will engineering graduates going into the next millennium need?

Our first response was that the professional formation of engineers must continue to be based on a sound technical education. Nothing in this book should be seen as a detraction from this fundamental requirement or from our recognition of its importance. Of course, what a "sound technical education" means will change over time. Even in the short term we can look to increasing emphasis on systems approaches and more extensive use of computers in both synthesis and analysis.

The social dimension of engineering activity is also of central importance. The current challenges of professional practice require engineers to have breadth as well as depth in their education. Our book is designed to help meet this objective by introducing readers to the background and the present context for engineering. It draws significantly on a successful Australian book *Engineering and Society: An Australian Perspective*, but has been extensively reworked, and contains new chapters. Most of the case studies were specially written for this new work.

Our primary aim is to provide a textbook for college students in engineering and related technologies. It may be suitable for some freshman courses but is likely to be more relevant for senior students. However, no specialist knowledge is assumed; the book should interest both practicing engineers (whose original professional education is unlikely to have included much of this material) and the general reader with an interest in this area. The insights into how engineering has developed and the prospects for engineering in the future should be useful to a wide audience. We have deliberately drawn our illustrations from all the main branches of engineering.

Professional engineers spend much of their working lives managing the development and implementation of technology, so they need to understand how technology is developed and transferred. They must also be sensitive to its impacts on the society and environment in which they live and work. Engineers should be prepared to promote and defend their work, both within the profession and to the community at large.

There has long been concern by engineering employers that new graduates in engineering lacked skills and competence in communication and industrial relations, as well as in people, costs, and resources management. They are seen as insufficiently aware of the broad social context of engineering. Some of these issues probably reflect the fact that we lack a well developed philosophy of engineering. Without a philosophy of engineering there is not even a consensus about what problems should be addressed in the preparation of engineering students, let alone how the problems should be addressed. We have tried to shed light on these issues in this book.

During the 1990s a series of major reviews of engineering education around the world, including two each in Canada and the United States, have drawn attention to the changing content and context of engineering work. The American Society for Engineering Education report (ASEE 1994) asserted that:

> engineering education should be... *relevant* to the lives and careers of students, preparing them for a broad range of careers, as well as for lifelong learning involving both formal programs and hands-on experience; *attractive* so that the excitement and intellectual content of engineering will attract highly talented students with wider backgrounds and career interests...; *connected* to the needs and issues of the broader community through integrated activities with other parts of the educational system, industry, and government.

In an address to the U.S. National Academy of Engineering, Charles M. Vest noted the importance of preparing students for the international environment in which engineering is already practiced. He went on to spell out the need to:

> ... de-emphasize narrow disciplinary approaches ... pay more attention to the context in which engineering is practiced [and] ... educate students to work better in groups ... because of the complexity of the tasks that engineers and their colleagues deal with today. (Vest 1995)

The Accreditation Board for Engineering and Technology (ABET) is recognized in the United States as the sole agency responsible for accreditation of educational programs leading to degrees in engineering. One of our aims here is to respond to and anticipate ABET requirements on the teaching of social context and professional ethics. In its Engineering Criteria 2000 requirements, ABET places increasing emphasis on social and environmental areas (ABET 1998). *Criterion 3, Program Outcomes and Assessment* requires engineering programs to demonstrate that their graduates have:

(a) an ability to apply knowledge of mathematics, science, and engineering

(b) an ability to design and conduct experiments, as well as to analyze and interpret data

(c) an ability to design a system, component, or process to meet desired needs

(d) an ability to function on multidisciplinary teams

(e) an ability to identify, formulate, and solve engineering problems

(f) an understanding of professional and ethical responsibility

(g) an ability to communicate effectively

(h) the broad education necessary to understand the impact of engineering solutions in a global and societal context

(i) a recognition of the need for, and an ability to engage in life-long learning

(j) a knowledge of contemporary issues

(k) an ability to use the techniques, skills, and modern engineering tools necessary for engineering practice. (©1994 by the American Society for Engineering Education)

This book, *Engineering and Society*, is primarily concerned with components (f), (h), (i), and (j), thus augmenting the typical engineering curriculum, which would adequately cover the other components. The ABET criteria for accreditation go further. *Criterion 4, Professional Component* includes the requirement that:

> … Students must be prepared for engineering practice through the curriculum culminating in a major design experience based on the knowledge and skills acquired in earlier coursework and incorporating engineering standards and realistic constraints that include most of the following considerations: economic; environmental; sustainability; manufacturability; ethical; health and safety; social; and political. (©1994 by the American Society for Engineering Education)

It is just these considerations that underpin this text that provides a wealth of material that can help engineering schools meet ABET accreditation requirements.

STRUCTURE AND CONTENT OF *ENGINEERING AND SOCIETY*

Engineering and Society begins by recognizing that our present technologies are the outcome of a long and exciting history of creative effort. Chapter 1 takes us through some highlights of this process. Chapter 2 carries the story into the present and on to the new millennium. These two chapters provide a basic historical framework to which later chapters add.

The social and political contexts in which engineers practice are outlined in Chapters 3 and 4. We see the practice of engineering as central to the generation of wealth in modern societies, raising questions that are explored in economic terms in Chapter 5. Chapter 6 describes how engineers create new products, processes, and systems.

Engineers need to work in diverse teams with other professionals, both to make the best use of their technical capabilities and to enhance the efficiency and effectiveness of productive enterprises. Engineering leadership and management are discussed in Chapter 7.

Economic development and the choice and transfer of technology are explored in Chapter 8, and its case studies demonstrate the increasingly global character of engineering practice.

Challenges in reconciling development and long-term ecological sustainability are discussed in Chapter 9, with particular attention to the most fundamental resource, energy.

The distinctive character of engineering is discussed in Chapter 10, which suggests the possible nature of a philosophy of engineering. Chapter 11 is a capstone chapter that considers ethics and the profession of engineering, including some of its prospects and challenges.

∎ Multidisciplinary Approach

The theory of multidisciplinary analysis developed by Kline (1995) and others indicates that analytical approaches adopted within any single traditional discipline will not provide a sufficiently comprehensive representation of the systems in use in modern industrial societies.

Accordingly, we have used a multidisciplinary approach in this book. It draws on those disciplines we judge most likely to be interesting and relevant to engineers and offers insights into both professional and personal development issues.

Continuity across the wide canvas we attempt to cover is provided by three central themes:

1. the nature of technology and its relationship to engineering;
2. the nature of development and its relationship to engineering;
3. the nature of professional engineering practice, and in particular, the roles it plays in the development of technology and the sustainable creation of wealth.

We explore what engineers do, the education, knowledge, and skills they need, and their roles and responsibilities. We discuss how these aspects of engineering are changing, and consider present and possible future norms for how engineers ought to behave, including codes of ethics.

▌Key Concepts and Issues

We have tried to deal openly and constructively with a range of issues, some of which may be either unfamiliar to many readers or somewhat controversial. We highlight central questions in boxes showing "Key Concepts and Issues." These include basic definitions and descriptions, state major assumptions underlying our approach to a topic, and (if appropriate) outline the variety of views and the main areas of disagreement.

▌Case Studies

The case studies written for this book show some of the ways engineers work. As indicated more formally in the Acknowledgments, we are indebted to faculty and students in Stanford University's long-standing Science, Technology, and Society (STS) Program for their contribution towards these case studies, some of which include reflections on their own experience of engineering. Readers might develop their own case studies along similar lines.

▌Discussion Questions

The discussion questions at the end of each chapter should help to test the reader's experience and ideas against our ideas and interpretations, as well as those of other writers. They are also intended to stimulate further exploration of the issues. At the end of each chapter there is also a list of suggested readings and other useful material such as films and videotapes.

▌Personal Attitudes and Values

The practice of engineering is a social as well as a technical activity. It draws on the creativity and the values of the practitioner. Most engineering is now a large-scale, often corporate endeavor, but we all constantly make value judgments in our work. To understand the choice, development, and application of a particular technology, it is necessary to take account of the underlying values, assumptions, and attitudes of those who propose, develop, and apply it.

Our values, as authors, are inevitably reflected in this book. So are the values of those who contributed case studies. We make no apology for this; indeed we hope that this variety of views and attitudes will challenge and stimulate our readers to clarify their own thinking on the issues we raise. We do not ask you to agree with our views, but we do ask you to respect the values that underpin them and to reflect on how and why they may differ from your own. The authors share a belief that the only firm basis for moral and ethical action by the engineering profession is the informed moral autonomy of its members. This book is intended to challenge present and future engineers to think issues through and to come to their own conclusions.

Before you read on, you might care to think about values you hold strongly. How can you draw on these values to shape and influence your professional career, and to direct it into directions that you will find fulfilling?

References

ABET (Accreditation Board for Engineering and Technology) 1998, *Engineering Criteria 2000*, at: <http://www.abet.org/>.

ASEE (American Society for Engineering Education) 1994, *Engineering Education for a Changing World*. A joint project by the Engineering Deans Council and Corporate Roundtable of the ASEE. ASEE, Washington, DC.

Kline, S. J. 1995, *Conceptual Foundations for Multidisciplinary Thinking*, Stanford University Press, Stanford, CA.

Vest, C. M. 1995, "US Engineering Education in Transition," Address to the annual meeting of the National Academy of Engineering, published in the Winter 1995 issue of the NAE journal *The Bridge* (Vol. 25, No. 4).

Contents

Preface vii

CHAPTER **7**

Management in Engineering 333

CHAPTER **8**

Development and Technology Transfer 397

CHAPTER **11**
Ethics and Professionalism

551

Index

603

Engineering and History

To know nothing of what happened before you were born is to remain forever a child.

— *Cicero (106–43 B.C.) Roman orator, philosopher*

hroughout this book we are encouraging readers to question why, as engineers, we think and behave the way we do. The very first question we have to anticipate is: Does history really help our understanding? A prominent , American engineer, Henry Ford, once said "History is more or less bunk," and many engineers might still agree with that sentiment. Engineers are rarely encouraged to reflect on the history of their own profession or how engineering relates to broader human history. So why set ourselves and our readers to this task?

History is both a series of events and the discipline that studies these events. The work of professional historians involves "reading" the past, seeking to understand *why* things happened, rather than producing straight chronologies of events. Good historians also recognize an obligation to retell the story in ways that will appeal to their audience. Making history fascinating and useful depends on reaching out and engaging the interest of the audience.

We all want to present the past in a way that makes more sense of the present. We all see the past through the eyes of our own times. In this and the next chapter we will look at history, and the history of engineering in particular, through the eyes of engineers living in a Western society at the end of the twentieth century. This affects both our choice of relevant historical "facts" and the way we interpret them. By selecting an important historical theme, such as human technological development, to provide the connecting thread, we can give some structure and continuity to what appears to be at first glance, a daunting sweep of time and a multiplicity of apparently unrelated events.

The different disciplines presented in subsequent chapters—sociology, economics, and so on—use a variety of conceptual frameworks to approach such questions as: What is the connection between engineering and other basic aspects of human society? These approaches often rely on our having some grasp of the time dimension of modern engineering. Thus our history of the development of engineering serves several purposes: it celebrates past achievements, improves our practice through insight into the society in which we work, and focuses our vision of how to make the world a better place for all.

The Technological Formation of Human Society

If engineers can be persuaded that history is relevant to their profession, it will be because it sheds light on such basic questions as:

- When did engineering really begin?
- What is the connection between engineering, technology, and science?
- How do these activities affect other members of society?
- How do these activities affect you as an engineer?

Historically, we can demonstrate that *engineering, science,* and *technology* are three distinct but related concepts. All three can be defined as areas of human endeavor and forms of human cultural activity. A great deal of human effort has been invested in them; they are enterprises that engage whole societies, although the degree of involvement by individuals varies greatly. It is also important that we do not confuse or blur them. They have different purposes, methodology, and content. They are not simply technical knowledge or activity that exists independently of human social, cultural, economic, and environmental systems.

Technology, science, and engineering emerged as distinct areas of human endeavor at different periods in human history. The relationships between the activities associated with them have changed and will continue to evolve over time (McGinn 1991, 1997). Chronologically, we will begin with technology.

Technology

Technology has both technical and social aspects and is an important part of human culture. The main features of modern technology—systems, modeling, and design—can be discerned even in very early technologies.

Technology has five major characteristics:

1. It is a form of human cultural activity.
2. It is essentially for practical ends and purposes.
3. It involves exercising human freedom and responsibility, particularly in choosing problems and in design approaches; that is, it involves making choices in response to normative values, such as those derived from a belief in God.
4. It ultimately involves forming and transforming the material world and not primarily the sphere of ideas, thoughts, or symbols.
5. It is typically done with the aid of tools and procedures.

Technology is about solving problems and meeting needs in the real world. We favor a very broad definition that includes products, processes and systems, and takes into account the environment required for their development and use.

▪ Early Technologies

The origins of technology date from the Stone Age. Purposeful, systematic technological action was a distinctive characteristic of our earliest human ancestors. They were hunter-gatherers in the savannas of East Africa some two million years ago, living in groups and moving around a territory to take advantage of seasonal variations in plant and animal life. They were also tool makers. They developed simple, portable tools made from wood and stone; the commonest tools that have survived are stones that had been fractured to produce a sharp edge for cutting. Other tools were grinding stones, the wedge, the carrying yoke, the ax, and the spear. Important early technological developments included the controlled use of fire, the fashioning of clothing and utensils for domestic use, and early art forms.

The development of agriculture was a critical turning point in human history. It was accompanied by the establishment of an agrarian society. Fertile river valleys provided favorable conditions for crop growth: a warm climate, an annual cycle of flooding and alluvial soil deposition, grasses that yielded enough seed to be worth cultivating, and animals that could be domesticated for food and/or as beasts of burden. One such area was Mesopotamia, the Fertile Crescent of the Tigris and Euphrates valleys. The valley of the Nile was another; there were others in India and China.

Key Concepts and Issues

Sociotechnical Systems

A critical issue in considering early technology is the extent to which innovations were not simply random, but were captured and perpetuated in human society. Stephen Kline (1995) suggests that successful innovations were incorporated into what he called "sociotechnical systems."

Sociotechnical systems are complex systems of linked *social* and *technical* behavior. Humans continue to devise and operate such systems, primarily to control our environment and to carry out tasks that we could not do otherwise.

In small-scale societies, the key sociotechnical systems, which include the social context (households, social groupings, and culture) as well as the activities directly involved, are:

◆ Sociotechnical systems of *production*—ways of making tools and other products.

◆ Sociotechnical systems of *use*—ways in which tools are used and products consumed.

Kline discussed the way these sociotechnical systems together supported a pattern of practice that included design, manufacture, use, maintenance, reflection and modification.

He called this the "basic pattern" and saw it both as a characteristic of early human technological activity and as a factor that eventually greatly increased its effectiveness.

These systems included more than just the tools and techniques that resulted from exercising emerging human design capabilities. They involved social arrangements, such as the education of the young to preserve innovations that the community accepted as desirable. For techniques to be taught and learned effectively they had to be standardized by tradition within any given society.

In large-scale modern societies, we would add:

◆ Sociotechnical systems of *distribution*.

◆ Sociotechnical systems of *research and development*.

(After Kline 1995)

Settlement facilitated the widespread adoption and progressive improvement of tools and implements, such as a primitive sledge for transporting crops. The lever was in use in Egypt around 5000 B.C. in the balance beam for weighing, and subsequently in the *shadoof* for raising water. The wheel is arguably the most important mechanical invention of all time. It first appeared around 3500 B.C., in Mesopotamia. Although logs had been used as rollers (and continued to be used for very heavy loads), the critical part of the wheel is the axle and bearing. The first application may have been the potter's wheel, but the wheel was quickly applied to transportation, in four-wheeled chariots. Associated with these developments was increasingly sophisticated smelting and working of metals (see box The Beginnings of Metallurgy). Most communities were dispersed and rural but some urban communities emerged. Large-scale control of water resources was fundamental to river-based societies. The administration of large settlements required record keeping and eventually, around 4000 B.C., writing. In Europe and the Middle East we can trace a succession of societies with progressively more sophisticated technologies. However, change was rarely a smooth or steady process; some technologies were lost when the societies using them collapsed as the result of internal and/or external pressures.

Building technology was initially constrained by locally available materials and minimal transport facilities, but it advanced quickly from around 3000 B.C. The Mesopotamians built in brick, but in Egypt stone was the favored material, used for constructing large and beautiful pyramids and temples. The pyramids, tombs for the Pharaohs, were built between 2650 and 2190 B.C., with stone brought large distances along the Nile. The Great Pyramid of Gizeh was commissioned by the Pharaoh Cheops around 2600 B.C. Over 230 meters square at the base and 147 meters high, it was originally faced with glistening limestone and remains one of the world's great building feats. Its construction involved moving thousands of massive stone blocks, averaging two tons but as heavy as thirty tons, many miles along the Nile and then shaping and assembling them with an accuracy that remains impressive today. The Greek culture that followed built, almost literally, on Egyptian foundations.

▮ Classical Antiquity

The Greeks laid the first philosophical and scientific base for knowledge, starting in the sixth century B.C. with Thales of Miletus, who attempted to explain the world on the basis of rationality rather than myth. Socrates (ca. 470–399 B.C.) and his student Plato (ca. 428–348 B.C.) also made

The Beginnings of Metallurgy
in the Ancient Middle East

The smelting of metals was a remarkable technological breakthrough, probably associated with the firing of clays in pottery kilns.

Unsmelted ("Native") Copper

9500 B.C.: The earliest known metal artifacts were made in Iraq.
5000 B.C.: A copper mace head was cast in a mold in Turkey.

Copper and gold were smelted at an early date, but their softness limited their effectiveness for tools, and their use tended to be ornamental.

Bronze

3500 B.C.: A technological revolution occurred with the discovery of bronze in the Middle East. Bronze was discovered independently in China, and very early bronze artifacts have been found in China and Thailand. A hard, tough alloy of copper, typically with 10 to18 percent of tin, bronze was probably discovered after careless reduction of ores. It proved so important that the subsequent period is known as the Bronze Age. Weapons and vessels were cast (often by the lost-wax process, still used today to make such complex shapes as microwave guides and aircraft engine blading). The shortage of tin for making bronze encouraged the development of exploration and trade, and led to a search for substitutes.

Iron

1500 B.C.: Usable quantities of iron were first smelted somewhere south of the Caucasus. Iron was not technically superior to bronze, but its ores were more widely distributed. The Iron Age followed the development of effective ways of working the metal.

Iron melts at a much higher temperature than bronze and the smelting and working of iron presented major technical difficulties. Iron technology developed relatively slowly.

The quality of iron products gradually improved with the development of the blacksmith's arts of quenching and tempering, but temperature restrictions limited the quality of iron work.

The Greeks used charcoal fuel, with a foot-operated bellows for draft, to produce small, spongy balls of iron that they hammered into wrought iron.

The casting of iron was not achieved in Ancient Greece or Rome, but case-hardening and tempering produced steel suitable for knife and sword blades.

major contributions to philosophy. Around 385 B.C. Plato founded the great Academy of Athens, effectively the first university, to which Aristotle (384–322 B.C.), also belonged. However, Greek technological achievements were less spectacular. Mechanical arts were considered to be less important and prestigious than liberal arts.

Although scholars such as Archimedes created some of the scientific basis for the later engineering profession, in general scholars disdained practical craft skills. Greek mathematics appears to have developed out of legal and political discourse. Great thinkers such as Euclid, Pythagoras, and Plato struggled to find the philosophical and theoretical basis for natural phenomena and to find humanity's place in the universe. Increasingly, however, the debates were closed. From Euclid (fourth to third century B.C.) onwards, the Greek attitude toward mathematics was essentially that: "mathematical truths exist ... eternal, unchanging and divine." (Hoyrup 1985) Plutarch suggested that Archimedes (287–212 B.C.) "looked upon the work of an engineer and everything that ministers to the needs of life as ignoble and vulgar." Archimedes used mechanical models to arrive at mathematical results, but discarded them in his proofs. In the Greek colony of Alexandria, later famous for its library, some ingenious mechanical devices were invented; but these were mechanical toys, never put to any practical purpose.

With the rise of the Roman Empire, engineering practice became more clearly recognizable. Rome (by tradition founded in 743 B.C.) became increasingly important from the third century B.C. onwards. The Romans inherited Greek philosophy and mathematics but did not encourage abstract science. They did, however, produce engineers who were thoroughly competent in contracts, specifications, and costing. Road building, with proper attention to alignment, foundations, drainage, and wearing surface, was a key military technology. It made possible rapid communication and troop movement throughout the Empire, greatly extending the area that could be centrally controlled. The arming of the Roman Legions also took full advantage of the technology of the time. Roman cities, and buildings like the Colosseum, remain impressive today. The Romans excelled in hydraulics, bringing water to the towns in large stone aqueducts and reticulating it in lead pipes (plumbing is from the Latin word for lead, *plumbum*). (Like engineers everywhere, the Romans had their occasional failures. The dry ruins of a proposed Roman aqueduct in Leicester, England, show where an intended source of water for the town turned out to be lower than the proposed destination.)

Rome, like classical Greece, became technologically stagnant. Productivity in both societies was limited by the widespread use of slaves, which

Some Persistent Social Attitudes

To Work

The role of work in shaping human society is not fully understood, but some of its impacts can be traced from very early times. Improved agricultural productivity freed some people from farming and allowed them to specialize in crafts, administration, or religion (the latter often associated with the beginnings of astronomy).

There was increasing separation and specialization in the kinds of work people did, particularly division into "mental" and "physical" work, and "men's" and "women's" work. We should not underestimate the long-standing, entrenched nature of such ideas when we encounter them in the home or workplace today.

To Social Stratification

The early appearance of socially constructed ways of differentiating ourselves may be deduced from evidence found in human prehistory. As societies grew and became more stable, different types of work (deriving at least in part from different technological practices) brought differing rewards. Increasing social differentiation was linked to the accumulation of goods. As discussed in Chapter 3, social stratification became a permanent feature of large-scale societies, up to and including the present.

To Technology

The negative attitudes of Greek élites to technology reflected their experience of manual work being largely done by others, including slaves or women. Partly because of the continuing high regard for philosophers such as Aristotle and Plato, these disparaging views have been extraordinarily influential in Western nations, inhibiting technological development and undervaluing the importance of engineering.

To the Role of Women

Greek writings about women and their "proper place" in society have influenced attitudes in all subsequent Western societies. They have been used in comparatively modern periods to justify male domination of the professions, including engineering.

Some Engineering "Firsts'

Mechanika: Represented as a minor work of Aristotle, but more probably written in the second century B.C., it is arguably the oldest surviving engineering text.

Aeolipile: Hero of Alexandria (first century A.D.) produced a primitive reaction turbine and a windmill driving a trip hammer.

Ingeniator: The term *engineer* as a job classification dates from the Roman Empire. It derives from the Latin *ingeniator*, meaning an ingenious person. Roman engineers marched with Roman armies. Their job was to build roads and bridges and to solve technical problems quickly, with a minimum of materials and equipment.

devalued all labor and precluded sustained efforts to develop labor-saving technology. Roman engineering made advances only in the military and transport fields, closely connected with imperial endeavors. Indeed, from its beginning, major roles for the engineering profession seem to have been designing ways to fortify cities, and ways to invade them. In spite of their engineer's efforts, the Roman Empire was under pressure beginning in the second century A.D. and crumbled during the fourth and fifth centuries A.D.

∎ Other World Civilizations

Roughly contemporary with the earliest known stone building at Sakkara in Egypt and the move from digging stick to simple plow in Europe, on the Indian subcontinent the Dravidians had established an impressive civilization in the Indus Valley, with well laid out towns of mud brick. Evidence has been found of the use of bronze tools, bullock carts, and the first irrigated cotton farming anywhere in the world. Yet, by 1750 B.C. the towns had been abandoned in the face of Aryan invasions from Persia. Successive civilizations rose and fell. The rest of the world is deeply indebted to India for the later development of Buddhism and Sanskrit learning. One of India's greatest legacies was its numerical system (see box Early Technologies in Other Parts of the World).

Europeans saw India as significant not just in its own right, but also for its position on the route to the Far East. This was because of the more advanced cultural and technological position that China enjoyed, with the flowering of Chinese civilization from about 500 B.C. onwards. Confucius (ca. 551–478 B.C.) laid the moral foundations for civil institutions that were to serve China for another two millennia. By 700 A.D. China had the largest city in the world, Changan (modern Xian), with a population of over one million people. Buddhist monks fostered the spread of knowledge throughout the whole region. Many discoveries and inventions from China filtered through to Western Europe, including gunpowder and pyrotechnics.

The spread of Chinese civilization to Japan from about the sixth century A.D. ushered in several centuries of imitation of China. The eventual development of a distinctively Japanese script, forms of Buddhism, and political governance, resulted in a Japanese culture that was increasingly closed to outside influence. Prosperity was based on agriculture and trade; impressive architecture, the fine art of the swordsmith, and exquisite ceramics were distinguishing features of Japanese technological development prior to contact with the Portuguese. Through subsequent isolationist phases, the technological focus continued to be on religious, military, and artistic applications.

Islam had a major cultural impact on the technological formation of Europe and the modern world. Its influence was assimilated through several centuries of cross-fertilization of ideas during the *Reconquista* (the reconquest of Spain from the Muslims) and the Crusades. In technical areas, Muslim scholars interpreted Greek texts, explaining the practical relevance of the very abstractly expressed Greek works (for example, in astronomy). The great Arab contributions to mathematics, namely algebra and trigonometry, also reached the West through them, as did Chinese inventions and the Hindu (Arabic) numerals. Muslim influence on the West was one of the forces that brought about that remarkable flowering of European thought and creativity, the Renaissance. When Constantinople eventually fell to the Muslims in 1453 A.D., effectively closing direct overland routes, development of navigation was spurred by the demand for a sea route to the Far East.

The continent of Africa provides the longest continuous record of human habitation, and there the impact of changes to our environment has been most dramatic. Artifacts, including stones used for grinding grain, have been found deep in what is now the Sahara Desert, demonstrating early technological development there. These artifacts also show the extent of the quite rapid climatic change in northern Africa beginning around 2500 B.C. The land became drier and more inhospitable; savanna grassland, on which sheep, goats, and cattle were grazed, became desert.

The first known Black African civilization was Kush in Nubia, around 1000 B.C. This civilization was the source of much of the gold of the ancient world and in the eighth century B.C. was strong enough to establish Kushite pharaohs in Egypt. Two centuries later, possibly after exhausting local wood for smelting iron ore, the capital was moved three hundred miles further up the Nile, from Napata to Meroe, which had rich deposits of iron ore. Even today, huge slag heaps are evidence of the scale of mining and smelting operations. Meroe became a cultural center that flourished for nearly 500 years. By 500 B.C., population pressure along the Niger and Nile led to Africans who had learned how to mine and work metals, especially iron, spreading south and east, taking this knowledge with them.

Early Technologies in Other Parts of the World

India

◆ Metallurgy—in 300 B.C. the best steel was produced in India, using a crucible process.

◆ Building technology—including the seventeenth century Taj Mahal, built for Shah Jehan.

◆ Indian number system—first used in Hindu temples 1500 years ago. The number system incorporated both the actual symbols, 1 through 9, in use today (commonly known as Arabic numerals) and the concept of the zero.

China

◆ Power generation—water-driven rice mills and water-powered bellows for furnaces were being used as early as 100 B.C.

◆ Textile technology—in the first century B.C. the Chinese had silk-weaving machinery and a rotary winnowing machine.

◆ Printing technology—invention of paper, block printing; Buddhist monks printed the first books.

◆ Transport and military technology—the compass, gunpowder and pyrotechnics, the horse collar, the stirrup, and the stern-post rudder were all Chinese inventions. Construction of the strategic Great Wall of China was one of the world's great engineering achievements.

◆ Ceramics—as early as 5000 B.C., temperatures around 1,400°C had been sustained in the kilns that were to give China mastery in pottery.

◆ Metallurgy—evolved from pottery. New types of bellows, of Chinese invention, were finally used some two thousand years later in the West, in blast furnaces that made possible the economic production of cast iron.

(continued on next page)

Africa

- ◆ Metallurgy—Kushites mined iron ore around their capitals, Napata and Meroe, and worked it for weapons and tools; Yoruba produced fine bronze busts using the "lost wax" process; Benin also developed bronze sculpture as a high art form.
- ◆ Building technology—extensive ruins of Kush civilization show architectural evidence of transfer of technology across much of the known world of the day; great ruined mosque and palace of Hausa city of Kilwa still stands; impressive granite ruins at Zimbabwe (nineteenth century Europeans, committed to white supremacy, refused to accept such evidence of an important African civilization).

The Americas

- ◆ Building technology—the earliest American pyramid built by the Olmecs about 1200 B.C. at La Venta (like some Egyptian pyramids, American pyramids were stepped, with flat tops); the Moche built extensive canals and aqueducts in Peru. Huge stone pyramids and associated urban structures were built throughout much of the Americas.
- ◆ Ceramics—widespread, high-quality manufacture, frequently elevated to an exquisite art form, but without the use of the potter's wheel.
- ◆ Mayan calendar and number system—developed independently of outside influence. Their complex system in effect incorporated a zero—a millennium before the corresponding Indian innovation became generally accepted in Europe.

Successive states in Axum (later Ethiopia) and Nok (west Africa) and the ancient kingdom of Ghana gave way in turn to powerful empires such as Mali. Medieval Africa saw the establishment of the Hausa city states and Muslim states along the east coast.

Much of the wealth of the coastal city states came from further south, in particular, gold from Zimbabwe. The level of technological development among this and other African civilizations is perhaps most clearly seen among the Yoruba at Ife and the Edo at Benin (both in what is now Nigeria). The kingdom of Benin flourished from the fifteenth to seventeenth centuries. It was a walled city several kilometers across: a Dutch visitor in 1600 A.D. compared it favorably with Amsterdam. Yet successive European incursions and three centuries of slave trade dramatically altered the course of history in the African continent.

The Americas demonstrate a similar pattern of flowering of several civilizations prior to sustained European contact. Olmecs, who by 1500 B.C. were well established on the coast of the Gulf of Mexico, were probably the "mother civilization." The impressive architectural achievements of the Zapotecs at Monte Alban (100 A.D.) were rivaled by the great urban center of Teotihuacan in the Valley of Mexico. The Maya built a civilization in the highlands of Guatemala and Honduras between 330 A.D. and 900 A.D. Their intellectual stature is demonstrated by a hieroglyphic system of writing and a complex calendar and number system. Moche warriors established a highly organized society on the north coast of Peru. There is monumental evidence of the architectural abilities of many other peoples in the Americas. Impressive stone structures were built with astonishing precision, but without the use of the keystone arch. The Toltecs learned to smelt metals around 950 A.D. The Incas used their engineers to construct roads and forts for administering their Andean empire, much in the manner of the Romans, but they lacked wheeled vehicles.

A number of important pre-Columbian civilizations are thought to have declined long before European invasion because they destroyed the natural environment on which they depended for subsistence. Even so, the lure of treasure they offered was still very powerful. The first Europeans in the Americas in the fifteenth and sixteenth centuries found bronze weapons and lavish use of gold and silver for utensils, building decoration, and personal adornment.

■ Craft-Based Technology in Europe

Although it took centuries to emerge, Europe's technological superiority was eventually to prove devastating in its contacts with the rest of the world. Relative chaos followed the collapse of the Roman Empire, but an expanding population provided the impetus for clearing and cultivating forested lowlands. Tilling the heavy soils of northern Europe required new agricultural technology, and the difficulty of turning the new plows encouraged land division into very long and narrow fields.

By the ninth century A.D., a new social formation, feudalism, was emerging. Kings and powerful nobles gave grants of land (tenancy, not ownership) to members of their retinues, in return for their promise to render military service. Peasants had become serfs and were legally bound to the land. They were required to provide labor on the manor, but were also free to work their own strips of communal land. This allowed some degree of primitive accumulation of capital and eventually gave rise to a class of middle peasants. In return for their labor, the serfs "enjoyed" the military protection of the lord of the manor, their liege lord. Political sovereignty was not focused on a single center, but fragmented. Such fragmentation allowed the growth of free cities.

In towns there was the beginning of commodity production and the rise of the medieval craft guilds, which planned production and supervised, trained, and even employed workmen. The diffusion of technology during this period was slow, but an economy based on serfs rather than slaves did encourage the increasing use of labor-saving technology, since serfs and free craft workers in towns could themselves benefit from such technological developments.

In 1348 A.D., the Black Death started to spread through Europe, the first recorded outbreak of bubonic plague in the West. Over two generations it caused a net 40 percent loss of population, shaking confidence in existing social, political, and religious authority. The resulting labor shortages led to better working conditions, with the introduction of the clock to the workplace. There was also a move away from feudalism. By the time the population started to recover, arable land was becoming scarce. This was due partly to past overworking, but also due to the enclosure of land for sheep to supply the woolen industry, particularly the large-scale industry in Flanders (modern-day western Belgium and the adjoining areas of France and the Netherlands). Technical barriers were limiting expansion of agriculture and mining. It was not possible, for example, to sink deep shafts to obtain silver, because the mines flooded.

Major social changes began sweeping through Western Europe, accompanied by (and often arising from) technological change (see box Some Medieval Technologies). Power was increasingly centralized in monarchies, creating the absolutist state. Land became absolute private property; but power based solely on land holding was increasingly being challenged by town-based merchants. Suspicious of the feudal barons, the new monarchs introduced standing armies, a permanent bureaucracy, national taxation, codified law, and the beginnings of unified markets, changes reminiscent of earlier Mesopotamian and Egyptian states. Only monarchs backed by merchants, or powerful republics such as Florence, were rich enough to take advantage of the new forms of social organization and military technology that helped to establish favorable conditions for the rise of merchant capitalism. The amount of money in circulation was increased. Ships such as the Portuguese caravels, that could sail into the wind as well as with it, opened the way for the discovery of the New World of America.

The rediscovery of ancient Greek and Roman culture and the discovery of the New World brought new ideas and knowledge to Europe. Poets, painters, and sculptors developed new perspectives imbued with a humanistic outlook, far removed from medieval religious symbolism.

The first universities in Christendom were founded in the twelfth century in Salerno, Bologna, and Paris, and subsequently in Oxford and Cambridge. They were centers of learning, especially in theology, law, and medicine. Direct involvement in natural sciences and technology was to come much later, but the beginnings were there for that vital phase of "rebirth" or "revival" in Europe known as the Renaissance.

Some Medieval Technologies

Agricultural and Power Technologies

Advances included: sturdy iron plowshares and wheeled plows, three-field crop rotation, modern horse harness, the whippletree (which distributed the draft load between two horses), nailed horseshoes, and the spinning wheel. Water mills were used for grinding foodstuffs, for processing cloth, and other functions. In 1086 A.D. there were 5,642 water mills south of the Trent and Severn Rivers in England. Water mills and windmills had to be made and serviced by people who understood gears, a task beyond most village smiths. With mills came the millwright, the first "mechanic" in the modern sense of the word.

Military Technology

At the Battle of Hastings in 1066 A.D., King Harold and his Saxon warriors used their horses simply as transport, but the Normans had a technological advance, the stirrup. Introduced from the East and expertly used by Frankish cavalry since the late eighth century A.D., it gave the rider a much more secure seat. The horse could be used as an elevated fighting platform. This technology made a significant contribution to the Norman victory over the Saxons (and arguably altered the future course of the history of English-speaking peoples).

By the early 1500s A.D. new technologies were revolutionizing warfare. These included the musket and the cannon, used by well-trained infantry forces. Cannons could batter down the land-based aristocrats' castle walls. Disciplined infantry, armed with muskets and pikes, could stand against charges by heavily armed knights. Any well-trained yokel with a musket could kill a knight, and the feudal order started to crumble before the power of the emerging nation state.

Somewhere in Germany or Austria an unknown artisan produced the rifled gun barrel. The spin it imparted to the projectile stabilized its flight, increasing its accuracy and range.

(continued on next page)

Printing

By 1456 A.D., Johannes Gutenberg of Mainz (ca. 1400–1468) had introduced movable type for printing. With the increasing availability of paper (also from the mid-fifteenth century onwards) printing became a practical technology and spread rapidly through Europe.

Printing is a powerful example of a social and organizational need making use of, and further developing of, a technical device. When the Bible started to be printed in the local languages, its increased accessibility provided part of the basis for the Reformation: "each man his own priest." Later, secular material was increasingly printed. By 1500, over eight million books had been produced for a European population of seventy-five million.

A major technical book, *de Re Metallica (On Metals)* was published by Bauer in 1566, using the pen name Agricola. It had superb drawings of mining equipment, including windlasses, basket chains, pumps, and hammer-mills, mostly driven by water wheels, although windmills were also in extensive use in Europe by this time. Bauer also described some of the environmental impacts of mining.

Diffusion of Technology

As with later communication/media technologies, the impact of printing on other technologies was profound. Before printing, techniques were traditional, handed down from master to apprentice. Printing greatly accelerated the transmission of ideas. It made it possible, and then necessary, for craftsmen to be literate. Literacy provided an important basis for further social and technical change. By the middle of the eighteenth century, two-thirds of the adult population in England could read, setting the stage there for further technological breakthroughs.

Leonardo da Vinci (1452–1519 A.D.) is probably the best known example of what later generations have idealized as the Renaissance man. Like most engineers before the Industrial Revolution, his technical role was essentially military. On a practical level he analyzed fluid flows, drew the trajectories of cannon balls, and inspected fortifications. He produced sketches for submarines, helicopters, and elaborate gear trains, but his notebooks were not printed at the time and therefore not widely available.

More generally, observation and experiment started to present successful challenges to authority and to accepted dogma, and a new view of nature began to emerge. The Renaissance ushered in what has been called the Scientific Revolution.

THE SCIENTIFIC REVOLUTION AND THE RISE OF CAPITALISM

By the beginning of the seventeenth century, the foundations of the old order of society were shifting. Ideas that had been accepted for centuries could be challenged. This upheaval can be seen in the life of Galileo Galilei (1564–1642). Galileo developed the telescope and did experimental work on dynamics and the strength of beams. Unlike Leonardo da Vinci (but like most modern academics) he was quick to publish his findings. The official view at the time, endorsed by the church, was that the Earth was the center of the universe, and that all other celestial bodies rotated about the Earth. Galileo discovered that Jupiter had satellites, and concluded that the Earth revolved in the same way around the sun. These ideas were inevitably denounced by the Church as heresy. Galileo was imprisoned by the Inquisition and forced to recant, but the spirit of scientific inquiry was not so quickly quenched.

Science has been of inestimable importance for modern engineering. However, as discussed in more detail in Chapter 10, we are concerned during the course of this historical overview to differentiate engineering from both applied science and technology. The Scientific Revolution is a label that was applied by historians to this period only comparatively recently. The term "science" was not in common use until the end of the eighteenth century; the term "scientist" was coined only in 1840. Given its significance for engineering, it is appropriate that we provide a working definition of science at this point (see Key Concepts and Issues).

One of the first major figures in modern science in Britain was Francis Bacon (1561–1626). Bacon was an enthusiast for industrial science, and emphasized practical issues. His approach was essentially inductive, working towards conclusions from a mass of experimental data. He was obsessed with institutionalizing science. His dream of a college for inventors, with laboratories, workshops, furnaces, and a library, laid the foundations for the establishment of the Royal Society.

René Descartes (1596–1650) was another major figure of the Scientific Revolution. Descartes emphasized the deductive approach, and his work on mathematics was very influential. His development of coordinate geometry united the previously separate areas of algebra and geometry. Working in Catholic France, Descartes avoided the fate of his Italian contemporary Galileo by espousing a division between the purely mechanical animals

Science

Science refers to a body of a certain sort of knowledge—organized, well founded, testable knowledge about natural phenomena. More broadly, science, like both technology and engineering, is:

- a general area of human endeavor;
- a distinctive kind of human cultural activity;
- a total societal enterprise with an increasingly global character.

Science is essentially a complex of knowledge, methods, techniques, and materials and their outcomes. Its purpose is to discover, create, confirm, disprove, reorganize, and disseminate statements that accurately describe some portion of physical, chemical, biological, or social nature.

Over the last three centuries, science as defined here has become so powerful and respected that occasionally exaggerated and inappropriate claims are made for it, such as that it can cover *all* matters that are of vital concern to humans. If we consider for a moment values such as "beauty", or "good", we can see this claim is inappropriate.

Modern engineering often, but not always, draws heavily on science (and vice versa), but they have distinctly different purposes, methods, and outcomes.

(After Kline 1995 and McGinn 1997)

(including man), operating according to physical principles, and the rational will and spirit dwelling within, which was the unchallenged province of the Church. This division avoided a confrontation and allowed science to advance. It also allowed scientists to sidestep moral and ethical issues, a legacy we are now having to address (most directly in this book in Chapter 11).

In the seventeenth century, the rise of capitalism brought about renewed interest in the practice of the trades of the artisan, and a more practical orientation towards knowledge. As Robert Boyle (1627–1691) put it: "experimental philosophy may not only itself be advanced by an inspection into trades, but may advance them too." (Armytage 1976, p. 78) However, at least until the end of the seventeenth century, science gained far more from its contacts with practical work than it contributed in terms of improvements in technique. Among other pioneers in Britain was

Robert Hooke (1635–1703), the first curator of experiments for the Royal Society, which was granted its Royal Charter in 1662.

Intellectual assumptions and attitudes toward knowledge from the Middle Ages were yielding to a new view of the world. Knowledge was much less controlled by the Church. The Earth-centered cosmos of Aristotle gave way to a sense of a solar system. The universe was seen to obey Newton's Laws: knowledge became a means of controlling Nature. The contributions of many great scientists were summed up in *Mathematical Principles of Natural Philosophy*, published in 1685–1686 by Isaac Newton (1642–1727). By the end of the seventeenth century, the new scientific world view had triumphed.

■ Exploration, Empire, and the Accumulation of Capital

With superior navigational aids and ships, and an advantage in arms, Europeans had set out to explore and conquer foreign lands and seize new resources. The main European economic and commercial centers moved north from the Mediterranean to the ports of the Atlantic seaboard. Truly global empires would be realized for the first time in human history.

In the sixteenth century, Spain and Portugal led the way in overseas conquest. Gold, silver, and other resources poured into their coffers. By the early seventeenth century, despite being at war with Spain, the Netherlands dominated world trade, art, and science, and founded extremely profitable empires in the East Indies (now Indonesia) and the West Indies. However, the large, well-established Dutch cloth industry met increasing competition from subsidized English linen. Britain used its large coal deposits to drive its industries; in time, it surpassed the Netherlands and established a vast colonial empire.

Revolutions in England, Europe, and America in the seventeenth and eighteenth centuries were as much the result of economic tensions as they were of religious and/or political tensions. Monopolies in mining, manufacturing (including basics such as soap and salt), and foreign trade had routinely been granted by absolute monarchs to their court favorites and to merchant guilds. In Britain, the royal power had been sharply reduced by Parliament in 1624. The "middling" sorts of men who formed the core of the increasingly important British middle class—freeholders, merchant employers, and nascent manufacturers who had risen from the ranks of the craftsmen—struggled for access to these areas. The patent system protected the entrepreneur, and the new era saw markets open up and new sources of capital for investment throughout the economy. Land holding

The "Technology Gap" and Colonization

From about 1500 A.D. onwards, there was a growing "technology gap" between European powers and the rest of the world. European ventures into other world civilizations had increasingly adverse outcomes for the non-Europeans, viz.:

◆ Contacts with Africa led to the slave trade, colonial partition of virtually the entire continent, and widespread destruction of African society.

◆ In India, technological efforts for centuries had been channeled into military and religious activity. Beginning about 1600, growing trade contacts with the West led to multiple colonial incursions and altered the entire course of Indian history.

◆ China was technologically ahead of Western Europe in most fields until the end of the fourteenth century. However, innovation then seems to have been discouraged by a powerful Chinese bureaucracy concerned with maintaining social harmony and keeping the merchants in their place. Technological change was ultimately stifled by a series of barbarian invasions. Even so, Chinese authorities showed remarkable awareness of the significance of contact with Western cultures. China resisted European influence until the mid nineteenth century, after which it was forced to come to humiliating terms with the West.

◆ Japan shunned contact for as long as possible. Interaction with Europe was carefully mediated. Although contacts were socially disruptive, Japan avoided being colonized.

◆ In spite of the obvious prosperity and social stability of many of the civilizations in North, Central, and South America, they were isolated from the rest of the world and their technological development was uneven. Continued use of stone-age tools was widespread, use of bronze was lim-

under the feudal system was officially abolished, and many of the large estates of the defeated Royalists were sold.

Some of the merchants who bought these estates had the considerable capital needed for increasingly scientific large-scale farming methods. These were required to satisfy the expanding market, created in part by the growing population. More effective and efficient farming practices led to increases in productivity of the land; taken together, the changes

ited, and iron and steel had not been discovered. The principle of the wheel was known to the Aztecs, but they lacked draft animals to pull carts or plows, and their use of wheels was limited to children's toys. They had no mechanical devices using wheels, gears, or rotary motion.

◆ Conquest, decimation of the population, and long-term impoverishment followed European invasion.

◆ While Europe languished in its Dark Ages, the Arab culture of the Middle Ages was one of the world's most advanced cultures. However, after the Arabs' devastation during the Crusades and their suppression by the Ottoman and other empires, the Arabs fell increasingly further behind Europe in technological development. Today, the Arabs of the Middle East tend to resent the technological, as well as cultural, dominance of the West over their own culture.

How important was the technology gap to these outcomes? Were some technologies more important than others? Does the use of more complex technologies make a population more "developed"?

The advances the European nations had made in military technology and social organization certainly helped them to impose their will on the inhabitants of the Americas, Asia, Africa, and Australasia. However, the destruction by colonizers of indigenous societies, often well-developed cultures with sophisticated technologies, more commonly resulted from introduced diseases (particularly smallpox) and disruption of society and culture than from direct military action.

There is considerable argument as to why and how colonial empires arose when they did. We look in Chapter 8 at the extent to which plunder from the colonies helped to finance the European industrial take-off, and how continuing exploitation of colonial resources helped sustain industrial growth in some countries but left a legacy of underdevelopment in others.

have been described as an Agricultural Revolution. Peasants could now leave the land; indeed, many were forced off because of new farming methods and because land was being enclosed and cleared to raise sheep to supply the growing textile industry.

By the early eighteenth century, urban merchants with sophisticated banking and exchange practices had gained control of commodity production, which was still at the handicraft and domestic stage. State institutions became increasingly secularized, official, and visible in their rule over civil society. Capitalists were defining their social identity as a class, distinguished by their disproportionate control of privately owned capital. The open public debate they started was based on a commitment to secular enlightenment, and was aggressively rationalist and antitraditional. It challenged religious authority and hereditary privilege and advanced the vision of a new constitutional state. This emphasized liberty, particularly the freedom of the individual to invest capital and to trade as he saw fit.

The Changing Organization of Production

The Industrial Revolution transformed the way goods were produced. Three major stages of development of the production process in Britain can be identified:

1. *Handicraft production*, essentially the guild system of the Middle Ages, with a master and a few journeymen each making complete articles. This patriarchal (or occasionally matriarchal—women could become "masters") arrangement persisted until the middle of the sixteenth century.

2. *Handicraft manufacture*, where a master paid money wages to a number of workers. Although there was a substantial division of labor in the making of each article, manufacture was still done by hand tools, using human skills.

3. *Modern industry*, with improved, power-driven machines reducing the need for skilled workers.

THE INDUSTRIAL REVOLUTION

These profound social, political, and economic transformations, the Scientific Revolution of the preceding years, and countless small changes and improvements in mechanical devices, dramatically increased productivity. In England, the weakening of the guilds by antimonopoly legislation had curbed guild restraints on innovation (by contrast, in some European countries such restraints continued to be significant well into the nineteenth century). These were among the factors that went into making the Industrial Revolution and go some way towards explaining why the Industrial Revolution started in the mid-eighteenth century in Britain, rather than elsewhere.

Abraham Darby (1677–1717), a Quaker like many of the innovative new factory owners, discovered how to use coal to smelt iron in 1709, although problems with quality continued until the end of the century. Transferring technology from the brewing industry, Darby built large blast furnaces running on coke. Ultimately, his plant at Coalbrookdale in Shropshire had sixteen steam engines, eight blast furnaces, nine forges, rolling mills, foundries, and thirty-two kilometers of railways (along which horses or people pushed the trucks). Coalbrookdale was a center for innovation for almost 100 years in pottery as well as ironmaking (see Case Study 2.4.)

Adam Smith (1723–1790), in his *Wealth of Nations* (1776), described the division of labor in pin making, with one worker drawing out the wire, another straightening it, a third cutting it into lengths, and others making the heads and fitting them, sharpening the points, whitening the pins, and packing them. This fragmentation of work greatly increased the productivity of the workers, but it was only the first step. Once production processes had been standardized and broken down into separate elements, a machine could be used for each element. The development of these machines completed the transformation to modern industrial production.

In popular imagination the period is seen as "the Machine Age" because of the great profusion of new machines and the mechanization of so many tasks. Machines were now used for spinning and weaving and to smelt iron for yet more machines and their accessories. Machines would soon power ships and trains: in short, machines revolutionized many areas of production and everyday life. Beginning in the middle of the eighteenth century an increasing proportion of production was done by machine, with journeymen reduced to machine minders. These developments ushered in the Industrial Revolution and changed the way most people lived. Enclosure of common land and elimination of jobs in agriculture forced people into towns, where they were packed into crowded city tenements. Time took on a different meaning; lives were regulated by the factory clock rather than by the hours of daylight and the rotation of the seasons. People survived by selling their raw labor power to the owners of the new factories and mills.

Power to drive the new machines was critical. Before the Industrial Revolution, productive activity was driven by animal and human power, along with windmills and waterwheels. This kind of power was adequate for what we would describe today as cottage industry, but it clearly had limitations. Just as our society today runs on oil, early eighteenth century Britain ran on wood, consuming so much that its limited supply became a problem. Timber was burned as wood, or reduced to charcoal for making glass or smelting iron ore to make cast iron, wrought iron, and steel, all increasingly widely used. The blast furnace had been developed; it burned charcoal and was driven by waterwheel. Coal from the surface had long gone; mines were getting deeper and wetter. Better ways were needed for raising coal, water, and miners out of the mines. This energy crisis was a major spur to invention. Some of the essential background scientific research had already been performed on atmospheric pressure, and vacuum and pressure vessels, but metal working machinery was still rather primitive.

The dramatic developments in steam power are summarized in the accompanying box. Starting with textile mills, steam engines drove the new machinery that took British industry to world leadership. James Watt's successes with the steam engine were not an accident but the outcome of systematic and informed investigation and development. His approach foreshadowed what Whitehead (1938) suggested was "the greatest invention in the nineteenth century...the invention of the method of inventions."

Watt's steam engine was the basis for the rise of the profession of mechanical engineering. Beginning with the period of the Industrial Revolution, modern engineering, as we chose to define it here, becomes increasingly differentiated from other technological and scientific activities. It is now useful to suggest a definition of engineering (see Key Concepts and Issues).

The ability to move people and goods quickly, reliably, and economically was essential to large-scale production and mass marketing. Improvements to transport, communications, and production technology, and wealth from the New World, were all important in the transformation of Britain into an industrial society. By the eighteenth century, the road network in Britain was in a worse state than it had been when the Romans had left, over 1,300 years before. Better transport was needed, particularly for the increasing quantities of heavy goods.

Works to make rivers navigable had started in the latter part of the seventeenth century. By 1758, over 1,600 kilometers (1,000 miles) of river improvements had been made in Britain, preparing the way for the canal boom. Canals had long been basic to the great irrigation-based societies. In Egypt, in the third century B.C., a canal over 100 kilometers long was built from the Nile to the Red Sea. The Grand Canal in China, almost 2,000 kilometers long, was completed around 600 A.D. In Europe, the 238-kilometer Canal du Midi, including 100 locks, was completed in 1681. The Duke of Bridgewater was so impressed when he saw it in 1753 that he employed James Brindley (1716–1772), a millwright, to build him

Innovators and Steam Power

Thomas Savery (ca. 1650–1715) invented and patented a primitive steam engine. In 1698 Savery went into partnership with Thomas Newcomen (1663–1729), a Dartmouth (U.K.) blacksmith, who supplied iron tools to the tin mines in Devon. By 1712 they had developed a more effective product operated by atmospheric pressure on top of a piston in a cylinder. In the lower part of the cylinder, steam was condensed to form a partial vacuum. The piston was connected to one end of a rocking beam, the other end of which drove a pump or other mechanical device. Although inefficient, Newcomen's engines were in great demand, and saved the tin and much of the coal mining industry by keeping the deep mines dry. The largest Newcomen engines had a bore of 6' 8" (nearly two meters) and a stroke of 10' (three meters).

James Watt (1736–1819), an instrument maker in a Glasgow University laboratory, was asked to repair a Newcomen engine; instead, he transformed it. Watt realized that alternate heating and cooling in the same cylinder was wasteful. He recognized the need for a separate condenser that would be maintained permanently at the lower temperature. This was possibly the most important breakthrough in mechanical engineering since the discovery of the wheel. The efficiency, in other words the ratio of the mechanical energy output to the chemical energy in the fuel burned, was improved dramatically, from less than 0.5 percent to around 2 percent. Watt pioneered many other improvements. In 1782 he adapted the steam engine to produce rotary motion, allowing it to be applied to the general mechanization of industry.

Matthew Boulton (1728–1809) was the wealthy backer Watt needed ("I sell what the whole world wants—power"). By the end of the eighteenth century, 500 Boulton and Watt engines were in use in Britain. Watt's engines operated on steam pressures above atmospheric, but, aware of the danger of explosions, he was extremely cautious in raising the pressure.

Difficulties in containing high pressures were overcome by Richard Trevithick (1771–1833) in England in 1801, and by Oliver Evans (1755–1819) in the United States in 1802. Evans promoted the use of scientific principles to improve existing technologies, publishing The Young Mill-Wright and Miller's Guide in 1795 (Layton 1991, p. 233). Trevithick's advance was based on the use of a new boring machine developed by Wilkinson, which permitted a much tighter-fitting piston. Trevithick developed the large beam pumping engine known as the Cornish engine, which continued to be used well into the twentieth century by some mines in Cornwall. In 1804, Trevithick built the first steam-powered locomotive, but he used a brittle cast iron track that broke under the load.

Engineering

Engineering has been central to the economic growth characterizing the rise of industrial capitalism. The best known definition of engineering is perhaps still the one that was written for the Institution of Civil Engineers by Thomas Tredgold in 1828:

"... the art of directing the great sources of power in nature for the use and convenience of man."

Such a definition would be unacceptable today. Despite its limitations, however, it captures some of the idealism and positive social purpose that imbue the best engineering enterprises. Engineering should be, in the broadest sense, at the service of humanity.

A more formal definition of engineering would recognize a number of meanings. Thus engineering can be viewed as:

- ◆ A specific area of human endeavor, defined by McGinn (1997) as:
  "that highly specialized and professionalized branch of technological activity which is devoted to systematic design, production, and operation of technics and technical systems to meet practical human needs under specified constraints [time, money, performance, reliability, etc.]." (The range of knowledge involved encompasses many distinct areas of human endeavor, tasks and competencies, including civil, mechanical, electrical engineering, etc.).

- ◆ A kind of human cultural activity (including design, research, development, manufacturing, testing, operation, maintenance, etc.).

- ◆ A total societal enterprise, with significant influences on all aspects of human life and a major role to play in moving the world towards particular goals (e.g., wealth creation, improved quality of life, sustainability).

A modern statement of the purpose of engineering, which we endorse, has been proposed by the Institution of Professional Engineers New Zealand:

Engineers will translate into action the dreams of humanity, traditional knowledge and the concepts of science to achieve sustainable management of the planet through the creative application of technology. (IPENZ 1993, The Pathway to the Future, Wellington, NZ.)

a canal from his coal mines to their market. Completed in 1761, the canal halved the price of coal in Manchester. The Duke was eventually involved in over 365 canal projects.

By 1858, when the last canal tunnel in Britain was completed, there were hundreds of canals with a total length of over 6,800 kilometers, including 72 kilometers of tunnels. Their downfall was that in order to maximize short-term profit they were built narrow and shallow. As a result, British canals were too small to meet the long-term demand they generated. From the 1830s, railways began to replace them.

The railway age took off following the introduction of wrought iron rails in 1825, and the success of George Stephenson's Rocket in 1829. Once railways were commercially competitive, the growth of their installation around the world was very rapid. In 1869 the Union and Central Pacific crossed the American continent (see box, Some Pioneer Engineers in Britain). In Britain and Europe the impact of railroads was perhaps less dramatic, but equally inexorable. Their speed changed peoples' sense of place and opened up broader markets for goods and services. The Iron Way meant transportation, exploitation, profit, investment, and more profit. The building of railroads was a major factor in the colonization of much of the rest of the world.

The Industrial Revolution was in full swing. The Agricultural Revolution preceding it had vastly increased agricultural productivity. The Industrial Revolution would provide employment for all those displaced from the land, and many more. The social and economic impacts of the Industrial Revolution were spectacular. Between 1750 and 1800, the population of England and Wales rose by 50 percent, from about six million to nine million. In the next fifty years it doubled, to eighteen million. The population of Europe, excluding Russia, rose from around 120 million in 1750 to 210 million in 1850.

Population growth was both an underlying cause and a fuel for the Industrial Revolution. It was due, at least initially, to a decline in the death rate, a decline in which engineering works played a central role. Better transport made fresh food more available and affordable. Public health was greatly improved by the provision of clean drinking water, and by better disposal of waste, particularly sewage. The rapid growth in population, in turn, provided labor for the new industries and markets for their products.

Many great engineers contributed to the Industrial Revolution (see box, Some Pioneer Engineers in Britain), often running extremely successful businesses. The versatility of some of the great engineers was astounding. Isambard Kingdom Brunel and his father designed and built almost everything—from production lines for ships' pulley blocks to tunnels, railways, and ships. However, as Pacey (1983, p. 95) pointed out,

Some Pioneer Engineers in Britain

John Rennie (1716–1821)

Rennie built canals, docks, and bridges. He was unusual for his time in having studied at Edinburgh University, working as a millwright between university terms. More typically, pioneering eighteenth- and nineteenth-century British engineers were either wealthy enthusiasts or poor men who learned the job from the bottom up.

John Smeaton (1724–1792)

Smeaton was elected a Fellow of the Royal Society in 1753 at the age of twenty-nine. He built a lighthouse on the Eddystone reef and went on to build bridges, engines, windmills, and other works. He was the first Englishman to describe himself as a *civil engineer*, a title indicating that he could be engaged to carry out nonmilitary engineering work.

Thomas Telford (1757-1834)

Telford, the first president of the Institution of Civil Engineers, founded in 1820, was apprenticed at fifteen to a mason. He went on to build canals throughout Britain. In Scotland he built the Caledonian Canal across the country, as well as extensive harbor and other works. In eighteen years he transformed the highlands, building over 100 bridges and 1,500 kilometers of new roads, and training over 3,200 workmen.

John Loudon McAdam (1756-1836)

McAdam was a wealthy enthusiast who spent £5,000 of his own money in experiments on improving road construction. His techniques were increasingly adopted from 1815 onwards. By 1823, his three sons were responsible for over three thousand kilometers of turnpike (toll) road, built using his system.

there is already a dilemma here. Should limits be placed on creativity, or on the pursuit of technical marvels? As with many more modern projects, Brunel's emphasis on technical virtuosity, often regardless of cost, consumed resources that could have been more directly used to benefit people. An emphasis on virtuosity in technology may be symptomatic of a lack of concern with economic and social reality. Much military and aerospace engineering today certainly raises these sorts of broadly political concerns, as discussed in Chapter 4.

Sir Marc Isambard Brunel (1769–1849)

Born in France, Brunel fled the Revolution in 1793, going first to the United States, where he was chief engineer of New York for some years, and where he designed a gun foundry and the defenses of Long Island and Staten Island. Brunel went on to Britain, where he pioneered mass production with his machinery for making ship's blocks, the pulleys for handling the ropes that controlled the spars and sails in the sailing ships of the time. Ten semiskilled men using his machines produced as many blocks as 110 skilled workers using the best previously available techniques and equipment. Brunel solved many of the problems of underwater tunneling in building the Thames Tunnel from Rotherhithe to Wapping.

Isambard Kingdom Brunel (1806–1859)

Sir Marc's son, Isambard, born in England, worked with his father on the Thames Tunnel. Isambard also designed the *Great Western (1838)*, the first steamship to cross the Atlantic; the *Great Britain* (1845), the first ocean-going steamship to use a screw rather than a paddle wheel; and the *Great Eastern* (1854–1858), then the largest vessel ever built and, at 18,915 tons, six times as large as any ship afloat. He was engineer to the Great Western Railway, and built many other civil engineering works.

George Stephenson (1781–1848)

Stephenson, the great railway engineer, was a selfeducated engine-wright. In 1847 he accepted the invitation to become the first President of the Institution of Mechanical Engineers, founded partly because of the conservatism of the leaders of the Institution of Civil Engineers and their reluctance to recognize his expertise. Many notable engineers followed Stephenson as President, including his son Robert, and Sir Joseph Whitworth, who pioneered the standardized system of screw threads named for him.

THE CHALLENGE FROM THE UNITED STATES

The pace of the Industrial Revolution was set by Europe, with France and Germany rapidly catching up to Britain. Across the Atlantic, approaches to education and knowledge in the United States contrasted sharply with those in Britain. Settlers brought with them extremely utilitarian attitudes: often dissenters from established churches, they saw knowledge as a commodity to be used to satisfy human needs. The scarcity of labor also encouraged a frontier spirit of inventiveness. Yankee ingenuity and "know-how" became legendary.

Some pioneer engineers in America included Robert Fulton (1765–1815), best known for his pioneering work on steamboats, and Eli Whitney (1765–1825). A successful manufacturer at age sixteen, Whitney went back to school in his twenties and graduated from Yale in 1792. In 1794 Whitney patented a simple cotton gin to separate fiber from seeds. By increasing the productivity of this stage of production fifty-fold, the gin transformed the economics of slave-operated plantations, but the planters pirated the invention and Whitney did not share in the benefits. (The role of Whitney's employer, Catherine L. Green, in the conception and development of the cotton gin remains unclear. Stanley [1983, p. 10] notes that Whitney paid royalties to her. At the time it would have been socially unacceptable for Green to have patented it herself.) Whitney pioneered quantity production of muskets, using machine tools to make precision, interchangeable parts. This "uniformity system" eventually revived his fortunes and was the forerunner of the modern assembly line. Whitney's first milling machine, with an automatic feed, was introduced in 1820.

The vast natural resources and economies of scale in production in the United States, associated with an acceptance of standard (rather than made-to-order) products, gave rapid results. The leaders of the young American Republic recognized the need to encourage domestic manufacture, particularly of basic military equipment. Work opportunities on the farms and in the forests attracted skilled labor away from the cities or increased its costs. American entrepreneurs responded by designing industrial equipment that reduced their dependence on labor, especially skilled labor. Their innovations were initially crude, inflexible, and wasteful of raw materials, but they were an effective response to local conditions.

The engineering profession that emerged in America in the nineteenth century was a blend of British and French traditions, with continuing tensions between advocates of "school" and "shop" (practically-based) education. French military engineers served in the American Revolutionary armies (1775–1783) and some of them emigrated to America during the French Revolution. American engineers went on to take the lead in machine shop techniques, developing effective grinding and milling machines. The first turret lathe, carrying eight different tools, was introduced in 1845.

The Crystal Palace Exhibition in London in 1851 was in many ways a celebration of engineers and engineering. It marked the high point of British industrial ascendancy. Even at this Exhibition, however, a distinctive "American system of manufacture" was recognizable. It was characterized by highly standardized products, made up from interchangeable parts, produced by specialized machines. These products were well suited to the demands of the rural middle class who were the dominant American consumers. The expansion of American railroads opened up national markets for machine tools, providing the basis for high levels of specialization. Tensions between the slave-owning South and the increasingly industrialized North contributed to the outbreak of the devastating American Civil War (1861–1865), the first to be fought with "modern" weapons on both sides.

Stagecoaches on land and steamboats on canals, rivers, and coastal routes had helped to build regional markets in the eighteenth century, but they were relatively slow and geographical isolation was still a problem. The cheap and effective U.S. postal services were probably more important in creating links within the new nation. Similarly, in the nineteenth century the railroads were symbolically and physically important in uniting the nation (see box, Early Railroads)—particularly important so soon after the bitterly divisive civil war. However, the blessings of the railroad were not universal. Some Native Americans had seen the railroad as a threat to their way of life, and had attacked survey and construction teams. Even in their wildest nightmares they could not have anticipated how quickly their world would change. In little more than a decade after the first trans continental rail link the great buffalo herds had been reduced to trainloads of skins, going east, with mountains of bones left behind. The way of life of the Plains Indians had changed forever.

The Spread of Industrialization

Britain had been a lucky country, rich and powerful without needing to examine and conserve the sources of her wealth. Following the Agricultural Revolution, the British aristocracy was essentially capitalist but lived on income from rents and bonds rather than from its own entrepreneurial or productive activity. It was relatively open, able, and willing to assimilate wealthy industrialists and remake them in its own image. Wiener (1985, p. 7) argues that the apparently peaceful and rather protracted transfer of power fostered a self-limiting element in Britain's development. The industrial revolution in other countries came at least partly from without and thus challenged and disrupted social patterns. In Britain, on the other hand, industrialization was indigenous; it more easily accommodated to existing social structures that, as a result, did not need to change radically, and have barely done so even now. Pressures for the adoption of the values and interests of industrialization were effectively resisted by the ruling élite.

Early Railroads

The first successful passenger railway, between Stockton and Darlington in England, started operation in 1825. Within a few years, the railway boom was on! During the 1840s and 1850s, thousands of miles of railroad track were laid all round the world.

There were four likely routes across the United States, and at least one across Canada. While still in his twenties, Theodore Dehone Judah had worked as an engineer on the Erie Canal and built bridges and railroads. In 1854 he built forty kilometers (twenty-five miles) of railroad from Sacramento to Folsom in California, and became passionately committed to the idea of building a railroad across the continent. In 1860 he surveyed a route and started to organize funds for his "Central Pacific Railroad" (CPR). In Sacramento he met with a hardware dealer, Collis P. Huntingdon, his partner Mark Hopkins, and an associate, Charles Crocker. They later brought in a grocer, Leland Stanford, making up the group that became known as the Big Four. The Central Pacific Railroad of California (CPR) was incorporated in 1861, with Stanford as president, Huntingdon as vice-president, Hopkins as treasurer, and Judah as chief engineer. When Judah fell out with the group, Crocker became construction superintendent.

The first trans continental railroad was built from both ends towards the middle. The CPR started from Sacramento in January 1863, building east, and the Union Pacific Railroad started from Omaha, Nebraska in December, building west. Government subsidies to the builders included 180,000 square kilo-

Britain's loss of technical leadership from the 1850s onwards reflected complacency and conservatism in society at large, and in the engineering profession itself. The emphasis in universities on the pursuit of scientific knowledge for its own sake saw practical and commercial applications undervalued and neglected.

The rise of nation states in Continental Europe eliminated many internal tariff barriers. Rail transport increased the size of potential markets, even for heavy equipment such as machine tools. In the United States in particular, customers were prepared to buy standard items, creating the opportunity for mass production. Rosenberg (1976) argues that British preoccupation with the individually fitted final product, rather than with the production process, discouraged standardization, and was an important element in Britain's relative decline. The dominance of the steam engine delayed the development of other forms of power in Britain. Restrictions were placed by the railway lobby on any kind of road trans-

meters (forty-five million acres) of land beside the tracks, and made the Big Four a major force in Californian life.

Early progress was good, but the CPR faced a chilling barrier, the Rocky Mountains. Progress was difficult. Large, well-coordinated teams were needed, and the sorts of men who had come west were unsuitable. The problem was solved by bringing in teams of Chinese laborers direct from China, complete with a clan leader who coordinated them. They needed courage, since the work was hard and dangerous, especially in winter. Some whole teams were trapped and frozen to death. Nevertheless, their coordinated efforts drove the western road forward. Eventually the CPR employed 15,000 Chinese.

Once through the mountains, progress was rapid. On one record-breaking occasion Crocker's teams laid ten miles of track in a single twelve-hour day. Competition between the two railroads was so fierce that eventually a Congressional order was issued that the lines would meet at Promontory Point, north of the Great Salt Lake in Utah. With much ceremony, the last piece of track was laid on May 10, 1869. Leland Stanford drove the last spike (of solid gold, using a solid silver hammer) and the continent was joined. (Stanford later founded Leland Stanford Junior University as a memorial to his son.)

In the 1870s and 1880s, four other transcontinental links followed. Their demand for materials gave tremendous impetus to the steel industry. By 1890 there were 265,000 kilometers (165,000 miles) of railroad tracks. Altogether the railroads received land grants of 710,000 square kilometers (180 million acres), an area one and a half times the size of Texas.

port (including the law that, until 1896, required a man with a red flag to walk in front of any road vehicle). In this climate, France and Germany developed the internal combustion engine without real competition.

Rosenberg (1976) described developments in the United States in the second half of the nineteenth century as *technological convergence*. By this he meant that a range of key metal-working industries emerged, with *common technological needs*. The same machine tools were used across a range of industries, and machine tool companies became a focus for technological change. Over several decades within the same manufacturing enterprises, there was a progression of products from guns to machine tools, to sewing machines, bicycles, motorcycles, and finally, to automobiles. There was a comparable progression of production machinery. Automatic lathes were operating in the United States by 1873. The innovative development of machine tools, and their use to produce standardized interchangeable parts, paved the way for Henry Ford to use moving assembly-line techniques for producing automobiles. This was the beginning of our modern industrial era; the changes in the organization of work are discussed in Chapter 2.

Between 1860 and 1900 manufacturing replaced agriculture as the leading source of economic growth in the United States. From a nation typically composed of farmers in small rural communities, the United States became a nation of industrial workers and city dwellers (Nash and Jeffrey 1998, p. 615ff). Manufacturing industry has been equally important for most of the other nations in the world (although they have experienced very different rates of industrial development).

As manufacturing transformed America in the second half of the nineteenth century, manufacturing industry was itself transformed. Before the Civil War, manufacturers concentrated on such areas as textiles, clothing, and processing agricultural goods. From about 1870 onwards the shift was to goods intended for other producers, rather than for consumers. Heavy industry—steel, iron, petroleum and machinery—grew rapidly, boosted by a number of technical innovations. When Andrew Carnegie introduced the Bessemer steelmaking process at his plant in the mid 1870s, the price of a ton of steel halved to $50 a ton; by 1890 the price had fallen to $12 a ton. Cheaper steel led to new goods and new markets, and it stimulated new demands and further technological change.

In Germany, only when Bismarck became Chancellor of a unified nation-state in 1871 was there a concerted drive towards industrialization. Bismarck's creation of a state-owned national railway system helped open up the internal market. There was also a successful push in Germany to expand foreign trade, with an emphasis on exporting industrial products. German governments at all levels took strong roles in promoting education relevant to industrialization. In a development that was later widely imitated by large organizations in the United States, German manufacturers in emerging science-based industries set up their own closely integrated industrial research laboratories (McGinn 1991, pp. 25–26). By the early 1880s, the United States and Germany were well on the way to overtaking Britain's commanding industrial lead, especially in fields that required university-level technical expertise. By 1900, Germany had a highly integrated industrial sector, and had surpassed Britain as an industrial power (Vernon 1989, p. 18).

New sources of power facilitated the conversion of industry to mass production. In the United States in 1900, steam engines fueled by cheap coal accounted for 80 percent of the total energy supply to the nation (Nash and Jeffrey 1998, p. 617), but electricity soon began to make inroads as a major power source. Although the railroads were the initial big industry, by the last quarter of the nineteenth century the textile, metal, and machinery industries were of equal size. A trend toward large enterprises had emerged. In 1870, U.S. iron and steel firms typically employed fewer than 100 workers. By 1900, 450 American factories employed more than 1,000 workers, and more than 1,000 factories had labor forces between 500 and 1,000.

The increasingly global character of engineering became evident from the latter part of the nineteenth century onwards. Case Study 1.1 (at the end of this chapter) on the construction of the Panama Canal is a good example of this trend, which has accelerated during the course of this century. The political implications of such large-scale engineering projects are already apparent in this case study.

Japan was a latecomer to industrialization. Commodore Matthew Perry arrived with his "Black Ships" at what is now Tokyo Bay in 1853. He brought a demand from the U.S. president for Japan to trade with the West. Perry's flotilla of warships included the first steamships ever seen in Japanese waters, and one of his aims was to extend their range by gaining the right to use Japanese coal. The external threat that Perry demonstrated and the unequal treaties imposed on Japan over the next decade forced political acceptance among Japanese of the need for rapid technological and social change. The Meiji Restoration of 1868 dismantled the feudal system in Japan.

Throughout the world, rapid population increase continued. From 210 million in 1850, the population of Europe (excluding Russia) rose to 393 million in 1950. A major effect of industrialization was that the gross national product in Europe grew even faster than the population. From 1850 to 1900, average income per head in England increased by more than 2 percent per annum. World population has continued to grow, from perhaps 5 million in 8000 B.C., to 750 million in 1750, 2.52 billion in 1950, and an expected 6 billion by the year 2000.

In tracing the development of engineering through the Scientific and Industrial Revolutions, we have outlined some of the reasons why the Industrial Revolution started in Britain in the mid-eighteenth century, rather than elsewhere in Europe, and why Britain was eventually eclipsed by other nations. Initially craft-based and craft-focused, as industrialization spread through Western Europe and North America the Industrial Revolution drew increasingly on the broadening base of the physical sciences. Capital accumulated from merchants' profits funded manufacturing industry, mining, and new means of transport and communication. The steam engine was important, but what really drove the Industrial Revolution was the interplay between the new means of production—machinery, engines, materials, and technological expertise—and the growing requirements of industry and commerce.

Some International Engineers

Ferdinand de Lesseps (1805–1894)

While serving as a diplomat in Egypt from 1832 to 1837, de Lesseps, a French Engineer, recognized the need for and then designed a canal across the Isthmus of Suez. His idea was eventually accepted. Begun in 1859, the Suez Canal was formally opened on November 17, 1869, earning de Lesseps numerous honors. Success with the Suez Canal led to his being chosen president of the French company that worked on the Panama Canal from 1881 to 1888 (Case Study 1.1). Shortcomings in their management of the Panama project destroyed the reputations of both Ferdinand de Lesseps and his son, Charles.

Sir Henry Bessemer (1813–1898)

Bessemer, an amateur scientist but a professional inventor, revolutionized steel production with the invention in 1856 of his converter, an innovation that brought the first mass-production technique to the steel industry. Investors in his works enjoyed an effective dividend of 600 percent per annum for fourteen years!

Sir William Siemens (1823–1883)

Born in the German state of Hanover, Siemens was one of the family that founded the great German manufacturing firm that still bears their name. Sir William settled in England in 1844 and later adopted British nationality. An enthusiastic inventor, consulting engineer, and steelmaker, he was elected Fellow of the Royal Society in 1862. He was in turn President of the Institution of Mechanical Engineers, the Institution of Electrical Engineers, the Iron and Steel Institute, and the British Association. As well as making improvements to steel production Sir William was a great believer in the future of electricity and invented the arc furnace and the electric pyrometer.

Gustave Eiffel (1832–1923)

Born in Paris into a family that included artisans and timber and coal merchants, Eiffel was educated at the Lycée Royal in Dijon and the College Sainte-Barbe in Paris. He graduated from the École Centrale des Arts et Manufactures in 1855. Eiffel founded his own company in 1866, specializing in wrought-iron structures. In 1884 Eiffel constructed the Garabit viaduct in France. For a time the highest bridge in the world, it earned Eiffel's company a worldwide reputation for excellence in craftsmanship and design. Eiffel designed the superstructure for Frédéric-Auguste Bartholdi's Statue of Liberty, which was dedicated in New York City in 1886. Soon after he began work on his greatest project, the three hundred meter high Eiffel Tower, for the celebration of the centennial of the French Revolution of 1789. An immediate success with foreigners, many notable Parisians hated it, and it was nearly torn down in 1909. The French government finally preserved it on the basis that the Tower's antenna was essential for telegraphy.

Jean Lenoir (1822–1900)

In the nineteenth century, coal gas became readily available, and efforts were made to use it for driving engines. In 1860, Lenoir designed and built the first successful gas engine. Lenoir's engine was similar to a double-acting steam engine, but was especially designed to run on a gas-air mixture. The mixture was ignited on alternate sides of the piston, using an electric spark. Some four hundred Lenoir engines were built, with outputs between 0.37 and 2 kilowatts. They powered pumps, looms, machine tools, and other machinery.

Thomas Alva Edison (1847–1931) (See also Case Study 6.1)

Edison pioneered the modern approach to invention and innovation, with his well-set-up Menlo Park and West Orange laboratories and his organized, intensive approach to research and development. He was keenly aware of the need to turn individual inventions into complete systems. Thus, in addition to his incandescent electric light bulb, he developed complete electrical lighting systems, including dynamos, transmission lines, switchgear, and even domestic light fittings. Edison's timely encouragement of Henry Ford in his early struggles to develop an economical automobile was critical to Ford's eventual success. Edison's encouragement of Ford led to a lasting friendship and eventually to Edison's Menlo Park laboratory being preserved at the Henry Ford Museum in Dearborn, Michigan. In conjunction with the national markets provided by national transport and other infrastructures, systematic approaches such as Edison's helped build the foundation for the eventual emergence of the United States as a superpower.

Cornelius Lely (1854–1929)

Lely, a Dutch engineer, graduated from Delft University, the largest and oldest university of technology in the Netherlands (or "low countries"). The Netherlands had always also had to battle the sea, with dikes (embankments) as their main line of defense. During the thirteenth century, ocean encroachment created a vast inland sea (known as the Zuyder Zee or Southern Sea) in the southwestern part of the Netherlands. In 1886, as a humble thirty-two-year-old *Waterstaat* engineering assistant, Lely started on a plan for draining the Zuyder Zee. He proposed building a series of meters-high dikes between the highest parts of the land under the Zuyder Zee and then draining the areas enclosed. The idea was eventually accepted, and he supervised its implementation. Today, protected by Lely's dikes, hundreds of thousands of people live and work on the bottom of the former Zuyder Zee.

Charles Algernon Parsons (1854–1931)

Parsons' father, the Earl of Ross, had been president of the Royal Society. Parsons was an unusual nineteenth-century British engineer, in that he came from the upper classes and studied mathematics at Cambridge. He learned the practical side of engineering at Armstrongs'' Works in Newcastle-upon-Tyne. (British experience during the Industrial Revolution had encouraged a severely practical approach to engineering education and a denigration of the role of theory.) Parsons'' unusual background, with its combination of theory and practice, placed him in a unique position to develop a device that could not have been developed on the basis of practical talent alone. Michael Faraday (1791–1867) had discovered the relationship between electricity and magnetism in 1831. This led to the invention of the dynamo, which needed a source of driving power at high rotational speed. Parsons set out to provide this source. In 1884 he built and patented the first practical steam turbine, fulfilling an ambition that dated back two millennia to Hero of Alexandria. Parsons fed high-pressure steam through a turbine with alternating rotor and stator rows. Almost all of our electricity today is produced by sophisticated developments of this invention. The jet engine can also be seen as a spin-off of Parsons' pioneering work.

ENGINEERING EDUCATION AND THE RISE OF THE PROFESSION

From the middle of the eighteenth century, formal and informal associations of practicing engineers, manufacturers, and scientists played an important role in the exchange of ideas and experience and in the general development of the engineering profession. What sort of people were the engineers who managed these changes? One early leader of the profession was John Smeaton, the first Englishman to differentiate himself from the traditional engineering role attached to the military (hence a "civil" engineer). An important informal engineering association, the Smeatonian Society, grew up around him. Armytage (1976) described the men (they were all men) and their period in detail (see the box, Some Pioneer Engineers in Britain, above).

Through the whole period of the Industrial Revolution, formal training as an engineer in Britain normally involved paying a substantial fee for a five-year pupilage in an engineer's office. The first engineering schools in Britain, at London and Glasgow Universities, were not established until 1840. Failure to recognize the need for systematic science-based education for technologists played a part in allowing other countries to overtake Britain.

Other major powers had taken a much more structured and centrally directed approach to training engineers. As early as 1676, France set up a specialized army corps of engineers, recruited largely from the nobility and upper middle class. In 1747, France established the first professional engineering school, the École Nationale des Ponts et Chaussées. This school was an outgrowth of a project to create a national road system, recognized as essential to the centralized control of the country. It was followed in 1794 by the École Polytechnique (Reynolds 1991, pp. 7–8). The engineering profession in France continues to enjoy high social status.

The U.S. Military Academy at West Point, established in 1802, was the first American engineering school. Claude Crozet, an early graduate from the École Polytechnique, was the first engineering professor in the United States. He taught at West Point before accepting the position of state engineer for Virginia in 1820 (Reynolds 1991, p. 12). The first non military engineering school in America, the Rensselaer Polytechnic Institute, was established in 1823 in Troy, New York (then a major industrial center). Rensselaer was so successful in attracting students that it forced several established U.S. universities to introduce technical courses.

However, British rule had heightened American antipathy to strong central government. Apart from military engineering, patronage for major engineering works and for engineering schools had to come from state governments and private enterprise. One result was that although some strong engineering schools were set up, for most of the nineteenth-century American engineers tended to be practically rather than "school" trained. This is perhaps most clearly seen in civil engineering. The construction of the first major American canal project, the Middlesex Canal (1793–1803) in New Jersey, drew heavily on the experience of a British engineer, William Weston, but foreign-born engineers were quickly replaced by native talent. New York State built the 365 mile long Erie Canal to attract trade away form Pennsylvania and it was during its construction (1816–1825) that the most widely used system for practical training of civil engineers emerged. Potential engineers worked their way up to head survey teams, learning the technical essentials on the job. The best then became assistant engineers. When the Erie was complete, its engineers took the training system on with them to other projects (Reynolds 1991, p. 13).

Germany established a number of polytechnics, the first at Karlsruhe in 1825. The example of the University of Berlin, founded by Wilhelm von Humboldt in 1809, was also important. Berlin became the prototype for the modern research university, with a new academic rigor in scholarship and laboratory experimentation replacing conjecture. Berlin also pioneered modern standards of academic freedom.

In Japan the new Meiji government took a leading role in setting up the physical and educational infrastructure for industrialization. The Imperial

Who Is an Engineer?

In the nineteenth century a confusion in the English language, apparently dating back to Chaucer, emerged as a problem for the identity and status of the profession. "Engineer" was being used to describe both those who operate engines as well as the ingenious people who devise them.

By 1850, the gulf between the two was widening rapidly. The intellectual preparation of the latter group (and those who designed other technologies) would increasingly include a grounding in mathematics and engineering science, as well as a lengthy period of practical training, justifying the title of "professional engineer." Even so, the confusion with the skilled operator of that supremely impressive creation of the engineering profession, the railway locomotive, continues in English-speaking countries to this day.

THE PANAMA CANAL

The Panama Canal is an early example of global engineering. Begun by France, it was completed by the United States. The construction workers came from more than ninety-seven different countries. Today, the canal continues to allows ships to pass across Central America between the Atlantic and Pacific Oceans. It operates in a politically neutral fashion, but its development was highly political. First thought of in 1517, in the following centuries Spain, Portugal, England, the United States, and France all considered proposals for a canal that would service their respective colonial empires. The first attempt at construction was made by France in the late nineteenth century.

French Construction

In 1876, Lieutenant Wyse led a survey team into the isthmus of Panama, in hopes of discovering the best canal route. He found fifty-one miles of undulating hills draped in dense rain forest, swamps, and unpredictable rivers. Wyse failed to survey a route, but he did negotiate successfully with the Colombian government (Panama was a part of Colombia until 1903) for permission to build the canal.

Ferdinand de Lesseps, who had built the Suez Canal, headed France's canal effort. He formed a construction company in 1879 and by 1881 it had plunged into the construction phase. However, de Lesseps' company failed to raise the projected 400 million francs—the project was undercapitalized and began with just 30 million francs. De Lesseps hired the engineering firm Couvreux & Hersent, which had worked with him on the Suez Canal project, to head con-

struction. These experienced engineers quickly discovered that their equipment, which had been suitable for the sands of Egypt, was too light for the hard rock of the isthmus. Couvreux & Hersent pulled out of the project after eighteen months, leaving it in the hands of subcontractors.

Unforeseen difficulties arose. The near-vertical walls of the canal were unstable and prone to landslides, which sometimes undid days of excavation. Some sections of the canal had to be dug nearly five times as wide as initially projected in order to have stable side walls. Unpredictable floods and raging rivers meant that much of the excavation had to be done underwater. The canal's projected path crossed six major faults, skirted five volcanic cones, and passed through seventeen fundamentally different types of rock, each of which required different drilling techniques. Yellow fever claimed the lives of an estimated 22,000 workers during the first seven years of construction, whereas malaria debilitated nearly all of the rest.

As technical difficulties and pressure from shareholders, the French government, and public opinion increased, de Lesseps' company was reduced to bribery and deception, paying magazines and newspapers to support the canal project: a report to shareholders even claimed that malaria and yellow fever were nonexistent.

In 1888, after excavating fifty-five million cubic meters of earth, the French company went bankrupt and excavation halted. In 1894 another company was formed in Paris to continue construction, but in the next ten years it only succeeded in excavating another six million cubic meters. The canal area fell into disarray: broken machinery and unemployed workers were the legacy of twenty-five years of French endeavor.

American Construction

The national interests of the U.S. Government were more directly involved than those of France. The United States realized its own need for an inter-ocean channel in Central America during the Spanish-American war. In 1898, the U.S. battleship Oregon left San Francisco, sailed 21,000 kilometers around Cape Horn, and arrived at Key West—just as the war was ending! Passage through a canal (a journey that today takes just eight hours) would cut approximately 17,700 kilometers from the journey from the Pacific to the Atlantic.

Aware that the United States was considering a Nicaraguan route for its canal, the French company persuaded the American Senate to purchase France's Panama route for $40 million, contingent on the Colombian government's approval. Colombia refused to ratify the treaty providing for the canal sale, but after considerable political upheaval, the newly seceded Panamanian government ratified the treaty in November 1903. This allowed the United States to build and administer a canal for an annual fee of $250,000.

John Wallace, the Chief Project Engineer hired by the military-organized U. S. Isthmian Commission, was apparently technically competent, but a poor leader. Like their French predecessors, the Americans plunged ahead with excavation without detailed specifications for the canal, and even before they had decided what type of canal (sea-level or lock-type) to build. The commission suffered from the same inefficiencies and lack of leadership that had destroyed the French. Some work teams nearly starved to death while food sat undistributed in warehouses. An outbreak of yellow fever and malaria in 1904 brought construction to a near stand-still.

Early in 1905 President Roosevelt dismissed the First Commission and set up a civilian-staffed Second Isthmian Commission, headed by Theodore P. Shonts. John F. Stevens, a former railroad official, replaced Wallace and created an effective railway system to distribute supplies and equipment. Sanitary housing was provided for all workers.

Stevens also recognized the work of William Gorgas, Chief Sanitation Officer. Gorgas, a doctor who had helped discover the cause of malaria and yellow fever (carried by mosquitoes), had been appointed in 1904 by the First Commission; however, it had dismissed his mosquito theories and denied his repeated requests for screens, sulfur, quinine, and other supplies. Finally provided by Stevens with these materials, Gorgas launched an attack on the epidemic by draining or filling in swamps. In just a few years, he largely eradicated yellow fever and malaria from the Panamanian jungle.

Pressed by Stevens to decide on the specifications, President Roosevelt formed the International Consulting Board of Engineers in June 1905. It was instructed to keep speed and feasibility as the prime considerations. Yet, despite a projected three extra years of construction time, the board ultimately recommended a sea-level canal. Why the selection of a sea-level canal? It appealed

(continued on next page)

idealistically, bringing to mind "dividing" a continent, while a lock-type canal would require that ships rise twenty-six meters and sail "over" the continent. Known as the "American" style canal, lock-type canals somehow sounded inferior to the "European" design. Roosevelt, however, suspected that this was poor advice. Stevens and several Commission members lobbied against the board's decision, and in June 1906, Congress voted for a lock-type canal.

Success at last

With disease curbed, a solid development plan, and efficient materials transport, construction of the Panama canal moved ahead. It opened on August 15, 1914 with the United States controlling and operating the canal, as well as a five-mile border on either side of the canal, known today as the Canal Zone.

The American effort had very nearly ended as the French one did—in ruin. Similar difficulties plagued the Americans: a bureaucratic supply system, poor leadership, and overseas management that failed to appreciate local problems. The United States had the advantage of developments in equipment (rock drills were faster; steam shovels and trains had larger capacities), but both projects were jeopardized by poor planning. In the rush to "make dirt fly," both made poor technical decisions. Disease nearly destroyed the Americans, not for a lack of medical technology, but because of bureaucracy and politics.

Why did the Unites States succeed where France failed? The gifted leaders President Roosevelt appointed in 1905 made the difference. Shonts eliminated red tape, improved organization, and devised a solid development plan. Stevens provided a disease-free, properly fed and housed workforce, a railroad system to distribute equipment and supplies, and more advanced construction tools. Well resourced, equipped, and led, they were finally successful.

Sources

Based on work by Jennifer Gadda, Stanford University, 1997, and:

Busey, J.L. 1974, *Political Aspects of the Panama Canal*, University of Arizona Press, Tucson.

Cameron, I. 1972, *The Impossible Dream: The Building of the Panama Canal*, William Morrow & Company, Inc, NY.

Encyclopedia Americana, 1990, Vol. 21.

College of Engineering was established in Tokyo in 1873 with a Scottish director, Henry Dyer (1848–1918). The College strongly emphasized practical as well as academic skills. Another very effective foreigner hired by the College was William Edward Ayrton, probably the world's first Professor of Electrical Engineering. In 1885 the College merged with Tokyo University, and by the end of the Meiji era in 1912 there were four Imperial Universities, all with engineering faculties. Military training centers, including the Japanese Naval Academy established in 1870, were also important for engineering education (Morris-Suzuki 1994, pp. 80–82).

The Scientific Foundations of Modern Engineering

In the period we have covered in the latter part of this chapter, technical innovations led to major advances in science. The role of science in technical matters had previously been mainly to explore, explain, and systematize what engineers had done. However, by the mid-nineteenth century, craft skills and technical know-how had taken engineering practice as far as they could. Profits from the high productivity of industrial achievements and the proceeds of colonial and imperial ventures financed the next stage of technological development, which required much closer interaction between science and engineering.

The leaders of the engineering profession in Britain were slow to recognize the importance of mathematics and applied science. The steam engine had provided the motive power for industries that, although changed by factory-based production, were built on traditional techniques. By the end of the nineteenth century, major new industries were emerging, based on science and on systematic research. Individual inventors and engineers were starting to give way to scientists and industrial researchers, and in time to large-scale research and development departments, often integrated with major industrial enterprises.

In the interplay between science and engineering, pivotal roles were shifting. The ongoing relationship between the two will be explored in later chapters of this book.

The history of the engineering profession itself has reflected the transformations wrought by technological change. In the seventeenth and eighteenth centuries civil engineering was increasingly differentiated from military engineering. Mechanical engineering came with the steam engine and the railroads in the late eighteenth and early nineteenth centuries. As engineering practice has developed, many other specializations have been added to these basic disciplines.

In the next chapter we focus on a small number of key technical areas, and explore the way engineering itself is changing. Chapter 2 takes the history of engineering into the next millennium, foreshadowing new challenges to engineers that we examine in more detail through the balance of the book.

Discussion Questions

1. Why study the history of engineering?

2. Roman roads enabled speedy communication and transport throughout the Roman Empire. What are some of their modern equivalents? What Empires do they serve?

3. Muslim influence contributed to the European Renaissance of the fifteenth and sixteenth centuries. Discuss how the blending of Eastern and Western cultures is shaping change in modern times.

4. How important were technological factors discussed in this chapter in the social and economic changes in Europe and America?

5. Why was Britain the first country to go through an Industrial Revolution? Why not the Netherlands?

6. What are some of the things missing from the history we have presented here, and why do you think they have been omitted? For example, how would you add a female or a minority perspective to the history of technology?

7. Imagine and describe the thoughts, feelings, and emotions of a nineteenth century craftsperson who has been made redundant by a machine. Pinpoint modern equivalents, and imagine them, too.

Further Reading

Billington, D. P. 1996, *The Innovators: The Engineering Pioneers Who Made America Modern*, Wiley, NY.

Burke, J. 1985, *The Day the Universe Changed*, BBC, London. This book, of the television series of the same name, deals with a number of profound changes in the way people saw or understood the world.

Davidson, B. 1991, *Africa in History*, Simon & Schuster, NY.

Diamond, J. 1997, *Guns, Germs, and Steel: The Fates of Human Societies*, W.W. Norton, NY. Diamond and Landes offer competing, perhaps complementary, and very readable accounts of the dynamics of the period discussed in this chapter.

Landes, D. S. 1998, *The Wealth and Poverty of Nations: Why Some Are So Rich and Some So Poor*, W.W. Norton, NY.

Merson, J. 1990, *The Genius that was China: East and West in the Making of the Modern World*, Overlook Press NY. The book and television series vividly demonstrate the development, transfer and control of technology.

Pacey, A. 1976, *The Maze of Ingenuity: Ideas and Idealism in the Development of Technology*, MIT Press, London. Discusses the period treated here with much more emphasis on the technological aspects.

REFERENCES

Armytage, W. H. G. 1976, *A Social History of Engineering*, Faber and Faber, London.

Hoyrup, J. 1985, Varieties of mathematical discourse in pre-modern socio-cultural contexts: Mesopotamia, Greece, and the Latin Middle Ages, *Science and Society*, XLIX, pp. 4–41.

Kline, S. J. 1995, *Conceptual Foundation for Multidisciplinary Thinking*, Stanford University Press, Stanford, CA.

Layton, E. 1991, *Science and Technology in 19th-Century America*, in T. S. Reynolds (ed.) *The Engineer in America: A Historical Anthology from Technology and Culture*, University of Chicago Press, Chicago.

McGinn, R. E. 1991, *Science, Technology and Society*, Prentice Hall, Englewood Cliffs, NJ.

———, 1997, Lecture notes for STS 101, *Science, Technology and Society Program*, Stanford University, Stanford, CA.

Morris-Suzuki, T. 1994, *The Technological Transformation of Modern Japan: from the Seventeenth to the Twenty-first Century*, Cambridge University Press, Cambridge.

Nash, G. B., Jeffrey, J. R. 1998, *The American People: Creating a Nation and a Society*, Longman, NY.

Pacey, A. 1983, *The Culture of Technology*, MIT Press, Cambridge, MA.

Reynolds, T. S. 1991, *The Engineer in America*, University of Chicago Press, Chicago.

Rosenberg, N. 1976, *Perspectives on Technology*, Cambridge University Press, Cambridge. Analyses the development of modern industry in the US.

Stanley, A. 1983, Women Hold Up Two-Thirds of the Sky: Notes for a Revised History of Technology, in Rothschild, J. (ed.) *Machina ex Dea*, Pergamon, NY.

Smith, A. 1776 (reprinted 1986), *An Inquiry into the Nature and Causes of the Wealth of Nations*, Penguin, Harmondsworth.

Vernon, R. 1989, *Technological Development: The Historical Experience*, Economic Development Institute of the World Bank, Washington, DC.

Whitehead, A. N. 1938, *Science and the Modern World*, Penguin, Harmondsworth.

Wiener, M. J. 1985, *English Culture and the Decline of the Industrial Spirit 1850–1980*, Penguin, Harmondsworth.

2

Engineering in the Modern Era

All progress is initiated by challenging current conceptions,
and executed by supplanting existing institutions.

— *George Bernard Shaw (1856–1950) Anglo-Irish dramatist*

By the end of the nineteenth century, engineering had emerged as one of the most potent forces in western society. It is now something of a cliché to state that engineers have played a major role in shaping the modern world but in this chapter we see just how tremendous its impact has been. The time-line lists some important milestones, but cannot really convey the sheer pace of invention and innovation in the twentieth century.

Britain provided the first model of an industrialized nation, the United States of America offered a different model, and others have since emerged. However, the basic level of technological development we associate with a "modern nation state" is much the same: access to natural resources, competitive local industries, a degree of political stability, a positive financial investment climate, and a labor force with an internationally competitive level of education and technical expertise. In such conditions engineering has continued to flourish. Dramatic social change has commonly been associated with technological change, often in turn encouraging further technological development. At the end of the twentieth century industrialization has reached into every part of the globe, not always as a blessing.

Technologies pioneered in the aftermath of the Industrial Revolution, including those associated with the internal combustion engine and electricity, are integral to our modern quality of life and standard of living. In this chapter we look briefly at some of the most important and exciting areas of engineering activity, including the automobile industry, infrastructure, flight, electronics and computing, and manufacturing. We also discuss some of the implications of newer engineering disciplines that are carrying us forward into the next millennium, such as systems engineering and bioengineering.

THE AUTOMOBILE INDUSTRY AND CHANGES IN PRODUCTION

The last years of the nineteenth century saw what was virtually a "bicycle revolution." This energy-efficient technology, based on the pneumatic tire and the roller chain, greatly increased the mobility of many millions of people. It also created a continuing attachment to relatively cheap, independent personal transportation. It was one reason the automobile became such a pervasive technology in the twentieth century.

We can use the technology associated with the automobile to demonstrate the fundamental importance of a series of questions that engineers should pose about new technologies (see Key Concepts and Issues). Like ripples spreading out from a stone tossed into a pond, the impact of any technology depends on the time elapsed and the vantage point of the observer. The development phase of the automobile is summarized in the accompanying boxes and discussion in this chapter; the transfer of the technology is dealt with separately, notably in several case studies in Chapter 8, but our main focus here is on the remaining questions we have asked about the impact of the automobile and the industry that arose from it.

■ Ford and the Early Automobile Industry

During the early years of the industry, automobiles were assembled by highly skilled craftsmen. Among other problems, parts had to be hardened after machining, causing distortion, which then had to be corrected by the assemblers filing the parts to fit together. There were no standard models; automobiles were made to individual customer order. One result of the inexact processes of manufacture was that no two automobiles were identical, even if they had been ordered to the same specification.

1850	Crystal Palace Exhibition 1851
	Union Pacific Railroad across United States 1869
	Suez Canal opened 1869
1875	
	Invention of Telephone 1876
	First Internal Combustion Engine 1876
	Beginning of Electric Lighting 1878
	Hargrave's work on Flight 1880s
1900	Invention of Radio 1900
	Wright Flyer—Heavier than air flight 1903
	Theory of Relativity 1905
	Ford Moving Assembly Line 1913
	Panama Canal Opened 1914
	Splitting the Atom 1919
1925	Invention of Television 1926
	Gas Turbine Invented 1930
	First Electronic Computer 1942
	First Atomic Explosion 1945
	Invention of Transistor 1947
1950	
	Discovery of DNA 1953
	Sputnik Launched 1957
	First Integrated Circuits 1958
	First Satellite TV Broadcasts 1962
	First Man on Moon 1969
1975	
	First Personal Computer (Apple II) 1977
	Three Mile Island Nuclear Leak 1979
	Compact Disc 1982
	Bhopal Disaster 1984
	Challenger Disaster 1986
	Chernobyl Disaster 1986
	Nuclear Weapons Limitation 1987
2000	English Channel Tunnel 1990

Figure 2.1 Timeline of Modern Technology.

Henry Ford (1863–1947) introduced the moving production line in 1913, using a common gauging system throughout the manufacturing process. The recent development of high-speed steel for machining pre-hardened metal parts helped Ford by eliminating distortion problems. His standard automobile, the Model T, was specifically designed for ease of assembly.

Ford's moving production (or assembly) line is said to have been adapted from the chain conveyor systems used in Chicago meat works in the 1890s. The line enabled Ford to almost double labor productivity, partly

Technology: Some Basic Questions

- ◆ How and why was this technology developed?
- ◆ How has it been transferred?
- ◆ What are the "technology drivers" in this area?
- ◆ How have the goals of technological development in this area changed over time and how might they best be achieved?
- ◆ What are the roles and responsibilities of professional engineers with respect to this technology?
- ◆ What have been its broad social, political, economic, and environmental impacts?

because it brought the work to the worker, and partly because it enabled Ford to control the pace of work. Ford effectively eliminated the craft assembly of automobiles, replacing skilled work with jobs that could be learned in minutes by workers just off the farm. To start with, annual labor turnover was as high as 380 percent. When Ford doubled pay rates, to $5 a day, he held on to his workers, but there was little intrinsic satisfaction in the work (Mathews 1989, p. 25).

Mass production was strikingly successful in achieving Ford's central aims of increasing productivity and reducing manufacturing costs. Its negative implications for work and workers of early approaches to mass production are discussed below.

Although Ford put his stamp so dramatically on the early automobile industry, other players soon emerged. Alfred P. Sloan, Jr., another engineer, became president of General Motors (GM) in the early 1920s. His first job was to consolidate a number of previously independent automobile manufacturers into a single corporate structure.

Sloan created decentralized divisions, each serving a sector of the market, and managed as separate profit centers from a small corporate headquarters. The high point of U.S. mass production was 1955, the year Sloan retired. More than seven million automobiles were sold in the United States that year. Three giant enterprises, GM, Ford, and Chrysler, sold 95 percent of them, with just six models accounting for 80 percent of sales! However, other challengers, with different approaches to manufacturing, were emerging.

Social and Economic Impacts

Our experience with the automobile shows that a technology needs to be considered in a wider social context. It also illustrates the difficulty of anticipating and controlling the social consequences of new technology. Collingridge (1980, pp. 16–18) probably overstates the case when he argues that the initial understanding of any new technology is so limited that control can only be arbitrary. Major early concerns about the social and environmental impact of the automobile were that it would raise excessive amounts of dust from unsealed roads and frighten horses; it was even feared that it would stop cows producing milk! Steps were taken to control the dust problem, but by the time a more serious issue, the road toll, became clearly evident, the technology was well developed and widely diffused. By then the range of control measures was rather limited.

The automobile shows how much a single technology can change the way we live and work. The difference between compact Old World cities and sprawling New World ones highlights the automobile's impact on our lives. Automobiles have also changed the way we die. By the year 2000, they will have killed twenty-five to thirty million people, approaching the death toll in both world wars.

Manufacturing industry from the nineteenth century onwards was the source of immense economic growth. The industries that grew up around the automobile are still arguably the most important in modern manufacturing. However, one of the long-term legacies, in which the automobile played a big part, has been widespread pollution of the environment. Air pollution is a major hazard, with exhaust emissions of lead, carbon monoxide, unburned hydrocarbons, and oxides of nitrogen. Engineers have only started to address this problem seriously in recent decades.

Other challenges (some of which took many years to become apparent) include traffic congestion, the need for alternative energy sources, disposal of nonbiodegradable materials such as plastics, and depletion of natural resources including oil and metals. Some of these problems relate to the adequacy of provision of infrastructure, discussed later in this chapter. Another whole cluster of societal issues are highlighted by the development of the automobile industry: it has been the site of profound changes in work practices and important interactions between employers and organized labor.

Scientific Management

In developing his form of mass production for mass consumption, Henry Ford pursued organizational ideas that came to be known as "Scientific

The Early Automobile

The development of the automobile started in the nineteenth century. Attempts to develop new power sources by applying steam engine practice to internal combustion engines produced noisy machines that ran erratically. The four-stroke cycle was conceived in 1862 by Beau de Rochas and first made to work in 1876 by Nikolaus Otto (1832–1891). Over a century later, this engine still provides motive power for the ubiquitous automobile.

A prime requirement was a suitable fuel. Gas produced from coal was beginning to be piped around many cities for cooking, heating, and lighting. It was convenient for stationary engines, but impractical for vehicles. Towards the end of the nineteenth century, experiments began on using oil as a transport fuel. When crude oil is distilled, the first or lightest fraction to be vaporized is called gasoline or petroleum, then comes kerosene, then medium, and finally heavy oils. Development of engines to work on gasoline and medium oils proceeded simultaneously.

In 1885, Gottlieb Daimler (1834–1900) and Karl Benz (1844–1929) produced four-stroke engines that used gasoline. In the same year Daimler produced the first motorcycle and Benz the first automobile. Their early methods of fuel injection were quite crude, and their engines were greatly improved by the carburetor, invented in 1893 by William Maybach. He used the principle of the venturi, in which a narrowing in the air passage increased the flow velocity, reducing the air pressure. This reduced pressure was used to draw in and atomize a jet of liquid fuel. The basic configuration of the carburetor has changed relatively little since the 1890s.

Starting gasoline engines by hand was slow, awkward, and potentially dangerous. The electric self-starter made gasoline the fuel of choice for

Management." The person most responsible for developing and promoting Scientific Management was Frederick Winslow Taylor (1856–1915). A member of a leading Philadelphia family, Taylor was an engineer who had worked as a laborer to improve his health. His methods (summarized in the following Key Concepts and Issues: Taylor's Scientific Management) aimed at maximizing both labor productivity and management control.

Because of the way Taylor's work was actually applied, there is a danger of crude oversimplification in discussing his approach. In *Principles of Scientific Management*, (1967, p. 128) Taylor argued that workers should be encouraged to suggest improvements in both methods and equipment.

automobiles. It was first fitted to an Arnold car in Britain in 1895. An improved version, developed by Charles F. Kettering (1876–1958), was introduced into production automobiles in 1912 by the American company, Cadillac. Until this invention made starting gasoline-engined vehicles quick, clean, and safe, they held only 22 percent of the market. Steam power had 40 percent, but steam took time for pressure to build up. Electric battery-powered cars made up the other 38 percent, but their range was limited, and recharging was slow.

German technical knowledge and skill had produced a working internal combustion engine, and it was German thermodynamics education that laid the foundation for the compression ignition ("diesel") engine. From 1897, Rudolph Diesel (1858–1913) worked on the concept of a compression-ignition system designed to burn heavier fractions of oil. Diesel had a strong background in science and mathematics, and was so confident about his new engine that he called it a "rational engine." All his confidence was required in overcoming the early difficulties of fuel injection, inefficiency, and mechanical integrity under high pressure. (One of his prototypes produced an explosion that destroyed the prototype and nearly killed him.)

Progress on the diesel engine was slow until fuel injection processes were perfected in the 1920s, but such engines did successfully power ships and submarines in World War I. Although most private cars are still powered by gasoline engines with maximum efficiencies around 25 percent, many trucks, trains, and ships use diesel engines. Small to medium-sized modern diesel engines are commonly turbocharged and have overall efficiencies around 40 percent. The efficiency of very large, slow, constant-speed diesel engines in ships approaches 50 percent.

These suggestions were to be carefully tested, and adopted if markedly superior to current approaches, with the worker given full credit and paid a cash premium. However, this more participative approach was not widely applied.

In the first decades of the twentieth century, "Scientific Management" became the basis for mass production of many standard items. Control over the organization and planning of work was taken away from the shop floor and from the people actually carrying out the tasks. As long as manual manipulative labor was a key production cost, the logic of this approach appeared compelling, despite its obvious drawbacks of degrading

Taylor's "Scientific Management"

The principles Taylor enunciated as Scientific Management were essentially:

- division of labor to a level of minute detail, allowing just the necessary amounts of strength and skill to be purchased as were needed for each of the subdivided elements of the overall task;
- separation of planning and execution, with planning removed entirely from the shop floor;
- time and motion study to find the best and quickest way of doing work and to allocate a time for doing it (by combining the times for its component motions or by timing a worker);
- payment by piecework; later replaced by machine pacing.

Rosenbrock (1990, pp. 145–151) argues that Taylor significantly overstated the likely benefits, disguising much of the real thrust of his work. He summarizes Taylor's principles:

- Improvements in working, and in machines, that offered the benefit of greater production for an unchanged human effort...
- A number of techniques to allow the employer to redefine the contract of employment to his own advantage...
- The withdrawal of control and initiative from the lower levels of an organization, and their concentration, to the greatest possible extent, in the higher levels. This entails the simplification and precise definition of tasks, which assists in subsequent mechanization.

work and demotivating workers. Although it was immensely successful in increasing productivity, the critics of scientific management have stigmatized this approach as tending to take the human purpose out of work, often reducing it to an essentially alienating activity.

Individuals such as Lillian Moller Gilbreth (Case Study 7.2), who had a detailed understanding of production issues and who worked with Taylor and her husband, Frank Bunker Gilbreth, did much to reduce alienation in the workplace.

In the period between 1880 and 1930, automobile manufacturing and much of the rest of industry in America were restructured in accordance with Taylorist principles. The world crisis of the late 1920s and the 1930s was

Mathews (1989, p. 23), gives more attention to Rosenbrock's third point—de-skilling work and maximizing management control—as the essence of the Taylorist approach. Mathews calls this approach Fordism, and sees the process as made up of three steps:

1. Dissociation of the labor process from the skills of the workers: appropriation by management of all the workers' traditional knowledge and skills and reducing this body of knowledge to systematic formulae and rules.
2. Separation of conception from execution: removing all possible brain work from the shop floor and relocating it in a planning or layout department.
3. Using this monopoly over knowledge to control each step of the labor process and its mode of execution, dictating how each job is to be done through complete written instructions. These describe in detail the task, how it is to be done, and the time it is to take.

Some General Observations

The use of the term "scientific" in this context is interesting. It has been argued that the failure by the advocates of scientific management to test or even identify their underlying assumptions led to subjective, value-laden material being presented as scientific. Taylorist philosophies still underpin many approaches to job design today, but they are not consistent with the level of worker involvement needed for modern quality production.

a result of the production of more goods than those who could purchase them were prepared to buy. The crisis was only resolved with the diversion of productive capacity into preparations for what became World War II.

At first his production lines essentially assembled parts made by other suppliers, but Ford moved rapidly to complete vertical integration, even adding a steel mill and glass factory to his Highland Park plant. The associated bureaucracy and Ford's centralization of decision-making in his own hands nearly bankrupted the company.

Ford's focus had been on maximizing the effectiveness and efficiency of manufacturing to produce a uniform product, a rather basic transport vehicle. But by the 1930s this was no longer enough; market demands

How the Automobile has Changed

In the century since it was invented, engineers have played major roles in automobile development. The major technology drivers have been reliability, comfort and convenience, performance, safety, and most recently, energy efficiency. Changes include:

Power

Power to weight ratios, both for the engine itself and for the overall vehicle, have risen sharply. Higher compression ratios and turbocharging have increased efficiency. However, the associated higher combustion temperatures have increased production of nitrogen oxides, contributing to formation of photochemical smog. Catalytic converters have been introduced to clean up the exhaust gases.

Fuel

As compression ratios increased, tetra ethyl lead was added to increase the octane rating of fuel to prevent "knocking" (detonation rather than steady combustion of the air/fuel mixture in the cylinder). Concerns about the toxic effects of lead, especially on young children, and the fact that it would poison the catalyst in catalytic converters, led to a change to unleaded fuel.

Body

Improving fuel economy meant losing weight. For "full-size" U.S. cars, the change from a separate chassis to a stressed skin (monocoque) body reduced weight

---

were for a more varied and sophisticated range of products. Ford's company survived by reorganizing along GM lines.

■ The U.S. Automobile Industry and Organized Labor

A typical union reaction to Scientific Management was fragmentation of job classifications, making for demarcation problems and a very inflexible workforce. In Britain and Australia, with their tradition of craft rather than industry unions, the difficulties were compounded by the presence in the workplace of numbers of separate, even mutually antagonistic, unions.

When automobile makers in the United States laid off workers during economic slumps (treating wages as a variable rather than as a fixed cost),

from around 2,300 kilograms to 1,400 kilograms. Replacement of steel by plastic and aluminum made further savings.

Brakes

On early autos, mechanical brakes worked on only two of the four wheels. The change to hydraulic drum and then disc brakes improved stopping performance amazingly. Recent anti-skid technologies sense the rate at which wheels are slowing and ease the brakes off to maintain control and prevent skidding.

Some Other Changes

All-independent suspension and better braking offered increased comfort and better handling, improving "primary safety" (the likelihood of avoiding an accident). Changing from rear to front wheel drive increased interior space by eliminating the transmission tunnel.

Nader's dramatic 1965 exposé of the U.S. auto industry helped move the emphasis from styling to engineering. Safety started to be marketed. Systematic crash-worthiness testing, improved seat belts, air bags, and more robust bumpers have improved "secondary safety" (making crashes more survivable). Compatibility between vehicles is important for secondary safety, and the height difference between standard automobiles and recreational vehicles is a serious concern.

tensions between labor and management occasionally erupted into battles between strikers and company police. A 1936 Ford strike was settled with a contract that gave workers a seventy-five-cent minimum wage, good overtime pay, rules respecting senior workers, and a 20 percent reduction in assembly line speed. In January 1937, the National Guard was deployed to protect union strikers during a strike against GM. On February 11, William Knudsen, the president of GM, agreed to recognize the United Auto Workers (UAW) as the sole bargaining agent for workers at nearly twenty GM plants that were on strike. The union signed a four-page contract that outlined bargaining procedures, standardized work schedules, established a seniority system, instituted a grievance procedure, and set a solid minimum pay. In 1939, the entire GM corporation signed an agreement with the UAW. However, the U.S. Government remained wary of organized labor, and after the war passed laws aimed at limiting its power.

By contrast, in Occupied Japan in 1946 the Japanese government strengthened union rights, with American support. Japanese workers were looking to lifetime employment, and they refused to accept mindless jobs. They went on to make a major contribution to what became known as the Toyota Production System, or more generally as *lean production*, discussed in Chapter 7. For employers, one implication of lifetime employment was that continuing employee training and development made excellent economic sense. A different approach by the state to industrial rights has led to a very different workplace culture in Japan, one that supports flexible, high quality manufacturing.

Is there an optimum level of mechanization and automation? The answer probably depends on the breadth of the design criteria. Increasing mechanization has been typical for automobile assembly lines; in 1952 the Ford Motor Company began using automatic drilling machines in an engine plant and found that 41 workers could now do a job that previously had taken 117 workers to complete (Nash and Jeffrey 1998, p. 928). The trend towards machines replacing workers has continued, but very highly automated plants lack flexibility. As labor costs become less significant overall and flexible manufacturing and marketing of individualized products become more important, workers' skills are increasingly valued.

ENGINEERING AND INFRASTRUCTURE

Infrastructure is the modern term for a very traditional focus of engineering activity. It covers all the basic facilities and systems that allow a large-scale society to function effectively, including transportation and communication systems, energy and water supply, and waste removal. Engineering remains central to the conception, design, and implementation of these systems.

Infrastructure has also been the site of one of the most dramatic developments in modern times: the extension of the professional engineering role into the political arena. This has opened up a wider debate about the nature of engineering work, including the role of nonengineering inputs. As the twentieth century progressed, broad political issues (as defined in Chapter 4) began increasingly to affect decisions on infrastructure. It is now recognized that choices about the location of a road, bridge, or dam are likely to be an advantage to some members of a community and be a disadvantage to others. The size of projects and their impact on society and the environment varies tremendously. During this century, the trend has been towards large-scale projects. The larger a project is, the wider its effects are likely to spread, and the less straightforward it is to avoid "secondary," unintended consequences (as Case Study 4.3 dramatically illustrates).

Typically it has been politicians, rather than engineers, who have had to confront such questions as: What are the options? How much will they cost? Who pays? How far is it acceptable and desirable to charge those who benefit, and compensate those who lose from this development? In more recent times, engineers, however reluctantly, have been drawn into these debates. Preferred technological responses are increasingly being mediated by other considerations. The answers often depend more on the political culture and power relations in the society than on what engineers might consider to be "best practice." Some of the current problems and issues are summarized in the Key Concepts and Issues box on Infrastructure (p. 64).

▌Roads and Railroads

"Walking cities" were the usual human experience until well into the second half of the nineteenth century. Problems with dust, mud, and poor sanitation were common on unpaved streets; overcrowding and congestion became increasingly serious as the population grew. The building of the railroads in the nineteenth century (see Chapter 1) and of road systems that allowed the automobile to become the dominant mode of transport in the twentieth century enriched urban life and lessened rural isolation.

In the United States the railroads were the pioneers of big business and can be seen as a major modernizing force in American society. The scale of the railroad enterprises (with huge amounts of capital for construction and on-going maintenance, large work forces, and very complex operations) required new company structures and management techniques. An engineer and inventor hired by the Erie Railroad, Daniel McCallum, emphasized division of responsibility and a regular flow of information. His system became a model for other businesses. The steel industry, closely dependent on the railroads, adopted their practices, and other large-scale businesses followed suit. They emulated the behavior of the railroads in other ways, with cutthroat competition, underpricing, attempts to cut workers' wages, mergers, and so on. The "Age of the Robber Barons" grew out of the original expansion of the railroads (Nash and Jeffrey 1998, pp. 618–620).

Great trains such as the Twentieth Century Limited rivaled the legendary British and European trains, and the romance of steam, with its visible, almost hypnotic power, appealed to generations of enthusiasts around the world. The locomotive whistle and the hiss of steam inspired poets and singers in a way that the diesels, which took over in the 1950s and early 1960s, could not do, but competition from airlines and from vehicles using the interstate highway system finally overtook U.S. railroads.

The ascendancy and then decline of the railroads raises issues about the finite life of a technology and the extent to which obsolescence is a social rather than (or as well as) a technical construct. With steel wheels on steel rails, the railroad remains the most energy-efficient way of quickly transporting large loads overland. It ought to remain highly competitive with road and other modes of transport, particularly if operated as part of an integrated transport system. Proper comparison of the relative costs of road building with heavy or light (streetcar) rail requires sophisticated analysis, as Case Study 2.1 shows.

By the 1930s, the inadequacy of the U.S. road system was very apparent. The German system of autobahns greatly impressed U.S. military personnel at the end of World War II. The interstate highway network in the United States took forty years to build; it was the largest engineering project ever undertaken, bigger than building the pyramids in Egypt or the Great Wall of China. The cost was around $50 billion. Modeled significantly on the autobahns, it involved the construction of 40,000 miles of high-speed divided highway. In the process, it changed the way most Americans live. The promise of the new highway system was that one would be able to get anywhere quickly. Seventy-two-hour cross-country trips became a rite of passage for young people, but in the cities traffic congestion seemed, if anything, to get worse. The politics of the interstate highway system, including its very uneven social impacts and the eventual rise of concern about the effects of running highways through cities, are discussed in Case Study 4.4.

The Interstates were a remarkable engineering achievement, the envy of most of the rest of the world, but they raised issues that have still not been addressed effectively. Perhaps most serious is the long-term U.S. dependence on cheap oil. Targets have in the past been set for vehicle fleet fuel efficiency, but then allowed to slide. This has sent the unfortunate, indeed regrettable, message to industry and the public that the issue is not really serious. The boom in fuel-thirsty "off-road" vehicles and light trucks, used essentially for single-occupant commuting, reflects this perception.

Regardless of the "correctness" or otherwise of past decisions, we now need to deal with today's and tomorrow's realities. Since World War II, the population of the United States has doubled, from 133.4 million in 1945 to over 270 million in 1998. The country's housing stock, indeed way of life, has been substantially relocated to suburbs designed for automobile access. Goods transport, even over long distances, depends increasingly on trucking. The huge public and private investment in these systems cannot be ignored and must not be wasted.

Traditional transport infrastructure development was focused on mass-transit installations; the twentieth century saw the emergence and eventual dominance of internal-combustion-powered road vehicles. Despite their tremendous flexibility, the inherently low-passenger density of automobiles with a single occupant presents problems. At the century's end, there is a return to more intensive systems, often including underground passenger rail transport (subways or metros), to encourage and support socially rich and compact city centers. Case Study 2.1, below, and recent studies like those reported in Diesendorf and Hamilton (1997) that take into account whole-of-life costs and sustainability issues, suggest that the overall costs of road transport have been seriously underestimated.

▪ Other Infrastructure Issues

For a modern, productive, and congenial society, there is no question as to the importance of reliable water and electricity supplies, safe and effective waste treatment and disposal, and good communications facilities. Case Study 4.2 describes the work of the U.S. Army Corps of Engineers, which has a major role in managing U.S. water resources. In the 1990s, decisions were made in developed countries around the world to dismantle monolithic utility companies and open electricity and communications markets up to competition. To a significant extent, and certainly in the case of AT&T, the U.S. telephone company that was broken into smaller companies that came to be known as the "Baby Bells," this move was seen as an antitrust, antimonopoly measure. It may in part have reflected a perspective of utility companies as protected from and unresponsive to public concerns, and even perhaps overcapitalized and inefficient. All the implications of the changes are not yet clear. There was certainly a belief that in an open and deregulated market costs to consumers would fall. Does this seem likely to be realized?

A parallel and increasingly urgent question is: How can national infrastructure be made more sustainable? The scale of the problems is immense. Hull (1990) suggests that it will cost $75 billion just to upgrade and replace municipal sewage treatment plants in the United States, and the expense of protecting the environment could easily be hundreds of billions a year. The American Society of Civil Engineers (ASCE 1998) has become so concerned at the situation regarding U.S. infrastructure that its world wide web site emphasizes the need to renew and extend it (see Key Concepts and Issues box on infrastructure below). Of particular concern

CONVENTIONAL VERSUS COMPREHENSIVE ANALYSIS OF TRANSPORTATION INVESTMENTS

The following case study describes a situation that occurred in a particular North American community, although it could have happened almost anywhere in the industrialized world:

The main highway connecting a fast-growing city with a nearby suburb was increasingly congested during peak periods. Regional transportation planners considered various transportation improvements to address this problem. There were two main options:

- widening the highway by one lane in each direction to accommodate two thousand to four thousand additional peak-period vehicles, with cost approximately $250 million; or
- a light-rail option to accommodate the same number of trips, with cost approximately $300 million.

The highway-widening option was recommended by transportation engineers and selected for funding by government agencies. However, these estimates only considered construction costs directly borne by the transportation agency. A number of important additional impacts of the two thousand to four thousand additional vehicles traveling into the city were overlooked, including:

- Additional parking spaces for these vehicles. There was already insufficient city parking capacity. As the highway project was being completed, the city government was giving away a $1 million parcel of downtown land to subsidize construction of another 275-space multistory parking structure.

- Increased traffic congestion on city streets, with associated future costs to address the problems. Traffic congestion on arterial roads presents increasing problems for both the city and the suburbs. If the same number of trips had been made by transit, this would have saved tens of millions of dollars for local government, businesses, and consumers.

- The fixed cost to consumers of owning an automobile. Vehicle operating costs and transit fares were considered in the analysis, but analysts assumed that each traveler had an automobile that would simply sit unused if they rode transit. In fact, high-quality public transit allows some households to defer the replacement costs or reduce their automobile ownership.

- Environmental and social benefits and costs. Public transit serves people who are unable to drive an automobile due to economic, legal, or physical constraints (including those too young or too old to drive). Transit provides environmental benefits by reducing air and water pollution, reducing the amount of land that is paved for roads and parking, and by discouraging urban sprawl. These additional benefits of transit are commonly overlooked.

Recent research indicates that rail transit and separated busways reduce congestion on parallel highways, because whenever congestion delays increase for automobile users, a portion shift to transit. Transportation engineers often underestimate conges-

Highway vs. Transit Investments Improving Mobility to Suburban Communities	
Estimated costs to accommodate two thousand to four thousand new trips:	
Light Rail:	$300 million
Highway Expansion:	$250 million
Highway Option Net Benefits:	$ 50 million
Costs Not Considered: (Litman 1997)	
Parking (assuming three thousand urban parking spaces with average cost of $10,000 each)	~$30 million
Surface street traffic congestion (assuming three thousand additional vehicles traveling ten kilometers per day, three hundred days per year on surface streets during peak periods, with an average cost to road users of $0.20 per kilometer, over twenty-five years with a 7 percent discount rate)	~$35 million
Vehicle Ownership Costs (assuming 20 percent of users save $2,500 annually)	~$29 million
Environmental & Social Benefits	? (probably substantial)
Uncosted Highway Expansion Factors	? (probably substantial)
Highway Option Costs not considered	−$94 million
Transit Option Net Benefits	**$44 million++**

tion reduction benefits that a dedicated corridor for public transit can provide to automobile users.

Another important factor often ignored in transportation investment analysis is the effect of generated traffic. Generated traffic is the increase in vehicle traffic resulting from a roadway improvement that would not otherwise occur. Recent research indicates that 70 to 90 percent of the increased capacity on congested urban highways is filled with generated traffic within five years (Hansen and Huang 1997). Failure to take into account the effects of generated traffic tends to overstate the benefits and understate the full costs of highway investments.

This is not to say that highway projects are always the wrong choice, but it does show that significant impacts are frequently omitted from engineering analyses. Planners and policy makers have an obligation to take into account all possible costs and benefits, regardless of who bears them. In this case, a more comprehensive analysis of costs (like those presented in the box above) would probably have resulted in a different outcome.

Source

Contributed by Todd Litman of the Victoria Transport Policy Institute, Canada.

References

Hansen, M., Huang, Y. 1997, Road Supply and Traffic in California Urban Areas, *Transportation Research*, Vol. 31A, No. 3, pp. 205-218.

Litman, T. 1997, *Transportation Cost Analysis; Techniques, Estimates and Implications*, Victoria Transport Policy Institute, (Victoria, Canada; www.islandnet.com/~litman).

NHI (National Highway Institute) 1995, *Estimating the Impacts of Urban Transportation Alternatives*, Federal Highway Administration (Washington, D.C.), Publication No. FHWA-HI-94-053, December.

Infrastructure

What is Infrastructure?

Look around. Infrastructure is everywhere ... You use a subway system, or drive to work each morning on roads and highways ... Infrastructure is often underground and out of sight, like pipes and sewers ... [Infrastructure is] that vast network of structures and systems civil engineers design and build to improve everyone's quality of life.

A Sound Infrastructure:

- ◆ Creates jobs: Every $1 billion of federal investment in infrastructure creates 47,000 construction and service-related jobs, and nets local governments $80 million in vital tax revenues from contractors, consultants, and construction payrolls.
- ◆ Enhances public health and safety: Poor roads are a factor in 30 percent of all highway accidents. Modern water treatment plants make our water safe to drink.
- ◆ Increases productivity: Modernizing our highways, for example, lets businesses sell products faster and at lower prices.
- ◆ Promotes economic development: States and communities with strong infrastructure attract businesses seeking to increase productivity and better compete in U.S. and world markets.

What are the Problems?

- ◆ Aging Infrastructure
- ◆ Lack of Maintenance
- ◆ Lack of Funding

are the condition and long-term adequacy of water and sewage services, waste disposal, roads, bridges, and airports.

The leaders of the ASCE could argue that their Code of Ethics requires them to bring infrastructure issues to public attention. Concerns are raised in Case Study 2.1 and later in this book about the adequacy of automo-

About 234,500 miles (nearly one of every five miles) of interstate roads are in poor condition ... [costing the average motorist] more than $112 annually on wasted fuel, added tire wear, and extra vehicle repairs ... More than forty-five million Americans drink water from public water plants that have contained cryptosporidium, a microorganism that caused some 100 Milwaukee residents to die and more than 400,000 others to become ill in 1993. One of every four of America's 800,000 miles of wastewater pipelines needs immediate repair or replacement ... severe congestion at seventeen major airports by the year 2002 [is predicted] unless we add new runways.

Neglect Skyrockets Future Costs

In 1992, Chicago failed to invest $10,000 to repair a small underground leak that caused a rupture that cost taxpayers and companies more than $1 billion in emergency response, lost business, and property damage. More than 200,000 people were evacuated when millions of gallons of water flooded a tunnel underneath the downtown district.

Developing Workable Solutions

Shrinking federal budgets have encouraged community and state officials to find innovative [technical, financial and organizational] solutions ... better infrastructure can happen only with a strong partnership among the American people, civil engineers, lawmakers, and businesses.

You Can Make a Difference ... Learn about infrastructure needs in your area ... Share your concerns with officials and your Congressional representatives ... Inform your community ... Support bond issues to improve your infrastructure ... With your help we can rebuild America.

From the American Society of Civil Engineers (ASCE) Web Site at:
<http://www.asce.org/govnpub/rightwro.html>

---

biles for inner city transport and the need for more sustainable mass transit. We need to look carefully at the sorts of important issues raised by the ASCE. How can adequate infrastructure provision be reconciled with moves towards "smaller government?" Such questions involve political as well as economic and technical decisions.

FROM EARLY FLIGHT TO THE SPACE PROGRAM

Engineering, science, and technology in the modern era have tackled some major challenges to human ingenuity. Flight has always been a human dream. Who could watch birds soar and not yearn to emulate them? Unfortunately, flight was not simply a matter of scaling up birds' wings and fitting them to humans, as envisaged in the Greek legend of Icarus. For humans, heavier-than-air flight would need a different approach. There were three central problems: power, lift, and control. The resolution of these problems would draw on major scientific and engineering advances over several centuries. Gunnery had given rise to the science of ballistics and

Early Flight

Balloons must be lighter than the air they displace. Dry air at standard temperature (15 °C) at sea level has a density of 1.225 kilogram per cubic meter, so balloons need to be large. The first successful human flight was in France in 1783, in a hot-air balloon built by the Montgolfier brothers. Later developments led to large, rigid, hydrogen-filled airships, with substantial range and lifting capacity. The flammable hydrogen was a continuing hazard, but helium was hard to get and expensive. The Hindenburg disaster and the failure of the British airship program saw balloons lose out to heavier-than-air machines.

Heavier-than-air flight needed a real understanding of the technical problems combined with systematic experimental development of the means to resolve them. Sir George Cayley (1773–1857) provided an essential scientific insight. He sketched the forces that would act on an aircraft in flight; in applied mechanics we would now call it the free-body diagram. Cayley showed that the weight of an airplane had to be supported by the lift generated by airflow over the wings, and that the drag, the resistance to the plane's motion through the air, had to be overcome by the forward thrust provided by some sort of propulsion device. In working aircraft, thrust would first be provided by a propeller or airscrew driven by a relatively light internal combustion engine, later by jet and/or rocket.

In a glider, thrust is provided by the forward component of the weight as the glider descends. Soaring requires finding the updrafts of air that occur over particularly warm areas of land or where winds are pushed upward by hills or cliffs. Cayley built working gliders, and in 1849 an unknown boy from his estate was the first person to fly in a heavier-than-air machine. Cayley had developed the theory of lift using cambered wing surfaces. In 1884 the first great

the steam engine had given rise to the science of thermodynamics. In turn, the thermodynamics of Carnot, Joule, and Clausius laid the basis for the internal combustion engine and, through it, for powered human flight. However, it still took many decades of efforts of aviation pioneers around the globe (see box, Early Flight) before December 17, 1903, when, in the Kill Devil Hills south of Kittyhawk in North Carolina, Orville and then Wilbur Wright made the first powered, sustained, and controlled flights in their twin propeller pusher biplane.

By January 1905, six years after they started work on the project in earnest, the Wright brothers had built their Flyer III, really the world's first practical aircraft. They offered the machine to the U.S. War Department, which rejected it. An approach to the British War Office was

aerodynamicist, Horatio Phillips, patented the theory of lift based on a differential camber between upper and lower wing surfaces.

The German pioneer, Otto Lilienthal (1848–1896), recognized that more was needed than slavish copying of birds. He sought to analyze the way birds fly and apply the same principles to a structure that would support the loads involved. He died after the monoplane hang-glider he was flying stalled and crashed, but he had developed a practical and controllable machine and his efforts inspired many others.

By the late nineteenth century, experimenters around the world had started to realize that curved airfoil surfaces generated high lift with a minimum of drag. In Australia, Lawrence Hargrave (1850–1915) designed and built tethered box kites that generated enough lift to raise him off the ground. His "braced biplane" arrangement was widely adopted for aircraft design. Wilbur Wright and his brother Orville drew on this and other work for the shaping of the wing surfaces of their Flyers. Over several years they carried out extensive and systematic tests on a series of gliders. The Wright Brothers built one of world's first wind tunnels, which they used in 1901–1902 to compare their experimental data with Lilienthal's. Testing also confirmed that they could control their machine in roll (about the axis parallel to the length of the airplane) by warping the wings.

The next critical step was a light but powerful engine. The first person to make a powered take-off (down an inclined ramp) was a Frenchman, Felix du Temple, in 1874, using a steam-powered machine of his own design. However, steam engines were too heavy for extended flight. No suitable propellers or internal combustion engines were available, so in 1902–1903 the Wright brothers built their own and made the first successful powered and controlled flight.

equally unproductive. Thoroughly discouraged, but still aware of the commercial and military potential of their machine, the Wrights locked the Flyer III in a shed to protect their ideas and stopped work on flight for the next three years (Jakab 1990).

Eventually, in 1908, Wilbur demonstrated the Flyer in Europe and Orville began acceptance trials for the U.S. Signal Corps. Tragically, during the trials a propeller blade cracked, and in the ensuing crash the passenger, Lieutenant T. E. Selfridge, was killed and Orville was injured. Even so, aircraft were moving rapidly towards acceptance as practical (and controllable) machines. Bleriot's flight in a tractor monoplane across the English Channel on July 25, 1909, confirmed this view and foreshadowed the increasing strategic significance of aircraft.

By the outbreak of World War I in 1914 there were well-designed and well-built aircraft, with increasingly competent pilots. By the end of the war, the aircraft was an accepted item of military equipment. War-surplus aircraft were used to establish the feasibility of commercial routes. Pioneering flights included the first nonstop Atlantic crossing by Alcock and Brown in 1919 and Captain Charles A. Lindbergh's solo flight from New York to Paris in May 1927.

Aviators like Sir Charles Kingsford Smith, who opened up air routes from Europe to Australia, became popular heroes. Amy Johnson in Britain and Amelia Earhart and Jacqueline Cochran in America were important in popularizing and demystifying flying (Chant 1978).

Between the wars, the leading technologies changed. Speed and maneuverability are critical for fighter airplanes and they continue to drive their development. In a quest for speed, all-metal monoplanes had been built for the Schneider Trophy races. They inspired the development of that classic airplane, the Supermarine Spitfire. Sir Frank Whittle patented the turbojet engine in 1930, but it was not tested until April 1937. A parallel German effort with strong government support was led by Hans von Ohain. Ohain produced a working turbojet a month after Whittle, and the first jet-powered aircraft to fly was the Heinkel He 178, on August 27, 1939, less than a week before the outbreak of World War II.

By this time military doctrine around the world recognized a range of essential roles for aircraft and the critical importance of achieving air superiority. After Japan and America entered the war, major battles in the Pacific were fought by carrier-based aircraft, with opposing naval task forces never in sight of one another. Jet- and rocket-powered aircraft were developed during World War II, but were not decisive; indeed the early units were more likely to kill their pilots than the enemy.

As operating speeds increased, aircraft began to experience fierce buffeting as they approached what came to be known as the *sound barrier*. As an aircraft approaches the speed of sound, the disturbances set up by, for example, the nose of the aircraft, form a shock wave. The attachment or detachment of this shock wave causes severe turbulence where it intersects other parts of the aircraft. It was found that wings for supersonic flight needed to be very thin to minimize this turbulence. The rocket-propelled Bell X-1 aircraft in which U.S. Air Force Captain Chuck Yeager broke the sound barrier on October 14, 1947, had thin straight wings. As speeds rise the shock front streams back in a cone from its point of formation. It then becomes desirable to sweep the wings back so that they stay within the conical shock front propagating back from the aircraft nose.

The first jet airliner to enter commercial service was the de Havilland DH 106 Comet in 1952. The prospect of commercial supersonic flight emerged with the unveiling of the Soviet Union's Tupolev TU-144 in 1968 and the Anglo-French Concorde in 1969. The latter went into regular service in 1976. Fuel costs and limitations on operating over land because of the sonic boom associated with supersonic flight made the economics of these elegant airplanes rather marginal. More socially and economically significant was the opening up of the mass air-travel market, particularly associated with the introduction in 1968 of the Boeing 747, powered by high by-pass-ratio turbofans. The introduction of the 747 was an example of "technology push" (discussed in more detail in Chapter 6) rather than of market demand. Designed for but rejected by the military, Boeing recovered its development costs by adapting the 747 for airline use.

The helicopter, and vertical take-off and landing airplanes generally, are important areas of flight that we do not have space to deal with here. The Gossamer Condor and Gossamer Albatross, discussed in Case Study 6.2, illustrate development in a different direction, human-powered flight.

An aviation gas turbine works by compressing air to several atmospheres pressure, injecting fuel (typically kerosene) into this air, burning the mixture in a combustion chamber, and then using the expanding hot gases to drive a turbine that powers the compressor. The surplus energy from combustion of the fuel can be used simply as a high-speed jet of hot gases to provide thrust to propel a turbojet airplane. It can, through the compressor, drive a fan around the engine, giving more thrust at lower speed than with a pure jet, as in the turbofan engines used on the Boeing 747, the Lockheed C-5 Galaxy, and the Antonov An-225 Mirya. The An-225, which entered service in 1989, is the world's largest and heaviest conventional aircraft, with a loaded weight of 600 metric tons (1,322,750 pounds), almost half as much again as the 409 tons (902,000 pounds) of the 747. For even more thrust, even more slowly, the compressor can drive a propeller, as in an older (1957) turbo-prop military transport plane such as the Soviet Union's Ilyushin Il-18. Engine noise constrains airport operation, so a continuing aim in jet engine development has been to reduce the noise associated with the turbulence of the hot exit gases. Major design efforts have reduced the intensity and lowered the pitch of this noise.

Rocketry came of age with the World War II German V2 rockets, and went on to the Space Programs that started with the launching of the U.S.S.R.'s Sputnik in 1957. The Soviet success shook U.S. confidence in its technical superiority. The wide-ranging response included shaking up technical and scientific education (including engineering) at all levels and allocation of substantially increased resources to technical programs. The Space Program reached an emotional peak with the Apollo moon landing in 1969, though it went on to work on the Space Shuttle programs and to increasing commercial application of associated technologies.

Long-range rocketry became a significant basis for projection of military force and attracted major national military funding. It is also the basis for increasingly important military and commercial systems of communications and navigation satellites.

It is chastening to reflect on the pattern of technological development revealed by the aerospace industry. Outstanding engineering effort realized the long-standing dream of human flight. Flight was then used first for military applications and to enhance national prestige, second for competitive applications (racing, conspicuous consumption), and only third for practical applications for the benefit of humanity. A similar argument could be made about the space program and its commercial spinoffs. Originally focused on national prestige and military capability, space-based communications and geographical positioning systems are increasingly being used for civilian applications. An apologist for engineering might argue that the demanding requirements of military and prestige applications, and the associated availability of almost unlimited resources, were necessary to develop the technology to a level where it could be commercially successful.

THE ELECTRONICS INDUSTRY AND THE DEVELOPMENT OF THE COMPUTER

The changing roles of science and engineering in the modern era are well illustrated in the electrical industry. Electricity had been known for centuries as a scientific phenomenon. The first device for storing electricity, the Leyden Jar, was invented in 1745. Discharging the device through the body was experienced as a violent electric shock and electricity and electric shocks became the new social rage. However, it was the electromagnetic theories of Faraday and Maxwell, arrived at by planned, systematic scientific research, that gave rise to electric lighting and power systems.

When problems of design and production arose within the new electrical industry, engineering practice was built upon and in turn enriched the scientific base. The integration of the electrical industry with academic physics brought about a new applied science: electronics.

Electronics was first used in industry in radiotelegraphy (1896), then radiotelephony (1906), and radio broadcasting (1920). In 1905, Lee de Forest invented the vacuum tube triode amplifier, by which electronic signals could be amplified and regenerated. In general use by 1913, it was eventually important in the development of the electronic computer.

The modern digital electronic computer and the various other uses of digital data transmission, processing, and storage on optical and magnetic discs, and magnetic tape illustrate the growing significance of *technological convergence*, first mentioned in Chapter 1. Initially developed for the military, equipment for communicating, processing, and storing data (words, sounds, numbers, and images) in digital form is used for a wide range of applications, from news and entertainment to business and engineering. The area is known collectively as Information Technology or IT.

As with the introduction of printing, many of the impacts information technology has had on our lives were not foreseen. Intended uses like electronic commerce are still the subject of heated debate. The accuracy, integrity, and security of database records involve important political and privacy issues as well as technical ones. The future commercial and social importance of electronic transmission of data is clearly immense.

Mathematicians such as Blaise Pascal (1623–1662) and Gottfried von Leibniz (1646–1716) had tried to solve the problem of carrying out tedious calculations by mechanical means. Charles Babbage (1792–1871) designed a series of analytical engines, and the method of programming them (using cards like those first used in 1804 by Joseph Jacquard to weave patterns in textiles) was devised by Lady Ada Lovelace (1815–1852). Lovelace, a pioneer woman in what was already a male-dominated sphere, was probably the first person to do what we would now recognize as computer programming. These early efforts were constrained by mechanical limitations.

The first person to process numerical data successfully was Herman Hollerith, who built a calculating machine to process the 1890 American census figures. His machine used a series of metal pins, placed in contact with cards that had been punched to record data. Where there were holes in the cards, the pins went through and completed a circuit by touching a pool of mercury. The company that built the machine went on to become IBM (International Business Machines). In 1885, William Burroughs marketed the first recording adding machine. At about the same time the National Cash Register Company introduced the cash register. All these calculating machines required user intervention after each operation.

The first machines to be programmed to carry out a series of operations without intervention were the "differential analyzers" developed by Vannevar Bush and others at the Massachusetts Institute of Technology (MIT), from 1925 onwards. The MIT machines were analog computers that used electrical devices such as motors and relays to actuate the mechanical parts: the gears, levers, and cams. An improved model where the angle through which a gear that had turned was measured electrically was built in 1942 and was used to compute firing tables for gunnery during the war.

The design and construction in 1942 of the first electronic computer, the Atanasoff Berry Computer (ABC), by John Vincent Atanasoff, is described in Case Study 2.2. Although Atanasoff solved the essential technical problems, he was not in a position to develop and promote the machine effectively.

In 1944, the first fully automatic digital computer, the IBM Automatic Sequence Controlled Calculator, Mark I, began operating at Harvard University. This machine was the result of a joint project between the University and IBM. It still used mechanical parts, but operated electrically. A more specialized computer had been introduced in England in 1942 for decoding wartime signals, but it could not be used for general purpose computation.

Not until the ENIAC (Electronic Numerical Integrator and Computer) did electronic tubes entirely replace mechanical parts. Patented by John W. Mauchly and J. Presper Eckert, ENIAC was the result of a project at the University of Pennsylvania, backed by the U.S. Army Ordnance. It was begun in 1943 and completed in 1946. Vacuum tubes, when used as relays (a switch operated by an electromagnet) had a switching speed of less than a microsecond, and ENIAC could complete the multiplication of up to twenty-three digits in seconds. It took 130 kilowatts to run its 18,000 vacuum tubes, and on average at least one of these failed every few hours. Despite these limitations, it was a general purpose machine that laid the basis for further development. Mauchly and Eckert, its creators, led the way in much of the further work, although their patent was overturned in 1973 when Atanasoff's earlier contribution was finally acknowledged.

It took almost a century to progress from Babbage's analytical engine to ENIAC. Evolution from ENIAC to high-speed digital computers took about fifteen years, and involved breakthroughs like the transistor (invented in 1947 but not commercially available until the late 1950s), the adoption of the binary system of notation, and developments in storage and logic structure.

Development of the transistor into the large-scale integrated (LSI) circuit, including capacitors and resistors, made the pocket calculator possible. Very large-scale integration (VLSI) gave us silicon chips powerful enough to drive computers. The Apple II, the first commercial personal computer, was released in 1977, followed in 1981 by the IBM PC and its clones. They led the way towards desktop workstations. The role of Silicon Valley as a focus for many of these breakthroughs is discussed at the end of this chapter in Case Study 2.4.

Although it was common for these major technological breakthroughs to originate in the United States, often in work financed by the military budget, by the 1970s and 1980s it was the Japanese who were excelling in turning technological breakthroughs into desirable consumer products. In the postwar period the Japanese had learned remarkably quickly, moving from being producers of low-cost, poor-quality, imitative products to a position where their products commanded a premium for quality and design. Part of the basis of their success was that they recognized that the future lay with quality products, and that the route to quality lay through continuing product and process innovation. The involvement of production workers themselves and the high level of technical support available at the shop floor in Japanese manufacturing enterprises are discussed in more detail in Chapter 7.

Throughout the industrial world the introduction of computers has revolutionized the way offices operate, sharply reducing white collar work forces in manufacturing, banking, insurance, and retailing. However, the character of the gender division of labor in these areas (discussed in Chapter 3) remains largely unchanged. Routine, repetitive tasks such as data entry remain "women's work," whereas the more interesting and challenging areas such as programming, systems analysis, and management have remained largely male preserves.

The computer is profoundly changing the practice of engineering. High-speed data processing is central to computer-aided drafting/design, computer-aided manufacturing, computational fluid dynamics, and finite element stress analysis. Computer workstations, with high-definition color screens and powerful software packages, are becoming standard equipment for professional engineers. Their power makes it more important than ever that engineers appreciate the limitations of the physical and computational models they use.

MODERN MANUFACTURING

The fundamental importance of engineering for manufacturing industry has been noted in the discussion of the automobile (above). The sheer scale of this area of modern human endeavor warrants further consideration.

ATANASOFF, FORGOTTEN FATHER OF THE ELECTRONIC COMPUTER

John Vincent Atanasoff (1903–1995) was born in Hamilton, New York. He was a 1925 electrical engineering graduate of the University of Florida and received a master's degree in mathematics from Iowa State University, where he taught for fifteen years. He received a doctorate in physics in 1930.

Atanasoff said the idea of a digital electronic computer had come to him over bourbon and water in a roadhouse in Illinois in 1937. While driving he had been thinking about the computing devices he had been working on since 1935. He needed an automatic machine that could solve the sets of linear equations that he and his graduate students had been working with using hand calculators.

Atanasoff's ideas for his computer included a number of modern computing concepts. It would be electronically operated and would use base-two (binary) numbers instead of the traditional base-ten numbers. It would use capacitors for memory, along with a regenerative process to prevent data loss due to electrical failures. It would have fifteen-significant-figure decimal precision, perform linear algebra operations with parallel multiply-adders, and be targeted at solving problems in structural analysis and quantum theory.

Work began in earnest in 1938. Assisted by a talented graduate student, Clifford Berry, Atanasoff developed and demonstrated a breadboard model in 1939. A working prototype of his electronic computer was completed in 1942. Although it used a mechanical clock system, the computing was electronic, using several hundred vacuum tubes. There were two rotating drums containing capacitors that held the electrical charge for the memory. Data were entered using punch cards. For the first time, vacuum tubes were used in computing. The machine was dubbed the Atanasoff Berry Computer (ABC).

The ABC weighed 750 pounds. Its memory consisted of a mere three thousand bits. It "printed" the results of its calculations by using a five-thousand-volt spark to burn holes in paper cards. The project, which cost about $1,500, was detailed in a thirty-five-page manuscript. University lawyers sent a copy to a patent lawyer, but the patent was never filed, and the ABC was eventually stored in the physics building's basement at Iowa State. By the time the computer industry was off and running, Dr. Atanasoff was involved with other areas of defense research and was out of touch with computer development. The prototype was gradually cannibalized without his knowledge while he was working for the Navy during World War II.

In the mid-1940s, John W. Mauchly and J. Presper Eckert were the first to patent a digital computing machine, which they called the ENIAC (electronic numerical integrator and computer). Mauchly and Eckert developed the ENIAC at the Moore School of Engineering at the University of Pennsylvania. Completed in 1946, it was much larger than the ABC and was a general purpose machine, not restricted to solving sets of equations. Eckert and Mauchly said they had worked out the ENIAC concept over ice cream in a restaurant. However Mauchly, while he was a physicist at Ursinus College near Philadelphia, had met Dr. Atanasoff at a conference and then visited him at Iowa State during the summer of 1941. Mauchly had stayed several days at the Atanasoff home, where he was briefed extensively about the computer project and saw it demonstrated. He left with papers describing its design.

Mauchly and Eckert patented the ENIAC and went on to found their own computer company, incorporated in December 1948. Eckert and Mauchly sold the company to Remington Rand in February 1950. In June 1951, Remington Rand delivered its first computer, the UNIVAC, to the U.S. Census Bureau.

Due to their patent of the ENIAC and their high visibility in the computing industry, Mauchly and Eckert have been seen as the fathers of electronic computing. However, Atanasoff became the court-recognized "father of modern computing" in 1973 when, in a patent infringement case brought against Sperry Rand by Honeywell, a federal judge voided Sperry Rand's patent on the ENIAC. On October 19, 1973, Atanasoff finally obtained official recognition for his creation of the world's first digital electronic computer. U.S. Federal Judge Earl R. Larson ruled: "Eck-

ert and Mauchly did not themselves first invent the automatic electronic digital computer, but instead derived that subject matter from one Dr. John Vincent Atanasoff." The decision made news in the industry, but Dr. Atanasoff, by this time retired, lived on in relative obscurity. Mauchly and Eckert continued to be credited with inventing the electronic digital computer.

John Atanasoff died in 1995 at the age of ninety-one, without ever making money from his invention. However, in 1990, President George Bush acknowledged Dr. Atanasoff's pioneering work, awarding him the National Medal of Technology. Other honors bestowed on Dr. Atanasoff included five honorary doctorates, the Navy's Distinguished Civilian Service Award, the Computer Pioneer Medal of the Institute of Electrical and Electronics Engineers, the Holley Medal of the American Society of Mechanical Engineers, and the Distinguished Achievement Citation of Iowa State University. He is a member of the Iowa Inventors Hall of Fame.

Sources

Campbell, J., *Atanasoff: Inventor, Problem-Solver, and Inventor of the Real computer*, The Shore Journal.

Do, H. C., *John Vincent Atanasoff*, *http://ei.cs.vt. edu/~history/do-Atanasoff.html*.

Gustafson, J. L., *An FPS Forerunner; The Atanasoff-Berry Computer*, Ames Laboratory, Department of Energy, ISU, Ames, Iowa.

Snow, J. A. 1995, *Looking at the Future with John Atanasoff*, Commentary aired on WOI Radio on July 14.

Table 2.1

The Role of Manufacturing in Selected Economies (1980–1990)				
Indicator	United States	Germany	United Kingdom	Japan
Share of manufacturing in total employment	20.0	34.7	25.0	27.2
Share of manufacturing in GDP	20.1	31.4	21.6	29.0
Share of manufacturing investment in GDP	7.9	8.5	8.2	11.2

After UNCTAD 1996, Table 29.

Manufacture is defined by Webster's Dictionary as:

1. the making of goods or wares by manual labor or machinery, especially on a large scale;
2. to make or produce by hand or machinery especially on a large scale;
3. to work up (material) into form for use.

A central characteristic of the Industrial Revolution was a long period of high growth in productivity and in gross national product (GNP, defined in Chapter 5). This is how rich countries came to be rich, whereas poor countries that had low economic growth stayed poor. Manufacturing has played and continues to play a key role in the process, as Table 2.1 shows. We look in Chapter 8 at the sources of the resources that were used to finance growth, including the extent to which they came from the countries that, at least partly as a result, stayed poor.

Another characteristic of the Industrial Revolution was the emergence of a pattern of economic "boom" and "bust," where workers were laid off periodically. Industrial workers, dependent solely on their wages for livelihood, had become an increasing percentage of the American workforce. Millions suffered real hardships in widespread depressions, even before the Great Depression of the 1930s. Political instability was associated with widespread unemployment. Following World War II, manufacturing had a high priority in public policy-making in both the developing and the developed worlds because of its importance in generating national wealth and its role as a major provider of employment.

The effects of World War II on manufacturing varied widely. Countries that became battlefields, or were systematically bombed to destroy their military capability, lost much of their prewar productive capacity. To the extent that they were able to replace that lost capacity with modern equipment their productivity increased dramatically. Other countries that were involved in the fighting but were not themselves attacked in any major way had a mixed experience. Many former colonies had some industrial capability before the war but had been encouraged to depend on the colonizing power for manufactured products. Wartime conditions interrupted the flow of manufactures to the colonies and encouraged them to develop an independent technical capability, including in manufacturing.

The United States emerged from World War II as the world's major productive force, so powerful that it was able to assist countries that had been on both sides of the conflict to rebuild their industries. This was widely seen not only as humane, but also as politically astute, helping to avoid national and international instability. Britain, on the other hand, was exhausted. Her factories were worn out and outdated and the euphoria of surviving undefeated did not provide the much-needed spur to modernize.

A central feature of the rapid postwar industrialization of Japan was the high profit (gross operating surplus) on investment in manufacturing. This high level of profit provided both the necessary cash flow for further investment and the encouragement needed to make that investment. During the 1960s gross operating surpluses were around 55 percent in Japan, compared with 35 percent in West Germany and 25 percent in the United States and United Kingdom.

Table 2.1 shows the more recent role of manufacturing in the economies of four key countries from 1980 to 1990. During this decade, employment in manufacturing fell slightly in the United States and Germany, fell sharply in the United Kingdom (27 percent), and rose in Japan (14 percent).

Manufacturing continues to be an important source of employment, although the actual percentage of the workforce involved can be expected to continue to decline. However, the general use of advanced manufacturing technology is still likely to continue to increase output and the role of engineers continues to be crucial. A reduced work force will need to be highly skilled and motivated. Experience around the world also shows the role of governments to be critical to the success of manufacturing because they are responsible for strategic direction, that is, for setting the context in which manufacturers operate. A healthy manufacturing industry seems to be an essential aspect of being a developed country. Governments cannot ensure the success of manufacturing enterprises, but they can destroy them by inappropriate policy settings. Potential roles for government and their effects on engineers and engineering are discussed in several of the following chapters.

◼ From Mass Production to Flexible Manufacturing

Mass production was a major factor in postwar recovery, but even by the early 1960s there was increasing overcapacity in manufacturing industries in Western countries. Cutting the cost of labor was not an effective competitive strategy for manufacturers in the West because Third World wages were already lower. Widespread use of manipulative practices like high-pressure salesmanship had also started to raise concerns. Bodies such as the Consumers Union emerged, concerned with informing themselves about products and preventing rip-offs. Increasing government control reflected rising awareness of environmental problems and rising dissatisfaction with the failure of suppliers to regulate their activities in the public interest. Implications for engineers of consumer protection legislation are discussed in Chapter 6.

It became clear that manufacturers needed to tailor products more closely to the wants and needs of consumers and needed to keep up with rapid changes in demand. The focus in manufacturing moved to flexibility and to reducing the time needed to bring new products to market. Simultaneous engineering is one approach to reducing product development time. Product design is done by a team that stays together from the conception of a product through to its arrival in the marketplace. Aesthetic, ergonomic, and manufacturability issues are addressed in parallel.

Computer Aids to Flexible Manufacturing

The strategy for profitability in modern western manufacturing firms is to focus on variety. Flexible manufacturing and information-based, computer-aided technologies allow dramatic changes in work organization. Worker commitment and responsibility are critical.

Maximizing the benefits of computer assistance requires vertical integration of the whole process, from conceptual design through to manufacture. The combination of computer-aided drafting and design (CADD) with stress and failure analysis leads to computer-aided engineering (CAE). Advances in computer drafting have led to computer modeling systems, with objects built up from three-dimensional elements rather than from two-dimensional lines.

An effective approach to meeting staffing needs for CADD has been to train and promote skilled machinists who can apply their manufacturing knowledge to the design process. Firms that attempt to deskill the design process and simply replace professional engineers with technicians, fail to reap the full benefits of the technology.

SYSTEMS THEORY AND SYSTEMS ENGINEERING

To minimize the unintended consequences of technology, engineers in the second half of the twentieth century have increasingly emphasized systems approaches. An early example was in electricity supply and use. The pioneers associated with commercial distribution of electrical energy soon realized the need to understand better the increasingly complex systems they were creating. Edison himself recognized the importance of going beyond the invention of the incandescent lamp. He developed a complete electrical power distribution system, including prime movers, generators, switchgear, wiring standards, and so on, through to well-designed domestic light fittings. Similarly, the emerging telephone network highlighted the need for systems approaches. The American Telephone and Telegraph's Mechanical Department was set up in 1883: in 1925 it became part of the Bell Telephone Laboratories Inc. One of the two major divisions of the Bell Labs was Systems Development. It pioneered the discipline that emerged after World War II as systems engineering.

Systems engineers work on projects that are too complex for a single engineer or other specialist, such as projects that require a team of specialists working together and pooling their expertise to solve the overall problem or to design the system. "Hard" or technical systems are usually

---

Design is a difficult and challenging process; most of the potential for maximizing performance and quality and for minimizing cost are set at the design stage. Design should be seen as the major opportunity to add value, rather than a chance to cut corners. Leaving design problems to be addressed in production is disastrous; it is both expensive and time-consuming.

Integrating CAE with the programming of production machinery leads to computer-integrated manufacturing (CIM). This typically includes computer numerically controlled (CNC) machining centers, integrated with materials handling equipment into flexible manufacturing cells. Linked cells can become *flexible manufacturing systems* (Mathews 1989, p. 127).

Flexibility and quick response to changing demands depend on skilled operators who can use equipment to its full potential. Labor costs are no longer the central issue. Warnecke (1993, pp. 210–211) argues that increased attention to people is central, and that a revolution in corporate culture is needed for future manufacturing. He cites an example where human resources development costs (education and training) made up 20 percent of the investment budget in the updating and restructuring of one German manufacturing operation.

---

dealt with by engineers within the discipline of systems engineering. Where human behavior is involved (as in the sociotechnical systems of production and use we discussed in Chapter 1) the problems become much too complex to solve completely. To deal with these systems we need to use multidisciplinary, soft systems approaches.

The development of modern systems theory has been a major intellectual and technical achievement this century. Kline (1995) and others maintain that the use of the system concept is the single way in which post-Newtonian science most sharply differs from ancient approaches. A systems approach assists us to think through and address, in a structured and organized way, *all* the aspects of a problem or application, not just one or two aspects. In many fields of engineering and science, the system concept is used as the starting point for the analysis of every problem.

Key Concepts and Issues

Systems

Although we can treat anything we choose as a *system*, there is no point in using a systems approach unless the object of study is reasonably complex.

For the object of study to be a formal system, it needs to meet specific criteria; for example, it must:

- ◆ have purpose and performance;
- ◆ include a decision-taking process;
- ◆ have components that are themselves systems that are connected and interact;
- ◆ exist in wider systems and/or environments with which it interacts;
- ◆ be bounded;
- ◆ have resources and some degree of continuity and stability (Checkland 1981).

Most engineering systems do not include a decision-taking process, but the other characteristics would normally apply. The complexity of such systems implies that they will, for example, have *emergent properties*, that is, properties that could not be forecast from a knowledge of the system components, but result from the way the components interact. Case Study 2.3 (at the end of this chapter) illustrates the importance of emergent properties on the modes of failure of complex technical systems and emphasizes the need for a more sophisticated understanding by engineers (and others) of uncertainty and risk.

Human inputs to system operation are critically important. Maximizing the benefits from computer-based equipment requires broadly trained staff who understand the way the technology operates and who are committed to making it work effectively. In banking, for example, the integrity of the data entered is of critical importance. The systems used are extremely vulnerable to error, so clerical staff need to feel responsibility for results. The systems appear increasingly abstract to the operators, so intellectual mastery of them is essential. Tasks become increasingly interdependent because there is no longer a processing chain in which errors can be picked up by staff further down the line (elaborated in Thompson 1967). Effective operation of these computer-based systems requires workers who are involved with their work, who have become part of a team during the process of being consulted, and who are involved in the process of selecting and implementing the new technology (Mathews 1989, p. 211).

In studying a system, we draw the system boundary, and we choose what may pass through this boundary. We make the system and its representation as limited or as extensive as we want, depending on where we draw the boundary.

If we focus, for example, on the automobile, and we draw the system boundary around the vehicle itself, we can use a systems approach to study the operation and control of the vehicle, including its dynamics and handling characteristics. We address a very different set of problems if we draw our system boundary around the road transport system as a whole. As Case Study 2.1 shows, we deal with a different system problem again if we focus on how to move people, rather than vehicles. As usual, to get good answers we have to ask good questions and draw good system boundaries, and we need to do this before we can usefully begin modeling and analyzing the system of our choice.

In systems work it is essential that we carefully distinguish the actual system from the pictures, equations, mental images, or models we make of it for the purposes of analysis. Kline (1995) calls these abstractions *system representations* or *sysreps*. In the process of simplifying the real system to a level of complexity that we can analyze, we lose important, perhaps essential, detail. The degree of loss depends on how complex the original system is and how adequately we can model it. Kline emphasizes the complexity of real systems, particularly those involving human beings. Kline also highlights the limitations on our ability to deal effectively with these systems, particularly if our approach is limited to a specific discipline.

Another example of a systems approach is in the use of free-body diagrams to solve elementary statics and dynamics problems in engineering mechanics. We draw a system boundary around a single body or group of bodies, and we let forces pass through this boundary, for example the forces exerted by the supports. Our system representation or *sysrep* commonly simplifies the problem by neglecting friction, and perhaps also the self-weight of some or all of the bodies. When we use this approach to model a real system, such as a simple truss or machine, we need to be aware of the effects of differences between the representation and the actual system. The fact that there are differences between models and reality may seem blindingly obvious, but it is easy to become so involved in the modeling process that its outputs seem real and unchallengeable. In engineering there is an additional problem, in that the detailed realization of a project may subtly violate the design assumptions.

▪ Developments in Systems Engineering

Two recent descriptions of Systems Engineering are offered in the following box. Why are systems issues important for engineers? These descriptions give us some clues. Systems can be assembled from smaller, less complex elements. Testing at each level of assemblage is necessary because complex systems have *emergent properties*, that is, their behaviors are likely to be different from those we would predict simply from a study of the elements that make them up. A corollary is that complex systems can fail in ways that can only be discovered by very extensive testing. In real life, this means that complex failures are likely to occur only after a system has gone into service (see Case Study 2.3 at the end of this chapter).

The collapse of the atrium walkways in the Hyatt in Kansas City in 1981 illustrated the gap between model and realization. Because the load paths in the system as built differed from the design model, two crowded walkways collapsed onto a crowded floor below, killing 114 and injuring almost 200 more (Petroski 1985, 1994).

Approaches based closely on systems design have also been used in the design of social systems. Early approaches along these lines were generally ineffective. Commonly authoritarian and technocratic, they failed totally to recognize the additional complexity of social systems over technical systems. Later approaches moved through what Benathy (1991) called "designing for" people, towards a recognition that effective, sustainable,

Systems Engineering

An interdisciplinary approach and means to enable the realization of successful systems ... a design and management discipline that is very useful in the designing and building of large or complex systems ... [It was] developed to counteract the difficulties encountered in the engineering of increasingly large, complex, and inter-disciplinary systems. However, this discipline has also been evolved to aid in the design of many other types of systems.*

Systems engineering is at once all-encompassing and non-specific. Strategies may involve either breaking down a complex system into smaller and smaller component subsystems, or integrating low-level components one at a time, testing at each stage so that bugs are discovered quickly.

Systems engineers must understand what needs to be done ... how each small piece fits into the big picture, and how to break large conceptual procedures or processes into smaller, more easily manageable chunks ... user input is essential, so engineers in this field must possess excellent communication skills ... conveying technical details to non-technical personnel is a constant challenge. Systems engineers must also be flexible, since the day-to-day work often depends on the project at hand ... The nature of the job is designing new systems ... not simply working with existing technology.**

---

*The International Council on Systems Engineering (INCOSE) (www.incose.org/intro.html).
**Graduating Engineer & Computer Careers, September 1997, p.18.

human systems were inherently "open, complex, indeterminate and self-organizing," with a nature that was "value-laden, purposeful, and even purpose-seeking." We will look more closely at such systems in Chapter 8 when we consider the differences in approach between *technocentric* and *high-performance* (or skill-based) technology design.

A total system test is essential to confirm that the completed system will work effectively. Panitz (1997) comments that, "to save money," NASA skipped this step before launching the $1.6 billion Hubble Space Telescope. As a result, they failed to discover that the primary mirror had been ground to the wrong shape. Fixing the problem in orbit cost NASA millions of dollars and a three-year delay in bringing the telescope into full service.

Complex Systems Failure—The de Havilland Comet

The difficulty of fully analyzing a system, in this case the service stresses on an airframe, was demonstrated by the problems experienced by the first commercial jet airliner, the de Havilland Comet. It entered service very successfully in 1952, but in 1954 mysterious crashes began. It was as if the plane was being blown apart in midair. The first two incidents were over water, but investigators eventually found the tail section in the second case. Even so, it was difficult to work out what had gone wrong.

Investigators took an actual Comet, submerged it in a tank of water, and simulated flight conditions by pressurizing and depressurizing the fuselage and hydraulically simulating flight loadings on the wings. After about three thousand simulated flights a crack appeared at a corner of one cabin window. It spread rapidly when the cabin was pressurized. The investigators' analysis focused on the fact that the cabin cross section was elliptical, rather than circular. Each time the plane went up, internal pressurization bent the fuselage along its length, changing the cross section towards a more nearly circular shape. However, the window corners were square, not rounded, with rivet holes nearby (we would now recognize this as poor detail design), significantly raising the local stresses associated with pressurization effects. Eventually fatigue cracks propagated far enough along the fuselage from the window corners that, as the test showed, failure of the structure was inevitable.

Airworthiness certification now requires two fuselages to be tested to destruction in a rig that simulates service loading. Because of their level of complexity, we now recognize our inability to analyze completely the loading of aircraft structures (although modern computer-aided design tools bring us much closer to this goal). Full-scale testing increases the likelihood of finding problems before they endanger the lives of passengers or aircrew.

System Design

Systems Engineers start with an iterative, top-down (or hierarchical), interdisciplinary approach to design. A complex system is designed by breaking it down into its component subsystems. The process is then repeated for each subsystem until easily designed or off-the-shelf components are all that remains.

Systems engineers are also responsible for integration from the bottom-up. Large systems are made up by taking the lowest level components and putting them together, one level at a time. Each stage is tested to make sure that it works properly. In this way even very large systems can readily be built, with problems found and resolved before they are buried too deeply. Greater use of outsourced components, potentially from anywhere on the globe, makes this integration capability increasingly important, and it is essential for the large-scale projects that include design, construction, and commissioning of facilities.

The reality is usually more complex than this ideal model. A number of iterations of the top-down/bottom-up analysis may be needed to arrive at the most effective approach.

The first step, *requirements analysis*, is a detailed definition of the problem the system is to address. It involves composing specific, clear, and unambiguous system requirements. This calls for excellent communication skills on the part of the systems engineer—talking through their needs with the system stakeholders and exploring the "user perspective" on what the system must (and must not) do, how well, and what are the cost and scheduling requirements. It is essential that every type of user of a potential system (operator, maintenance, management, etc.) be involved in specifying system requirements. This process is essentially the development of the *functional specification*, described in Chapter 6 in the context of engineering design.

Even after the system requirements have been determined, it is essential for users to be involved in the detail design. After all, who knows better than the user what feedback and controls they want, what production support they need, and what output quality they are capable of? Supporting users in optimizing new system development is a challenging exercise that calls for a high level of empathic and communication skills. Human factors issues, including interfacing between individuals, machines, and society, are increasingly central.

■ Soft Systems and Multidisciplinary Approaches

Engineers and engineering are involved with both hard and soft types of system. Their analysis requires inputs from science and technology ("hard" discipline inputs) and from the social sciences ("soft" discipline inputs). We emphasize soft inputs here because engineers tend to focus on technical matters and to neglect or even dismiss social issues.

Sociotechnical systems have very substantially reduced the amount of hard and unappealing physical labor we need to do to survive, and have also greatly extended human powers. Developments in transport, communications, computing, and military technologies demonstrate this, sometimes all too clearly. Why do problems involving sociotechnical systems such as these require different approaches? Because complex systems, particularly those involving humans, require going beyond the reductionism that has been so effective in much of science and technology for the last few centuries. (*Reductionism* is the theory that every complex phenomenon can be understood by analyzing the simplest, most basic mechanisms involved. It does not address the emergent properties of complex systems.)

As we have indicated, the essential complexity of sociotechnical systems makes it impossible for us to produce adequate models of them. No individual professional discipline offers all the conceptual and analytic tools to model sociotechnical systems adequately. Take, for example, the process of innovation, which has conventionally been defined in economic terms as "the introduction into the market of a new product." This definition ignores *process* and *system* improvements, both critical technical aspects of innovation, particularly in production systems. Economics ignores these aspects because economists treat technology as something inside a black box, the details of which they do not need to concern themselves with. We will show in Chapter 6 that similar limitations on other individual disciplines involved with innovation allowed both an inadequate definition and an inadequate model to continue.

A more satisfactory definition of innovation, based on a multidisciplinary approach, is: "Any improvement in the sociotechnical systems of manufacture and/or use which increases performance as perceived by customers" (Kline 1995, p. 180). When we start to think in detail about making such innovative improvements, we recognize the relevance of social issues. As discussed in Chapter 7, a key factor in the successful emergence of Japan as a manufacturing power was the social organization of Japanese production facilities as learning organizations. Their approach is based at least as much in the social as in the technical sphere, with deliberate planning to encourage incremental process innovation, as well as associated product innovation (redesigning the product to make it easier/cheaper/more cost effective to produce).

Taking the Engineering Profession into the New Millennium

This chapter has ranged very widely in topics but has left untold many of the fascinating engineering episodes of the last century or so. Some of our omissions may turn out to be very significant indeed. Historians recognize that they often need some "distance" from recent events to see them in perspective. We may not yet be able to distinguish the truly "defining points" in modern technological development, but they will inevitably set the stage for the engineering profession in the new millennium. The pace of technological development has continued to accelerate in the late twentieth century, and we know that engineering in the new millennium will be constantly changing. By now, many of the future directions in technology are discernible, yet our own recent experience suggests we will be unlikely to anticipate the most dramatic changes.

Just as, in the seventeenth and eighteenth centuries, the (newly defined) civil engineers designed roads, bridges, ports, and other such works, and mechanical engineering emerged with the steam engine and the railroads in the nineteenth centuries, so too the engineering profession today continues to reflect transformations wrought by technological change during the modern era. The twentieth century saw the widespread introduction of electricity and electrical engineering, followed by aeronautical (now aerospace) engineering, chemical engineering and production, industrial and manufacturing engineering. The process continues with computer engineering, telecommunications engineering, and systems engineering. Materials engineering is increasingly important. Other recent additions are environmental engineering (commonly based on civil engineering but with a strong environmental and systems emphasis) and, possibly the most significant for the new millennium, bioengineering (see box).

In new engineering disciplines, especially those clearly oriented to issues of social concern, women and minority groups have begun to assert their right to be a part of the profession. Engineering has lagged behind other professions in supporting and encouraging this change. Some of the reasons for this and the ways engineering needs to continue to change in order to be more broadly welcoming and socially inclusive are discussed in Chapter 3.

Bioengineering

An important branch of engineering to emerge since the 1960s is bioengineering (or biomedical engineering). The rise of bioengineering as a strong new discipline reflects the emergence of biology as a foundation science, alongside physics and chemistry.

Bioengineering is the increasingly broad field of engineering that applies engineering principles and practice to problems in biology and medicine. Initially dominated by researchers with the advanced skills needed to pioneer the field, it has expanded to include those with bachelor's degrees. Probably because of its clear social relevance, bioengineering attracts a high percentage of women (Panitz 1996).

A close association between engineering and medical schools seems essential. The scope of research activity is indicated by the focus areas at two of the pioneers:

- At Johns Hopkins University, current research topics include: biomaterials; biomedical imaging systems; biomedical sensors and instrumentation; cardiovascular systems physiology; molecular and cellular systems physiology; systems neurobiology; and theoretical and computational biology.

- At Georgia Institute of Technology, the Institute for Bioengineering and Bioscience listed the following research topics: biomechanics and tissue engineering; bioinstrumentation and medical imaging; and medical informatics and telemedicine.

We only need to look at one of these areas, tissue engineering, to appreciate the possibilities. Researchers have already grown skin cells in a layer that can

A central challenge for engineering in the new millennium will be contributing to the move towards global sustainability as population continues to increase. The United Nations suggests that, barring global disasters, world population will level off at about 9.8 billion in 2040-2050. Almost 80 percent of the world's present population, 4.7 billion, live in what is known, perhaps over-optimistically, as the "developing world." Only 1.2 billion live in the "developed world." We will look at the social dimension of this division and its implications for engineering in Chapter 8.

The progression from the mechanical steam engine to the electronic, and perhaps eventually the biologically based, computer, and from machine-aided craft work to automatically controlled mass production, is not just a matter of clever engineers and the triumphant march of tech-

be placed onto burns or ulcers to promote skin regeneration. Other tissues being researched include parts of the nervous system, eyes, liver, pancreas, blood vessels, cartilage, bone, and muscle.

The First International Conference on Ethical Issues in Biomedical Engineering, held at Clemson University in South Carolina in 1997, spelled out some important issues:

"Biomedical engineering has been largely responsible for many of the advances in modern medicine. Biomedical engineers have developed medical devices and implants such as pacemakers, total joint replacements, and various imaging modalities (e.g., computer-assisted tomography (CAT) scan, magnetic resonance imaging (MRI), and ultrasound machines), which in turn have significantly improved our quality of life and our life expectancies. However, these technological advances have also created new moral and ethical dilemmas ... allocation of scarce resources, clinical trials of new devices and implants, confidentiality, genetic engineering and testing, conflict of interest, and animal experimentation."

Bioengineering is not limited to medical applications. It includes some or all of agricultural engineering, bioprocess engineering, food engineering, and even environmental engineering. Some engineering educators are starting to argue that all engineering students should learn some basic biology, to help them to deal with problems associated with living systems. It is not surprising that this proposal is meeting resistance, given the legendary overload in the engineering syllabus, but if bioengineering becomes as important as its promise suggests, room may have to be found.

nology. These huge changes have not come easily or without costs to society, as well as costs to many of the individuals affected. They have involved changes in education (craft apprenticeship to tertiary training); in political structure (absolute monarchy to liberal democracy); in social structure (peasant to industrial laborer, merchant to capitalist); and in the economy (family firm to joint stock company to multinational). The role of governments has been more clearly recognized. The rise of fascism in Europe in the 1930s drove many intellectual leaders to emigrate to the United States, including brilliant scientific and technical professionals like Albert Einstein and Edward Teller. Following World War II, the GI Bill dramatically increased and broadened recruitment to the professions, particularly to engineering and science. Continued systematic large scale funding of scientific and technological development, proposed by Vannevar Bush in 1945 and adopted by President Truman, carried the United States to scientific and technological leadership on the world stage.

There have been, and will continue to be, international, national, and local struggles about which technologies should be adopted and which should be rejected. (This applies particularly, but not only, to the military use of technologies.) There have often been bitter struggles in the workplace for control of the production process. Along with changes in technology, it can be argued that the most critical skills in engineering of design and production have moved from those associated with empirical development, through the capacity (and determination) to resolve problems at the design stage, to skills associated with meeting the demands of global markets. Most recently the focus has been on design for assembly and even disassembly (to facilitate recycling), and on innovation in the production process itself, including improvements in quality and inventory control. The progression reflects changing social and political circumstances at least as strongly as it does the internal logic of the technologies involved. This view is supported by the fact that the leadership position has passed from the Britain of the Industrial Revolution to Germany, the United States, and in some areas on to Japan.

Other issues for the future need to be borne in mind. Aspirations of individuals and their communities are orders of magnitude higher today than they were even a few generations back. However, the long-standing trend towards "unfettered" development, the impetus to seize every opportunity and exploit every breakthrough in order to maximize profit, is no longer the only goal. Real choice has become a possibility. Notions of "the greater good" (sometimes even for "the greater number") have come to hold increasing sway.

It is undoubtedly true that engineering has shaped the modern world, but will it continue to do so into the new millennium? Already we can see that, as well as being an agent for change, in the process engineering has been directed into some unexpected paths. Technology, including engineering, has itself increasingly been shaped in ways that were not conceivable before the twentieth century. Pulitzer Prize winning author Paul Starr suggests that modern technology has been given form and direction not only by engineers and technologists who see it as their preserve, but also by the "constitutive choices" of political and business leaders and, at times, by popular decision (P. Starr, Personal Communication). He argues that these choices have had far-reaching consequences. They are associated with an increasing recognition of the need to shape technology to meet human values, including protecting our environment, rather than uncritically accepting whatever technologies the engineers and marketers may care to present.

Transfer of technologies, including to what are now known as newly industrializing countries (NICs), has been important in broadening access to the benefits (and drawbacks) of modern industrial society. Convergence of the technologies associated with digital data processing, transmission, and storage continues to transform engineering work, with computer-aided engineering now including drafting, design, and integrated manufacturing. Engineers have been and are important in all these processes.

In this chapter we have looked both backwards and forwards. There is, of course, a lot more to the history of engineering than we could present here, especially in the twentieth century. The reader is strongly encouraged to expand this narrative, concentrating on areas of particular personal interest.

Engineers have developed transport to take humankind to the moon and back; they have also given us the means for our own destruction. With the end of the Cold War, the prospects for world peace are tantalizing, and there is already some beating of swords into plowshares. The new challenge for engineers, discussed in Chapters 8 and 9, is to find sustainable ways to help to address the needs of the less economically developed countries while maintaining quality of life in the industrialized countries and conserving the Earth's resources.

Discussion Questions

1. Why should we study the recent history of engineering?
2. Why were internal combustion engines developed in France and Germany and not in Britain?
3. Many people assert that the arts are only possible because of technology. Discuss the validity of this statement. What is technology?
4. Some engineers see governments as grossly inefficient bodies that merely move resources from one sector to another, almost always with a net loss of jobs. What other agency might provide essential infrastructure? What sorts of social implications would your proposals have?
5. Why did the ABC computer (Case Study 2.2) fade from the historical record?
6. What are some of the things missing from the history we have presented here, and why do you think they have been omitted?
7. How would you have added a female or a minority perspective to the history of technology?

8. What differences (if any) will it make to the way engineering is practiced when all the population at large, including women, is equally represented in the engineering profession?

9. Imagine and describe the thoughts, feelings and emotions of a twentieth-century craftsperson who has been made redundant by automation. What choices does such a person have for remaking a full, rich life?

10. Many people talk about the computer revolution. What similarities and differences can you see between this period in America and that of the Industrial Revolution in Britain?

11. Chose a technology not discussed here (e.g., modern printing technology or television) and using the "Basic Questions about Technology" provide a brief assessment of the "technology drivers," the role of engineers in developing the technology, and its significance for society as a whole.

Further Reading

Crompton, M. 1996, *Airframe*, Ballantine Books, NY. Crompton tells a good story but offers a rather stereotypical view of engineers.

Meadows, D. et al. 1972, *Limits to Growth*, Universe Books, NY. This is the classic early systems study of the Earth.

Petroski, H. 1996, *Engineers of Dreams: Great Bridge Builders & the Spanning of America*, Random House, Inc., NY. Petroski is relevant, readable, and infectiously enthusiastic.

Shute, N. 1954 *Slide Rule: the Autobiography of an Engineer*, Heinemann, London. An aeronautical engineer turned novelist, Shute is interesting, informed and readable.

References

ASCE (American Society of Civil Engineers) 1998, World Wide Web site at <http://www.asce.org/govnpub/rightwro.html>

Benathy, B. H. 1991, Comprehensive Systems Design in Education: Who Should be the Designers? *Educational Technology*, Sept., pp. 49-51.

Bush, V. 1945, *Science—the Endless Frontier, A Report to the President on a Program for Postwar Scientific Research*, OSRD, Washington, D.C.

Chant, C. 1978, *Aviation: An Illustrated History*, Orbis Publishing, London.

Checkland, P. B. 1981, *Systems Thinking, Systems Practice*, Wiley, NY.

Collingridge, D. 1980, *The Social Control of Technology*, Frances Pinter, London.

Diesendorf, M., Hamilton, C. (eds.) 1997, *Human Ecology, Human Economy: Ideas for an Ecologically Sustainable Future*, Allen & Unwin, St. Leonards, Australia.

Hull, C. W. 1990, Engineering in a Global Economy: A United States Perspective, *Technology in Society*, Vol. 12, pp. 107-120.

Jakab, P. L. 1990, *Visions of a Flying Machine: the Wright Brothers and the Process of Innovation*, Smithsonian Institution Press, Washington, D.C.

Kline, S. J. 1995, *Conceptual Foundations for Multidisciplinary Thinking*, Stanford University Press, Stanford, CA.

Mathews, J. 1989, *Tools of Change: New Technology and the Democratisation of Work*, Pluto Press, Sydney.

Nash, G. B., Jeffrey, J. R. 1998, *The American People: Creating a Nation and a Society*, Longman, New York.

Panitz, B. 1996, Bioengineering: A Growing New Discipline, *ASEE Prism*, Nov., pp. 22-28.

———, 1997, Training Technology's Maestros, *ASEE Prism*, Nov., pp. 18-24.

Petroski, H. 1985, *To Engineer is Human: The Role of Failure in Successful Design*, St. Martin's Press, NY.

———, 1994, *Design Paradigms: Case Histories of Error and Judgment in Engineering*, Cambridge University Press, Cambridge.

Rosenbrock, H. 1990, *Machines with a Purpose*, Oxford University Press, Oxford. Quotation reproduced by permission of Oxford University Press.

Taylor, F. W. 1967, *The Principles of Scientific Management*, W.W. Norton, NY.

Thompson, J. 1967, *Organizations in Action*, McGraw-Hill, New York.

UNCTAD (United Nations Conference on Trade and Development) 1996, *Trade and Development Report*, United Nations, NY.

Warnecke, H. J. 1993, *The Fractal Company: A Revolution in Corporate Culture*, Springer, Berlin.

EXPERTS, COMPLEX SYSTEMS, AND CHALLENGER

Much that has been written in the engineering community on the NASA Challenger accident (January 28, 1986) has focused on the question of engineering ethics.

Did Boisjoly and Thompson, engineers at Morton Thiokol Inc., manufacturers of the rocket boosters, argue strongly enough that the temperature at the launch site was too cold for the O-rings? Should they have "blown the whistle" when they were overruled?

Did the engineer-managers act unethically when they recommended a launch in spite of the dangers?

In short, would a better appreciation of and adherence to engineering ethics have saved the lives of those seven astronauts? Although the direct answer to this question is probably yes, we would miss some important lessons about the nature of technology and the implications for a broader view of engineering ethics if we were to stop our analysis there.

A closer look at the events and problems surrounding the accident indicates that many things were wrong. The launching of a space shuttle involves the operation of a very complex system, a system that includes not only complex hardware but also the cooperation of a wide range of people. Furthermore, the even larger sociopolitical context often has a direct impact on the system. In complex systems, single-component failures can occur in unanticipated ways, can lead to other failures, or can work in conjunction with other failures to produce major accidents. A narrow view of engineering ethics focuses on single failure problems; a broader view is needed to deal with the possibilities of multiple failure problems. A whole range of problems and failures surrounded the shuttle accident, although not all were immediate issues with a direct connection to the disaster.

Let's look at the issues one by one:

Cold Temperatures

The temperature at the launch site was expected to fall to 10 °F (−12 °C) at dawn on the day of the launch, creating several problems: the O-rings in the field joint of the booster rockets might be too stiff to seal properly; the water in the trough used to absorb the shock wave would be frozen; and the pressure transducers in the nose cone were not rated to give reliable results at that temperature.

Responses to these problems varied: the possible effect on the O-rings was ignored (as discussed later); the troughs were doped with antifreeze and the water was left to run in the hoses to avoid freeze-up; and the temperature criterion for the transducers was rewritten twice, first to 28 °F (−2 °C) on the Monday and then to 10 °F (−12 °C) on the actual day of the launch. Allowing the water to run in the fire hoses created a new set of problems: the drain pipes froze, allowing water to flow all over the launch structure and create sheets of ice and large icicles. This was potentially dangerous because ice could have been sucked into the engine during take-off or could have fallen off and damaged the sensitive tiles on the shuttle.

Field Joints

Because the failure of the O-rings in the field joint was found by the Warren Commission to be the primary cause of failure it has received a lot of media coverage. I will review only a few of the central issues. Erosion of both the primary and secondary O-rings had become so common by the eighteenth flight that it was considered acceptable and did not require review after each flight. The manufacturer had the problem closed because they were "working on it." New designs were being considered, but very little had been done to correct the fault. One of the changes that did occur was to increase the pressure during ground testing. This is now believed to have made the problem worse.

Hatch Closing Indicator

A faulty microswitch (indicating that the shuttle door was not closed) eventually led to an earlier launch being scrubbed. What went wrong was a micro-example of failure in complex systems; it was Murphy's Law in action. In order to fix the microswitch the handle of the hatch had to be removed. However, the threads of the bolts were stripped and could not be removed without drilling. A battery-powered drill carried by the technicians for this environment was found to have a dead battery. Of the nine batteries eventually sent up as replacements, only one actually worked. The metal was eventually found to be too hard to be drilled out, so the bolt was cut off. All this took hours. By the time the problem was fixed, the wind had picked up and the launch had to be postponed.

Wind Shear

Although the temperature problem has received a lot of attention, the importance of a large wind shear that day was not recognized until much later. Some calculations indicate that the wind shear was large enough to exceed the structural capacity of the rocket, fifty-eight seconds into the flight and just before the fatal explosion. Although it is clear that the O-rings did fail at the moment of ig-

nition, the failure did not lead to the fatal accident at that point. It is likely that propellant oxides resealed the joints. The Challenger may have made it had the shear forces not been so powerful one minute later.

Production Pressures Within NASA

On record NASA had said that their highest priority was to make the shuttle system "fully operational and cost-effective in providing routine access to space." The goal was at least fifteen flights a year in 1986 and twenty-four by 1990. However, this was unrealistic; NASA had until then completed only four shuttles and had enough spare parts for two. Only half of the $200 million cost per flight was being recovered from fees. Each minute of a shuttle mission requires three person-years of preparation, with 14,000 people involved. The pressure to keep on schedule was very high and felt by everyone; many worked long work weeks with no breaks.

Production Pressures Within Thiokol

Up to this point Thiokol was the sole source of the booster rocket. There was pressure for open competition for the contract. Thiokol could ill afford to admit problems in design and could not afford a redesign.

Storm in the Atlantic

A major storm in the Atlantic had forced the recovery ships for the booster rockets to leave their post. This meant a loss of $50 million in hardware. Why was the launch not postponed? This becomes clear when we consider some of the political factors.

Political Pressure/Involvement

This was present in a number of ways. There is some evidence of conflict of interest and political connection in the awarding of the contract to the company who designed and

(continued on next page)

built the shuttle. Other designs did not require a low-lift space glider that had to land on its first try.

The decision not to launch on Sunday, January 26 had been made because although Vice President Bush would have been able to be present at the launch there was fear of bad weather that would result in a cancellation, with associated bad publicity. It turned out to have been a perfect day for a launch. January 28 was the last possible day to launch if the next shuttle flight was to go before March 6. This next flight was to carry a special project to view Haley's Comet. The politicians wanted to scoop the Russian Vega 2 that would send back its pictures on March 9.

One other political connection involved the President's State of the Union Address. The speech was to mention NASA in connection with the Teacher in Space program. NASA wanted to have the shuttle in orbit to capitalize on the administration's emphasis on education.

THE NATURE OF EXPERT DECISION MAKING

People who get involved in projects of this nature often demonstrate a very high level of technical skill and tend to have a "can do" attitude. In some ways they get wrapped up in the task for the sake of the task. A sound level of skepticism does not always accompany these virtues; there is a tendency to take risks and perceive risks differently from the general public. This is often linked to overconfidence and blind spots due to previous successes. We see this in how the O-ring erosion problem eventually became "acceptable." Sometimes there is long agonizing over whether it is safe to fly with a certain known risk, and the decision to go is made and nothing goes wrong. There is then a feeling that the decision to go ahead was not so risky after all, and next time it will not be so much of a concern.

Sometimes experts have difficulty in handling uncertainty. Evidence is not always conclusive and there is always need for judgment. If launches were postponed every time any engineer had some doubts they would probably never happen. After a while engineers get used to making decisions with uncertainty. The concern about the O-rings was another case of uncertainty. Had the managers not been engineers and had consequently become accustomed to ignoring some degree of uncertainty, they might not have been so quick to overrule their engineer subordinates in the case of the Challenger Shuttle.

In order for the final launch to occur there was a lengthy prelaunch certification process involving several levels of command. Often the upper levels were more skeptical than the lower levels. Theoretically this is a good procedure. In reality, however, the procedure was flawed. Lower levels could waive constraints, the details of which were not communicated to the upper levels. For this particular launch, for example, Thiokol's hesitation and the concern about the effect of the cold temperature was not communicated between Levels III and II in the launch chain of command.

The above analysis has shown that the accident was due to a number of factors: if only the weather had not been so cold; if only the design of the field joint had been changed in time; if only the production pressure was not there; if only there was not pressure for the March 6 flight; if only the managers had not been engineers; if only they had had more experience with failure; if only the concern had been passed up the chain; and if only there had not been strong shear winds that day. Charles Perrow (1984) argues that multiple failure systems accidents are normal for complex, tightly coupled technologies; the failure is to a large degree inherent in the technology. Perrow distinguishes between component failure accidents, where failure is linked in an anticipated sequence, and system accidents, where there is an unanticipated interaction of multiple failures that nevertheless start with component failures.

Space missions are more complex than chemical plants but are less tightly coupled because of the number of built-in redundancies (back-up devices that provide an alternative action if something fails). There are, however, over seven hundred critical items for which there are no redundancies: failure in any one of these would lead to a loss of a mission.

Was there a failure due to one or more persons failing to act ethically? All of this analysis is not to excuse unethical behavior, but to illustrate the need for a concern wider than individual ethics. A broader view would consider not only the human operators, but also the nature of the technology and the nature of the process of design and operation of the technology.

Concerning the nature of the process, engineering design and operation are very much human activities. Manager engineers need to be aware of, and know how to handle, the pressures that affect their work. These pressures can be political, economic, social, or even personal. Multiple lines of communication would help here.

More deeply, however, as engineers we need to recognize the blind spots inherent in being an expert. Our perception of risk and uncertainty is affected by our past successes and by the distance we can take from the consequences of our projects.

Concerning the nature of technology, we need to consider the characteristics that get built in. The tendency for highly complex and tightly coupled systems to behave in totally unexpected ways has been demonstrated many times. This is the nature of the system we deal with, so we need to design policies and procedures accordingly. In the operation of these systems, if anomalies occur we should be careful not to ignore them or rationalize them away; they may be unexpected interactions, common to complex systems.

In this regard it is important to encourage people to talk about problems they experience; there is a tendency to suppress unpleasant information. In the risk analysis of complex systems we need to widen our worst-case scenarios to include three or four simultaneous component failures.

Source

Case contributed by Robert Hudspith, Professor of Engineering, McMaster University, Hamilton, Ontario. Originally published in IEEE Hamilton (On) Newsletter, April 1988.

Further Reading

Feynman, R. L. 1989, *What Do You Care What Other People Think*, Bartan, NY.

This engaging and very readable memoir by a physics Nobel laureate includes an account of his role in the investigation of the Challenger disaster. The book is essentially an extended reflection on Feynman's work and illustrates the value to a professional of such a reflection.

References

McConnell, M. 1987, *Challenger: A Major Malfunction*, Doubleday & Co, NY.

Perrow, C. 1984, *Normal Accidents: Living with High Risk Technologies*, Basic Books, NY.

FROM COALBROOKDALE TO SILICON VALLEY

What conditions are conducive for innovation and what makes an area thrive? Since the dawn of history, valleys have been associated with technological innovation. Natural resources have played an important part in this. In Egypt, the Nile provided easy bulk transport and a usually reliable food supply. In later centuries a free-flowing river was used to drive a water mill, provide water for material processing, and eventually was used to condense steam used for power generation. Industrial developments were inevitably concentrated in the more significant river valleys. This was certainly the case in the development of the Severn Valley in England, where in the eighteenth century the Coalbrookdale area saw the birth of the Industrial Revolution.

Some two hundred years later, in the new industrial revolution of the "information age," suitably educated and trained people (Peter Drucker's "knowledge workers") are increasingly recognized as the essential resource. Silicon Valley in California is the most obvious and successful example of their success. The technologies on which Silicon Valley is based are not directly dependent upon a good supply of water or on river transportation. Is the fact that both these revolutions, in industries so very far apart, are based in valleys merely a coincidence? Or are there more salient connections? What can we learn from them about the common factors that will be essential to the success of future industrial developments?

Coalbrookdale

In the coking process, sulfur is eliminated from coal, greatly increasing its usefulness for most industrial purposes. In England, patents for the use of coke to smelt iron had been granted as early as 1620 but no-one had managed to make the process work in practice. Pacey (1991) suggests that the Chinese may have used coke for smelting iron several centuries earlier, but there was no knowledge of this in Europe. By 1700 the rising demand for timber in England and the resultant high cost of charcoal were limiting the availability of iron.

Abraham Darby was the son of a farmer who also worked as a blacksmith; he grew up knowing about forges and wrought iron. He was apprenticed to a copper foundry, gaining experience of the use of coke in copper smelting, and in 1707 patented a method for casting iron pots in sand. In 1708 he moved to Coalbrookdale and took over a site with an existing blast furnace. He seems to have experimented almost immediately with the use of the local coal; transferring technology from the brewing industry, he built blast furnaces running on coke and succeeded in producing good quality iron in them.

The iron that Darby produced in a coke-fired furnace was ideal for casting the cooking pots and utensils that were Darby's main business, but it was not immediately accepted for conversion to wrought iron. Acceptance did not come until the price advantages of the use of coke over charcoal became significant. In the United States, with its profusion of timber, the acceptance of coke for smelting was to take even longer. But Darby's breakthrough was so profound that it is widely accepted as the start of the industrial revolution.

The plant at Coalbrookdale grew rapidly; Darby developed additional blast furnaces and added forges, rolling mills, foundries, and thirty-two kilometers of wagon track. The site had been well chosen. The general area was one of high ground sloping down to the River Severn, at that time England's busiest river system. The furnace site was halfway down a tributary stream, below a number of large coal mines. The stream was dammed and the furnace blast was produced by bellows, powered by waterwheels. The obvious scheme was for the coal to be carted downhill to the furnaces in wagons. The cast iron products could then be readily carted further down to a dock on the river for shipment.

Dependence on water power restricted the use of the furnaces in summer and this, and the relatively weak blast available, limited the output of iron. In a nice symbiosis, Coalbrookdale produced cast cylinders for the Newcomen engine, which had been introduced in 1712. The installation of a steam engine at Coalbrookdale in 1742 allowed year-round running of the furnaces, with a much stronger draft. Output increased dramatically. Steam engines were becoming an integral part of both the iron industry and coal mining.

The Darby family were Quakers and the attitudes associated with their religion clearly had a major influence on their work. It was said that they did not really separate the two and that they chose to work with people of the same faith. Quakers were favored for employment at Coalbrookdale and the prevailing atmosphere was one of friendliness and mutual respect. The surrounding community was transformed from poverty to enlightenment and health. Other local industries, such as the Coalport China Works, benefited from very direct transfers of technology, and the potteries actually employed more local residents than the iron or coal industries.

Silicon Valley

Formally the Santa Clara Valley, Silicon Valley is the industrial concentration, consisting largely of electronics and information corporations, located between Menlo Park and San Jose in the San Francisco Bay Area. In global terms these corporations are usually considered to be the leaders in the "Information Revolution."

Since before World War II, fundamental research on semiconductor technologies has been conducted at Stanford University, which was keen both to spin-off "start-up companies" and to keep in close touch with them. In 1937, with the encouragement of their Stanford electrical engineering professor Frederick Terman, Stanford graduates William Hewlett and David Packard started a company. A similar course was followed by Charles Litton and the Varian brothers. When Professor Terman returned to Stanford as Dean of Engineering after the war, he realized how far the West Coast had fallen behind the East Coast in innovation. Terman dedicated himself to developing Stanford and the industry in tandem as a "community of technical scholars."

In 1954 a Stanford graduate, William Shockley, who had invented the transistor while at Bell Labs, returned to the Bay Area to found a company. Two years later eight of his leading engineers broke away to found the Fairchild Semiconductor Corporation, in a pattern that was to be repeated many times as Silicon Valley became the home of the transistor industry (Case Study 6.3). Thirty-one semiconductor companies were

(continued on next page)

started in the Valley in the 1960s, and the majority traced their lineage to Fairchild.

The approach to industrial organization at the "Fairchild University" was egalitarian. Saxenian (1996) quotes Tom Wolfe as commenting: "It wasn't enough to start up a company; you had to start a community, a community in which there were no social distinctions." The prevailing culture was a complex mix of social solidarity and individualistic competition.

Silicon Valley survived a setback in the 1980s, when the industry treated semiconductors as commodities but became vulnerable to Japanese competition. Fortunately the Valley's innovation culture was sufficiently strong that new start-ups continued to be born, and the emphasis shifted to application-specific chips and computer systems. By 1992, some 113 technology companies in the Valley were reporting revenues in excess of $100 million. Most had their headquarters in the region and had been started in the last twenty years.

In "Regional Advantage," Saxenian contrasts the system in Silicon Valley with the more rigid structure of corporations around Boston's Route 128. She concludes that regional networks are more flexible and technologically more dynamic than arrangements where experiment and learning take place within individual firms.

Common Threads

Although Coalbrookdale and Silicon Valley enjoyed their extraordinary success a quarter of a millennium and half a world apart, there are striking similarities. What factors were strong enough to produce not just successful technological innovation but a whole technology revolution? What is it about a valley that is important?

Cincinnati's machine tool industry may provide guidance on some important linkages. In the last quarter of the nineteenth century, the initiative in U.S. manufacturing shifted from the east coast to the Midwest. Despite the obvious transportation advantages of Cleveland and Chicago, the machine tool industries flourished along Cincinnati's Mill Creek (hardly a great river valley!) Scranton (1994) describes how a web of actions and contexts generated a technological momentum that was "open, non-systemic, locally determined, and as much dependent on social relations as on engineering refinements." The resulting rich and robust industrial legacy is carried on by such large high-technology enterprises as GE Aircraft Engines.

In both Coalbrookdale and Silicon Valley, the informal channels of information exchange seem to have been as important as formal mechanisms. Successful innovation certainly required the knowledge and the strategic decision to exploit it, but beyond that it required a web of action and context. It was important for the technologies to be developed independent of large bureaucratic structures. Supportive local social frameworks and value systems, simultaneously competitive and cooperative, appear to have been essential for success.

The structuring of the organizations that knowledge workers inhabit is critical. Over the last few decades we have seen a flattening of formal organizational structures, with moves towards smaller operating units, networked flexibly together. As discussed in Chapter 7, this is consistent with recent management thinking. Moves dictated by quality considerations are aimed at empowering workers, giving them authority to organize their work more effectively and responsibility for ensuring appropriate outcomes.

Dedicated research facilities, well networked with educational institutions, seem an important ingredient, as does the attitude to knowledge within the academy and the broader society. The American attitude has traditionally followed that of the religious nonconformists who were a significant proportion of the immigrants. Their belief is that learning carries a responsibility for its application. This attitude was as manifest in the work of Dean Terman at Stanford as it was earlier in the work of the Darby family at Coalbrookdale.

We are well used to valleys being a focus for innovation and this historical perspective dates from the establishment of civilizations after the last ice age. Because most valleys look towards a river, with its potential advantages for the flow of goods and materials, we tend to focus on these physical attributes to explain technological success. Certainly the Coalbrookdale example shows that they can be very important in fostering innovation.

However Silicon Valley is not really a valley at all in the conventional sense. No great river bisects it and clearly its location was not based on any particular local supply of silicon nor on easy transport of finished goods to the market (although its proximity to San Francisco, a major transport hub, is certainly helpful). So how important is the "valley" to the success of innovation in Silicon Valley? The answer to this question may be an important one that the more obvious physical attributes of valleys in the past have masked.

Saxenian shows the way with her interesting comparative study of the rise and fall of the Route 128 industries, which had no perceptible "valley" effect, and the rise and then renewal of the Silicon Valley industries. This comparison prompts the question: is there something special about the relative intimacy of a valley that supports the favorable social climate that Saxenian observes there?

The Cincinnati example points to the sense of community that a valley fosters. What else characterizes a valley other than its river? Possibly the most important aspect is that it inhibits the spread of the population. If you work in Silicon Valley, it is most convenient to live between the Bay and the Santa Cruz mountains, or possibly in one of the cities east of the Bay that are generally considered to be extensions of Silicon Valley and are even more constrained. This imposes a certain concentration of population and inhibits urban sprawl. The result is a much greater likelihood that the people you work with will also be the people you socialize with. An underpinning common purpose may develop—in a real sense you will work for the valley. This probably puts people in touch with some of their most basic communal feelings. These effects were almost certainly present in the successful industrial community at Coalbrookdale, but would have been masked by the more obvious attributes of the place. The valley physically locates and concentrates people, the most important resource in innovation.

References

Pacey, A. 1991, *Technology in World Civilization*, MIT Press, Cambridge, Mass.

Raistrick, A. 1951, *Dynasty of Iron Founders: The Darbys and Coalbrookdale*, Longmans, Green, London.

Saxenian, A. L. 1996, *Regional Advantage: Culture and Competition in Silicon Valley and Route 128*, Harvard University Press, Cambridge, Mass.

Scranton, P. 1994, "Determinism and Indeterminacy in the History of Technology," in *Does Technology Drive History*, Merritt Roe Smith and Leo Marx, eds., MIT Press, Cambridge, Mass.

3

Sociological Insights

History is, strictly speaking, the study of questions;
the study of answers belongs to anthropology and sociology.

— W.H. Auden (1907–1973) Anglo-American poet

Engineers have a major role in shaping the society in which they live, particularly the structure of workplaces and production relationships. To lead and manage technological change, and to carry out their work effectively and ethically, they need to understand their society and the way they interact with it. The discipline that deals with how society works is sociology.

Sociology has been described as the history of the present. Just as we gain valuable insights from a historical perspective on the engineering profession, a systematic study of engineers and the place of engineering in contemporary society is also very relevant.

In this chapter we are concerned with engineers as a social group with a distinctive, often unique profile. We also offer a broad-brush sketch of American society as a whole; this provides the context for engineering activity and identifies areas of common interest between engineers and other groups in our society. Sociology sheds light on aspects of modern society that constitute some of the major challenges for professional engineering practice. Sociology addresses such questions as:

Engineers and Sociologists

Sociologists can suggest answers to all sorts of queries that, as engineers, we might have about our society and about our own place and prospects in it, such as:

- ◆ Does the work of engineers influence how opportunities and resources are distributed in our society generally? Who gains? Who loses?

- ◆ Are engineers in control of technological change and its on-going impact on society? Or are they more like conscripts, responding to management imperatives to develop and implement technological change that can cause major social dislocation? To what extent are they subject to the imperatives of capital accumulation by others?

- ◆ Given the changing nature of technical work and whole industries, what are engineers' expectations? Are engineers moving up or down the social scale?

- ◆ Do engineers mostly align themselves with those who enjoy substantial rewards and greater autonomy in the workplace and high status in society at large? Or do their interests really lie with low-to-middle wage earners?

- ◆ Where do engineers see themselves? Where does the rest of society locate them in the overall scheme of things? Why might this matter?

The answers to such questions typically draw on a *sociological* analysis. Sociologists use a range of concepts and methodologies unfamiliar to most engineers. Even where the focus is on issues that are associated with or derive specifically from engineers' own technical expertise, that expertise alone does not supply the answers to some very crucial questions.

- ◆ How is a society constructed and how does it operate?
- ◆ Why are some groups in society more powerful than others?
- ◆ What causes social change?
- ◆ Is society normally in orderly balance or conflict?
- ◆ What is the relationship of the individual to society?

In seeking answers to such questions, sociologists have developed a number of conceptual tools. Rather than total numbers, they concentrate on structures and on groupings of people. The focus is on broad patterns in society and how these are changing. For instance, stratification models (such as those based on social class, gender, ethnic, and racial differentiation, discussed below) look at people's opportunities in life and how they relate to one another. Other groupings often used by sociologists are occupational categories, age cohorts, religious or political affiliations, or simply population breakdowns based on where and with whom we live. Some of these groupings reflect the important role of institutions (such as marriage or the church) in our society. Such concepts are central to understanding the basic functioning of any society. Engineers should be familiar with the various tools employed by sociologists and the insights they offer.

An area of focus for sociologists that is of particular interest to engineers is "work." This is one of the most important activities in which members of human societies engage. As engineers, not only is our own work important, but we also have major responsibilities for the ways in which work is organized for others. The technology developed by engineers impacts on workplace relations and the characteristics of work that make it interesting, safe, and satisfying. Another area that draws heavily on sociology and is of increasing importance for engineering is marketing. Marketing demands systematic social analysis, rather than decision making based on superficial or anecdotal evidence. These are some of the ideas we will explore in this chapter.

Sociology is "the scientific study of the social behavior or social action of human beings" (Gould and Kolb 1964). Like most definitions, this one raises questions as well as answers them. In particular, what makes sociology scientific? Its claims rely on an organized, disciplined approach to research in which proposed theories and concepts are open to testing and should be able to be disproved. Individual sociologists perceive society in different ways; they have different values and beliefs and these underlie different theoretical approaches. Some schools of sociology focus on sources of conflict within society, such as social class. Others emphasize institutions that tend to draw society together. Another emphasis, bordering on psychology, is on exploring the ways people experience and make sense of their experience in society. Detailed descriptions of these various perspectives are beyond the scope of this book, but we should note that none of them is "right" or "wrong." Human society is so complex and diverse that all of them offer useful insights. This wide divergence of approaches within a subject that calls itself a science may seem strange to engineers, who come from a technical background. It reflects the fact that we cannot really stand outside human society because we are part of it. This is also consistent with the systems approach described in Chapter 2. Humans, and human behavior, are so much more complex than inanimate natural phenomena that it would be absurd to expect the same level of predictability from them.

THE SOCIOLOGICAL IMAGINATION

We use sociology to help us to stand back and look at everyday events and relationships in new ways, giving us a clearer perspective on how societies function and on the tensions within them that can lead to social change. The ultimate test of the usefulness of sociology is whether it offers a better basis for understanding what is going on and for the effects of our actions, than is provided simply by common sense.

Mills (1959) described sociologists' analysis and interpretation of the present as *the sociological imagination*. He suggested that it could help people become aware of the connections between their own experiences and the changes going on in the world at large. He argued that this involved people seeing their period not as something static but as a moment in the flow of history.

An essential part of developing this sociological imagination is differentiating between personal issues that reflect individual characteristics and circumstances and public issues that arise from the social situation and affect large numbers of individuals similarly. The sociological imagination enables us to shift from one perspective to another, from the economic to the political to the social, to draw on a novel or a film to illuminate the human effects of (say) economic policies. It helps make connections between different styles and approaches to reality, so as to build up a rounded picture. Perhaps its most valuable aspect is that the sociological imagination can help us to keep a hold on what is really important, the fully human, spiritual dimension, and to see how this is affected by what we do and what is happening in the larger world around us.

When applied through social surveys of the engineering profession, the sociological imagination may provide helpful insights for professional bodies or engineering educators. On a personal level, awareness of what professional engineers generally do, how they view their work, what they value, and how much they are paid is of interest to most of us. It can also help us to assess our own prospects realistically. Some stocktaking may be required. How quickly are educational requirements changing? What skills are in demand? We address these questions by looking at some of the raw information drawn from census and similar data by sociologists, and then go on to models of social structure and social differentiation.

U.S. Households and Families

One area of human behavior that has been studied closely by sociologists covers pairing, parenting, and domestic arrangements. A brief look at trends in household and families provides us with useful information against which to test our own assumptions about norms and social stability.

The U.S. Bureau of the Census figures for 1996 show a total U.S. population of just over 265 million people, living in almost 100 million households. (It defines a household as comprising all persons who occupy separate living quarters; it includes all related family members and all unrelated persons). The average number of people per household was therefore 2.65 (U.S. Census Bureau 1997, pp. 1, 60).

What is a family? For statistical purposes, it may be defined as: two or more persons, one of whom is at least fifteen years of age, who are related by blood, marriage, or adoption, and who are usually resident in the same household. Separate families are identified for each married couple and for each one-parent family in the household.

In most Western societies, including America, there has been an expectation that "normal" living and reproductive arrangements are based on the nuclear family—mom, dad, and the kids all living together in the family home. This notion has been closely scrutinized by sociologists over the past few decades. They point out that it is by no means the universal experience, and certainly not over a whole lifetime! This model appears to be under serious threat. The makeup of U.S. households is shown in Figure 3.1.

Households in America, 1996

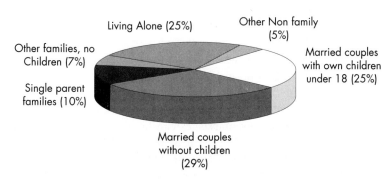

Figure 3.1 Households in America, 1996

Married couples, with or without children, make up 54 percent of U.S. households (59 percent of White households, 34 percent of Black households, and 57 percent of Hispanic households included a married couple). During the two decades from 1970 to 1990, the proportion of all U.S. households accounted for by married couples with children fell from 40 percent to 26 percent. Our society is thus less likely now to conform to the traditional nuclear family stereotype, but the decline does seem to have "bottomed out"—from 1990 to 1996 it only fell by another 1 percent.

At least half of all children will experience living with only one parent at some stage before they reach eighteen years of age. The rise of single-parent families (mother or father but not both parents present and mostly headed by women) has become a widespread phenomenon. It has slowed very sharply, but still represents nearly one third of all families with dependent children. The numbers of births to unmarried mothers and to teenage mothers (traditional cause for concern as regards social stability) is declining. New trends are emerging: there are now at least one million gay or lesbian couples who are raising one or more children (Macionis 1997, p. 476).

Of course, most of us do not define "family" in cold statistical terms. Our view of the family is more likely to be influenced by our own experience at Thanksgiving or other holiday get-togethers. (It is likely to be different at weddings, and different again at funerals.) There are movements in and out of family households, and flows of support (financial, practical, and emotional) between different households within the same family. These are common and very significant but may tend to blur the outlines of a family. Such flows of support are not universal, and life for families without them can be bleak.

We should not lose sight of the fact that over twenty-five million people live alone. This is one household in four, up from only one household in nineteen in 1950. The statistical data do not tell us whether this represents a striking increase in loneliness, or is the outcome of a higher value being placed on economic independence and self-reliance.

With greater social and economic equality between the sexes, divorce has become more common. Women are less likely to be trapped in an unsatisfactory relationship by an inability to make an independent living. People expect more out of marriage, and are more prepared to divorce if their expectations are not met. According to government estimates, about 40 percent of new marriages can be expected to end in divorce (for African-Americans, this figure is 60 percent). This is more optimistic than the five out of ten previously predicted.

Taking into account separation without divorce, and possible death of a partner, it is estimated that couples who marry today when the wife is age twenty-five have a 53 percent chance of being together thirty years later. This is still an improvement on the prospects one hundred years ago when divorce rates were very much lower, but at that time there was only a 46 percent chance of both partners still being alive thirty years later! People do not break up relationships lightly: divorce is the most stressful event in life, apart from the death of a spouse or a child, and individuals commonly take at least two years to get over it.

All this social data is by no means simply of academic interest. These figures have implications for a range of issues (some of which are discussed below) including wealth and income distribution in America, budget allocations for schools and health care, size of living units, changes in residential housing developments, and so on. They are reflected in policies espoused by the major political parties, such as those relating to state-subsidized child care and tax penalties faced by married couples. Information on families and households is used in a variety of ways. For instance, it can be vital in assessing markets for different sorts of products, which is in turn crucial for industry sectors such as manufacturing and communications, where levels of consumption are of direct interest to engineers.

We should also be aware of any shifts in human behavior, as these shifts may impact on our work in all sorts of otherwise unexpected ways, especially if they reflect underlying shifts in values taking place in our society.

Social Stratification

Societies differentiate between their members on the basis of whole sets of criteria that are socially constructed (although there may appear to be a biological or other basis). The distinctions can have a significant effect on the probabilities of people achieving any given social outcome: life expectancy; having a satisfying job; graduating from university; owning their own home. The three most socially significant types of differentiation are stratification, gender differentiation, and ethnic or racial differentiation.

Stratification involves social ranking of people by economic position, prestige, and/or power. Economic position, determined basically by property ownership and income, is commonly referred to as class position. Prestige is status or social honor; social power and the way it is applied in the United States are discussed below. There are several types of stratification models: societies may show features of more than one model, but for Americans, social class is the most important.

■ Social Class

A "common sense" view of American society might doubt that the class system is all that significant in the United States today and might dispute the notion of a "ruling class." Yet the proposition is readily accepted that status is related to possession of income or property. What is the connection? Notions of stratification may be intuitive but can also be quite rigid. Certainly, considerable social mobility is possible, but it may be limited in practice by ethnic/racial and gender differentiation, as well as by a tendency for classes to reproduce themselves. These are issues we wish to explore.

For reasons of status and prestige, people are often inclined to identify with those above them on the socioeconomic ladder. There is thus commonly a difference between a person's objective class position, based on occupation and income, and their subjective class (the one to which they consider themselves to belong). Mobility between social classes is

Marx and Class

Marxist definitions of class emphasize the relationship to capital.

As discussed in more detail in Chapter 4, the predominantly private ownership of capital characterizes modern Western industrial societies such as the United States. The traditional Marxist approach emphasizes an inherent conflict between two dominant classes:

1. the capitalist class or *bourgeoisie*, defined by its ownership or control of capital, particularly in the form of factories and other workplaces; and

2. the working class or *proletariat*, the group that has to sell its labor in order to survive.

The middle class or *petit bourgeoisie* lies between these two and plays an important role in cushioning their differences. It includes shopkeepers and self-employed people (and by no means corresponds to what most Americans would consider the "middle class.").

This framework has been considerably modified by later sociological definitions of class. These are more relevant for our purposes and are discussed in the Key Concepts and Issues box.

recognized as more widespread in mature industrialized ("open") societies such as the United States than, say, traditional agrarian societies. For instance, it is estimated that about 40 percent of the sons of U.S. blue collar workers show some upwards mobility and 30 percent of the sons of white collar workers move downwards on the social scale, compared with their fathers (Macionis 1997, p. 274). Nevertheless, although social class may be most readily defined in terms of wealth (as discussed below) it should be stressed that it is birth into a particular family that most affects our social standing for the rest of our lives.

So how best to define the class system in the United States today and why is social class relevant? Some caution should be exercised in attempting any class analysis. Wealth, defined in terms of ownership of property, is still the best starting point, although income may be more relevant in certain contexts. In 1993, the richest 20 percent of U.S. families owned approximately 80 percent of the country's entire wealth. The least wealthy 60 percent owned less than 6 percent of the wealth and the poorest 40 percent have virtually no net wealth, and their debts commonly exceed their assets (Macionis 1997, p. 263).

Descriptions of social class are of interest to a wide audience. Journalists write stories for popular consumption based on sociological studies. There is no reason why engineering students should not find such material equally interesting. However, sociologists draw certain inferences from such data. Our purpose here is to reflect on some of the implications, so that we may become more sensitive to the society in which we live and which we serve as professional engineers. In particular, a grasp of the realities of how wealth and power are distributed is of practical importance to engineers involved in developing project proposals or product design and marketing. For instance, social awareness and a sensitivity to local community needs was conspicuously absent from much of the engineering input to the building of the Interstates, discussed in Case Study 4.5.

■ The Widening Income Gap

One of the messages that at least some sociologists continually stress is that marked social stratification may give rise to a range of social problems that are eventually everyone's concern. The health of a whole society is jeopardized if, for instance, the numbers of its homeless people grow too great, or if people's faith in the possibility of upward social mobility (if not for themselves, then for their children) is threatened. Insofar as social class, with all its attendant cultural baggage, is at least partly dependent on income, then wide disparities in income are cause for concern. Many

Social Class in the United States

A social class is generally taken to be a category of people who share a common social position, economic situation, and cultural attitudes.

Upper Class

It is generally accepted that the upper class comprises the top 5 percent of the American population, and that this group owns more than half the nation's property. The "super rich" comprise the 1 percent of U.S. households that own one third of our nation's privately held resources.

The origins of the wealth of the most affluent group are illuminating. Unlike "old money" families such as the Rockefellers and the Vanderbilts, who accumulated their wealth in earlier generations and thus are viewed as sort of American aristocracy, the "new rich" are likely to be the top level of managers and entrepreneurs. The emergence of a global economy has created extraordinary opportunities for "upstart" financiers to accumulate vast fortunes. As a result, the number of super rich Americans has exploded in recent years. At the end of World War II there were 13,000 millionaires; by 1976 there were 250,000 (Appelbaum and Chambliss 1997, p. 208). In 1995–1996 the figure was approximately 1.8 million, including 128 billionaires. In 1997 Microsoft's Bill Gates was the richest individual in the world.

Thus the upper or most affluent class in America consists mostly of heads of large corporations, people who have made large amounts of money through investment or real estate, those fortunate enough to have inherited great wealth from their parents, highly successful movie and television stars, a small number of professional athletes, and a handful of others (Appelbaum and Chambliss 1997, p. 208). There are strong connections between rural and urban élites. Education, family background and social acceptance are important, but the major factor determining membership of this class is property.

Middle Class

"Middle class" describes about 45 percent of the American population. This grouping is less firmly structured than the previous group. In other societies and/or in previous times it was separated from the upper class by its lack of income-earning property, and from the working class by education, income, and attitudes, but these distinctions no longer hold good in modern American society. Many subcategories can be identified. Occupations range from what some analysts, such as the Marxists, would describe as "nonmanual working class," through to small business people and upper professionals, the Marxist petit bourgeoisie, people with enough capital to be self-employed, and perhaps to employ a small workforce.

In terms of cultural differences and for practical purposes, such as targeting groups for advertising, it has been common to divide the middle class into at least two groups, including upper middle and average middle class. The former class, largely self-employed people, managers, and professionals, makes up about 5 percent of the population. Its members are affluent but self-confident and hard for marketers to manipulate. Two thirds of upper middle class children go to college and many go on to high prestige professions (physicians, lawyers, engineers, and accountants or business executives) (Macionis 1997, p. 269). The lower middle class have substantial disposable income, but are less secure in their status, and therefore are seen as easier for marketers to manipulate (Kotler et al. 1989).

Working Class

The manual working class in America (sometimes called the lower middle class) includes the 30 percent who are in or have retired from manual work and their families. Although they may have higher incomes than many that might be considered middle class, they commonly work longer hours under poorer working conditions, with fewer fringe benefits. This class is often subdivided into an upper, skilled group, and a lower, unskilled group.

Lower Class

There is also a lower class in the United States, comprising the remaining 20 percent of the American population. Some 39.3 million people or 15.1 percent were classified as poor in 1993 (Macionis 1997, p. 271). They are substantially excluded from participation in the economic and social life of the community. Social security payments are too low to lift people out of poverty. Some of the sick, single parents and the unemployed may, with difficulty, move out of this class if their circumstances or opportunities improve, but poverty can make it harder to find paid work, perpetuating unemployment. Experience in the United States, illustrated vividly by the 1992 Los Angeles riots, is that people in this group can move from being temporarily unemployed to permanently alienated. Well-directed welfare payments are not wasted public spending—they may be a preferable alternative to increased spending on policing, prisons, and emergency health care.

problems are associated with poverty, including the strength of social institutions such as the family, health standards, crime rates, and consumer demand; indeed the whole stability of the nation and its economy are associated with poverty.

U.S. society is very unequal in terms of income distribution. Sociologists compare different societies using a standard analytical approach. The ratio of the average income of the wealthiest quintile (the top 20 percent of households or families) to the poorest quintile can be compared across societies, even where real income levels are very different. In the late 1980s the ratio for the United States was about 12 to 1, followed by France and Canada (tied for second, but well below the United States, on 8.5 to 1), and Britain (8 to 1). It is interesting that Japan, until recently one of the strongest economies in the world, had one of the lowest rates of inequality (4 to 1). Socialist countries also had very low rates of

---

In California the Poor Get Poorer

California has long basked not only in its sunshine but also in its reputation as a booming land of opportunity. The latest picture is not so rosy. The Center on Budget and Policy Priorities (a Washington D.C. "think tank") reported on changes in income for Californian families from 1978 to 1996 (San Francisco Chronicle, 12/16/97, p. A3).

As is usual for income studies, the sample population was divided into five groups of equal size. The income for each of those five groups in any given year was then averaged. The differences in these average incomes can then be tracked and used as a measure of inequality over time. The data were drawn from annual surveys done by the federal government. The study found that California's poor and middle class families have experienced sharp falls in income over the last two decades or so, whereas the richest families are nearly 30 percent wealthier.

Families in the top group, whose incomes averaged $98,023 in 1978, had incomes of $127,710 in 1996. (The 1978 earnings were inflation adjusted to their 1996 value.) Meanwhile, the average income of the bottom fifth plummeted from $12,298 in 1978 to $9,033 in 1996.

Over the same period, even middle-class families mostly lost ground, especially those with earnings under $60,000 annually. The shift in incomes is shown in Figure 3.2.

The Executive Director of the California Budget Project, which collaborated with the Washington organization, commented:

inequality, although it should be noted that their base incomes were much lower, and the picture may have changed in the last decade (Appelbaum and Chambliss 1997, p. 215).

Some of the root causes of inequality are discussed in later chapters in this book. However, engineers should at least be aware that inequality has a well-established connection with technology and level of industrialization, although the precise, long-term interaction is still debated. Sociologists and economists can point to evidence that inequality increases during the early stages of industrialization and then declines and eventually stabilizes at a relatively low level (Appelbaum and Chambliss 1997, p. 204). However, the fact that levels of social inequality appear to be increasing in a mature industrialized society such as the United States suggests that the links with technology may be more complex than is currently appreciated.

"It isn't good when the poor are getting poorer but you have to be really concerned when middle class families aren't strong. If people start believing that working hard and playing by the rules doesn't get them anywhere, you've got a formula for problems; but these numbers show you can't count on upward mobility here."

There is a general impression of a growing economy, especially in California, where Silicon Valley has been a pace-setter and "help wanted" signs are ubiquitous. However, as one of the co-authors of the report pointed out, not everyone is sharing the gains. She found that incomes have been uneven and generally lagging for most Californians (and most Americans over the same period) due to:

- loss of unionized jobs;
- falling pay rates for people with only a high school education;
- movement of jobs offshore;
- lack of capital to invest in financial markets.

The main reason the wealthiest fifth have enjoyed soaring incomes is because of their ability to make money with money through investment in the stock market and financial products. However, stagnating incomes will need to be addressed, perhaps by measures such as tax cuts, if California's income gap is not to widen still further.

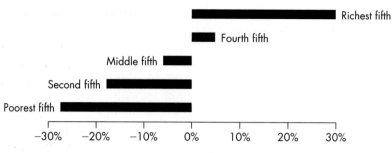

Figure 3.2 Change in Average Income in California (1978–1996)

Every society on earth lives with a degree of tension between its rich and poor, but sociologists see the wisdom in monitoring even small shifts in the pattern of "haves" and "have-nots" as the box suggests. If we are indeed becoming a more unequal society and people are discontented with their financial position, this may have unexpected outcomes.

■ Engineers and Social Class

Where do engineers themselves fit in this pattern of social stratification? The question of the social class or classes to which engineers belong is one factor in determining their status and public influence. This will be discussed from a different perspective in Chapter 11.

The status of engineers varies from country to country. In France and Germany, engineers have generally enjoyed high social status. However, in Britain and Australia particularly, the status of engineering has been affected by surprisingly widely held prejudices against applied technical activity of any kind. Fortunately this is generally not the case in America, where traditionally engineers have enjoyed high esteem in the community. One indication of the social status of the profession comes from "prestige scores" in numerous social surveys conducted over the years. A series of U.S. surveys suggests a spectrum that ranges from "physician," the most prestigious occupation (with a score of 86), through to the lower end of the scale (occupations such as "janitor," with a score of 22). The respondents rated "aerospace engineer" level with "dentist" (at a score of 72); "electrical engineer" (at a score of 64) was grouped with "registered nurse, accountant, economist, high school and elementary school teacher." (Macionis 1997, p. 265)

However, a degree of community ignorance still exists about what it is that professional engineers actually do; this may reinforce any prejudices regarding manual versus nonmanual work. The location of engineers in a class society is complicated by the existence of a continuum of engineering activity, from engine driving and fitting and turning, through paraprofessional activity, to high-level intellectual activity. High-level activity includes analysis, research, design and development, and technical and general management, as well as the most senior levels of management in large corporations and government agencies.

A significant number of professional engineers are self-employed, operating their own businesses or as consultants (approximately 67,000 engineers in the United States were self-employed in 1994, together with some 49,000 computer professionals). Still, the majority of engineers, 96 percent, are employees (U.S. Bureau of the Census 1997, p. 610). Engineering workers are spread out across the whole spectrum of class, but with only a very small proportion in positions of very high social status and authority.

To which social class do engineers typically belong in terms of income levels? Although the data is somewhat dated, the graph in Figure 3.3 gives a good idea of relativities.

The boundaries between classes may not be sharply drawn. For example, the authors of this book do some independent engineering consulting work, which is middle class entrepreneurial activity. Further, our ownership of shares, both privately and through superannuation funds, gives us an interest in the success of a range of business enterprises in our society. However, we are also employees, selling our ability to labor to survive, so that economically we could perhaps be described as non-manual working class. Yet the fact that we are university professors immediately gives us a different social status. Our current income levels would probably put us in the upper middle class, but after retirement our incomes and social position could slide.

When we consider the employment hierarchy in which engineers typically work, we have to ask ourselves how far up people have to be before their interests really lie with the management and owners rather than with other employees. Professional engineers almost invariably see themselves as part of management, regardless of the actual responsibilities they bear. Although part of the work force, they certainly have a great deal of control over the pace and direction of their work. They may, however, be less secure in their jobs in the event of a disagreement with the boss or a business downturn than employees backed by strong union organizations and working for wages. High levels of unemployment among immigrant engineers from the former Soviet Union to Australia in the early to mid-1990s suggest that, with globalization, even engineers will sometimes find the labor market quite precarious.

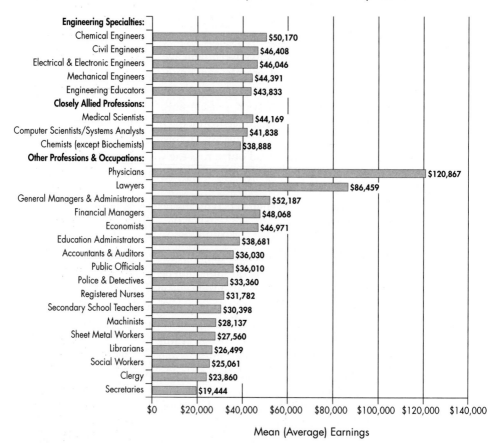

Figure 3.3 1989 Average Salaries of Engineers. *Source:* Graphic by John Wayland; Engineering Workforce Commission Reports.

Stratification and social class is one form of differentiation in our society but there are other, equally powerful ways in which we tend to separate ourselves into groups.

GENDER DIFFERENTIATION

Although a person's sex is a matter of anatomy and physiology, gender is essentially socially constructed. Gender differentiation is related not to sex differences, but to the way people of different sex are treated in our society. In a multicultural society there will obviously be variations between cultural groups with different traditions, but there are broad general patterns of gender differentiation in the United States.

There are statistically significant differences between the opportunities and experiences of males and females across the whole range of social life. Many of these reflect differences in the ways males and females are socialized. These differences appear to start from infancy: when people pick up a baby, the way they hold it is influenced by whether they believe it to be a boy or a girl. Girls tend to be held more gently and more closely to the body than boys.

In 1997, there were 82 million men and 89 million women over the age of twenty-five in the United States. Differences in levels of educational attainment between men and women are shrinking but they are still present. In 1997, 32 percent of men and 35 percent of women had only a high school education, and 26 percent of men and 22 percent of women had received either a bachelor's, master's, doctoral, or professional degree (March 1997 Current Population Survey, U.S. Census Bureau).

In employment, gender differentiation is manifest in a number of ways. Of the 102 million women age sixteen and older in the United States in 1994, 60 million were labor force participants (in paid employment or looking for work). Women made up 46 percent of the total workforce. Forty-one million women worked full time (thirty-five or more hours per week). Nearly 16 million, or 28 percent of all women workers, held part-time jobs. Two-thirds (67 percent) of all part-time workers were women. Some (but not all) women, particularly those with young children, prefer this status.

Women have made substantial progress in obtaining jobs in virtually all managerial and professional occupations. In 1984 they held 42 percent (10.3 million) of these higher paying jobs, a figure that rose to 48 percent (16.3 million) in 1994. However, women are still over represented in lower-paying jobs. Two-fifths (24 million) of employed women work in technical, sales, and administrative support jobs. Even though the earnings gap between men and women is slowly closing, median weekly earnings of all full-time workers in 1996 indicate women earned only 76 percent of male earnings. Engineering is one of the best-paid occupations for women, along with computer systems analyst, scientist, lawyer, physician, and university and college teachers.

Women are clustered in a much narrower range of relatively lower paid jobs, and part of the reason for the pay disparity between men and women has been that the generally patriarchal leadership in the trade union movement has neither attracted women nor represented them effectively. Deliberate efforts are now being made by many trade unions to address this situation. Their narrower occupational base and the vulnerability of women's jobs in a period of rapid technological change add to the problem. In practice, low wages for women may not actually represent a net saving for their employers, since low wages may cause an increase in labor turnover and its associated costs.

At work and away from work, differences between men and women tend to be perpetuated by social expectations. Members of both sexes accept gender differentiation in an unquestioning way. They commonly fail to recognize that a host of apparently minor differences add up to lasting intellectual and behavioral effects, and ultimately very different life outcomes.

Gender differentiation is also complicated by racial and ethnic differentiation in many societies, including the United States.

Racial and Ethnic Differentiation

Physiological differences (particularly skin color) and cultural differences are also used to differentiate individuals and groups in America. The U.S. Bureau of the Census acknowledges this, utilizing broad categories— White, Black, Native American, Asian, and Hispanic (the latter, as it points out, can be of any race)—in one set of approaches to presenting social data. In spite of some blurring and blending, and the importance of one's own sense of identity, birth into a particular family is most important. From then on, differentiation is perpetuated in many ways in our society. What is particularly significant is that these differences commonly translate into different and markedly unequal outcomes for minority groups (Figure 3.4).

Some key indicators of social status and social disadvantage are:

- ◆ income;
- ◆ education;
- ◆ unemployment; and
- ◆ single-parent families.

Then it is possible to obtain a series of quick snapshots of how ethnic and racial differentiation manifests itself in American society in the 1990s. It should also be emphasized that, although separate figures are given in Table 3.1 for each of these indicators, the effects of multiple disadvantage, especially where clustered within families, greatly exacerbates the situation for both individuals and whole communities.

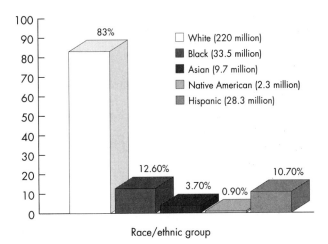

Figure 3.4 U.S. Population 1996—Race and Ethnicity. *Source:* U.S. Bureau of the Census, 1997 Statistical Abstract of the United States (various tables).

TABLE 3.1

Minority Groups and Social Disadvantage				
	Black	Native American	Hispanic	White
Income, 1995				
Median family income	$25,970	$21,619	$24,313	$42,646
Families below poverty level	26.4%	27.2%	27.8%	8.4%
Persons below poverty level	29.3%	31.2%	30.7%	11.2%
Educational Attainment, 1996				
Elementary school	9.3%	N.A.	12%	7.8%
High school graduate	73%	66%*	53%	84%
College-4 yrs. or more/ Bachelor's degree or higher	13.6%	9.4%*	9.3%	24.3%
Unemployment Rate, 1996	10.5%	N.A.	9.3%	4.7%
Single-Parent Families, 1996†	64%	27%*	37%	26%

N.A., not available.
* 1990 figures.
† % of all families in minority group with dependent children.
Source: U.S. Bureau of the Census, 1997 Statistical Abstract of the United States (various tables).

Significant aspects of the overall picture presented in Table 3.1 include:

1. *African-Americans*

   It is encouraging that the annual high school dropout rate for young Blacks is declining (from 11 percent in 1970 to 5 percent in 1993). By 1993, there was no statistical difference in the annual high school dropout rate of Blacks and Whites and of males and females. African-Americans have closed the historical gap and, consistent with lower dropouts, a growing proportion of all Blacks have at least a high school education.

   However, the civilian unemployment rate for Blacks was more than twice that of Whites and continues to be cause for concern. Seven percent of Black families received public assistance in 1993, in contrast to 1.2 percent of non-Hispanic White families.

2. *Hispanics*

   In 1993, the proportion of Hispanics twenty-five years and over with less than a fifth grade education was about 12 percent, down from 16 percent in 1983, but still more than fourteen times greater than that of non-Hispanic Whites (0.8 percent).

3. *Native Americans*

   In 1990, 66 percent of Native Americans twenty-five years and over were high school graduates or higher, up from only 56 percent in 1980. Despite the advances, the proportion was still well below the total population.

Of the four indicators of social disadvantage discussed above, poverty is the starkest measure of inequality in our society. When poverty levels are correlated with race and ethnicity, they show a grim picture.

In 1995, three out of ten Black persons were poor: there has been virtually no improvement since 1979, when the proportion was 31 percent. More detailed 1993 data reveals that nearly half of all poor Blacks were children under eighteen years old. Among poor persons fifteen years old and over, 35 percent of all Blacks and 42 percent of all non-Hispanic Whites worked. Compared with Black men (0.8 million), nearly twice as many Black women (1.5 million) were among the working poor.

About 30 percent of Hispanic people lived below the poverty level in 1995, compared with 11.2 percent for non-Hispanic Whites. As a result, although the Hispanic population was only 10.7 percent of the total population, in 1993 more than 1 in every 6 persons (18 percent) living in poverty in the United States was of Hispanic origin.

In 1990 (latest figures available), when the national poverty rate was about 13 percent, 31 percent of Native American persons lived below the poverty level. In 1989, 50 percent of Native American families maintained by females with no husband present were poor, compared with 31 percent of all families maintained by women.

There is also a great deal of data that correlates race, poverty, and a range of adverse outcomes, including increased incidence of serious illness and higher mortality rates. The individual tragedies behind the dry statistics are one aspect of the situation, but no one should lose sight of the fact that higher medical costs are borne by the whole of society. Fortunately sociologists are able to devise measures that quantify even an area as seemingly subjective as social disadvantage in ways that make it possible for policy makers to target assistance programs. Just as welfare relief measures are targeted at low income-earners, other programs focus on, say, expanding the employment options for Native Americans on reservations as one means of improving relative rates of pay, or adult education initiatives for Hispanic immigrants to reduce unemployment.

Is there a role for professional engineers in remediation of any of these social problems? The complexity of the issues involved should not be understated, but nor should it act as a deterrent to efforts to improve the life chances for everyone, especially young people who are at risk in our society. One program with which some engineering educators have been involved in recent years is discussed in Case Study 3.1.

DIVERSITY IN THE ENGINEERING PROFESSION

How do we react to the sorts of data presented so far in this chapter? The long-standing public debate has alerted most people to gender issues; and although we may be as far from consensus on many of these issues as we ever were, public opinion has inexorably shifted in the direction of more equality of opportunity. Race and ethnicity have long been emotionally charged subjects in American communal life, and the spectrum of views, whether privately held or publicly espoused, is very broad. However, Americans are used to the notion of themselves as an ethnic "melting pot" or a "mosaic," and over the years much of the rhetoric associated with this theme (from politicians, educators, social commentators, and the like) has welcomed this diversity.

Yet diversity should be much more than a pious hope or platitude; it ought to be the source of great creative tension, strength, and resilience in

SUPPORT FOR SUCCESS:
THE MESA WAY

Disadvantage, especially where young people experience it, is a continuing focus for activists and concerned citizens in American society. One concerted effort to overcome it is the Mathematics, Engineering, Science Achievement (MESA) program, started in California's Oakland Technical High School in 1970 with twenty-five students. By 1993, the number of California secondary school students participating in MESA had reached nearly seven thousand, and the program influenced another five thousand youngsters in elementary and junior high schools. More than 40,000 precollege students took part in MESA programs from 1970 to 1992. More than half of these participants have graduated from four-year colleges and universities. More than half of the MESA precollege program graduates are women.

MESA is one of the country's oldest and best-known programs that assist minority students in pursuing careers in technology; it has been profiled in *Science* magazine (Gibbons 1992) as one of the top programs in the nation that is successfully producing science professionals of color. MESA intervenes in the lives of average, often uncommitted students who are educationally disadvantaged in engineering and science. It nurtures them. It rewards them for success.

Today, MESA works with over 21,000 students throughout California from elementary through university levels, serving educationally disadvantaged students and, where possible, emphasizing participation by students from groups with low eligibility rates for four-year colleges. It is a program of the University of California.

For most MESA students, participation begins in ninth grade. Educationally disadvantaged youth already enrolled in, or willing to be enrolled in, algebra are recruited. They then agree to take geometry, trigonometry, precalculus, chemistry, and physics during their remaining time in high school. They must also voice an interest in a career in science or engineering. MESA then provides a rigorous learning environment that includes group study sessions, career exploration, academic advising, parent involvement, cash incentive awards, field trips, and summer jobs.

The California State legislature, corporate contributions, and various grants fund MESA. Corporations such as Bank of America, Chevron, Hewlett-Packard, IBM, Lockheed, McDonnell Douglas, Pacific Telesis, Toshiba America, and Toyota provide funding and in-kind contributions, and offer scholarships and summer jobs for students. Many of these corporations participate in MESA advisory groups.

MESA oversees four subprograms. The MESA Schools Program (MSP) includes twenty centers that serve close to three hundred elementary, junior, and senior high schools. The Success Through Collaboration (STC) program operates in eleven sites, bringing MESA to rural and urban Native American students. The MESA California Community College Program (CCCP), located on eleven campuses, is committed to increasing the number of math, engineering, and science majors who successfully transfer to and receive degrees from four-year institutions. The MESA Engineering Program (MEP) supports engineering and computer science students in twenty-four California colleges and universities.

The success rate of MESA students is outstanding. Some 97 percent of MESA high school graduates in 1994-1995 went on to a college or university. And MEP students earned over 80 percent of all engineering bachelor's degrees that went to under represented students in the twenty-four Universities of California, California State Universities, and independent California universities that have MESA programs. Several features have contributed to this success. The program is academically based. MESA does not provide remedial aid. Students with aptitude are identified and given the tools to help them excel. Each of the involved layers, public schools, higher education, and industry, assists the layer below it to achieve its goals.

The MESA Engineering Program began in earnest in 1982 when the California State legislature appropriated funds to establish twelve MEP sites on University of California and California State University campuses. Later, other public and private colleges and universities joined the program. MEP focuses on increasing the retention rates of educationally disadvantaged students majoring in engineering. MEP resources provide counseling, summer job placement, student study centers, financial aid, and other services. Students are also encouraged to join and become active in student chapters of professional engineering societies.

Among all freshmen (of any color) who entered University of California engineer-ing programs in 1982, 47 percent were still enrolled after three years. In comparison, 60 percent of MEP students remained enrolled. While 64 percent of African-American MEP students remained enrolled after three years, only 13 percent of those African-Americans not within MEP remained in engineering school. Among Mexican-Americans, the retention numbers after three years were 57 percent among those in MEP and 21 percent among those not in MEP. Data for other participating colleges and universities indicate similar trends.

The MESA Engineering Program at the University of the Pacific

The University of the Pacific (UOP) is one of the twenty-four California universities that have a MEP program. The School of Engineering began its involvement in MEP in the fall of 1990. Within four years, UOP saw its underrepresented engineering enrollment rise from 10 percent to nearly 20 percent.

The program offers seminars, midterm progress evaluations, and scholarship assistance. Students participate in UOP's professional student societies, including the National Society of Black Engineers (NSBE), the Society of Hispanic Professional Engineers (SHPE), and the Society of Women Engineers (SWE). The students also have access to the Student Study Center, which is centrally located in the School of Engineering and is equipped with computers and comfortable surroundings for studying.

Because UOP's School of Engineering is the only engineering school in the West that includes industrial cooperative education

(continued on next page)

(practical) experience as a requirement for graduation, its MEP students are better prepared for their respective engineering professions. MEP students at UOP also benefit from the school's association with the Pacific MESA Center Industry and Education Advisory Board. This Board, with nearly fifty members, offers personal, in-kind, and financial support to the program. Board members help coordinate an annual MEP Career Faire that exposes students to potential career opportunities. They also sponsor special recognition awards banquets.

The past seven years have changed the face of the School of Engineering and increased the morale and achievements of the students supported by MEP efforts. The successes have been recognized by MEP students and their parents, by the Dean of the School of Engineering, by the University, and most importantly, by the industry and community representatives who recruit and hire our MEP students.

California's MESA program has now become a model for similar programs in other states. At least twelve other states have established MESA-based programs. A number of states have combined their resources to form MESA U.S.A., a national organization with standardized goals and programs.

References

Gibbons, A. 1992, Minority programs that get high marks, *Science*, 258 (13), November, pp. 1190-1196.

Somerton, W., Smith, M., Finnel, R., and Fuller, T. 1994, *The MESA Way, A Success Story of Nurturing Minorities for Math/Science Careers*, Caddo Gap Press, San Francisco.

our society, enriching every facet of American life. It also cannot be simply taken for granted. We cannot assume that diversity, whether in gender, ethnic, or racial terms, naturally evolves or is somehow spontaneously generated. All too often, still, we observe differences in outcomes for different minority groups that are the basis for continuing exclusion and inequality. It is important to recognize just how differentiation manifests itself in microcosms, and in particular, within the engineering profession.

The Science and Engineering Equal Opportunities Act, passed in December 1980 by the U.S. Congress, declares that:

> . . . it is the policy of the United States to encourage men and women, equally, of all ethnic, racial, and economic backgrounds to acquire skills in science, engineering and mathematics, to have equal opportunity in education, training, and employment in scientific and engineering fields, and thereby to promote scientific and engineering literacy and the full use of the human resources of the Nation in science and engineering (Sec. 32(b)).

According to a 1994 report of the National Science Foundation, women constitute approximately half of the U.S. population and about 46 percent of the U.S. labor force in all occupations. Women make up 22 percent of the labor force in science and engineering occupations, compared with about 50 percent of social scientists.

The U.S. Bureau of the Census has provided the following figures on the 1.96 million engineering employees (including engineering technologists) in 1996:

- 8.5 percent female

  This is a significant increase from 5.8 percent in 1983 but still falls far short of ideal levels. Chemical Engineering (15.2 percent) and Industrial Engineering (13.2 percent) are the fields with the best female representation.

- 4.2 percent Black

  In 1983, Blacks made up only 2.7% of employed engineers and they are still seriously under represented in the profession. Again, Chemical Engineering (8.7 percent) is more inclusive than other areas of engineering.

- 3.8 percent Hispanic

  This figure is only a slight improvement on the proportion of employed engineers of Hispanic origin in 1983 (2.2 percent). Industrial Engineering currently has the best representation (5.9 percent).

Changes in the total makeup of the profession will be foreshadowed by trends in the composition of Engineering degrees conferred in America, in other words new entrants to the profession. The Bureau of the Census has provided the following figures for 1994: 78,225 Bachelors degrees (including Engineering Technology) were awarded, of which 14.9 percent were awarded to women. This makes for a dramatic comparison with the official figure of 0.8 percent in 1971, but we should bear in mind that we would need three times the 1994 figure if engineering were truly to reflect our society as a whole! (Corresponding increases have occurred in higher degrees: 15.5 percent of masters degrees were awarded to women in 1994, up from 1.1 percent in 1971, and 11.1 percent of doctorates.)

A more detailed analysis is presented in Case Study 3.2. The author of that case study also provided Figure 3.5, showing the trend towards engineering becoming a more diverse profession in the foreseeable future.

Case Study 3.2 adopts a fairly optimistic standpoint—other observers view the scene as still rather dismal. The trend towards breaking down the White male dominance of the profession is very slow.

Historically, and in spite of widespread discrimination, American society has undoubtedly been enriched by the talents of its men *and* women, by its Native American heritage, and by successive waves of immigration. However, there is little room for complacency. Whatever one's personal perspective, greater diversity within the profession would appear

THE AMERICAN ENGINEERING PROFESSION

As recently as 1970, more than 90 percent of American engineers were White males. The stereotype of a White, male-dominated engineering profession seemed unassailable. Women in the profession (also predominantly White at that time) accounted for less than 5 percent of the total.

Since then, the ethnic diversity of the profession has increased noticeably. By 1996 the proportion of non-Whites had risen to around 20 percent. However, it is mainly an influx of Asian Americans that has broadened the American engineering profession; minority representation still does not reflect the population as a whole. African Americans, Hispanic Americans, and Native Americans are almost 20 percent of the total U.S. population but are only 10 percent of the American engineering community. Asian Americans, less than 5 percent of the total U.S. population, are now more than 10 percent of all U.S. engineering graduates.

These numbers raise some interesting questions. Is the success of Asian Americans somehow occurring at the expense of other groups who are under represented in the profession? Or are they simply in a better position (in socioeconomic or cultural terms) to take advantage of opportunities to gain a professional engineering qualification (either here or abroad)? For instance, it would seem to be consistent with the idea that Asian American families place a higher value on a technical education and career than other ethnic groups. However, such cultural stereotypes should not be accepted unthinkingly and certainly should not be used to justify continuing inequities. Also, although Asian Americans are not now under represented in the engineering profession, it is recognized that they are still subject to discrimination in employment. Racial and ethnic differentiation in the community inevitably raises all sorts of issues for a subset, such as a professional group.

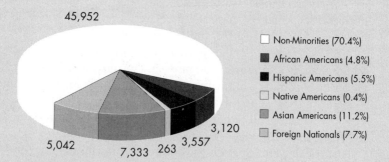

45,952

☐ Non-Minorities (70.4%)
■ African Americans (4.8%)
■ Hispanic Americans (5.5%)
☐ Native Americans (0.4%)
■ Asian Americans (11.2%)
☐ Foreign Nationals (7.7%)

5,042   7,333   263   3,557   3,120

Figure 3.2.1 1996 American Engineering Graduates Receiving Bachelor's Degrees—By Ethnicity

Female engineers are now better represented in the profession, particularly in the areas of environmental, chemical, industrial, and civil engineering. However, they are still markedly under represented compared with the American population at large. Moreover, the high concentration of women in certain specialties contrasts with the scarcity of women in mechanical, electrical, and computer engineering. Is it because these appear more socially detached and abstract? This unequal distribution of women between the different engineering specialties suggests that women are more likely to enter fields of study that are seen as more socially relevant.

The numbers of women and minorities in the engineering profession are still rising, but that does not necessarily mean that they will reach levels proportional to their population percentages anytime soon.

So what is the outlook for the profession overall? The engineering profession as a whole is slowly becoming more inclusive; numbers are growing, but there are shortages in almost all areas. The growth rate is not uniform between specialties. When the different engineering specialties are compared in terms of growth trends, it appears that a few specialties are growing at a very healthy rate whereas the rest are either growing slightly or even contracting. Computer engineering numbers have an annual growth rate of more than 6 percent (attributable to the enormous growth and popularity of the entire industry), whereas the next highest growth rate, in engineering management, is only slightly more than 2 percent. The intakes to petroleum and mining specialties are contracting. This is perhaps a result of the increased awareness that

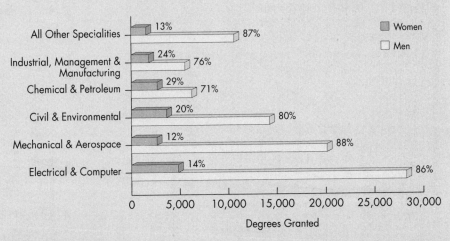

Figure 3.2.2 1996 American Engineering Degrees—By Discipline And Gender

(continued on next page)

our natural resources must eventually be depleted (although that conclusion might seem likely to promote more interest in better engineering in those areas, instead of a waning of interest).

In America, as elsewhere, the face of engineering is changing. The profession is finally starting to open its doors. Although the number of women and minorities in the profession is still low, that number is growing each year. As the White, male domination of the profession is slowly relaxed, it becomes more and more accepted (and possible) that anyone, regardless of ethnicity or gender, can pursue a career in engineering. This increased diversity is complemented by the diversity of engineering fields, as new specialties constantly develop and old specialties are often made obsolete. All these factors go into making engineering the dynamic profession that it is.

Source

Based on work by John Wayland, Stanford University 1997.

References

Engineering Workforce Commission reports.

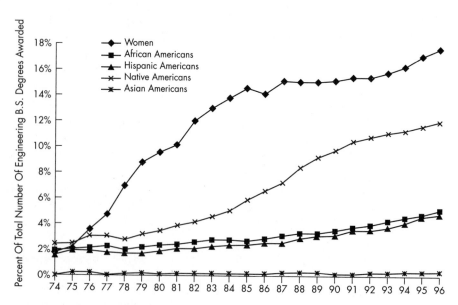

Figure 3.5 Trends In The Percent of Engineering B.S. Degrees Awarded To Women And Minorities, 1974–1996

to be in everyone's best interest. The United States is by no means alone in coming to this realization; a great many "women in engineering" programs around the world are attempts to achieve greater diversity.

Other goals of these programs include changing the culture of the engineering workplace to make it a more pleasant place for everyone involved. Case Study 3.5 shows the importance of such cultural change.

CONSUMERISM AND ADVERTISING

Detailed analyses of social stratification and gender differentiation, describing the basic characteristics of people's lives and incomes, also underpin modern marketing approaches that place product development, promotion, and selling effort in the context of consumer needs. Sections of the population are targeted for particular products. Commercial media outlets attract advertising by delivering access to those particular segments of the market (for example, by providing the sort of news and entertainment content that will attract the target audience). Advertisers are not generally interested in groups without much discretionary income (although targeting children continues to be controversial). Market segmentation requires a clear picture of the social structure of society and how it is changing. One useful way to do this is probably to use the class model discussed earlier in this chapter; another is age segmentation. Neither of these is indicated by a single factor, such as income. Wealth, income, education, occupation, and other variables contribute to lifestyle and taste as well as purchasing power.

We would suggest that the essence of consumerism is placing a high value on the whole process of consumption. Consuming becomes a need in itself for the individual and for the system, at times even a surrogate for mature human relationships. People value the novelty of a product or service, although this is often spurious (with some aspect of production emphasized to create the illusion of novelty). The most obvious example is fashion. Advertising plays a large part in this process, such as the transformation of the four-wheel-drive pick-up truck from a basic utility vehicle to a middle class status symbol.

A more fundamental critique of consumption driven by advertising can be made in terms of its impact on consumers. We can appreciate some of the needs and anxieties on which advertisers play. In our complex and compartmentalized industrial society it can be difficult for people to feel a sense of belonging. Advertisements often focus on this in a very unsubtle

Consumerism

The word consumerism comes from consumer, a purchaser or user of goods or services.

It is a mistake to see consumers simply as passive participants in the marketing arena. Consumers of all kinds are increasingly a force to be reckoned with.

Three different definitions are given for consumerism in the Webster's Dictionary. They are:

1. A modern movement for the protection of the consumer against useless, inferior, or dangerous products, misleading advertising, unfair pricing, etc.
2. The concept that an ever expanding consumption of goods is advantageous to the economy.
3. The fact or practice of an increasing consumption of goods.

The second and third definitions emphasize growth in the consumption side of the economy. The first suggests constraints on how consumption should be stimulated, but does not challenge the concept that the system needs continuing growth. Consumer movements, with their emphasis on education and protection of rights, are often a middle class phenomenon. The question they ask is: "Are we getting a fair deal?" Part of their importance may be that they give people a real sense of belonging to a group and help to define them in terms of their consumption.

way: if you drink a certain brand of beer, you will be better liked or will indicate your sophistication. Images and messages of intimacy in advertisements are all around us, offering us a sense of place, a consensus, a common code, and shared humor.

This process encourages people to allow themselves to be directed, rather than draw on their own sense of relevance and worth. How one is perceived by others becomes terribly important. It also becomes essential to keep up with fashion trends—which are set by other people. Producing artificial needs moves people towards a conformist mentality, through what becomes almost a process of self-hypnosis. The ultimate stage is the belief that possession of objects can transform our lives, a condition known as commodity fetishism.

Advertising is central to consumerism. It is also central to the profitability of the privately owned mass media. Questions that follow from these points are: What are our media like, and who owns them? What controls do government and the community exercise? Media advertising is big business. In 1996, a total of $175 billion was spent by advertisers in the United States, or almost $2 per day per American ($4.75 per day per household). Television advertising accounted for $42 billion of the total, newspaper advertising accounted for $38 billion, direct mail advertising consumed $35 billion, and radio advertising consumed $12 billion.

Sociologists would deny any complicity in advertisers' wide-scale manipulation of people for profit through using their analytical tools and the social data that they yield; nevertheless, the usefulness of sociological data to the advertising industry is quite apparent. This is another area where an informed engineering profession is less likely to accept uncritically its own association with the world of advertising and the potent messages it sends to a relatively unsuspecting public.

THE SOCIAL IMPORTANCE OF WORK

What do sociologists mean by work? We all have pretty clear ideas of our own, but sociologists would generally define it as "any human effort that adds something of value to goods and services that are available to others." It can be thought of as falling into three broad categories: the formal economy; the underground economy (which includes informal as well as illegal activities); and unpaid labor (Appelbaum and Chambliss 1997, p. 442). Sociologists and psychologists have long recognized that satisfying work is important for personal growth and for the fulfillment of the individual's potentialities as a human being. Productive activity helps establish relationships with the natural world and with other people in a society.

Individuals seek some control over the pace and general organization of the work systems that affect them. The social changes associated with the shift from small-scale handicraft production to modern industrial manufacture are discussed in Chapter 1. Karl Marx called the effect of these social changes alienation. In his view, the technology in use was not the central issue. He saw capitalist labor relationships, which deprived workers of control over the work process and the products of their work, as the major cause of alienation (Dickson 1974, p. 30). Other analysts have suggested that alienation might also be a consequence of the character of modern mass industrial society, reflecting both social fragmentation and work organization (Scientific Management is discussed in Chapter 2).

One approach suggests that alienation may be closely associated with the ways in which particular technologies are selected and implemented. Feenberg (1991) points out that an implicit criterion in the choice of technology in modern industry is the extent to which the technology assists in controlling the work situation. The classic example was Ford's mass production system. In this system, the hierarchical labor relationships, and the split between mental and manual labor, were effectively incorporated in the technology itself. The social and political dimensions of alienation were thus submerged into what was apparently simply a technical issue of how to maximize output. Chapter 7 discusses examples of renegotiation of the social relationships of the workplace.

■ The Quality of Work

Work is important to well being, but it must be recognized that many work situations fail to address the needs of the people employed in them. An environment that is so noisy that it prevents ready communication with nearby workers is very isolating. When the work is on a production line, carrying out a repetitive and in itself trivial task at a pace set by management, the situation can became very frustrating. When the pace of work is also so demanding that the worker cannot even stop long enough to go to the toilet without needing to have a supervisor replace him or her, the total effect is almost certainly alienating.

The seriousness of the quality of work problem has been reflected in absenteeism levels in the car manufacturing industry. Absenteeism remains a nagging, expensive problem. In America, a study of Detroit's "Big Three" auto makers found that, despite an all-out effort to get employees to show up consistently for work, at Chrysler Corporation the absenteeism rate rose from 5.2 percent in 1996 to 6.05 percent in 1997. The Chrysler numbers include excused absences, such as vacations, jury duty, and bereavement leave; however, unexcused absences account for half of the missing workers each day (The Detroit News, July 27, 1997). Industry experts say rates of absenteeism are generally equal or higher at Ford Motor Company and General Motors Corporation factories, although neither company releases these figures. In contrast, U.S. factories owned by Japanese auto makers, such as Toyota Motor Corporation and Honda Motor Company, have absenteeism below 2 percent. This lower rate of absenteeism says something about the way their factories are operated and gives them a big economic advantage over the American-owned plants.

Once we recognize the ways in which work contributes to people's happiness and well being we can begin to deliberately design work (and workplaces) to be more satisfying and fulfilling. At Ford's plastics plant in

How Much Can Workers Manage?

- How much control are we prepared to give workers over the production process?
- How will the factory of the future be structured and organized?
- What sorts of job structures and work organizations should engineers be designing so as to improve quality and encourage innovation?
- What roles will computers play? How important will computer applications be?
- Should the emphasis in designing computer systems be on offering intelligent assistance to operators, or on controlling their work?
- What contribution should we permit (or expect) production workers to make to the design process?

These questions involve engineers' support for quality improvement and innovation. Good answers will address social and cultural issues as well as technical ones. They will differ in different countries and, again, the sociological perspective will be very important.

Melbourne, Australia, the phasing-out of the assembly line and its replacement by worker teams that rotated jobs and managed their own work not only reduced absenteeism but also halved the number of production defects.

Engineers, historically involved in the design of these systems of work, need to appreciate how awful systems of work can be for the workers involved. The issues raised in the box above (How much can workers manage?) need to be addressed in workplace design. Modern approaches such as total quality management (TQM) show that around 85 percent of quality problems are in the design of systems, which can only be addressed by management (Sprouster 1987). However, there has been a tendency for management writers to focus on the shortcomings of workers as the reason for poor quality.

▪ What Makes Work Satisfying?

Generally, professional engineers are fortunate (and unusual), in that they like their work, finding it challenging and stimulating (if sometimes overly demanding), particularly in the areas of design, research, and development (Carter and Kirkup 1990). A U.S. survey applied some basic sociological techniques to the question of "What Motivates Engineers?" (see box).

What Motivates Engineers and Managers?

Engineers see "challenging work" and "opportunity for advancement" as the main ingredients of job satisfaction. A survey of engineers and managers in the United States provided some fascinating insights into engineers' attitudes to their careers. The survey questionnaire covered:

- ◆ Which job factors do you want more of?
- ◆ What do you like best about your job?
- ◆ Why did you leave your last job?
- ◆ What accomplishments are most important for you?

Job factors in order of importance were:

Challenging Work

In the survey, 100 percent of the engineers and 79 percent of the managers chose "interesting work" over "prestige and reputation" as their most important work requirement. It is also keeps them from moving to another job. Satisfaction and challenge are not the only reasons; professional status also plays a part. Much of an engineer's recognition from his peers depends on the importance and results of his projects. By "interesting work" respondents meant:

"...tackling complex designs; being able to handle a variety of engineering tasks; contributing to the technical decision-making on projects; working on their own ideas (rather than "knowing what is expected"); ...the sense of freedom when performing creatively."

Advancement

Asked to compare "opportunity for advancement" with "job security," 61 percent of engineers and 74 percent of managers chose "advancement." Increased opportunities are linked to higher salaries, but respondents valued "reinforcing their sense of self-worth and professional growth" most of all.

Money

Good salary is very important for morale and job satisfaction. In the survey it consistently ranked second. When asked what they wanted more of, a

These desirable qualities are virtually universal. Walshok (1990) made a study in the United States of skilled blue collar technical workers. She summarized the four key criteria they found central to their work satisfaction:

1. Accomplishment: a sense at the end of the working day of having done something significant.

good salary was the most frequently mentioned of all job factors (31 percent of engineers and 20 percent of managers) and 64 percent of engineers and 51 percent of managers chose a good salary above "recognition for a job well done."

Recognition

Aside from financial rewards, recognition in various forms is also sought after. This could include patent rights and royalties for new products; more generally, it is commissions and bonuses, recognition from top management, plaques, and write-ups in the company newsletter.

Working Conditions

A company that provided modern facilities had a distinct advantage, but this was not found to be a major motivating factor. Work space is particularly important: roomy and private is the ideal (in spite of the trend towards open-plan offices and workstations).

Fringe Benefits

These did not emerge as significant motivational factors, although they do represent extra cash (in some U.S. companies as high as 20 to 25 percent of the salary dollar).

Security

Only 11 percent of engineers and 5 percent of managers mentioned job security as something they wanted more of, but it was seen as more important than advancement by 39 percent of engineers. It is comprises four aspects: (1) personal (keeping one's own job); (2) related to the engineering department or group surviving within the organization; (3) one's employer succeeding in the marketplace; and (4) whole industry security.

Sources

(From Raudsepp, Eugene 1988, "Attitude Survey." *Machine Design* (Jan 21 pp. 84-87).
   Reprinted with permission from Machine Design, 1/21/88, a Penton Media publication.)

2. Challenge: a satisfying newness and variety in their work, a continuing need to solve problems.

3. Relevance: getting feedback and recognition for what they do, a feeling of being part of the mainstream of life, rather than stuck in a backwater.

4. Autonomy: control over the pacing of their work, discretion as to the sequencing and frequency of tasks, and feeling they define what is going to happen in their own day.

Thus job characteristics that are the common experience of most engineers are often an unattainable dream for most blue collar and many white collar employees, partly because of a continuing Taylorist emphasis on removal of employee control from the workplace, and an associated assumption that fragmentation of jobs is efficient. However, modern approaches to quality management are starting to improve the quality of work for many employees.

■ Unemployment

Having a job at all (even if it is hard, tedious, and badly paid) is important for the self-esteem of people in our society. This is not simply a matter of money. People define themselves by their work, and it provides them with broad social contact and interaction. Losing a job is a major blow to self-confidence, particularly during a recession, when getting another job may be more difficult than usual.

As Western economies become increasingly "efficient," they tend to produce at least temporary unemployment. The costs of unemployment are normally stated in terms of the loss of production of potential goods and potential income while the economy is not producing up to its capacity. This economic cost can be measured annually as the amount by which potential gross domestic product (GDP) exceeds actual GDP.

However, unemployment also results in serious personal costs. These include reduced income and standard of living, loss of personal self-esteem and self-confidence, loss of marketable skills, social isolation, and increasing poverty and ill health. These individual costs also have an impact on the community. The contributions people were able to make when they were employed to consumer spending, taxation revenues, and community savings are also lost. Although unemployment benefits exist to spread the cost of unemployment among the entire labor force rather than leaving that burden only on the unemployed, they are rarely adequate. As a result, demands on government and on private charities for support increase.

In spite of the booming U.S. economy in the late 1990s, unemployment will not be eradicated in the foreseeable future. The overarching goal of labor market policy should be to encourage the private sector to create jobs and to fill them quickly with the unemployed. The "creative destruction" that is at the heart of the market economy means that no job lasts forever, but it allows new jobs, professions, and even industries to arise continually. However, the total costs of unemployment provide a strong financial argument for governmental policies that are aimed at generating employment, even by reallocating existing revenues. With appropriate strategies, policy makers can keep the inevitable human and economic costs of unemployment to a minimum.

∎ The Dangers of Work

Although unemployment is conducive to ill health, it should also be remembered that work itself can be not only unpleasant but also dangerous. Thousands are killed and millions are injured at work every year. In 1996 private industry workplaces reported 6,112 fatal work injuries and 6.2 million injuries and illnesses (Census of Fatal Occupational Injuries, conducted by the Bureau of Labor Statistics, U.S. Department of Labor).

In 1996 highway traffic incidents and homicides led all other events that resulted in fatal work injuries, a continuing trend. These two events totaled over one third of the work injury deaths that occurred during the year. Work-related highway deaths accounted for 22 percent of the 6,112 fatal work injuries in 1996. Slightly over half of the highway fatality victims were driving or riding in a truck. (Maybe engineers can design better highways and safer trucks, which are today largely exempt from the safety regulations imposed on cars.)

Homicide, the second leading cause of job-related deaths, accounted for 15 percent of fatal work injuries in 1996. (There is probably little engineers can do about this, other than recognize the high levels of stress and tension that exist in some work environments, sometimes associated with the nature of the work.) Robbery was the primary motive of job-related homicides. Domestic disputes accounted for one-sixth of the workplace homicides for female workers.

Nine percent of the fatally injured workers were struck by various objects, such as falling trees, machinery, or vehicles that had slipped into gear, and by various building materials. The construction industry accounted for about 17 percent of the fatality total, about three times its 6 percent share of total employment. Other industry divisions with large numbers of fatalities relative to their employment include agriculture, forestry, and fishing; transportation and public utilities; and mining. Half of the fatal falls occurred in the construction industry.

The 6.2 million injuries and illnesses reported during 1996 resulted in a rate of 7.4 cases per one hundred equivalent full-time workers. About 2.8

million injuries and illnesses in 1996 were lost workday cases; that is, they required recuperation away from work or restricted duties at work, or both. Nearly 5.8 million were injuries that resulted in either lost work time, medical treatment other than first aid, loss of consciousness, restriction of work or motion, or transfer to another job. Injury rates were generally higher for midsize establishments employing 50 to 249 workers than for smaller or larger establishments, although this pattern is not universal.

In 1996 there were about 439,000 newly reported cases of occupational illnesses in private industry. Manufacturing accounted for three-fifths of these cases. Disorders associated with repeated trauma, such as carpal tunnel syndrome and noise-induced hearing loss, accounted for 4 percent of the 6.2 million workplace injuries and illnesses. Among goods-producing industries, manufacturing had the highest incidence rate of occupational illnesses in 1996 (10.6 cases per one hundred full-time workers), followed by construction (9.9 cases per one hundred full-time workers). Within the service-producing sector, the highest incidence rate in 1996 was reported for transportation and public utilities (8.7 cases per one hundred full-time workers) followed by retail and wholesale trade industries (6.8 cases per one hundred workers).

The Occupational Health and Safety Administration (OSHA) is the federal agency charged with improving worker safety in the United States. It has devised a number of strategies to assist with this objective (see box).

∎ The Future of Work

A major result of the industrialization of western society has been what Aungles and Parker (1988, p. 162) described as "the split between paid and unpaid work and between the public/rational world of work and the private/affective world of home." They described a division in paid employment between primary and secondary labor markets. The primary market includes the core activities of larger corporations and government. It offers stable employment with good pay and conditions. The secondary market generally involves smaller firms or peripheral or seasonal activities. It offers lower wages and little prospect of job training or security, and has probably contributed to the creation of the "lower class" described earlier.

What is the likely future of work in Western cultures? Paid employment is still seen as central to the lives of most workers. If significant levels of unemployment are to be accepted by policy makers and the public generally as inevitable, we need to recognize the implications. We must

OSHA

The mission of the Occupational Safety and Health Administration (OSHA) is to save lives, prevent injuries, and protect the health of America's workers. To accomplish this, federal and state governments must work in partnership with the more than 100 million working men and women and 6.5 million employers who are covered by its services.

OSHA and its state partners have approximately 2,100 inspectors, plus complaint discrimination investigators, engineers, physicians, educators, standards writers, and other technical and support personnel spread over more than two hundred offices throughout the country. This staff establishes protective standards, enforces those standards, and reaches out to employers and employees through technical assistance and consultation programs.

Despite a 57 percent drop in workplace fatalities since OSHA was created, the agency's job of protecting the health and safety of American workers has been expanding at the same time its resources have been declining. Protecting over 100 million workers in more than 6 million workplaces with 2,100 state and federal compliance officers is a daunting task. For example, it would take eighty-seven years for OSHA inspectors to visit every work site in the country.

Therefore, in 1994, OSHA set out to redefine itself. It established new tools, besides simply inspections, to enforce its regulations. These tools included incentives and partnerships with employers, trade associations, unions, and other governmental organizations. Partners would work together on the common goal of eliminating hazards to prevent injuries and illnesses. Traditional enforcement would focus on those who chose not to participate in the partnership efforts. Today, the modus operandi of OSHA has changed from mainly enforcement to increased partnership with industry and labor, and typifies the way the federal government is working

change our attitudes about what does and does not constitute "proper work," and we must guarantee adequate minimum incomes to all. The "Protestant Work Ethic" arose in a period when working conditions were often appalling, and people had to be given all the encouragement possible to become involved in it. How relevant is that today?

Further, what roles will the choice of technologies play in the type and amount of work available? Although telecommuting offers interesting possibilities, it does not appear to be a panacea. Effective networking, both within the office and to the remote site(s) is essential. Working from a cabin in the woods may be appealing at first, but not indefinitely. Social contact is part of the satisfaction people derive from work. A good pattern might be spending two or three days at the office, catching up on what is happening and collecting projects or assignments, then spending the rest of the week working from home, or at a telecommuting center close by. That degree of flexibility suits people with domestic responsibilities or those who want to (or must) live a long way from their nominal place of work. Many professionals already work rather like this, having discovered that it is the only way to put together big enough amounts of time to work through difficult material. The advantage for them of proper telecommuting is that the office and other information sources on which they rely may be fully accessible at the remote site. Problems for telecommuters are likely to be similar to those for any outworker. Will the employer exploit the employee by setting an unreasonably low rate for the job? Will the telecommuter set him or herself up and then discover that there is not enough work to cover outlays and provide a decent living?

There are still important productive activities that need to be attended to in our society, but because they are not being addressed within the paid work force they are regarded as marginal or less important. This attitude applies particularly (but not solely) to the work of women in the home; there is an urgent need to accord these activities the same status and prestige as paid work.

Social Power

As discussed in more detail in Chapter 4, power relations are fundamental to any society, and the way power is exercised in society is complex and subtle. In Chapter 4 we will look at power in the context of politics and the state, but we also need to explore it in a social context. The term "social power" is used here to emphasize that we are dealing with the exercise of power within an established society, and within existing social structures. There is no general agreement on a definition of social power: it operates on many levels. Our values and view of the world will influence our attitudes, particularly as to what we accept as *legitimate* exercises of social power. Examples of unequal power relationships in U.S. society are race relations, industrial relations, and gender relations. Here we look more closely at the last two of these.

■ Gender Relations

Gender as a social construct, and the concept of gender differentiation, were discussed earlier in this chapter. Biology is used to justify what are essentially power relationships. This process is played out in a wide variety of social situations, but is played out particularly graphically in the workplace. Discrimination is illegal, but women's median incomes in the United States for full-time work in professional specialties were still only 71 percent of men's in 1994 (Macionis 1997, p. 368).

The power structure in a number of work areas can be shown to be still essentially patriarchal. Patriarchy is defined as "a structure that gives some men power over other men, and all men power over women" (Game and Pringle 1983, p. 22). Some women may have power, but their levels of control and choice are constrained by social structures. Case Study 3.5 (at the end of this chapter) shows graphically how social power relations in the work place can flow from gender.

Men have not always dominated women in human societies. Societies based on male aggression seem to have come with agriculture, and by 1200 B.C. all the major cultures of the world were based on male domination. However, even in the medieval guilds women as well as men could be "masters." During the Industrial Revolution, manufacturers moved most paid work out of homes and into factories. In this process, women were excluded from entry into what became defined as men's jobs, particularly skilled trades. This paralleled increasingly dismissive attitudes to the roles and skills of women, particularly those not in the paid work force.

Beginning in the Middle Ages, women who attempted to compete with men in professional areas, particularly in medicine, were increasingly liable to be burned as witches. Allegations of witchcraft were effective, if appallingly drastic, social control measures. They supported patriarchal power relationships throughout society, not just in professional areas. In 1484, Pope Innocent VIII issued a papal bull against witches, and the number of (almost entirely female) deaths has been put as high as 2,000,000 (Bacon 1989, p. 84). Women continued to be tried for witchcraft in Britain until the early eighteenth century.

How are patriarchal relationships expressed in modern American practice? In the media (and often in the workplace itself), women's ambitions and their need to have both a family and a career may be extolled for a few "supermoms." More frequently, however, female ambition is depicted as selfish and unfeminine. Different adjectives are used for similar characteristics in male and female managers: "he is dynamic, she is aggressive."

For factory work in Australia, Game and Pringle (1983, pp. 28–29) showed a clear gender division in the workplace (in a pattern that is more or less replicated across the world):

> ... based on a series of polarities broadly equated with masculinity and femininity. The most obvious distinction is between skilled and unskilled work. The other main ones are: heavy/light, dangerous/less dangerous, dirty/clean, interesting/boring, mobile/immobile. The first of each of these pairs is held to be more appropriate for men, or men are assumed to be better at it.

Game and Pringle found that women were assumed "by nature" to be good at boring, detailed, and sedentary work. The particular jobs done by women and men varied from factory to factory, depending on the local availability of labor, but the interface between men's and women's work seemed to be set by the same criteria. However boring the men's jobs were, rather than pressing for job enrichment, they seemed to be reconciled by the fact that the women's jobs were worse.

Men's jobs have been protected by informal mechanisms (such as the way on-the-job training has been arranged), and by legal provisions such as weight limits on lifting that can be done by women and young people. It is notable, however, that it is men's jobs that are protected, not the male workers themselves. A lack of clearly defined weight limits for lifting done by men has been associated with a significant incidence of back injuries; an overall reduction of limits would make good economic as well as social sense. Social isolation and physical and psychological harassment (and even lack of toilet facilities) have been used to stop women filling "men's" jobs. The human impact of harassment is movingly described in Case Study 3.5.

A disturbing and widespread mechanism that is used to establish and reinforce patriarchal relationships in the workplace (and the classroom), is "joking" behavior, including low-level harassment and "hazing" of less dominant individuals. Supposedly in jest (and therefore more difficult to deal with), this behavior can chill an already unfriendly environment for people not conforming to established stereotypes. It is precisely the opposite of what is needed to attract a broader range of people to engineering (McLean et al. 1997).

Measures taken to attempt to address some of the problems discussed above, as well as other forms of discrimination in the workplace, include the passing of the U.S. Civil Rights Act of 1964. Their effective implementation is not only socially desirable but also important in increasing the effectiveness and international competitiveness of our economy. These measures are designed to increase the likelihood of the best-qualified person getting the job or the promotion. However, profound changes to organizational cultures are not made quickly or easily. They require real leadership at all levels of management, and solid commitment from all levels of the state. The tenacity of underlying power relationships that date back for millennia should never be underestimated. The seriousness of these issues was underlined by the size of the $34 million decree obtained in 1998 by the U.S. Equal Employment Opportunity Commission against Mitsubishi for sexual harassment of women at its Normal, Illinois plant.

▮ Industrial Relations

Relationships in the workplace clearly involve the exercise of social power. We discuss alternative approaches in Chapter 7, but the dominant model of enterprises under capitalism assumes that the management represents the people who actually own the capital. The resulting hierarchical structure of most organizations, with the workers (labor) on the lower layers, is then taken for granted. The relative power of management and workers depends on factors inside and outside the workplace.

One important element in worker bargaining power is the availability of other jobs. Industrial action tends to take place in relatively good times, when there are prospects of winning significant benefits. Political action is more common in bad times, when the whole system (rather than individual employers) may be seen to be failing people and therefore tends to come under challenge. From experience in the workplace we can see the changing relative strengths of employees and management reflected, for example, in changes in accepted work practices.

The relations between employers and employees are set in a legal, economic, and policy framework by the operation of the state, which has a long history of regulating conditions of employment. Early English factory laws dealt with the health, safety, and morals of child textile workers. In 1825 labor unions were legalized in Britain; however, agreements among their members to seek better wages and hours were punishable as conspiracy until 1871. In the United States early legislation was aimed at improving working conditions, but labor organizing was discouraged by the federal doctrine of conspiracy until state laws began superseding the doctrine in 1842.

The history of trade unions shows how their strength, and the acceptance of their legitimacy within society as a whole, evolved rather slowly. However, by the beginning of the twentieth century, it was clear that trade unionism had arrived to stay. American legislatures have enacted many relevant laws over the years, including laws prohibiting night shift work and work in hazardous occupations for women and children. Other laws have established minimum wages in certain occupations and regulated interstate commerce to discourage child labor and sweatshop labor. In 1916 Congress exempted unions from the antitrust laws. The Wagner Act of 1935 established the right of workers to organize and required employers to accept collective bargaining.

The defeat of widespread strike action by the Industrial Workers of the World (IWW) early this century confirmed the need for political as well as industrial action. In the United Kingdom and Australia this experience led to the union-sponsored formation of political parties, while in the United States it encouraged union support for the Democratic Party.

When the United Auto Workers signed agreements in the late 1930s with what by then had become the "Big Three" (Ford, General Motors, and Chrysler), the main issues were seniority and job rights, in a movement that was called job-control unionism (Womack et al. 1990, Ch. 2). In 1938 the Fair Labor Standards Act provided for minimum wages and overtime payments for workers in interstate commerce while leaving so-called states' rights intact. The outbreak of strikes after World War II, however, brought demands for more restrictive labor legislation, culminating in passage of the Taft-Hartley Labor Act of 1947. This Act banned union membership requirements for employment and secondary boycotts (where other unions take action in sympathy with the union actually on strike).

The Civil Rights Act of 1964 prohibited race-, sex-, and religion-based discrimination in employment, and age discrimination was barred in 1967. Concern over workplace safety led to the establishment of the OSHA in 1970. The Americans with Disabilities Act of 1990 barred job discrimination against otherwise qualified disabled individuals. In one way or another, all of these measures were efforts to "balance" the overt and covert exercise of power in the workplace.

Despite occasional excesses (on both sides) it is likely that the organized working class and its representatives can claim much of the credit for high standards of living for industrial workers and important aspects of political freedom in many western democracies. Having won so many benefits, unions now see their membership declining dramatically. In the decade 1985 to 1995, union membership fell in most of the ninety countries surveyed by the International Labor Organization (ILO). The decline was sharpest in Central and Eastern Europe; U.S. membership fell by 21 percent.

In practice, the role of unions is quite complex, and employers generally seem to take up pragmatic rather than ideological positions on this issue. An excessive emphasis on individualism may well inhibit moves towards organization of work based on self-managing teams. In Chapter 7 we describe the key role union leadership has played in close cooperation with management in changing the culture in this direction in some U.S. automobile plants.

REDEFINING ENGINEERING

Our introduction to the discipline of sociology has furnished some concepts and analytical tools that we can apply to our own profession. We have already attempted to locate engineers on income and social prestige scales and have explored other aspects, such as gender and ethnicity within the profession (see above). Significant changes towards making the profession more representative of the society as a whole are only likely to occur as a result of concerted effort.

But how best to direct such effort? In this section we suggest that a focus on the role of women in engineering may be particularly constructive in addressing wider concerns than simply the relatively low numbers of women engineers. The absence of women is important, and not just historically. It throws light on what sort of activity engineering is, how it developed, and where it should be going, as well as on other issues that engineers have habitually ignored. As one overseas engineer sees it: "Our biggest challenge is to portray engineering as a human act for humans, not as a cold technology."

▮ History of Women in Engineering

Very few women even receive a mention in the history of engineering, but we have tried to feature some in this book (see Case Study 7.2 on Lillian Gilbreth and Case Study 3.3 on Edith Clarke, below). Feminist critiques of the history of technology, like the ones by Rothschild (1983) and Pacey (1983), offer different and important perspectives on the apparent lack of a significant role for women. Overwhelmingly, it seems that the engineering profession and engineering education and training have been run by men, for men. Pacey (1983, p. 111), describing some nineteenth century British engineers, noted "a swashbuckling disdain for the social evils around them," and added that they had first denigrated the importance of sanitation schemes and then used them for "displays of technical virtuosity." For much of the period we identify with modern engineering, women were effectively excluded from it. Women's social position reflected the fact that they had little legal autonomy; they were

EDITH CLARKE,
PIONEER ELECTRICAL ENGINEER

Edith Clarke was born in 1883 in a small farming community in Howard County, Maryland. At that time, at most one woman had received an engineering degree in the United States. In the late nineteenth century any kind of higher education was unusual for a woman, and virtually no women considered engineering.

Clarke graduated from a boarding school for girls when she was sixteen. Her education was typical for girls at the time, including Latin and French, classic English literature, history, physics, chemistry, and astronomy. Fortunately for her eventual career, her boarding school gave her exceptionally good training in algebra and geometry.

Clarke's father died when she was seven. At eighteen, against the advice of relatives and friends, she decided to use her inheritance to attend Vassar College, a women's college in Poughkeepsie, New York. There, she specialized in mathematics and astronomy, graduating in 1908 with honors. After teaching for three years, she decided to try engineering, a radical step for a woman that time.

Clarke enrolled in civil engineering at the University of Wisconsin in the fall of 1911. During her first summer break, in 1912, she accepted a summer job with American Telephone & Telegraph (AT&T). There, she worked as a computer, that is, as a mathematician. She enjoyed the work so much that she decided to stay at AT&T, abandoning her plans to get an engineering degree.

In 1918, Clarke decided again to pursue an engineering degree, leaving AT&T to attend the Massachusetts Institute of Technology (MIT). This time, inspired by her work at AT&T, she majored in electrical engineering. After receiving credit for her work at Vassar and the University of Wisconsin, Clarke received her Masters Degree in electrical engineering in 1919. She was the first woman to receive an electrical engineering degree from MIT.

When she could not find work as an engineer, Clarke returned to work as a mathematician, this time at General Electric (GE). Her new work involved the management and training of a small team of women who worked on problems of vibration and stress in high-speed turbine rotors. While at GE, Clarke was granted

under the authority of their fathers and then their husbands. Around the world, political rights for women came later than for men (see Chapter 4).

Women who qualified in engineering began as a mere trickle in numbers in the late nineteenth century. The University of California claims it was the first in America to award an engineering degree to a woman, doing so in 1876 to Elizabeth Bragg. However, Iowa State College also claims it was the first, awarding its degree to Elmina Wilson in 1892. In Britain, the first woman member of the Institution of Electrical Engineers was Mrs. Hertha Ayrton, admitted in 1899. Two years later she almost went on to become the first woman to read a paper to the Royal Society, after demon-

two patents, the first for a graphical calculator, the second for a saturated synchronous condenser.

Thirsting for adventure and travel, in 1921 Clarke took a temporary position as a professor of physics at the Constantinople Women's College in Turkey. She returned to GE in 1922. Her new assignment was that of a salaried electrical engineer. At 39 years old, she had finally achieved her long-time goal.

Clarke spent the rest of her industrial career with GE, working in Schenectady, New York. She retired in 1945 at the age of 62. After two years in retirement, Clarke took a position as a professor of electrical engineering at the University of Texas, in Austin. She was the first woman to teach engineering there and perhaps the first to do so at any American university. She retired a second and final time in 1956, when she was 73 years old. Between 1923 and 1951, she was the author or co-author of nineteen technical papers. She also wrote a popular two-volume reference, *Circuit Analysis of A.C. Power Systems.*

Today, women in engineering are continuing to build upon the accomplishments of "technical grandmothers" like Edith Clarke. The engineering student population of women in the United States reached a plateau at around 16 percent in the late 1980s, but started rising again in the mid-1990s. Even so, fewer than 10 percent of the working engineering population are women. Steps are being taken to increase the numbers within engineering by introducing engineering to young women at an earlier age and providing better math and science preparation prior to college. Many engineering programs have programs specifically to help and recruit women in engineering. However, before women are represented in the profession in proportion to their numbers in the population, engineering will also have to change.

Further Reading

Brittain, J. E. 1985, "From Computer to Electrical Engineer—the Remarkable Career of Edith Clarke," *IEEE Transactions on Education*, Vol. E28, No. 4, November.

Goff, A.C. 1946, *Women Can Be Engineers*, Edwards Bros., Ann Arbor, Michigan.

Gusen, A. 1994, "Looking Back: Edith Clarke," *IEEE Potentials*, February.

LeBold, W. K., LeBold, D. J. 1994, "Women engineers—a 100 year look back and hope for a leap forward," *WEPAN National Conference.*

The Engineering Specific Career Advisory Problem-Solving Environment, <www.ecn.purdue.edu/ESCAPE/special/women/Histore/wiehistm.html>

strating her experiments on a Ladies' Night, but the members apparently did not yet feel able to compromise with tradition to this extent, so a male member read the paper on her behalf (Griffiths 1985, p. 57). One of the most renowned women in engineering in the United States was Edith Clarke, whose distinguished career is outlined in Case Study 3.3.

Engineering has been the large-scale, dramatic, visible, and prestigious public face of technology. Women have been involved in technology, but what they did was rarely treated as engineering. For instance, Florence Nightingale overcame the appalling complacency, conservatism, and incompetence of the army medical bureaucracy to reorganize the Crimean

War base hospital at Scutari. She went on to reform hospital design so that it was based on: "... consideration of what would best tend to the comfort and recovery of the patient ... [rather than] ... the vanity of the architect, whose sole object [in the design of the major new army hospital in England at the time] has been to make a building which cut a dash when looked at from the Southampton river" (Strachey 1973, p. 145). Her work, however, was called nursing—not engineering or even architecture. Her official recognition was not a knighthood or a peerage from Queen Victoria, but the gift of a brooch! Further, improvements in domestic technology, even where they had substantial social impact, were less dramatic than more public works, and have been ignored in most histories of technology.

Some women clearly did play outstanding roles in engineering, such as Dr. Lillian M Gilbreth (1878–1972), the subject of Case Study 7.2, who was central to the development of Scientific Management. Regrettably, their contributions tended to be ascribed to the males with whom they worked. In fact, Lillian Gilbreth pioneered the application of psychology to industrial management, and her early book, *The Psychology of Management*, is recognized as a critically important contribution to Industrial Engineering (Rothschild 1983, p. 23). Lillian Gilbreth was a mentor and teacher to numerous engineering students, both male and female, during her tenure at Purdue as a full professor of management in the School of Mechanical Engineering, as head of the Department of Personnel Relations at Newark School of Engineering, and as a visiting professor of management at the University of Wisconsin at Madison. Her presence for female engineering students must have been crucial in the 1930s to 1950s (Amy Chang, Stanford University 1997, unpublished paper).

Although the numbers of women engineers in many (former) Communist countries were much higher, in most Western nations the numbers continued to be very low until well into the 1970s. The steady increase over the past two decades is discussed in Case Study 3.2.

▪ Women and Technology

What effects has the history of exclusion of women from engineering had? It appears to be associated with the failure of the engineering profession, worldwide, to address the social effects of its activity. Another effect has probably been an emphasis on construction and the conquest of nature, whereas problems of operation, maintenance, and use of equipment have been relatively neglected.

There is a broad question of social control with respect to engineering that it is timely to recognize here. Does the way technology is defined control access to it within our society? Historians have looked at the ways in which women were systematically excluded from the professions which were then redefined by men. Since women were permitted into university courses for the professions from late nineteenth century onwards, they have become a major force in most areas, but not yet in engineering. Why have women been so slow to move into technical areas, particularly the physics and math-oriented ones? In the United States, although women make up 46 percent of the workforce, less than 3 percent of the top executives (chairmen, vice chairmen, chief executive officers, presidents, chief operating officers, and executive vice-presidents) in 1995 Fortune 500 companies were women. In spite of the fact that girls consistently match or surpass boys' achievements in science and mathematics as measured by scholastic aptitude tests, achievements tests, and classroom grades, 34 percent of high-school-aged girls reported being advised by a faculty member not to take senior math.

Prince (1992) discussed the influx of women into the Chemical Engineering program at the University of Sydney, Australia, starting in 1978 with 14 females in a first-year class of 70 students. By 1992, women totaled 120 of the 300 undergraduates. Among the innovations were subjects that introduced real engineering, especially laboratory work, into the first year of the course. In addition to putting the technical material into its professional context, these subjects gave students hands-on practical experience of dismantling, reassembling, checking, and operating basic process equipment such as pumps and valves—experience that many women (and some men) may not otherwise have acquired. The written, essay-based examinations focused student attention on the importance of developing their communication skills (one group of male students told Prince they assumed these would not be needed in engineering, but women students were generally comfortable with this requirement).

There is increasing evidence that differences in the ways girls and boys are socialized are relevant to women's view of technology and their participation in engineering. Prince (1992) suggested:

> Adolescent men are generally less easy about people relationships, and so may choose engineering, thinking that it is entirely concerned with "things," not people (they are wrong, but may not discover this until much later). Women, more confident about their interpersonal skills, then prefer people-oriented careers, and if they continue to see engineering as the men apparently do, will have a bias against us.

∎ Broadening the Appeal of Engineering

The ways technology and engineering are defined and practiced clearly discourage most women—and some men! Engineering, largely if unconsciously defined by men, is commonly an uncomfortable environment for women. The values of the workplace are all-pervasively male. Just as effectively exclusive can be the situation where all the real decision making is done over a drink in the local bar. A prime dilemma for all working women is overload, and women engineers probably find it worse than most because of the difficulties they have in maintaining networks of female friends. Their location in a nontraditional job surrounded by men isolates them from other women generally. Their life experience is often so different from that of the girls they grew up with that they lose touch with their early network of female friends. A critical mass of women is important in any profession. The engineering profession needs enough women in the engineering work force that their concerns and attitudes are specifically taken account of, rather than the women being treated as honorary males and simply being expected to adjust.

Hegarty-Hazel (1991, pp. 46–49) has identified five areas for particular action by engineering educators in her recommendations for change

1. Improving the advice available to potential students
2. Increasing the practical and social relevance of the course content
3. Providing an equitable, secure, and nondiscriminatory academic and professional climate
4. Enhancing staff support for students
5. Encouraging postgraduate study.

These steps are an essential part of the profession's planning for the future. They point to ways in which engineering educators can change the likely outcomes for *all* those students—women and men—who, because of their potential to enrich the profession, we particularly need to attract and retain.

Among many initiatives, around the world, that address these concerns is a manual, *Taking on Technology*, for use by girls at high school in Australia. It was developed by a Women-in-Engineering coordinator (Taylor and Giffard-Huckstep 1992). The manual seeks to "engender" the understanding of technology, encouraging girls to explore what technologies mean for them. The need for such a manual has been confirmed by broad industrial and academic support for its production, and its extensive adoption in schools. The manual emphasizes a theme that also runs through this book: that although engineering is a social activity, it is too often treated as if it existed in a social vacuum (Taylor and Johnston 1991). The very successful approach adopted by McMaster University in Ontario, Canada is described in more detail in Case Study 3.4.

A brief exploration of some of the directions in which engineering might be "redefined" are presented in the accompanying box.

"Male" and "Female" Values in Engineering

Studies quoted in Carter and Kirkup (1990) suggest that girls' socialization typically rewards

- recognition of systems and connectivity:
- knowing what others want, caring for them, and being attached to them; and
- awareness of context and consequences.

These skills and attitudes are precisely those needed for systems thinking, for recognizing linkages and relationships between new technologies and new tasks, and managing the process change.

Boys' upbringing, on the other hand, tends to encourage them to be much more object-oriented. The differences in boys' socialization affect experience, expectations, and learning styles. They also affect approaches to moral problems and conflicts, in which Carter and Kirkup argue:

Men are concerned with competing rights and exercising impartiality, [whereas] women see them as concerned with competing responsibilities. Men rely on fairness (objectivity) as their preferred way of resolving dilemmas; women rely on their understanding of relationships (subjectivity) to come to terms with them.

It is difficult here to formulate a vision for the future of engineering that takes adequate account of these insights. There is a real danger of presenting a caricature, rather than a well-rounded portrait. The issues involved are so important, however, that we believe they need to be raised, even at the risk of failing to do them justice. A strong case has been made (Kolmos 1987) that these differences in attitudes are reflected in the very aspects of technology that are considered to be relevant, and that are most valued in our current ideology. This argument relies in part on stereotypes of:

- *Male* qualities (dominating nature, linear, rational, objective, hierarchical, logical). These are contrasted with the virtual absence of:
- *Female* qualities (part of nature, iterative, caring, subjective, emotional).

However, it also points to the possibility of moving on to a new model or "vision" of technology, which would be characterized by:

- *Inclusive* qualities such as "respect for nature as the basis for human

ATTRACTING A DIFFERENT KIND OF STUDENT: McMASTER'S FIVE-YEAR ENGINEERING AND SOCIETY PROGRAM

The need for a more liberal and socially oriented undergraduate program is not a new idea to most engineering educators, nor to the profession. The difficulty has always been finding room to include a broader education in the already crowded degree program.

Undergraduate engineering programs are generally four years in Canada; McMaster University in Ontario has taken a different approach through its five-year Engineering and Society Program, started in 1992. By adding a year, the program maintains the same scientific and technical content as the traditional degree. It spreads the additional studies evenly over the upper four years, helping students integrate what they learn in the "Society" Program with their technical courses. Students join Engineering and Society and one of McMaster's nine specialized departmental programs in level 2, after completing the level 1 program common to all engineering students. Graduates receive a new and fully accredited degree, for example, a Bachelor of Civil Engineering and Society.

This new program enables engineering students to get a broader university education in two ways:

1. *Elective courses taken in the Faculties of Humanities, the Social Sciences, or Science.*

The elective part of the Program is used to develop the student's intellectual interests outside of engineering. The choice is up to each student, but must fulfill two conditions: it is focused on a subject or theme, and it adds a new dimension to their university education. The possibilities are almost unlimited. Some students choose subjects that expand on their engineering interests such as urban geography, biology, or environmental science. Others go further afield into such subjects as philosophy, psychology, sociology, music, drama, or art. In many cases students take enough courses to receive a minor in their area of interest. There are as many different sets of focused electives as there are students.

2. *A new set of core courses.* The seven core Society courses are interdisciplinary and all are concerned with relating engineering and technology to society. In the first three of these courses, students develop skills in formulating critical questions and carrying out all the phases of research into those questions, as follows:

◆ The first inquiry course introduces the skills and art of inquiry and uses the issue of sustainability as its focus.

- The second inquiry course is an introduction to the systems nature of many sociotechnical issues of today.

- For the third and final inquiry course, students choose their own area of investigation and find their own supervisor.

The other four core courses develop enquiry skills but are more focused in content:

- Case Studies in History of Technology is taken early in the program; cases are chosen to exemplify the complex and changing relation of technology to science and society.

- The Culture of Technology is taken with nonengineering students and examines the shared beliefs, attitudes, and values of our technological culture.

- Environmental Studies. This level four course takes on a positive orientation through the examination of examples of preventive engineering from around the world.

- The Social Control of Technology. The final course examines the various means used by society to regulate and steer technology, with a major focus on the role of the engineering profession.

Who is attracted to such a program? A few students have taken their previous "introduction to social responsibility" to heart and want to understand what this means in greater depth. However, most will admit to being attracted by the prospect of being able to study something outside of engineering. High school students have an image of engineering as being just science and math; this makes it difficult for many to choose engineering, thinking they have to abandon their other interests. It is not until this latter group has had some of the Society courses that they see why socially responsible engineering is so difficult.

McMaster University was confident that a significant number of students would elect to stay an additional year to broaden their education—largely because of the success of another five-year program at the university in Engineering and Management, a combination of the traditional engineering curriculum with several commerce courses.

The Engineering and Society Program has a steady-state enrollment in all levels of just over one hundred students, which represents about 7 percent of the undergraduates in the Faculty. All departments are represented with the majority coming from Chemical, Civil, or Mechanical. The Engineering and Society Program has an equal number of male and female students.

Source

Contributed by Robert Hudspith, Programme Director, Engineering and Society Programme, McMaster University, Ontario, Canada.

Other Issues for the Future

In this chapter we have considered possibilities for constructive changes that would make engineers more responsive to the needs of all sections of society and enhance the appeal of the profession for a broader cross-section of new recruits. The discipline of sociology gives all of us insights into patterns in our society and major social problems of which we might otherwise be unaware. It prompts us to reflect on ways that engineers could help to address some of those problems, including:

- designing quality workplaces and jobs;
- coping with the changes towards a postindustrial or digital information or services based or computer-aided society;
- finding ways to employ unskilled and semiskilled youth to stop them sliding into the underclass;
- meeting the needs of people with disabilities and an aging population; and
- job security.

The values and perceptions about themselves and their society that engineers bring to their work influence their designs, their judgments, and their choices. In the course of their work, engineers inevitably make decisions on the use of resources, in the face of considerable uncertainty and gaps in knowledge. It follows that whatever options they choose must be, to some extent, wrong. The plea for a new synthesis, a new assessment of what we are and how we work, is therefore very important.

The social sciences can make an important contribution to the natural sciences and engineering, in particular by reviewing the problems chosen for attention, and the way work is organized and paid for. These questions have important social elements. So do the ways that public debate about scientific and technical matters is conducted. Engineers must increasingly realize that they have a voice in the social debate. The inevitable political overtones of that debate are taken up in the next chapter.

Discussion Questions

1. Discuss the social structuring of the engineering profession. What sorts of people become engineers? What social groups are they drawn from and which do they enter?

2. What sorts of underlying attitudes and beliefs do engineers hold? What sorts of practical implications do they have for the way engineering is practiced in this country?

3. American society is a mixture of many cultural groups. Discuss the impact of this mixture of social heritages and analyze the probable effects of various immigration strategies.

4. Review some forms of unpaid work and discuss ways by which their status and prestige increased. What do you think about the proposition that "the work ethic is in decline"?

5. Discuss, preferably from your own experience, the effects on social power relationships of how a particular technology has been applied. What differences would there have been if one of the design criteria had been to maximize the control the people involved had over their working situation?

6. How would you go about setting up a new manufacturing line to minimize the alienation of the workers involved? How would you sell your approach to your senior management?

7. What are EEO principles? How would you implement them in either (a) a new engineering design office, or (b) an existing manufacturing firm? What major difficulties would you have to overcome?

8. Discuss the role of the media in image creation and marketing generally. Are the media a necessary evil? What effects do these activities have on the ownership and structure of U.S. manufacturing industry?

9. The antismoking lobby argues that the role of advertising extends beyond communication to augmenting and even creating demand. Is this true? How could it be confirmed or disproved? What do you think about the ethics of this sort of marketing? Is it freedom without responsibility, or are there other compelling views?

10. What differences would you expect it to make to the profession and practice of engineering if women and minority groups were present at all levels in similar proportions to their numbers in the community? Would the profession be more sensitive on environmental or other issues, or would those entering it merely be assimilated into the present culture? How will the culture of the profession change, for example, to embrace sustainability issues?

11. What do you see as the basic forces driving people to involvement in economic activity, such as going to work, or starting a small business? What are your views on the idea that "the consumer's avaricious pursuit of his own needs benefits society as a whole"?

12. The MESA and McMaster Case Studies (3.1 and 3.4) present models for approaching the problem of the under representation of women and minorities in engineering. Suggest some others, and research the extent to which they have been implemented.

FURTHER READING

Kass-Simon, G., Farnes, P. (eds.) 1990, *Women of Science: Righting the Record*, Indiana University Press, Bloomington, IN.

Vare, E. A., Ptacek, G. 1987, *Mothers of Invention: From the Bra to the Bomb, Forgotten Women and Their Unforgettable Ideas*, Quill William Morrow, NY.

Engineering groups that focus on broadening the experience of engineering education and attracting and supporting nontraditional entrants include:

- ◆ American Indian Science and Engineering Society
- ◆ National Society of Black Engineers
- ◆ Society of Hispanic Professional Engineers
- ◆ Society of Mexican-American Engineers and Scientists
- ◆ Society of Women Engineers (SWE)
- ◆ Women in Engineering Programs Advocates Network (WEPAN)

All can be contacted through their web sites or campus chapters.

REFERENCES

Appelbaum, R. P., Chambliss, W. J. 1997, *Sociology*, 2nd Ed., Addison Wesley Longman, NY.

Aungles, S. B., Parker, S. R. 1988, *Work, Organisations and Change: Themes and Perspectives in Australia*, Allen & Unwin, Sydney.

Bacon, E. E. 1989, Witchcraft, in *Encyclopedia Americana*, 29, Grolier Incorporated, Danbury, CT, pp. 83-84.

Carter, R., Kirkup, G. 1990, *Women in Engineering: A Good Place to Be?*, Macmillan, London.

Dickson, D. 1974, *Alternative Technology and the Politics of Technical Change*, Fontana/Collins, Glasgow.

Feenberg, A. 1991, *Critical Theory of Technology*, Oxford University Press, NY.

Game, A., Pringle, R. 1983, *Gender at Work*, Allen & Unwin, Sydney.

Gould, J., Kolb, W. L. 1964, *Dictionary of the Social Sciences*, Tavistock, London.

Griffiths, D. 1985, The Exclusion of Women from Technology, in Faulkner, W., Arnold, A., *Smothered by Invention*, Pluto Press, London, pp. 51-71.

Hegarty-Hazel, E. 1991, *Women in Science and Technology in Australia: Summary*, Report for DEET, AGPS, Canberra.

Kolmos, A. 1987, Gender and Knowledge in Engineering Education, *Fourth International Conference on Girls and Science and Technology*, July, University of Michigan (quoted in Carter and Kirkup 1990).

Kotler, P., Chandler, P., Gibbs, R., McColl, R. 1989, *Marketing in Australia*, Prentice Hall, Sydney.

Macionis, J. J. 1997, *Sociology*, 6th Ed., Prentice Hall, New Jersey.

McLean, C., Lewis, S., Copeland, J., O'Neill, B., Lintern, S. 1997, "Masculinity and the Culture of Engineering," *Proc., 3rd Australasian Women in Engineering Forum*, Women in Engineering Program, UTS, Sydney, pp. 32-41.

Mills, C. W. 1959, *The Sociological Imagination*, Oxford University Press, Oxford.

Pacey, A. 1983, *The Culture of Technology*, MIT Press, Cambridge, Mass.

Prince, R. G. H. 1992, "Peter Nicol Russell Oration," *National Engineering Conference*, IEAust, Newcastle, 29 March-1 April.

Rothschild, J. (ed) 1983, *Machina ex Dea: Feminist Perspectives on Technology*, Pergamon, NY.

Sprouster, J. 1987, *Total Quality Control: The Australian Experience*, Horwitz Grahame, Cammeray.

Strachey, L. 1973, *Eminent Victorians*, Penguin, Harmondsworth.

Taylor, E., Giffard-Huckstep, S. 1992, *Taking on Technology*, Women in Engineering Program, University of Technology, Sydney.

Taylor, E., Johnston S. F. 1991, "Why transforming engineering courses will attract a diversity of students—including women," *Proceedings, 3rd Annual Conference, Australasian Assoc for Engineering Education*, 15-18 December, Adelaide, pp. 239-244.

U.S. Bureau of the Census 1997, *Statistical Abstract of the United States 1997 (The National Data Book)* 117th Ed., Washington D.C. October.

Walshok, M. L. 1990, "Blue collar women," in A. H. Teich, *Technology and the Future*, St. Martin's Press, NY, pp. 288-296.

Womack, J. P., Jones, D. T., Roos, D., 1990, *The Machine that Changed the World: the Story of Lean Production*, Harper Perennial, New York.

A GOOD JOB FOR A WOMAN?
ONE ENGINEER'S EXPERIENCE
WITH HARASSMENT

Engineering managers have a professional responsibility to shape productive workplaces that are (at the very least) not damaging for those employed there. It is part of their social and ethical responsibilities, as spelled out in their professional Codes of Ethics. It is a prerequisite for the quality production that is now essential to the success, and even to the survival of, manufacturing and other enterprises. It is also a legal requirement, and failure to address harassment issues promptly and effectively can leave a supervisor open to immediate dismissal and even to civil lawsuits. This case is one woman engineer's description of her own experience, and her reflections on how the situation should have been handled.

Monday 10:00 P.M.

How many times have I driven this highway? Fifty? Sixty? As I drive, I begin to reflect on the events of the last few months.

Nine months ago, my supervisor proposed a new assignment. I wasn't too surprised. I never stayed very long in one position. Management was moving me quickly through various assignments and I should feel flattered. This would be the fourth assignment in four years, three of them in different departments. When my supervisor described it, however, I was a little overwhelmed. It involved a substantial increase in responsibilities and exposure. I would be design responsible for a major component in the redesigned version of the highest selling automobile my corporation manufactured. From a monetary viewpoint,

this represented $45 million per year at manufacturer's cost. The vehicle was being launched in a few months.

On the bright side, the component was very similar to a part I had worked on before. But when I took that assignment, the vehicle had already been in production for two years. The major problems were solved and I was fine-tuning the design. Now I would be responsible for taking the component into production. I needed to learn the design quickly.

I traveled right away to the main vehicle assembly plant to watch the last few cars of the pilot build. In total, three plants would be manufacturing this automobile. Two were in the Mid-west and one was in Mexico. The main or "lead" plant was only four hours away by car. Because flying basically took the same amount of time as driving, almost all engineers drove to the plant. Management also preferred us to drive as it was significantly less expensive.

The lead assembly plant was relatively new but was known throughout the corporation as the most difficult plant. Although labor union and management relations were improving in other areas of the corporation, this facility still had its share of problems. This was to be my first experience working with them and I was apprehensive. According to other engineers, not only was there an abnormally large rift between union and management, but the assembly plant relationship with central engineering was also strained.

I was very familiar with the other Mid-west assembly plant because I had worked there for a year and my last component was built there. I was glad I already knew those contacts and I enjoyed working with them. By and large, union, management, plant engineering, and central engineering enjoyed a good working relationship, and hierarchy was not an issue. The third assembly plant, in Mexico, built a number of different vehicles. Because of its location, the plant was largely self-sufficient. They were well known throughout the corporation as very cooperative and they usually tried to solve their problems before contacting central engineering. My first visit to the facility was scheduled a few months away.

As far as allocation of production volumes, the majority of the redesigned vehicles would be built at the lead assembly plant. That was the only product they would manufacture and they had the final say in any production or design decisions for all three assembly plants. In Mexico, the redesigned vehicle amounted to approximately half of their total production but most of their production was for the local Mexican market. I was told that central engineering in Mexico controlled the local market cars and I was responsible for the relatively few vehicles to be imported into the United States. Finally, in the second Mid-west plant, the redesigned vehicle was a small fraction of their total production, filling in capacity gaps.

In order to ease the burden on central engineering, the start-ups at the three facilities were staggered. First was the lead plant in the fall, next Mexico a few months later, and finally the second Mid-west plant in the spring.

The start-up at the lead plant was long and painful. A certain degree of "pain" is expected during all vehicle start-ups. I expected to work long hours and I wasn't disappointed. Starting in September, I would usually work Monday at the office, drive to the plant Monday night and work twelve to fourteen hours per day, Tuesday through Friday. Working Saturday was common. Every month, I would also travel to Mexico for a week. In Mexico, the start-up was progressing much more smoothly. Because the start-up in Mexico was later than the lead plant, Mexican engineers traveled to the Mid-west to watch and learn. This helped accelerate their learning curve. The start-up at the second Mid-west plant was not planned for another six months.

It soon became apparent that my component was causing installation difficulties. In an effort to empower the assembly workers, an Andon system or "stop-cord" was installed on the assembly line prior to the model change. If assembly workers experienced difficulty in their job or noticed a quality problem, they could pull the cord and stop the line. The area where my component was installed had the cord pulled frequently.

When I started to look at a redesign, I found I was at a disadvantage because I did not originally design the component. I had to find out why certain decisions had been made and investigate all the build variations of the vehicle. Fortunately, my supervisor asked another engineer in our group to help me; he was also responsible for a component in the same vehicle. We started sharing rides to the vehicle assembly plant. This made travel easier, but the extra responsibility of the redesign added Sundays to our work routine.

So many problems, and not enough time. Because the redesign would take many months, a short-term solution had

(continued on next page)

to be implemented first. There was considerable pressure on each engineer to fix their problems quickly because the assembly line was down more often than running. In addition, the slow start-up was starting to make national news. Consequently, I broke almost every purchasing rule in the book, chartered airplanes to fly material to the plant, and reworked material inside the plant. This relieved some of the build difficulties, but inherent design problems existed and a totally new design needed to be found. The months flew by.

There was also one other problem that was nagging me: the constant harassment at the vehicle assembly plant. Being a young, female engineer, I had anticipated this. In my previous assignments, I had worked at another Mid-west assembly plant for about one year. When I started there, "cat calls" were common, but as the months wore on and the assembly workers started to know me, the harassment eventually ceased. I found that my love of meeting people and talking to them helped significantly to combat the harassment. In the plant, I quickly learned that the people building the cars carried the essential knowledge. Whenever I heard about a problem from the field, I immediately went to the assembly line to talk to the workers. I respected them and they eventually began to trust and respect me.

But this approach wasn't working at this assembly plant. In fact, the harassment was getting worse. It was as if the entire plant was playing a cat and mouse game with me. When I walked with my male colleague, no one bothered me, but when I was alone the harassment was inevitable. The insults ranged from "cat calls" to one man that stalked and stared at me and a supervisor

that liked to touch me. I sat down and talked to my supervisor about the harassment a month ago. He suggested that I "yell back at them." Well, I was not going to even try this. Last week, I told my supervisor that yelling at people did not fit my personality. I couldn't do it. He then suggested that I ignore it.

I become conscious of driving again. My hands are gripping the steering wheel tightly. At least they aren't shaking. In order to support my extended working hours, my caffeine intake has increased significantly. That is why my hands are shaking.

I hate to talk about the harassment. I am receiving management praise for the redesign and I look like a hero. I am trying every trick I can think of to deter the harassment on my own: I wear absolutely no makeup, very baggy pants, shirts buttoned to the very top and my hair pulled back. I know I look terrible. In fact, my supplier asks me if I am sick. My supervisor also confirms that my dress should not invoke harassment. What else can I do?

3:00 P.M. Tuesday

I have only been in the plant for eight hours but I can't take it. Today three workers circled me and started barking like dogs. I left the plant floor and went to the plant offices. I couldn't stop the tears. I feel really embarrassed and I sense that the other engineers are embarrassed for me also. So I'm leaving. I'm going to just drive four hours back the office.

A Few Weeks Later

I'm driving again on this road to the assembly plant. I'm starting to hate this road. I talked to my supervisor again yesterday. I told him that I'm really tired. I can't get out of bed in the morning and my hands are shaking a lot. I frequently come in late to work. He said he doesn't mind. He said, "You get as much accomplished in half a day as my other engineers do in an entire day." I guess that should make me feel good. I'm a hard worker.

At the office, there have been a lot of complaints and rumors. Other engineers are starting to voice their frustrations concerning the strained working environment at the lead plant. This makes me feel a little better because I no longer feel like I am the only person experiencing abnormally high levels of stress. Also, it seems that all the engineering changes are not helping the line speed ramp-up at the lead assembly plant. People are starting to wonder if all the cord pulling is really necessary. I know that a lot of people were laid off at the lead plant prior to start-up because the redesigned model required fewer man-hours to build. Was the excessive cord pulling a labor initiative to pressure management to rehire some of the laid-off workers? On the other hand, the plant in Mexico was running smoothly.

Several Weeks On

I blew up at my director, my supervisor's boss, today. He asked why I've been so quiet. I said I didn't care if the lead assembly plant was blown into a million pieces. He had a look of shock on his face. I asked him if my supervisor told him about the harassment. He said no. How could he not tell him? I've been talking about this for months.

Afterwards, I went to talk to an acquaintance of mine. She was a manager in engineering but now she is a manager in an assembly plant. I thought she might be able to tell me what I can do differently to curb the harassment. Instead she was very upset. She said I should have said something earlier. I said I did.

What could I have done differently?

- If action wasn't taken immediately, talk to an engineering manager one level up about my supervisor.
- Talk to someone in the Personnel department.
- Not let it go on for so long.
- Reprioritized—put my health ahead of work.
- Trusted and listened to myself. What was going on was wrong and it was not my fault.

Update

Personnel was notified. For health reasons, I was transferred to a different position. Harassment training was instituted at the assembly plant. My supervisor was verbally reprimanded. The incident did not affect my career, as I had feared it might. A few years later, I received a corporate-sponsored graduate fellowship to pursue an advanced engineering degree.

Authors' Comments

Engineering managers have a professional responsibility to shape productive workplaces that are (at the very least) not damaging for those employed there. This is their common duty as civilized human beings. It is part of their social and ethical responsibilities, as spelled out in their professional Codes of Ethics.

There were clearly long-standing human relations problems in this plant, with its antagonistic, macho, even dysfunctional culture. Such problems may well be common in plants in old manufacturing areas that have seen jobs, factories, even whole industries, move away. What would it feel like to be a worker in such a plant, in such a city?

(continued on next page)

Although these issues in no way justify or excuse harassment, they may help us to understand it and to deal with it more effectively. Turning around the culture as a whole would require broad commitment to change, with cooperation between leaders in the community, the company, and the union. The future of manufacturing belongs to organizations that can learn to do new things and to work in new ways. Changing workplace behavior is a first essential step along a path to renewal, but without such change both the plant and the city are probably doomed.

What would you personally have done in this situation? How do you feel about the hours that the engineers were obviously expected to work? Were they reasonable under the circumstances? Would your nonengineer friends agree with your view? What do you think about the measures that were taken to address the harassment?

Source

Case contributed by Monika Ellis, Stanford University 1997.

Further Reading

Hamper, B. 1992, Rivethead: *Tales from the Assembly Line*, Vol. 1, Reprint Ed., Warner Books. A description of life on the car and truck assembly line that spells out problems of the life of factory workers in and out of the plant.

Walton, M. 1997, *Car: A Drama of the American Workplace*, W. W. Norton & Co. With a background in management theory, Walton spent two years in the Ford Motor Company researching this entertaining and critical account of the manufacture of the 1996 Taurus, and the relations between the company's engineering staff and its designers.

4

Politics, Power, and the State

Politics ought to be the part-time profession of every citizen who would protect
the rights and privileges of free people and who would preserve
what is good and fruitful in our national heritage.

— *Dwight D. Eisenhower (1890–1969) U.S. President and war hero*

The focus in this chapter is on the state as the formal organization of
civil power, on our political system, and on the various roles played
by engineers in this arena. Politics always involves questions of
power; it is about the different and often competing interests of
individuals and groups in society, about how decisions are made and conflicts contained. In this chapter we examine some of the ways in which capitalism as the dominant economic model in America influences interaction
between governments, agencies of the state, and engineers in a wider context
of power relations.

Engineers' work impacts on power relations within our society in a
variety of ways. Technological choices made by engineers in the workplace involve the exercise of power. They produce outcomes that are not
neutral, but that have different effects on groups within the nation as a
whole, advantaging some and disadvantaging others. The practice of engineering, from seemingly minor design decisions through to managing
mega projects, inevitably has political as well as technical implications. It
is important that engineers recognize that they will be politically involved,

both as citizens and professionals, throughout their lives and that this dimension of their work is of crucial importance.

Political Power and the Modern Nation State

Definitions of the word state in Webster's Dictionary include:

1. a politically unified people occupying a definite territory; a nation
2. (sometimes capitalized) any of the bodies politic which together make up a federal union, as in the United States of America
3. (informal) State, also called State Department—the [US] Department of State
4. the body politic as organized for civil rule and government (distinguished from church)
5. of or pertaining to the central civil government or authority

Professional engineers need to understand how the state works in order to be able to deal effectively with it. This chapter can only provide a brief introduction to the subject. Much has been written about the power and activities of the state in advanced capitalist societies such as America where the economic system is characterized by private ownership, but there is also significant public control of private corporations in advanced capitalist societies, for example through legislation covering labor relations, environmental protection, and antitrust measures.

The state is not static or immutable, but continues to change. One of the points that emerges from the development of the modern state is the importance of understanding the driving forces and the directions of change. For instance, "liberals," or "liberalists" (not to be confused with members or sympathizers of the American Democratic party—we use this term to refer to approaches that are market-oriented, rather than left-wing) may be devoted to the market mechanism and to private ownership and control of capital, but this devotion does not necessarily imply any commitment to democracy or equity. In the dominant model of the "liberal[ist] democratic" state, such attachments arise out of public political processes and are sustained by them.

A standard political science definition of the state emphasizes rule over society on behalf of some or all of its members. This includes a monopoly on the legitimate use of physical force within a given territory. Questions of legitimization are in fact crucial because they deal with the

The Right to Vote

Legitimization of political governance is most commonly achieved through voting by its citizens. Worldwide, the right to vote came slowly (even more slowly for women). Initially in the United States, women had the right to vote in their respective states. However, the 1787 Constitutional Convention placed voting qualifications in the hands of the states, resulting in women losing their right to vote in all states except New Jersey (in 1807, even New Jersey revoked the right). The fifteenth amendment of 1870, a consequence of the Civil War, extended the right to vote to African-Americans, to both ex-slaves and to those who had always been free. However, it was the Nineteenth Amendment, called the Susan B. Anthony Amendment, that granted the right to vote to women on August 26, 1920.

The right to vote is, by itself, a very limited measure of democracy. A more comprehensive model would require an informed and interested populace. It would also include direct decision making on issues rather than delegation of decisions to parliamentary or congressional representatives. More radical approaches would extend democratic decision making to industrial democracy—a direct voice for employees in the running of enterprises. All of these involve the exercise of political power. An aspect of voter participation that must be considered in the United States is the extent to which the electoral college system effectively silences the voices of many voters. In many countries there is provision for relatively small minority groups to be represented in the legislature.

Whether voting should be compulsory is an interesting question that the authors of this book do not agree on. The freedom to vote can be seen to imply a responsibility to do so. From that starting point, one position is that voting should therefore be compulsory and that citizens should be obliged to make up their minds and vote on matters that are seen by the community as being of general importance. Where people do not have a view they can spoil their ballot; but that represents a decision, not simply abstaining from the whole process. An alternative view is that any compulsion associated with the ballot process is inappropriate; coercion is undemocratic and leads to misleading or undesirable voting outcomes, where apathy can override informed choice. Provided there are no barriers to exercising voting privileges at the ballot box (e.g., physical distance or administrative "red tape") then it is argued that the state has met its obligation and it is then a question of individual rights. What do you think? How would you justify your position?

basis on which rule is justified and the extent to which rule is accepted by the members of the society. Government intervention in the economy, for instance, in the development and implementation of industry policy, may be limited by the degree to which such intervention would be accepted as legitimate by the various stakeholders.

The *modern nation state* is the term that best describes the form of governance that has emerged on the world scene over the last fifty years or so—the postcolonial era. It is not easy to define. However, most states are governed by a political apparatus that at least a majority of their cit-

---

Key Concepts and Issues

Politics

The typical dictionary defines politics only in terms of activities within the framework of the formal structures of the state. For example, the *Oxford English Dictionary* (1989) defines politics as:

> ... the science and art of political government; the science dealing with the form, organization, and administration of a state or part of one, and with the regulation of its relations with other states (hence imperial, national, domestic, municipal, parochial, foreign politics etc.)

Some definitions emphasize the amoral Machiavellian scheming and deviousness commonly considered essential to political success. (In the American context, where a single year's election campaign costs hundreds of millions of dollars, affluence and fund-raising skills may be equally important qualifications.) The broader definitions of politics used in the social sciences highlight other issues that are important for engineers. According to *A Dictionary of the Social Sciences* (Gould and Kolb 1964), for example:

> Politics denotes those processes of human action by which conflict—concerning on the one hand the common good and on the other the interests of groups—is carried on or settled, always involving the use of, or struggle for, power.

Power

The way power is exercised in society is complex and subtle. Power is not merely force. It is not just blatant control. It does not necessarily show in behavior. It is not always exercised consciously. To arrive at a satisfactory

izens regard as legitimate; that is, the use of power is recognized as rightful by those over whom it is exercised. Although states make continuing efforts to establish a common cultural identity (sometimes mentioned as a defining characteristic of the modern state), such an identity is perhaps the exception rather than the norm among the 180-odd nations in the world today. The development of modern nation states was important in the virtually universal spread of capitalism, which benefited from strong central governments and from legal systems that provided clear and predictable conditions for trade and commerce within and across borders.

definition, let us start with three propositions, and then see what insights they offer us, and how well they apply to practical situations:

◆ Power is not an abstract, but an aspect of social and political relationships in a society.

◆ Power is not random, but is specific to situations. It is an element of structural relationships in a society. Typical power relationships are between government and governed, between social classes, between different racial or ethnic groups, and between the genders. We should in particular look at how the choice of technologies influences these relationships (Feenberg 1991, pp. 71–88).

◆ Power relationships are not static, but are fluid and can themselves generate change.

Based on these propositions, a tentative definition of the way power is exercised might be as follows:

A power structure is a network of relationships that involves effective control by some people over others, and the ability of certain groups to organize political and social life to their own advantage or benefit.

One widely held (but rather too simplistic) approach, which can be characterized as the democratic-pluralist view, holds that power in Western societies is competitive, fragmented, and diffused; that everybody, directly or through organized groups, has some power and nobody has or can have too much of it (Miliband 1973, p. 4).

Later in the chapter we consider more realistic models, including the extent to which there is actually an identifiable U.S. "ruling class."

■ The State Apparatus in the United States

The formal political system in the United States is a federal republic, in which power is divided between a central governing authority and the individual states. The principal framework of government is the Constitution of the United States, drawn up in 1787. This constitution was initially modified by a group of amendments known collectively as the Bill of Rights. Subsequent Amendments have responded to changing social and political needs and attitudes. *Legacy* (Michener 1990) is a very readable account of the process of U.S. constitutional development.

The federal government consists of three branches: executive, legislative, and judicial. Executive power is vested in the President, who conducts the nation's administrative business with the aid of a cabinet comprising Secretaries (ministers) responsible for various areas of government. The Congress of the United States, the legislative branch, is bicameral, consisting of a Senate and a House of Representatives. The judicial branch is formed by a system of federal courts, the highest of which is the nine-member U.S. Supreme Court. For most people, the focus of American political life is the political parties and most Americans are committed to supporting one of the two major parties. The political differences between the Democrats and the Republicans have been less extreme than those between the major parties in many other Western democracies.

Engineers and the Political Process

Maintaining both quality of life and effective democracy requires (among other considerations) recognizing the political as well as the social and industrial implications of the creation and application of technologies. Historically, there have certainly been problems with technological decision making, partly as a result of widespread ignorance among our decision makers (including politicians) of the nature and importance of technology. There is also a lack of understanding of the importance of, and the process of, innovation; this is partly because political decision makers are overwhelmingly lawyers and economists. As we discuss elsewhere, it is also partly because the process of innovation and the relationships between science, technology, and engineering have not been well understood, including by many professional engineers, and the advice they offer the decision makers may well be unsatisfactory.

U.S. Political Parties

Democrats

Originally called the Democratic Republicans, the party was founded around Thomas Jefferson, who was elected the third President of the United States in 1800. The party originally emphasized personal liberty and the limitation of federal government. Democrat Andrew Jackson won the elections of 1828 and 1832, but arguments over slavery created splits within the party, and the Civil War all but destroyed it.

The Democrats, led by Franklin D. Roosevelt, again controlled the Presidency during the Great Depression and World War II. Harry S. Truman succeeded Roosevelt in 1945. The Democrats lost the Presidency to Republican Dwight D. Eisenhower in 1952, but regained it when Democrat John F. Kennedy narrowly won the 1960 election.

Jimmy Carter, a nuclear engineer, was elected President in 1976 and served one four-year term. Seen as somewhat lackluster at the time, his stature has risen in retrospect abroad, though perhaps not at home.

Today the Democratic Party (also described as the "liberal" party) consists of an alliance of labor, minorities, middle class reformers, and conservative Southern Democrats.

Republicans

Opponents of the extension of slavery into the Western territories founded the Republican Party in 1854. The election of its 1860 candidate, Abraham Lincoln, precipitated the secession of the Southern states and the Civil War. The Republicans then dominated national politics until they were perceived as unable or unwilling to take strong action to tackle the problems of the late 1920s. The Republicans, under President Herbert Hoover, were blamed for the Great Depression of the 1930s, and the party lost every presidential election between 1932 and 1952, when the war hero Dwight D. Eisenhower became president. In the 1970s the party's popularity suffered as a result of the Watergate affair during the administration of Richard M. Nixon. In 1980 a Republican, Ronald Reagan, was elected president. He was succeeded in 1988 by his vice president, George Bush, who was defeated by Democrat Bill Clinton in 1992.

The modern Republican Party is generally considered America's conservative party, representing business interests and middle and upper class voters.

All the case studies for this chapter and many others in the book demonstrate that there are political implications in a great deal of engineering activity. A project such as the building of a new dam or power line, or a highway system linking major cities, will have direct impact on people and the environment. It will benefit some people and disadvantage others. Politicians will invariably take sides. Although dealing with the technical questions might seem to be challenging enough for politicians who generally lack technical expertise, viewed from the other side of the political fence there are other concerns. Unless engineers are equally prepared to address the political issues, to appreciate what factors are likely to be weighed by politicians, and to appreciate what pressures they are responding to, everyone is likely to find that they have solved the wrong problems. Even the choice of a design specification has some political implications because it influences the resources committed to a project and involves an assessment of the seriousness of failure. The following case study discusses the lack of understanding of engineering in government and offers possible explanations for the low level of involvement of engineers in politics.

Wenk (1986) emphasized the extent to which engineering decisions may be political rather than technical. A former scientific adviser to the U.S. Congress, he pointed out the range of areas where politics and engineering intersect; for example, decisions on acceptable levels of risk have a political dimension. The political world in which engineers must often move demands social skills of quite a high order.

Wenk also observed that engineers' social attitudes reflect their need for a stable society in which to work. Engineers are typically employed by big firms or government bodies. They commonly work on large-scale and long-term projects that involve the mobilization of substantial economic and other resources. They therefore have a tendency to give uncritical support to powerful social institutions that provide the resources they need for their work rather than to the citizens over whom the power is exercised. This orientation towards large-scale systems may be more important than their social class position in determining engineers' attitudes to technology and politics.

Engineers who work for government agencies (see below) generally perform tasks similar to those they would perform outside the government. Most have little or no direct influence on major governmental decisions. Even those in management positions are often constrained by bureaucratic and political limitations. However, they may have important roles in framing and setting the parameters for the problems to be addressed, for example by ensuring that issues such as sustainability and whole life costing are considered.

Engineers responsible for developing new technologies need to keep the human and ethical purposes of those technologies clearly in mind. The Art of Inquiry, discussed in Chapter 6, can help with this. The choice of whether or not to proceed with powerful but potentially hazardous technologies (genetic engineering, for example) needs to be made in as public and informed a way as possible, taking into account moral and political as well as technological issues. Political issues need to be resolved in political ways. Many engineers appear to be uncomfortable with the political aspects of their work, and this can seriously limit their ability to take positive and constructive roles in political processes. On the other hand, as Case Study 4.1 shows, the political astuteness of some engineers has enabled them to rise to high leadership positions and succeed in them.

Case Studies 4.6 on the Tennessee Valley Authority at the end of this chapter and 4.3 on the Three Gorges Project in China show how much more critical public attitudes have become in recent years, even to relatively benign technologies like hydroelectricity. Many of the issues involved in large development projects, both at home and abroad, are clearly political, and the engineers involved find themselves (often reluctantly) at the center of the political stage. These are not isolated instances. Engineers in both public and private sectors make major inputs to decisions with significant political dimensions.

Engineers are increasingly likely to be held personally responsible for disasters such as the collapse of a bridge or a dam, or even for the sort of infrastructure failures described by the ASCE in Chapter 2. This is despite the fact that the causes may be not so much technical problems (that are present in any situation), but the failure to mobilize sufficient resources to resolve them.

Hudspith (1996) argues that engineers need to be more broadly educated, and need to be sensitive to political contexts and deal with them effectively and responsibly. This also implies their being more aware of the character and limitations of formal and informal power structures so that they can work out how and where to exert the necessary influences. Williams (1980) expressed a key concern about the way the dominant values of our society are put into practice, when he wrote:

> Ultimately this remains a world in which the political is paramount, for decisions great and small, public and private, are its very essence. And all decisions involving other people involve an exercise of political power, even when we attempt to pretend otherwise by invoking some technique or economic methodology.

Problems undoubtedly stem from reluctance on the part of many engineers to address the wider political impacts of their work. The education

ENGINEERS IN POLITICS

What do Yasser Arafat, Jimmy Carter, Jiang Zemin, and Boris Yeltsin have in common? Each of them has been an influential political leader whose decisions have had a significant national and international impact. They also share a common professional background that rather sets them apart on the political stage. Instead of treading the usual law or political science education path, these politicians received an engineering education. Although the relationship between politics and engineering may seem remote to most observers of the political scene, engineers are equipped with relevant skills and technological knowledge and have been significant as national policy makers and have served at many levels of government administration and political life.

The technological insights they bring to their political careers are very important. Unfortunately, such insights are still all too rare. Technology permeates our everyday life: in housing, communication, energy supply, health care, education, and entertainment. Its influence extends to religion, cultural traditions, and family life; it has also created a niche culture of its own. In terms of its most direct impact on policy making, technology presents a host of potential benefits, as well as consequences (often unforeseen) that may be detrimental to society, creating new issues and problems (for example, concerning the environment, energy sources, and security of information). Technology also plays an active and important role in generating wealth, as demonstrated by the vast information technology (IT) industry. Over the last five decades, American IT has grown to a $500 billion industry. Engineers are employed in great numbers in

this field, and in other fields outside of science and engineering.

Politics is one area where, in spite of a few illustrious role models, engineers rarely find their vocation. Yet the need for their expertise is increasingly apparent. Decision makers who may be technologically illiterate may not make the wisest choices; they may be confused or deliberately misled, and may be unable to see where the nation's best interests lie. The government should probably pay closer attention to technology and its impacts, but how well equipped is it to do so, given the human resources available to it? Engineering tools such as mathematical models and risk analysis are used to investigate policy options and to assist decision makers, for instance by calculating the costs and benefits of projects, but how widely are they understood? Ultimately, technology changes the focus of policies and the way issues are analyzed and presented; it may even determine national priorities and the way the government is run. Engineers would appear to have an important role to play in dealing with the large technical component of many issues.

How best, then, to draw on engineers' expertise? What is wrong with the present approach to tapping scientists' and engineers' particular strengths? Science policy advisers are currently available to investigate projects and provide feedback to the politicians who make the final decision. One problem arises when politicians do not achieve a reasonable grasp of the scientific point of view, or when their interests do not coincide with scientists and engineers' concerns. Legislators need to understand how science and technology are important to the country's future, but it may take congressional representatives who do not have an appropriate science background up to three terms (six years) to learn enough to engage at all knowledgeably in the relevant debates on the Hill. Yet, as of June 1996, half the House of Representatives had served three years or less, and most were attorneys with little knowledge of science or engineering. In addition, there are moves to limit politicians to only two terms (California already has such a provision). This has important effects in terms of budget allocation and government support. Decisions made by politicians who do not understand science and technology can affect the whole climate for growth. Tapping scientific resources gives experts in the field of study less control in decision making, and populist policies may be chosen over more effective ones.

Dealing with a nation's problems involves more than an understanding of technology. Engineers are equipped with problem-solving skills that can help in finding workable and practical solutions to national issues. Applying engineering economics to assess the risks and the potential effectiveness of policy shifts and new programs could help to channel limited resources more appropriately. Teamwork is increasingly needed in both the political and engineering workplace. Engineers often demonstrate leadership and management skills that are applicable to the running of a country.

So why do engineers remain "on tap" instead of "on top?" Most politicians receive a political science education, and hence are familiar with policies, laws, and government. The study of political science is tailored to provide politicians with sufficient knowledge to run a country, a background that engineers do not formally receive in school. Moreover, politics demands social

skills and involvement with a wide spectrum of society: "people skills." Engineers have a tendency to being somewhat socially detached. Professor Wenk, who served as a science advisor to Presidents Kennedy, Johnson, and Nixon, describes engineers in the following terms:

> ... many engineers approach the real world as though it were uninhabited ... [many] who are ... attracted to engineering are uncomfortable with the ambiguities of human behavior, its ... absence of predictable cause and effect, its lack of control, and with the demands for direct encounters with the public (Wenk 1997).

This is perhaps the most probable explanation why engineers shun politics. However, the situation can be improved by a range of measures. Introducing more humanities classes in engineering curricula, with a greater emphasis on social responsibility and commitment to social progress, is one approach. The academic discipline of political science will continue to play an important role in grooming our nation's future politicians; obviously we need people who are familiar with legislation and government. However, public policy makers with an engineering background have an extra edge. Engineering is the nation's second largest profession, but the number involved in politics is miserably small. What we need is an increase in the number of engineers willing to go into politics, to get involved in whatever capacity they can, to recognize this as an appropriate area of social responsibility and public service. It is they who can best represent the voice of science and technology that is rapidly altering the structure of our society.

(continued on next page)

Source

Based on work by Jaime Wong, Stanford University 1997.

References

American Society of Engineering Education. 1997, "What is the career outlook for engineers?" htpp://www.asee.org/precollege/html/questions_.htm#1, 18 Nov.

Lawless, T. 1996, "Educating Capitol Hill." *ASEE Prism* Oct. 12.

National Academy of Sciences. 1997, "Technology and the Nation's Future." http://www.nas.edu/21st/technology/technology.html, 20 Nov: 6.

Panitz, B. "Evolving Paths." http://www.asee.org/publications/html/evolving.htm, Nov. 1997.

Wenk, E., Jr. 1997, "Teaching Engineering as a Social Science." htpp://www.asee.org/publications/html/social.htm, 18 Nov.

most engineers receive leaves them unfamiliar with how power is exercised in society, and unaware of the political content of their activities. They may be uncomfortable in the hectic, cutthroat world of politics, but this can and should change.

STATE-GENERATED EMPLOYMENT OF ENGINEERS

According to the U.S. Bureau of the Census, (Public Employment, series GE, No. 1, annual), in 1992, the various levels of government in the United States were responsible for the employment of 18,745,000 Americans, representing about 16 percent of the workforce. Of these, 3,047,000 worked for the federal government, 4,595,000 worked for State governments, and 11,103,000 worked for local governments (counties, municipalities, school districts).

Because many engineers work for each of these levels of government, the state directly affects engineering activity in a wide variety of ways. Particular reference is made to legislation affecting engineering innovation in Chapter 6. The importance of national industry and technology policies is emphasized in Chapter 8.

According to the U.S. Bureau of Labor Statistics, in the United States slightly more than half of the nation's 1.2 million engineers work in non-manufacturing industries, primarily in engineering and architectural services, research and testing services, and business services such as construction design. Fewer than half of all engineers work in manufacturing industries, mostly in electrical and electronic equipment, industrial machinery, scientific instruments, aircraft and parts, motor vehicles, chemicals, guided missiles and space vehicles, fabricated metal products, and primary metals industries. Registration of engineers, and new national approaches to education and training, are discussed in Chapter 11.

Federal, state, and local governments employ approximately 14 percent of all engineers, around 180,000. The U.S. federal government deals with national and international activities, and encourages uniform legislation across state boundaries in areas where it does not have power. In 1994, the federal government employed 90,000 engineers, representing approximately half of all engineers employed by U.S. government agencies and 7 percent of all American engineers. Within the federal government, the Department of Defense is the largest employer of engineers, accounting for 60,000 engineers. NASA employed 12,000 engineers in 1991. Other federally employed engineers work in the Departments of Transportation, Agriculture, Interior, and Energy, as well as in technology research, the design and construction of highways and government buildings, and education. Engineers of all types work for the federal government.

State governments are responsible for the licensing of engineers in the United States. Most engineers in state and local government agencies work in highway and public works departments. Some engineers are self-employed consultants. Most of the engineers who work for the government are civil engineers, with nearly 40 percent of U.S. civil engineers working in federal, state, and local government agencies. A more detailed breakdown is given in Table 4.1.

TABLE 4.1

U.S. Engineer Wage and Salary Earners, 1994 (thousands)		
	Civilian Employment by Government Agencies*	Non-Government
Civil Engineers	70.8 (38.4%)	113.6
Electrical/Electronics Engineers	37.5 (10.8%)	311.1
Mechanical Engineers	13.1 (5.7%)	217.4
Total	180.6 (13.6%)	1,146.8

* Percentage of all engineering classification in U.S. Civilian Employment.
Source: U.S. Bureau of the Census, 1997 Statistical Abstract of the United States, p. 610.

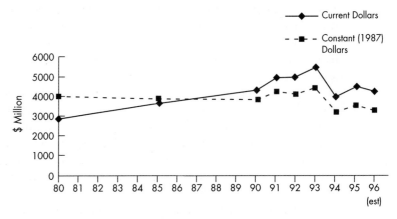

Figure 4.1 Federal Obligations for Engineering Research, 1980–1996

Direct employment is not the only way governments contribute to engineering activity. Figure 4.1 gives an indication of U.S. federal government funding for engineering research over the past two decades, carried out in industry, government laboratories, and higher education institutions. This funding, in turn, generates additional employment of engineers.

Two extraordinary institutions illustrate the pragmatism of the direct U.S. government use of engineering capability. Each has come under significant political attack, but has had broad enough support to continue in its clearly defined role. Neither has been accepted as an example for wider application. The first institution, the U.S. Army Corps of Engineers, had its origins in the American Revolution. It is discussed in Case Study 4.2. The second is the Tennessee Valley Authority (TVA), set up at the height of the Great Depression to improve conditions in a particularly depressed area. Its history and challenges are described in Case Study 4.6 at the end of the chapter.

CAPITALISM AND THE RELATIONSHIPS BETWEEN CLASSES

In societies today political power is derived from a variety of power bases and may be exercised in a number of different ways. One important power base derives from the ownership of capital: a countervailing force is that of organized labor. We are assuming an understanding of some basic concepts that are summarized briefly in the accompanying Key Concepts and Issues box (entitled Capitalism and Socialism).

The definition of capitalism given here is not intended to imply that workers have no personal property, or even that they do not own their own homes (as in fact, a majority of American workers do). Moreover, since this definition was written, in many countries there has been widespread privatization of formerly public assets. Many workers now own stock in companies; indeed a majority of American stockholders are non-degreed workers. In 1998 share ownership in the United States—including direct and indirect ownership through superannuation funds—involved 43 percent of the adult population. However, the question a political scientist would now pose is to what extent does this change the actual economic position of most workers? Do they still have to sell their labor-power to live?

We saw in the previous chapter that the poorest 40 percent of the U.S. population have no net wealth. The next 20 percent only own 6 percent of the nation's wealth. Therefore, broader share ownership seems unlikely to have made much difference to the lives of the poorest 60 percent. What affect has it had on the remaining 40 percent? For some, the change will add significantly to their assets and earnings, whereas for others it will make only a marginal difference. The situation is evolving, but it seems likely that the majority of the population will continue to depend on their employment (or their past employment, in the case of retired workers) for a livelihood.

There is controversy over whether capitalism involves exploitation of workers. Marxists argue that it does; that the price of workers' labor power, their capacity to work, is determined by the cost of reproducing and maintaining them, and is less than the value of the labor the worker does. They argue that the difference is the source of the capitalists' profit and hence of their political power. (There is an interesting qualification that may need to be added here. As the only remaining superpower, able to shape much world trade and other policy, the United States may be in a situation similar to that of Britain when it still had a mighty empire. It was suggested then that the living standards of the whole nation, including the working class, were inflated by the net inflow of resources from the rest of the world. This could now, to some degree, be the situation for the United States.)

Publicly owned enterprises have tended to be concentrated in areas that were less attractive to privately owned firms. Publicly owned enterprises can still be important for the functioning of the economy. They commonly include activities requiring a large capital base or having a long lead-time before they produce a return, and activities involved in the provision of infrastructure and essential services. Such activities are often engineering-intensive. There is no magic formula dictating which

THE U.S. ARMY CORPS OF ENGINEERS

The U.S. Army Corps of Engineers is an institution that is unique in American history. The Corps is of interest not merely because of its size (the largest engineering and construction organization in the world), but because it seems to present a significant anomaly: how could a country pervaded by a culture of "free enterprise" and civic autonomy sustain an arm of the military with substantial domestic responsibilities for activities of a social and economic character? The Corps does, in many respects, provide a "window" into the complex development of the United States itself.

The U.S. Army Corps of Engineers had its origins in 1775 as a vehicle for erecting the physical infrastructure of war. It was permanently organized in 1802 to provide such facilities on a continuing basis. West Point, also constructed at that time, provided the training for military engineers (it trained only engineers until after the Civil War)—de facto, the majority of engineers at the time. After 1824, the Corps' role gradually expanded into civil works projects. Internal waterways were the arteries of domestic commerce. Their spatial expansiveness and the multiplicity of their functions and associated problems gave them an intrinsically public character. Governments inevitably became involved, and gradually the federal government assumed wider responsibilities. As early as 1824, the Supreme Court declared that federal authority to regulate commerce extended to interstate navigation. The Corps was assigned responsibility for waterway management as a matter of course. The early role of French engineers in the United States was discussed in Chapter 1. Much early American engineering was "French" both in conceptual orientation and training (highly theoretical) and in its culture of civil works being overseen by state officials for the broader public good.

The massive Mississippi system presented the major problems and the major arena of activity (and of conflict over the Corps' role). Few nonengineers have ever confronted the scale and complexity of the technical problems that had to be solved to "tame" the Mississippi system for large-scale navigation and dense settlement of its hinterland.

The Corps' monopoly on advice and management of water resources was increasingly challenged in the latter half of the nineteenth century. The number of educational institutions and the number of formally trained engineers increased dramatically. Rapid industrial development fostered a growing private demand for engineers, a demand that was partly met by engineers with a British-style, "learning on the job" training. That was the background of James Eads, who staged a major battle with the Corps' Andrew Humphreys over enhancing the navigability of the Mississippi. This rising group of private engineers was not so much interested in reducing the government's role in such projects as in gaining the right to contract for public works themselves. Congress agreed with them. A steady stream of legislation required broader public involvement in water management, although the Corps continued to be the central instrument for this process. After 1910, the Corps was permitted to hire civilians to manage its growing list of projects, but carried over a strong elitist and meritocratic culture from its military origins.

Initially concerned primarily with survey work, the central formal responsibility of the Corps became the "navigability" of waterways. Inevitably, it acquired multiple inter-related responsibilities. With each step of the federal government's pragmatic acquisition of greater influence over the nation's economic life, the Corp's role followed. Important steps in this process were the post-Civil War Republican Party's assertive action on civil works programs to hasten industrial development; the influence of turn-of-the-century progressive movement politics that stimulated "multipurpose" waterway policies (flood control, hydropower, water supply, etc.); the post-World War I demand for electric generating capacity; and the 1930s propensity for public works projects for unemployment relief (see Case Study 4.6).

The threat of total war resulted in the Corps being handed the unenviable organizational responsibility for development of the atomic bomb. Inevitably, because of its size, experience, and military roots, the Corps also became an instrument of U.S. foreign economic policy, designing and overseeing construction projects overseas. It was involved in the building of the Panama Canal, described in Case Study 1.1.

In spite of this expansion, the relatively pedestrian matter of flood control came to be the Corps' dominant domestic responsibility. Expanded responsibility often came after a major disaster (as in 1927 and 1935) that stimulated a pragmatic Congressional response. By the 1940s, multipurpose dam construction was a major part of the Corps' activity. Despite conventional assumptions regarding the self-expanding nature of bureaucracy, the Corps has remained tightly project-oriented (ports, bridges, dams) rather than expanding into a program orientation (the control of toxic wastes).

Even in terms of flood control, the Corps long remained doggedly attached to levee construction and opposed to dams as the preferable form of control. This narrow focus has been the source of much criticism, but it has probably contributed to the Corps' survival. The Corps has survived the criticisms of regional planners, who claimed that river basins were an arbitrary basis for infrastructure development; of social planners who wanted to incorporate social indicators into development planning; of economists who wanted more rigorous cost-benefit evaluations of the planning process; and of political scientists who wanted to impose rigorous rational methods of administration. Continuity in focus is probably driven by a deeply entrenched "engineering" culture, but it has also been facilitated by a unworkably utopian dimension in the ambitions of its critics.

Post-World War II population growth meant continuing demand for Corps projects, and Congress used the Corps for developmental purposes. However, by the 1970s, Corps activity had been slowed—by procedural changes in project planning, by federal budgetary constraints that led to impasses over cost-sharing, and by community opposition to Corps priorities. There had always been community conflicts over land use, but environmental concerns now loomed large on the political agenda. Symbolic of the era was the 1969 National Environmental Policy Act, which required all federal agencies to consider the environmental implications of their activities. A representative conflict occurred in mid-Illinois. Long negotiations between the Corps and the town of Decatur to improve

(continued on next page)

the town's water supply led to a proposal for damming of the Sangamon River. However, the potential inundation of a woodland reserve area owned by the University of Illinois aroused organized resistance to the project. This resistance not only doomed the project, but also severely dented the credibility of the Corps' project evaluation techniques.

Nevertheless, the Corps appears to have responded positively to its critics, revising its decision making and consultative processes and its project evaluation techniques. Ironically, the Corps itself has become a key instrument in the furthering of environmental concerns. The River and Harbor Act of 1899 (the "Refuse Act") gave discretion to the Secretary of the Army (effectively the Corps) over the discharge or deposit of refuse. The Federal Water Pollution Control Act Amendments of 1972 reinforced this role; and several judicial decisions in the early 1970s effectively gave the Corps jurisdiction over the entire water mass of the United States, placing wetlands protection within the ambit of the Corps' responsibilities.

The use and control of water resources is an essential part of the nation's economic life and the Corps of Engineers has taken a central role in the development and management of those resources. In a country so profoundly imbued with an ethos of "free enterprise," it is salutary to discover that an arm of government has played such an important role in American development. That role has continued because governments and communities have continued to designate key resources as possessing a public or social character. The Corps as a public institution has retained legitimacy as a vehicle to develop and manage such resources.

Source

Case contributed by Evan Jones, Economics Department, University of Sydney.

References

Appelbaum, R. P., Chambliss, W. J. 1997, *Sociology*, 2nd Ed., Addison Wesley Longman, NY.

Mazmanian, D., Nienaber, J. 1979, *Can Organizations Change? Environmental Protection, Citizen Participation, and the Corps of Engineers*, Brookings, Washington, DC.

Moore, J. W., Moore, D. P. 1989, *The Army Corps of Engineers and the Evolution of Federal Flood Plain Management Policy*, University of Colorado, Boulder.

Reynolds, T. S. (ed.) 1991, *The Engineer in America*, University of Chicago Press, Chicago.

activities should be privately owned and which should be public. Although essentially a result of political choices, this division can have a major impact of the quality of life. Transport policy is one example of an area where large social savings could be made by policy formation that recognized all the costs. Case Study 2.1 illustrates the uncertainties in the calculations and some of the difficulties in ensuring that all the costs are taken into account.

Miliband (1973) suggests that the fundamental tension in a capitalist society is between two classes: on the one hand the class that owns and controls the workplaces (capital), on the other the class that has only its ability to work (its labor-power) to sell. The working class includes industrial workers, a group whose numbers are in decline. This class also includes increasing numbers in service industries, and a relatively smaller and decreasing group of agricultural wage earners. Miliband (1973, p. 16) asserts that the public political process is mainly about the confrontation between the two key classes, and is intended to officially sanction the terms of the relationship between them.

Intermediate groupings (between the capital-owning class and the working class) may include professionals, business people employing only a few workers, and the self-employed in small businesses. They are of great political and social importance, not least because they cushion the conflict between the two main groups (Miliband 1973, p. 17). White collar workers, although generally working class on the basis of Miliband's definition, commonly see themselves as middle class and reject working class values. However, economically they have much in common with the industrial working class. The attitudes and ideologies of members of the less affluent groupings do not necessarily correspond to their economic circumstances: this has tended to prevent them from forming effective political alliances.

▮ A U.S. Ruling Class?

The class structure of any society is important in the exercise of power. It has, for example, a significant effect on which views are heard and taken seriously. The levels of economic inequality and social stratification in the United States, discussed in the previous chapter, showed that the top 1 percent of U.S. households owned one third of the nation's wealth. Domhoff (1990) argues that a small group of the very wealthy and the managers who look after their interests can reasonably be characterized as a U.S. ruling class and he describes the ways in which they are interconnected. The processes of accommodation within the group and the way it establishes hegemony (sets the broad political agenda) have also been analyzed.

There have been important recent changes to the distribution of wealth within the most affluent groups in the United States. There is a rising group of "upstarts" (including the multibillionaire Bill Gates) who have become very rich as a result of their involvement in the global economy. Studies on the speed and extent of their accommodation with "old wealth" will be interesting.

What about the people who actually operate the state apparatus, the politicians, public servants (including roughly 14 percent of profession-

Capitalism and Socialism

Capitalism and socialism are the two major economic systems in the world.

Capitalism is characterized by private ownership of production and distribution systems, i.e., productive capital. Capitalist ideology generally emphasizes the concepts of individual initiative, competition, supply and demand, and the profit motive. A classic political science definition of a *pure* capitalist model, which today may need careful qualification, is:

> The economic system under which the ownership of the means of production is concentrated in the hands of a class consisting of only a minor section of society, and under which there is a property-less class for whom the sale of their labor-power, as a commodity, is the only source of livelihood (Penguin Dictionary of Politics, 1973).

The modern importance of capitalism dates from the seventeenth century, especially in Holland and Britain, when merchants, bankers, modernizing landowners, and, later, industrialists extended the influence of commercial imperatives on a predominantly subsistence-based society. By the early twentieth century capitalism had transformed many nations, making a significant social and economic impact and creating vast manufacturing and distribution systems. The pace of political change and democratization appeared to accelerate in association with capitalism.

However, not all citizens in capitalist countries benefited. There were large areas of the world either almost untouched by industrial capitalism, or where

al engineers), judges, the police, and the military? Where do they fit economically, and what are their social backgrounds and origins? How well do they understand the problems their fellow citizens face? With whom do their interests lie? Miliband (1973) argues that these people generally see their responsibilities as making the system work, and that government bodies, particularly at their higher levels, are still dominated by men (in the majority) drawn from, or co-opted into, the ruling class.

■ Other Models

The modern state is central to the operation of capitalism, even in countries most dedicated in theory to *laissez-faire* (noninterference by gov-

its effect was felt as colonization and exploitation by capitalist nations. To millions of such people socialism offered the only attractive alternative. Opposing capitalism, socialism is a political and economic approach based on collective or government ownership and management of the means of production and distribution of goods. A key concept is cooperation. Socialism developed in its present form in the nineteenth and early twentieth centuries. In large part, it was a reaction to the Industrial Revolution and the hardships caused by capitalism.

Historically, capitalism has been held responsible for various abuses, notably the exploitation of labor. The great British economist, John Maynard Keynes (1933), felt that capitalism "... is not a success. It is not intelligent. It is not beautiful. It is not just. It is not virtuous. And it doesn't deliver the goods." Broad controls on capitalist imperatives have included the passing of antitrust laws limiting the power of monopolies, legislation ensuring product safety, social reforms such as environmentalism, and the nationalization of certain basic industries in some countries.

Socialism has also been criticized as the source of a number of economic and social woes and, at the end of the twentieth century, is widely viewed as being in retreat. In spite of lasting legacies, including social welfare policies that are now taken for granted in a great many countries, its underlying principles and major social and economic goals are being explicitly rejected in many parts of the globe.

ernment in economic activity). In their discussion of a variety of different types of national approaches to the role of the state, Capling and Galligan (1992, p. 12) note that "liberal[ist] countries like the U.S.A. rely on market forces, and when those fail, adopt *ad hoc* measures to insulate threatened industries." This approach was illustrated by the rescue operation mounted in the 1980s by the U.S. government for the Chrysler Corporation, which included loan guarantees of billions of dollars.

As the scale of business within the military-industrial complex in the United States (Case Study 5.1) has dramatically illustrated, the state can also be the largest customer of the private sector. This creates important opportunities and responsibilities for government to encourage innovation in positive and beneficial directions, which should also increase the international competitiveness of the organizations involved. One dramatic example of this was the American government's Energy Star program for computers and their peripherals, which required them to have provision for a low energy consumption stand-by (sleep) state, reducing power consumption from around 200 watts to below 30 watts. This has become a standard feature and, over many millions of computers, the energy savings are significant.

Other countries have sustained a significant role for government. An effective alternative to the *laissez-faire* approach can be seen in countries such as Singapore and Japan, where the state takes a strong guiding role at the national level, shaping the directions of industrial development and promoting long-term structural change. Case Study 8.5 shows how Singapore has encouraged the development of the software industry there.

What sorts of constraints are there on the operation of the state apparatus? Checks and balances on power, and public accountability of those exercising it, are important in combating the tendency for power to corrupt. Just as significant for our purposes is to consider how the bounds of legitimate political action are defined in a capitalist liberal[ist] democratic state. What countervailing forces are there to the state apparatus, which includes the police, the judiciary, and ultimately the military? Some analysts (particularly Marxists) have seen the organized working class (i.e., labor unions) as the only alternative group with the necessary potential strength and coherence. This is discussed more fully below. Ironically, it was graphically illustrated by the success of Lech Walesa's Gdansk shipyard Solidarity movement in Poland *against* a communist government.

We have looked at power in the context of politics and the state, and in the previous chapter in both social and industrial contexts. An understanding of social power is particularly helpful in organizations that are in the process of rapid change, as, for example, with much of the manufacturing industry, which must continue to change rapidly if it is to survive.

In Chapter 11 we will look at professional groups and social power. For instance, to what extent (if any) are doctors in private practice in a position of exercising social power over the rest of the community? How do their circumstances differ from those of professional engineers? Engineers need to keep in mind the various types of power relationships in which they may be involved.

THE STATE TODAY

What changes are discernible in the institutions of the State today? How has the state changed between the end of World War II and the present? There have been major changes and challenges to the way the power of the state is exercised, include the following:

1. Deliberate identification of the state with the economy, so that the state puts itself increasingly at the service of private enterprise, providing not only education, training, and applied research, but also significant amounts of public funds for starting or modernizing selected industries. This has been important, for example, in maintaining the U.S. technical lead in the semiconductor industry.

2. The idea that smaller government is a good thing in itself. The associated enthusiasm for privatization has reflected the view that private rather than government ownership was a more efficient way of running enterprises. The aim of privatization has been to sell off those publicly owned activities that could be profitable. This has had a direct impact on engineers employed in a host of government agencies and enterprises. The 1993 UNDP Report (pp. 44–53) reviewed potential benefits and problems of privatization around the world, and offered guidelines for maximizing its social benefits.

3. The emergence of "postindustrial" nations. Britain, which led the whole world into the Industrial Revolution, may well be in the vanguard of countries emerging from it. A widely quoted estimate now puts the total employment and production of its traditional steel, shipbuilding, and coal industries at less than that of the nation's Indian restaurants! It is probably not coincidental that Britain is also seen as an attractive low-wage location for new manufacturing operations marketing into the European community.

■ Demilitarization and the Decline of Socialism

During the twentieth century a significant group of countries moved towards socialist relations of production, with substantial restrictions on the right to private ownership or control of capital. These countries were still caught up in the arms race. They had major problems with trade balances, national minorities, and self-perpetuating and selfish bureaucracies. The central planning process generally delivered on quantity, but failed to develop effective measures to encourage quality and to manage production effectively.

In the turbulence of recent rapid change it is easy to overlook the achievements of the socialist countries in education, health, and welfare. Despite their problems, they posed a serious challenge to the capitalist model of the state, particularly in times of recession. They also appeared to offer models for poorer developing countries in Latin America, Africa, and Asia. In terms of the affluence and standard of living of their citizens, most socialist countries were between the rich and poor countries of the West. Because their resources were more equally distributed, their citizens were generally better off than many of the poorest people in the most economically developed Western countries.

In the late 1980s the leaders and people of most of these countries lost confidence in their continued viability. At one point up to one million people at a time were involved in the clean up after the Chernobyl disaster, which may well have been the last nail in the financial coffin of the U.S.S.R. It was increasingly clear that attempts to open up part of the system would simply increase pressure on other parts. Mikhail Gorbachev tried desperately to open up the Soviet system and save its more positive aspects (*perestroika*). He achieved a more genuine thaw with the west (*glasnost*) but was not seen to be moving fast enough. Boris Yeltsin displaced him and has continued the transition to a market economy.

Despite the tragedy of Tiananmen Square, the Chinese leadership has generally kept its nerve, and has been able to draw on the skills and resources of local and overseas Chinese to open up the economy in a controlled and so far effective way. The ability of the Chinese government to undertake massive engineering projects (exemplified by Case Studies 4.3 and 4.4 and other case studies in this book) is indeed impressive, although there are serious questions as to the wisdom of committing such a high level of resources to a few giant undertakings, such as the Three Gorges Project discussed in Case Study 4.3. The long-term future of the few remaining socialist countries, including their influence on other states, is still unclear.

Western firms have made major efforts to get access to the large and expanding Chinese domestic market. China has capitalized on this interest by setting up joint ventures in key industries, applying considerable pressure for development of local technical capacity in these enterprises. This has included both transfer of technology and training of Chinese workers. Some of the political challenges such projects present for the Western engineers involved are illustrated in Case Study 4.4 on the Volkswagen plant in Shanghai. Associated technical difficulties are highlighted in Case Study 8.7.

For some four decades after the Second World War, real restraints were placed on aggressive behavior between states by the development of nuclear weapons and the competition between the Western and Eastern blocs. With the end of the Cold War in Europe, both sides moved towards reducing the burdens and risks associated with the more than 4,500 delivery vehicles and 24,000 nuclear warheads deployed worldwide. Under the Strategic Arms Reduction Treaties (START I and II), there was agreement to reduce the number of warheads to about 7,000 by 2003 (UNDP 1993, p. 9). This represented considerable progress.

Following the collapse of the Soviet Union, however, the picture has become less clear. There was a risk that many warheads would simply be moved to new sites, particularly in the North Pacific. There has also been serious danger of the creation of a number of new nuclear powers in a generally unregulated world power arena. Despite the ending of the Cold War, global stability has not improved. If anything, the trend has been to return to the anarchy of smaller, unstable, and/or hostile nation states, rather than two major power blocs, and more of them are now potentially nuclear-armed. There are still enough missiles pointing at cities to destroy most of the human race, and not enough has been done to eliminate the possibility of accidental nuclear war.

Reductions in military spending in the United States and the former U.S.S.R. created unemployment among engineers who were working on military projects that have been canceled. As indicated in Chapter 6, in combination with a worldwide recession, this highlighted the urgent need for industry policies that would generate useful work for engineers. As Markusen and Yudken (1992) point out, there is no shortage of important environmental and other problems for engineers to work on. However, mobilizing resources to address these problems requires national and international stability. In the next part of this chapter we look at challenges to the authority of the modern nation state.

THE THREE GORGES PROJECT IN CHINA

This case illustrates some of the issues and dilemmas for governments involved in very large-scale infrastructure projects. It demonstrates the capacity of the state to plan and carry out such megaprojects, which may address and raise both national and global concerns. It also highlights the importance of social and environmental issues and how these are being dealt with under a very different political system.

The Three Gorges Dam project is certainly a megaproject. The dam wall will be almost two kilometers long. It will form a reservoir 650 kilometers long, inundating 13 cities, 140 towns, 1,352 villages, and about 650 factories. It will raise the water level 175 meters, in the process submerging hundreds of ancient archaeological sites and flooding the base of the gorges that are among China's greatest tourist attractions. In most Western nations there would be doubts about maintaining the political will to undertake a project that would destroy national treasures and farming communities on such a large scale and disrupt the lives of millions of people.

The Chinese governing bodies that made the decision to proceed with the project were far from unanimous. However, in commending the project, the Chinese President, Jiang Zemin, claimed that the project "proves once again that socialism is superior in ... concentrating resources to do big jobs." The project may now be politically irreversible (Sly 1997).

Technically, the project is awe inspiring. When complete in 2009, the dam will be the world's largest. Its twenty-six generators are planned to provide 18,200 megawatts of clean and much-needed electrical energy, "as much as eighteen medium-sized coal or nuclear plants," for the Yangtze valley, where one third of China's 1.2 billion people live. The Chinese government is aware of increasing global concern with world carbon dioxide levels and the greenhouse effect. It sees a need to substantially increase the amount of electricity generated from hydroelectricity projects. It is also concerned at the extent to which industrial and commercial development is focused along the coast. It wants to encourage development in the interior, and this project will provide an immense boost to industrialization (Luk & Whitney 1993).

The site for the project is on the upper reaches of the Yangtze, the world's third largest river, where it runs through a 200-kilometer stretch of canyons with immense limestone cliffs. The three named gorges on the river—Xiling, Wu, and Qutang—are arguably the most scenic landscapes in all China. Locks around the dam will allow 10,000-ton vessels to sail 2,400 kilometers up the river from Shanghai on the coast to the city of Chongqing.

Dai Qing, a Chinese journalist who is one of the most outspoken critics of the dam, is quoted as calling it "the most environmentally and socially destructive project in the world ... The Three Gorges Project is not a hydroelectric engineering project. It is a political project exhibiting all the characteristics of a centrally controlled socialist economic system" (Sly 1997).

There is no question of the huge scale of the project. The flow rate where the Yangtze was blocked in November 1997 to initiate major works was almost 19,000 cubic meters (670,000 cubic feet) per second. This is more than twice the flow rate through Brazil's Itapu dam, the previous record holder. The estimated cost in late 1997 was $29 billion, and is likely to be exceeded.

The location is categorized as having a slight seismic risk. With 400 million people (50 percent more than the entire population of the United States) living in the valley below the dam, a breach could obviously be disastrous. Other concerns are that, as with the High Aswan Dam in Egypt, silt that used to be carried by the river and deposited on the fields downstream will settle to the bottom of the dam and that the lower reaches of the river will dry up and fisheries will be destroyed.

The Clinton administration strongly opposed the project and the U.S. Export-Import Bank refused to help finance the project because of environmental concerns. The bank's refusal effectively excluded U.S. companies, including Caterpillar, Rotek Industries, and Voith Hydro, from contracts on the project. By November 1997, a European consortium had a $740-million contract for initial work on the generators.

In most development projects, there are winners and losers. What is the situation here? The area to be flooded is one of outstanding natural beauty. Its 1.2 million peasants and fisher folk have already started to be displaced by the project. The environmental costs are likely to be significant.

The ecology of 3,200 kilometers of the river will be changed, probably irreversibly. The inland city of Chongqing already has a population of 15 million. If it continues to industrialize, and to pay insufficient attention to water treatment, as has historically been the case, water quality disasters could also be on a mega scale (Ryder 1990).

The availability of large quantities of clean electricity is certainly a bonus. What is not clear is the capital cost of that power. Cost accounting has not been a strength of socialist governments in the past. One major reason that China needs additional sources of electricity is that its heavy industrial electricity use is very inefficient by world standards.

The project has some resemblance to the U.S. Apollo moon project, or the Malaysian plan to build the tallest office building in the world, an understandable expression of national pride, the need to embody what President Jiang described as "the great industrious and dauntless spirit of China," and to display "the daring vision of the Chinese people for new horizons" (Sly 1997). Governments of all political persuasions still find that national pride is a spur to dramatic efforts and they frequently enlist engineers in such endeavors. China has had to put up with many insults to her pride since the colonial period, including the Opium Wars, the seizure of Hong Kong, the occupation of Shanghai by the Western powers, and the Japanese invasion in the period that led up to World War II. Such a grand achievement as this might be advantageous for both domestic and international reasons, providing the main aims could in fact achieved without serious unintended consequences.

Important questions are raised by this project. Is it likely to achieve its aims? How does one best balance the political aggrandizement of the nation and political concern for its citizens? In the circumstances, how adequate do you expect the environmental and other analyses to be? Is China getting value for the money? Might the project aims have been more effectively met by a series of less dramatic and (less costly) projects? In this situation, what would you have done if you were given the responsibility for deciding whether or not to proceed? What extra information might you have wanted? Would it all have been possible to get? What alternatives might you have considered?

References

Luk, S. H., and Whitney, J., Eds. 1993, *Megaproject: A Case Study of China's Three Gorges Project*, M.E. Sharpe Inc., NY.

Ryder, G., Ed. 1990, *Damming the Three Gorges: What Dam Builders Don't Want You to Know*, Probe International, Toronto.

Sly, L. 1997, "China begins work on world's largest dam," *San Francisco Examiner*, Nov. 9, 1997, p A-18.

VOLKSWAGEN SHANGHAI JOINT VENTURE

The automobile industry is one of the "pillar" industries that the Chinese government intends to be internationally competitive (discussed further in Chapter 8). Back in the 1980s, the industrial infrastructure for the local transportation industry was very poor, and even simple products were hard to procure locally. The government wanted to start modernizing the industry before China becomes a member of World Trade Organization. Because a national automobile industry was seen as an essential element of the country's industrial modernization, competition from imported cars was limited by huge import duties. An increasing gap was projected between supply and demand in the domestic market, where all models were completely outdated due to lack of access to international technology. In the twelve years from 1988 to 2000, production of passenger cars was planned to increase from 23,500 units to 1,800,000 per annum in order to substitute for car imports, and eventually to build an export industry. Moreover, the government was determined to develop a localized industry with a local supply base, rather than allowing foreign companies just to set up assembly operations. The emergence of a supply industry was seen as simply a matter of time.

Shanghai Volkswagen (SVW) became an integral part of Chinese Government planning. It is a Sino-German equity joint venture formed in 1984; the partners include the Chinese government-owned Shanghai Automotive Industrial Corporation (SAIC), China National Automotive Industrial Corporation, Bank of China's Shanghai Trust and Consultancy Company, together with Europe's largest car producer, the German Volkswagen AG (VW). The capital investment of about $40 million has been contributed on a 50:50 basis by the Chinese partners and VW.

Because of its lower costs, and its competitiveness with potential Japanese alternatives, VW decided to start with its sturdy Santana sedan (rather than the more modern Golf or Jetta models). As long as the domestic market was protected by huge import duties, the Santana would be profitable. The Chinese would have preferred to start with a relatively large and modern new vehicle developed specially for their market, and to rely on domestic suppliers. However, the German side ruled out that possibility, insisting that completely knocked down kits (CKDs) were the only possible way to start quality production of vehicles in China. The localization that the Chinese Government demanded was to be achieved in a gradual process that required a lot of assistance by SVW to local companies, as well as moving some Western part suppliers to China.

Government pressure to localize part procurement was high and was exerted in two main ways: a requirement to balance the foreign exchange accounts and a specified rate of increase of local content in the joint venture contract. The most difficult problems for VW were poor local supply, a shortage of foreign exchange for buying the CKDs, and the low technical capability of the Chinese partner. The Chinese workers had no experience in producing a modern car, and the quality of the manpower was low. Very few of the employees had significant technical training, so SVW had to set up extensive training programs.

On the other hand, marketing was simple, because influential institutions like the

Bank of China and the National Automotive Industrial Corporation were committed to the company, and the China Automotive Bureau, which completely controlled the car market in China, was involved!

Because of the low level of local manufacturing capability, in the first years there was little alternative but to depend heavily on imports of kits and components. However, in order to fulfill the requirement of balanced foreign exchange, something had to be exported to compensate for these imports. Because the Santana was not considered a competitive export product, VW agreed to export engines produced in the Shanghai joint venture to Germany. Later on, engines were also supplied to other VW joint ventures in China. By 1991 a local content ratio of 70 percent was achieved in the production of the Santana. In 1996 it reached 90 percent.

VW's Asia-Pacific Chairman has described the development of Shanghai Volkswagen as a four-stage process. In stage one, there was only assembly; in stage two, parts and components were partly produced locally; in stage three, the bulk of content was provided locally. In the fourth stage of development, in the late 1990s, design and development also started to be done locally.

Over the thirteen years from 1984 to 1997, SVW turned an old Chinese factory into China's largest car producer, with an annual output of around 200,000 units and a local content ratio of 80 to 90 percent. It had become a sizable engine exporter. Starting from a very low level, capability for manufacturing quality products had been developed within the local firm and its suppliers. A supply base of 190 local suppliers had been created, giving the backward linkages the Chinese Government wanted. The protected market made it an economically viable project, but the situation demanded a great deal

of adjustment by the foreign partner. VW's continued economic success depended entirely on the policies of a foreign government—even pricing decisions could not be taken autonomously by the firm.

Shanghai Volkswagen is a success story for the Chinese Government's industrial policy of using foreign capital and technology in joint ventures. The first steps in product innovation and local design have been taken. SVW is a good "corporate citizen," paying high wages and taxes, offering increasing employment, and with a policy of engagement in community issues.

Despite overall harmony, VW's Asia-Pacific head continues to stress the importance of the link between the product and the local supply industry, insisting that high local content levels cannot be reached with more modern models. The Chinese, on the other hand, are pressing for such models. SAIC recently signed an agreement with General Motors (GM) to produce more up-market Buick cars. The government is using this agreement to apply further pressure to VW. The project illustrates some of the technical and political challenges for engineers involved in such projects.

Source

Based on work by Dipl.-Kaufm. Philipp Riekert, Stanford University, 1997.

References

Business China 1996, "Part blessed, part stricken," November 11, pp. 1-3.
FT Asia Intelligence Wire 1997, "China Volkswagen to expand tech-centre," June 2, 1997.
Harvit, E. 1995, *China's Automobile Industry: Policies, Problems, and Prospects*, M.E. Sharpe, Armonk.
JAMA (Japan Automobile Manufacturers Association) 1995, "Vision is the key: An interview with Martin Posth," *JAMA Forum*, March 1995.

BEYOND THE NATION STATE

■ Supra-National Organizations

There are several important ways in which the authority of the nation state is being challenged. Formal supra-national authority is embodied, in different ways, in the United Nations Organization (UN), the North American Free Trade Agreement (NAFTA), the European Union (EU), and a number of others. For instance, the British Commonwealth of Nations is still one positive model for attempts to improve international understanding and cooperation.

The most important international organization is the UN, established in 1945 (replacing the ineffectual League of Nations, established in 1920). Formally a political body, the UN has inevitably acquired an interest in economic matters. The UN has acted as an umbrella organization, establishing subsidiaries (such as UNCTAD) or taking other institutions under its wing, such as the International Labor Office, the World Health Organization, and the International Court of Justice. The UN and its affiliates tend to be the bodies to which small and/or developing countries and their populations have appealed to balance their weakness in economic dealings.

The UN operates rather differently from the other organizations. As we will see in Chapter 9, over the last few years it has become clear that many problems of importance to individual states can only be effectively addressed at a global level. Working through its agencies, such as the United Nations Environment Program (UNEP), the UN has started to demonstrate some of the characteristics of a global government, coordinating action on issues that extend beyond national borders, such as acid rain, the greenhouse effect, and the depletion of the ozone layer (with some rapid and notable success on the last issue, see Case Study 9.1).

The NAFTA group of countries are signatories to a comprehensive trade agreement that simplifies virtually all aspects of doing business within North America, forming the world's largest free trade area, with 360 million people annually producing $6.2 trillion of goods and services and exporting and importing more than $1 trillion worth of goods. Within ten years of the January 1, 1994 implementation of NAFTA, all tariffs are to be eliminated on North American industrial products traded between Canada, Mexico, and the United States.

The EU is becoming a superstate, a federation of formerly sovereign states. It is taking over specified powers from its member nations and imposing international treaty requirements on them in areas such as human rights. In the EU, individual countries with proud traditions are expected to move to cede sovereignty in favor of a higher unity. By 1997, the EU had grown to include fifteen member states, representing 372 million people. The founding treaties had been revised three times: in 1987 (the Single Act), in 1992 (the Treaty on European Union), and in 1997 (the draft Treaty of Amsterdam). The ultimate goal of the EU is "an ever closer union among the peoples of Europe, in which decisions are taken as closely as possible to the citizen." The EU aims to promote economic and social progress that is balanced and sustainable, assert the European identity on the international scene, and introduce a European citizenship for the nationals of the member states. The EU has its own flag, its own anthem, and celebrates Europe Day on May 9. Its main objectives for the coming years are:

◆ implementation of the Treaty of Amsterdam (which contains new rights for citizens, freedom of movement, employment, strengthening the institutions);

◆ enlargement of the EU, to take in applicant countries from central and eastern Europe; and

◆ promoting of the Euro, a common unit of currency.

On a more informal level, the other main area of influence beyond the nation state is that wielded by the nongovernment organizations (NGOs). These are typically voluntary movements, including the international organization Greenpeace and aid organizations such as Oxfam.

In many areas of concern, the effectiveness of the UN and NGOs depends on their working together. NGOs generate ideas and contribute to the formation of a global political climate that encourages bureaucrats and politicians to work together. The UN/NGO interaction was clearly illustrated in the processes surrounding the Rio de Janeiro Conference on the Environment in June 1992. In parallel with the formal conference of 10,000 official delegates from 150 countries, a Global Forum of over 20,000 concerned citizens included representatives of a wide range of NGOs. Subsequent NGO pressures helped ensure that national governments gave serious consideration to the advisory as well as the potentially binding documents agreed to at the conference.

NGO communications facilities can assist with provision of information and the exchange of ideas. Computer networks offer significant possibilities for increasing the effectiveness of the democratic process (at least for the computer literate, which should include most professional engineers). They can improve the flow of information, for example by

Multinational Corporations

The development of capitalism has tended to extend the space over which goods and services are produced and sold. The simultaneous development of the nation state both facilitated and constrained the spatial extension of economic activity. Governments helped consolidate markets within a defined territory and protect markets outside their own territories. The first "multinational" corporations (MNCs) (such as the Dutch East India Company) were arms of the nation state. Since then, business corporations with a multinational focus have gradually increased in size and number. In particular, the scale of MNCs has escalated since the early 1980s; MNCs are linked to the globalization of financial markets and the privatization of government-owned infrastructure. Annual foreign direct investment by MNCs escalated from about $50 billion in 1982 to about $350 billion by the mid-1990s. The United Nations Conference on Trade and Development (UNCTAD) estimates that in 1996 one third of all world trade involved transfers *within* MNCs.

MNCs have become so powerful that governments' ability to control national economies has been significantly reduced. Some MNCs bargain aggressively with governments over the terms on which they will locate plants in a particular country: Ford Motor Co. and Intel, for example. Other MNCs are perennially moving plants in search of cheaper labor, especially in clothing, footwear, and electronics. Nevertheless, governments continue to exert substantial leverage over economic activity. Government policies, both defensive (such as tariffs) and proactive (such as the Singapore government's industry plans, Case Study 8.5), influence where MNCs will locate their activities. MNCs, to a large extent, still remain the creatures of a particular country (Ford is American, Ericsson is Swedish, Toyota is Japanese, etc.), and are given substantial assistance by that country's government. It is a rare MNC (though Rupert Murdoch's News Limited is an important example) that manages to manipulate many governments simultaneously.

The power of particular MNCs is partly a reflection of the power of countries in which they are based. Global competition involves competition between nation states as well as between companies. There is an international hierarchy of both states and business.

establishing "conferences," forums for the open discussion of issues. The computer networks provided by the Internet to university staff and students could play similar roles, allowing for continuing conferences on issues of concern. The essential contribution that NGOs add to information networks is their commitment to implementation of the conclusions arrived at through discussion. Although some people may feel they go overboard at times, one cannot dispute the effectiveness of organizations such as Greenpeace in putting issues such as pollution or whaling onto national and international agendas.

Historically the competition for territorial and economic advantage has led to both world war and global depression, and a number of multinational institutions have been created to attempt to stabilize international relations. A triad of significant international economic institutions were established after World War II: the International Monetary Fund, the World Bank, and the General Agreement on Tariffs and Trade (renamed since 1993 as the World Trade Organization). However, the structures and operations of these institutions have predominantly tended to serve the interests of the most powerful countries (especially the United States) and the most significant multinational corporations (MNCs). The tension between the pursuit of economic growth and the equitable distribution of its fruits is thus played out at the international as well as the national level.

Finally, any discussion of supra-national bodies would be incomplete without at least a brief mention of MNCs or "multinationals" (see Key Concepts and Issues box).

▮ Global Threats

There are many global threats that are beyond the control of individual nation states. These include international crime organizations; the Mafia and the Triads are by no means alone and there are now worldwide empires devoted to drug trafficking. International terrorism is still a real threat. However, the most serious concern must still be the prevalence of weapons of mass destruction (including nuclear, biological, and chemical weapons). If the survival of the planet, and particularly the human species, is placed at the top of a list of threats, then we must be concerned to ensure that such weapons are eliminated. The proliferation of nuclear weapons and their spread to new states, often poor Third World countries, is a critical global problem. The nuclear posturing of India and Pakistan in 1998 aroused worldwide concern. Although the threat of global nuclear war appeared briefly to have receded, it has by no means completely disappeared. Some commentators would rate biological weapons as an even greater threat due to the relative ease of manufacture, ease of concealment and use of this material, and the degree of potential damage that could be inflicted on the human race.

As discussed above, there is some encouraging progress, but the best strategies for achieving significant reduction in levels of international tension are still being debated. To some this entails an even more vigilant defense posture; to others it entails disarmament, or arms control, or the deliberate fostering of friendship between countries.

During the 1990s it has been argued that instabilities in regions as diverse as Bosnia, Iraq, Quebec, Ulster, North Africa, Rwanda, and Indonesia are leading to a resurgence of political divisions, essentially on ethnic and religious lines but often fueled by economic hardship or inequality. Kennedy (1993) suggested that the world was being moved by two currents. One, driven by technology and communications and trade, tended toward ever-greater economic integration. The second was the revived tendency toward ethnic separatism [and even tribalism?]

A new pluralism does appear to be emerging, with the nation state constituting merely one category in the spectrum from the tribe to the global agency. The "spaceship Earth" concept means that it is increasingly important in many situations to stop acting as separate nation states; international financial crises underscore this point.

■ New Political Challenges

At the national level, politics has increasingly come to mean the appropriation and short-term exercise of power. Effective action is too often curtailed by considerations of winning and maintaining office, not to mention the distractions of fighting internal party challenges and brush fires fueled by the media. Every political party is an alliance of more or less related interests. For a party to succeed it merely needs to become bigger than its opposition and stay bigger. One danger is that this exercise can descend to a cynical grab for the middle ground, in which ideals are left behind and the political process becomes an end in itself.

Great ideals with which we may all identify to some degree, including liberalism, conservation, and even broad social welfare, are in danger of being abandoned, but no others have taken their places. What cluster of issues could motivate a future populace to elect a government with a sweeping mandate for change? As South Korea and Indonesia demonstrated in the late 1990s, the combination of personal financial pain, a struggling national economy, limited democratic rights, and hurt national pride can still bring down governments, but the governments that replace them may not perform any better.

With the collapse of socialism in Eastern Europe and the often uncertain health of capitalism, the world faces something of a political vacuum. A vacuum reflects a dearth of good political thinking and a failure to identify clusters or alliances of interests. This has in the past been found to be a dangerous situation, an open invitation for demagogues and empire builders. So far, major new dangers do not appear to have emerged, but the potential for a nuclear build-up on the Indian Subcontinent raises growing anxiety. The next decade will be critical for the emergence of a stable new world order. The environment and its problems might conceivably provide a similar group of issues. The successes of Green parties and movements around the world have shown that they might constitute an even stronger international political movement.

Drucker (1993) argues that the world is presently nonsocialist and postcapitalist. He sees economic control of both the means and the tools of production as moving from a small ruling class to a rather larger group of knowledge workers. Drucker links this transfer of control to the increasing importance of the resources held by pension funds in national economies. This may be rather too simplistic, tending to confuse participation in the benefits of ownership with control of industrial and social outcomes, but the issues he raises are clearly significant.

IMPLICATIONS FOR ENGINEERS

In this chapter we have briefly considered the prevailing forms of political and social power in the Western world and in the United States in particular. In what the ways do they affect American engineers?

Drucker sees the postcapitalist phase as transitional, with knowledge workers (presumably including most professional engineers) assuming an increasingly powerful position. He argues that knowledge is the principal economic resource and that maximizing the productivity of that knowledge is the greatest economic challenge. The major social division emerging may be between knowledge workers and service workers, reflecting the increasing polarization of new jobs and the pay associated with them. Maintaining the dignity of service workers will be of major social importance. Is continuing conflict between management and knowledge workers as central to the new politics? Or are decreases in the power of the unions and the increases in cooperation between government agencies and business symptoms of a trend towards less conflict? There is some evidence of these tensions in advanced economies, and there have been some moves by knowledge workers to become unionized. In some countries, particularly the United States, this idea has not been acceptable, let alone attractive, to many graduates in technical areas.

With the dawn of the new millennium, we urgently need an increased global consciousness and awareness of our interdependence. Modern communications and transportation have greatly improved contact across the face of the earth, but we have yet to take full advantage of this. Technologies such as the modem, electronic mail, and the Internet have contributed tremendously to the development of fast, reliable worldwide communication networks. Along with the proliferation of efficient information storage and retrieval technologies, they are bringing about an age of global commerce and international corporations.

Whether we think globally or locally, professional codes of ethics explicitly require engineers to put community interests first. The brief discussion in this chapter of the formal political structures in the United States, with their explicit protection of private economic power, highlights potential tensions between community interests and the personal and career interests of professional engineers. Some of the issues involved, including the need for decision making and review processes that inspire public confidence, are discussed in Chapter 11. It is important that engineers take part in developing political mechanisms that involve the community effectively and constructively in setting design criteria and assessing proposed solutions. The following Case Study can be read as an object lesson for us all.

One obvious example of public consultation in local government is the use of precinct committees to advise councils on local traffic flow proposals. It is tempting and all-too-easy to dismiss these mechanisms as an unwieldy and time-consuming nuisance. They are! So is the democratic process generally; but it is also necessary for achieving solutions that acknowledge and address community aspirations.

Some of the implications for professional engineers include the following:

- ◆ Engineers need to be aware that the optimum outcomes for international business are not necessarily best for the quality of life of the majority of the people of the United States or the world. In their inputs to decision making, engineers have an obligation to take explicit account of public costs and benefits, as well as the profitability to the private company or government body involved.

- ◆ Engineers need to realize that maximizing short-term profit may be at the expense of long-term environmental or other costs. Sustainability must be an increasingly important criterion for new developments.

- ◆ The most technically challenging projects may not provide for the best use of resources or be most beneficial to society. They may address the wishes of the most economically powerful at the expense of the rest of the population. Challenging projects may divert scarce resources of capital and talent from fundamental needs, such as maintenance and improvement of basic infrastructure, to "gee-whiz" projects of little real long-term value. They may also turn out to be uneconomic and/or socially divisive. An assessment of the nuclear power industry, for example, has shown the damage that can be done to society and to the credibility of technology generally by failure to take account of these issues (Johnston 1986).

As engineers, proud of our profession, we need to recognize the importance of the political processes that help to ensure that the most socially desirable projects go ahead and that the costs and benefits of technology are equitably shared. Thoughtful involvement in the political process, both as professional engineers and as citizens, and at whatever level of government is appropriate is an important aspect of our responsibility to the society in which we live.

Discussion Questions

1. How well does the government structure in the United States match present needs? Discuss its advantages and disadvantages. Can you propose and justify a more appropriate structure?

2. In what ways are engineers in government agencies affected by changes in what are seen as the roles and responsibilities of the state?

3. The Secretary or Minister responsible for a major government agency announces that he or she is moving to change its organizational culture to make it more open in its decision making and more responsive to community demands. What implications does this have for the engineers it employs? How appropriate is this sort of political intervention in the working of such a body? What future developments would you anticipate?

4. In what ways might the opening up of the electricity supply industry affect you (a) personally and (b) professionally? On balance, will you be better or worse off?

5. How do you see the conflict being resolved between "the consumer's avaricious pursuit of his own needs" and the limited resources of our planet? How are the various sources of power discussed in this chapter likely to be involved?

6. Some of the discussion questions raised by the Three Gorges Project (and other megaprojects) include: What needs is the project aiming to address? How else could these needs be met? Is the project likely to achieve its stated aims? Could these aims have been met more effectively by a number of less dramatic (and less costly) projects? What would you do if you were given the responsibility for deciding whether or not to proceed? What alternatives would you consider?

THE POLITICS OF BUILDING
THE INTERSTATES

A thoughtful and carefully researched 1997 documentary film, *Divided Highways*, explores the genesis and the impacts of the U.S. Interstate Highway system. In the process, it reveals the politicization of the whole process, from government approval and funding, through planning, design and construction, to eventual outcomes, winners and losers. This case study is based largely on the film.

The importance of infrastructure issues, especially to engineers, was discussed in Chapter 2. The highway program was conceived as a "national system of interstate and defense highways"; the system was promoted with almost fanatical fervor in the 1950s at the height of the Cold War. Part of the rationale (which some saw as blatantly dishonest political propaganda) was the proposition that these highways would enable people to flee the cities in the event of a nuclear attack. In what now seems almost a blasphemous parody, the project was intended to create "one nation, indivisible, united by roads." The irony that such roads would also irreparably divide communities was not appreciated at the time.

Indeed, the uncritical acceptance by the American people of President Eisenhower's determination to build "the best system of highways in the world" may seem surprising to a more politically aware generation a half century later. Dazzled by the scale of these additions to their landscape and the convenience they appeared to offer, people clearly had no idea how much the highways would transform their countryside and neighborhoods. Their whole American way of life would change forever.

Yet the politicians of the day did not deliberately deceive their constituents; they were, as much as anything, responding to the prevailing mood of optimism, the belief that Americans could achieve even greater things in peace time than ever before in their history. The highway project did, to a very large extent, deliver—more than people had ever imagined!

Massive government spending (funded by citizen's taxes) was required. There were the initial construction costs and ongoing maintenance costs; in effect a huge government subsidy for auto transit and trucking over other forms of transport. The powerful highway lobby in Washington, D.C. had a large and influential constituency. It included not only the automobile manufacturers and the gas and oil corporations but also the construction industry, motels, and the entertainment industry. The real estate industry (housing developers were just one of the stakeholders in this whole exercise) recognized the potential for further suburbanization as middle class homeowners and firms of all descriptions moved out of the city centers. The only real opponents of the project were the railroads, but even they did not fully appreciate the threat. And by then they had insufficient "political clout" to be listened to.

Some stakeholders had no voice at all. As one commentator put it: "Nobody speaks for the non-driver, no-one speaks for the urban resident about to be displaced by some beltway ... they're non-players." The decision making process seems to have ignored them altogether.

Last but not least, the engineers were as enthralled as the politicians by the sheer scale of the project. When it got under way (the first section opened in 1956) the people building the highway system were the heroes who had won the critical production battles during World War II, designers and engineers who could do no wrong. That wonderful capacity to produce could now "live out a longer life." An engineer reflected later: "The engineering graduate in the 1950s wasn't thinking about the impact of the interstate road system on the country. I think we saw it as one mile at a time … we didn't sense we were going to change the fabric of the nation." They had no interest in, or even awareness of, the potential social, economic, and environmental impacts of building highways. At that time, government agencies didn't have to talk to anyone or consult with anyone. In what seems, in retrospect, a continuation of wartime "can do" attitudes, they just did what they considered necessary.

The highways were initially welcomed by most people. However, to minimize cost they went through the cheapest land. In rural areas this included swamps and waste land. The dilemma was where to locate them in cities? The planners were reluctant to cut a wide swathe through wealthy neighborhoods; the route was most likely to cut through urban slum neighborhoods where, coincidentally, a great many more people would be affected. A whole cluster of social issues including race, economic disadvantage, and urban decay surfaced in such situations. In the late 1950s and 1960s agriculture in the South was being mechanized, driving tenants and sharecroppers (Blacks, in particular) off the land into the centers of the cities, where housing was cheap. African-Americans were typically ineligible for the home loans that would have permitted them to move to the suburbs. At the same time, the vastly improved roads were encouraging industries to move out to the suburbs where they could expand, in the process reducing inner city job opportunities.

Overwhelmingly, it was the African-American communities who were affected: it seems in retrospect that the planners had an uncanny knack of picking out Black neighborhoods. In any number of Southern towns the Interstate went right down the Black/White divide, making integration that much more difficult ("you can't have the kids walking across the Interstate to school"). In St. Petersburg, Florida, ten African-American churches were picked up and moved for I-275; in St. Paul, the I-94 displaced one in seven of the city's Black residents; and "there were few Blacks living in Minnesota at the time but the road builders found them."

The effect of the elevated highway through the center of Overtown, a previously pleasant Black neighborhood in Miami, is perhaps one of the most dramatic examples. It was described in *Divided Highways* as "the political drive-by shooting of a town." Beginning in early 1960, thirty-day eviction notices began arriving. Over the next eight years, 30,000 of the 40,000 residents moved away. Overtown had been the vibrant hub of Black business, entertainment and civic life, but it became an empty shell. As one local explains it, "You have to understand that at that time … Blacks had no political power. [Overtown] was killed without any concern of those in political power at the time to save that community as part of the greater Miami community." Overtown's Black residents did not fight back. The destruction of the amenity of their neighborhoods may have seemed like just one more burden they had to suffer. People believed that their circumstances would improve with relocation to better housing.

However, within a few years residents in other cities did begin to protest about the destruction of their homes and neighborhoods.

(continued on next page)

The most striking example was in the mid-1960s in Boston, where an extension of the Massachusetts Turnpike into the city was planned. The proposal included the creation of an Inner Belt through the downtown area, complete with an eight-story interchange. The insensitivity with which the project was "sold" to the residents aroused strong resentment . Not only would their quality of living be adversely affected, so would their property values (always a touchy political issue). Local resistance manifested itself in unusual political coalitions: people coalesced around this particular issue in ways that cut across racial and ethnic lines. Eventually a large gathering of citizens came together in a protest rally literally across the street from the State House. As one analyst conjectured: "it may well be that, as the Governor looked out, he saw that these were very earnest, well-meant citizens with very serious concerns. And being as he was a smart politician, he decided he had better re-think [his support for the proposal] and listen to this constituency."

The Governor not only promised them that the Inner Belt would not be built in Boston, he pulled off a major political coup. He convinced Congress to break into the sacrosanct Highways Trust Funds and divert millions of dollars into an alternative mass transit project for the city of Boston. Funding for highway construction and for public transport had come from separate trust funds, with technically rather than socially focused guidelines for their spending. The question of how to move people most effective had not previously been asked. It had taken until 1973, seventeen years after the start of the program, for enough political muscle to be assembled to have Congress transfer the first money from highway funds to mass transit systems.

This was the first time that a proposed highway was scrapped and the money used for other modes of transportation. People now realized that their private misgivings about such projects could find a voice and that they would be listened to. It also was the start of attempts by the road builders, tentative at first, to consult with the people likely to be most affected by major construction projects. All the later interactions between citizens and governments grew out of what happened in the 1960s. Since then, the whole political landscape has changed.

The engineers and planners learned some important lessons. As they reminisce in the *Divided Highways* documentary, it is clear that the technocrats were totally baffled by the bitterness they aroused at the time. They had efficiently followed the design rules laid down for them, they had done everything "by the book." They could not understand why people did not admire them for the technically excellent work they were doing but rather complained about the social impact of their work. They felt they were being "dumped on by the public."

It is interesting to speculate on the extent to which their very narrow thinking, their lack of political sensitivity, might have been affected by having people of color, other minorities, and women among their ranks. Such groups might have been able and willing to bring different perspectives to the implementation of the highway construction program. The situation is improving, but these groups are still seriously underrepresented in the engineering profession.

References

Hott, L. R., Lewis, T. 1997, *Divided Highways: The Interstates and the Transformation of American Life*, Films for the Humanities and Sciences, Princeton, NJ. Video based on the book *Divided Highways: Building the Interstate Highways, Transforming American Life*, Tom Lewis 1997, Viking Press.

FURTHER READING

Feenberg, A. 1991, *A Critical Theory of Technology*, Oxford University Press, Oxford. A thoughtful and challenging perspective on the ways power relationships shape technology. Feenberg's web site includes a summary of his approach and a number of relevant papers, at: *http://www-rohan.sdsu.edu/faculty/feenberg/*

Poggi, G. 1978, *The Development of the Modern State: A Sociological Introduction*, Stanford University Press, Stanford CA. An important source for the treatment of the state in this chapter.

Many of the organizations mentioned in this chapter have their own web sites. So do some of their critics. Like all web material, it needs to be carefully and critically evaluated by the reader.

REFERENCES

ASEAN (Association of South East Asian Nations) 1998, *ASEAN Web*, web site at <http://www.asean.or.id/>

Capling, A., Galligan, B. 1992, *Beyond the Protective State: The Political Economy of Australia's Manufacturing Industry Policy*, Cambridge University Press, Melbourne.

Domhoff, G. W. (1990), *The Power Elite and the State: How Policy is Made in America*, A. de Gouyer, NY.

Drucker, P. F. 1993, *Post-Capitalist Society*, Harper Collins, NY. Probably the world's most perceptive management thinker, this is one of Drucker's most provocative contributions.

Gould, J., Kolb, W. L., 1964, *A Dictionary of the Social Sciences*, Tavistock, London.

Hudspith, R. C. 1996, The Art of Inquiry and Its Relevance to Preventive Engineering, in *Engineering Education in Transition: Proceedings of the 10th Canadian Conference on Engineering Education*, pp. 315-322, Queen's University, Kingston.

Johnston, S. F. 1986, Nuclear fuel, an incomplete cycle—Who pays for the power?, *Engineers Australia*, 16 May, pp. 43-8.

Kennedy, P. 1993, *Preparing for the Twenty-first Century*, Vintage Books, NY.

Keynes, J. M. 1933, *National Self-Sufficiency*, section 3; reprinted in Collected Works, vol. 9, 1982.

Markusen, A., Yudken, J. 1992, *Dismantling the Cold War Economy*, Basic Books, NY.

Michener, J. A., 1990, *Legacy*, Fawcett, NY.

Miliband, R. 1973, *The State in Capitalist Society*, Quartet, London.

United Nations Development Programme (UNDP) 1993, *Human Development Report 1993*, Oxford University Press, Oxford.

U.S. Bureau of the Census, 1997, *Statistical Abstract of the United States 1997*, Washington DC.

Wenk, E. 1986, Engagement of engineers in science policy, *Journal of Professional Issues in Engineering*, 112, (4), October, pp. 260-8.

Williams, R. 1980, *New Scientist*, 10 April, p. 66.

THE TENNESSEE VALLEY AUTHORITY

The Tennessee Valley Authority (TVA) is one of the best-known government-initiated engineering projects in the world. It inspired governments in many other countries to undertake similar large-scale, even visionary schemes to harness their natural resources. Along with such New Deal projects as the Hoover and Grand Coulee Dams, it captured the national imagination with its mythical dimensions, redirecting rivers and leveling mountains to build great dams and generate vast amounts of power. (The building of the railroads last century and the space program in the 1960s were similarly bold undertakings.) In the 1930s only national governments had the money and courage needed for such ventures; those were desperate times and a statesman of genius was at the nation's helm. The political will to overcome legal and bureaucratic obstacles and to overcome opposition from many different sources coincided with a particular set of circumstances and a climate of opinion that seems unlikely to return.

The TVA began during the Great Depression. It was a major part of President Franklin Delano Roosevelt's efforts to revive the national economy and provide employment. The Tennessee region was particularly depressed, with worn out soils and little industry. Roosevelt asked Congress to create "a corporation clothed with the power of government but possessed of the flexibility and initiative of a private enterprise." On May 18, 1933, Congress obliged. The TVA was to demonstrate an integrated approach to resource management. Each issue it faced—power production, navigation, flood control, malaria prevention, reforestation, and erosion control—was to be studied in its broadest context and weighed in relation to the others.

The TVA now supplies electricity over 27,000 kilometers (16,800 miles) of transmission lines to 7 million people, spread out over 200,000 square kilometers (80,000 square miles) in seven states. It is the nation's largest electric power supplier. Its generation facilities include 48 plants: 11 coal fired; 4 gas turbine; 3 nuclear; 29 hydroelectric; 1 pumped storage hydroelectric. In 1996, these plants supplied 134 billion kilowatt hours of electricity, and TVA revenues were $5.7 billion. The generation breakdown was 65 percent from fossil fuels, 24 percent nuclear, and 11 percent hydro, and TVA burned 40.7 million tons of coal.

Background

The TVA was conceived as a way of addressing two major concerns. The first was the management of the hydroelectric and related facilities created in northern Alabama during World War I. The second was to manage water flow along the Tennessee River to provide flood control and allow reliable navigation. In terms of stream flow the Tennessee is the fourth largest river in the United States and regularly caused serious flooding. At the same time a number of impassable shoals blocked navigation of the river's central channel.

Between 1918 and 1923 the federal government built two nitrate plants and the Wilson Dam on the rapids of the Tennessee River near Muscle Shoals, Alabama. They triggered major political debates over the river's potential as a source of hydroelectric power, the production of cheap fertilizer, and the possible industrialization of the Tennessee Valley. The ownership of the Muscle Shoals facilities came under the political spotlight: would they be publicly operated (as Nebraska Senator George Norris wished) or privately managed (as Republican presidents Coolidge and Hoover desired)? Political tensions arising from the different philosophies of the two political parties, and from the different philosophies of states rights versus federal intervention and control, were thus present from the beginning. After unsuccessfully fighting for congressional approval of a series of bills that provided for government operation (rejected by both Coolidge and Hoover, in 1928 and 1931, respectively), Norris continued his efforts. When Franklin D. Roosevelt became president in 1932 Roosevelt supported Norris, and in 1933 he introduced an expanded plan.

Roosevelt's plan resulted in the creation of the TVA, essentially, a government corporation in the hands of a board whose authority was subject to supervision only by the President and Congress. The three-man board was appointed by the President (with consent from the Senate) and one member was subsequently chosen as chairman. From the outset, the group was given notable independence in personnel and financial administration.

The TVA area included parts of seven states: Mississippi, Kentucky, Tennessee, Georgia, North Carolina, Alabama, and Virginia. The TVA Act of May 1933 identified its primary areas of action as:

- flood control
- navigation
- electric power generation
- reforestation
- proper use of marginal lands
- the socioeconomic well-being of the river basin residents.

For and Against

Like the nitrate plant developments, the TVA sparked immediate political controversy. Its constitutionality was questioned by those who viewed the TVA as government control encroaching on private interests. This attack was mainly from private power companies, and from conservatives generally opposed to government activity. However, in 1939, the U.S. Supreme Court confirmed the legality of the production and sale of hydroelectric power generated in association with measures for flood control and navigation improvement.

Associated with this debate was a government proposal to use TVA utility prices as a benchmark for assessing the fairness of private electricity rates. Needless to say, the private companies vehemently opposed this idea, claiming that it was an invalid comparison because the TVA was exempt from taxes, benefited from deficiency appropriations, and was able to borrow money at a lower interest rate than the private companies.

There was also pressure from Tennessee senators to grant them the authority to appoint TVA employees. Although their proposals were repeatedly rejected by Congress, these senators did get Congress to pass a bill requiring detailed accounting for TVA expenditures. The TVA was often accused of requesting unnecessary grants and subsidies while claiming questionable profits. It was even denounced as a government experiment in socialization and regimentation.

On the other hand, TVA supporters saw it as putting previously wasted resources to good use. Not only did it directly enrich a formerly depressed region, but it also helped development of the area by attracting businesses through provision of cheap electricity.

(continued on next page)

In agriculture, a (perhaps surprisingly) large part of the TVA policy was actually based on cooperation between the TVA and the states and localities. With the help of local authorities and citizen groups, the TVA provided significant assistance to local communities. As well as extensive rural electrification programs, it encouraged better farming methods, reforestation, and conservation techniques; it also promoted fertilizer research, phosphate production, and scientific agriculture on demonstration farms. In cooperation with the University of Kentucky College of Agriculture, it continues to operate a Rural Studies program.

The TVA's most widely known activity is in providing electric power. Here, the TVA largely ignores the states. It negotiates contracts directly with municipalities and cooperatives, providing them with electricity at wholesale prices. Cheap power is also supplied to large industries and various government agencies. The TVA sought to extend the use of electricity, developing an integrated power system with the lowest cost, highest per capita usage of electricity in the United States. Increasing demand during and after World War II led to the addition of coal-fired steam plants.

In an effort to enhance the socioeconomic well being of the area, the TVA set up programs to encourage sporting activities, including boating, fishing, and camping. During the 1960s, the TVA initiated the "Land Between the Lakes," a national outdoor recreation center between western Kentucky and Tennessee. Other auxiliary

TVA activity ranges from mosquito control and more advanced health services to funding libraries and educational programs.

Better communications are seen as attractive to business, and a recent focus has been improving communications in the Valley. With a number of partners including Bell South, the TVA has developed fiber-optic telecommunications infrastructure, and integrated services digital network (ISDN) access is now widely available throughout the region. With support from other sources, the TVA has built up an economic development loan fund that has provided tens of millions of dollars to assist the establishment of small businesses in the region.

Before 1959, construction projects for dams and power plants were primarily funded by appropriations, and TVA was often frustrated by political difficulties and delays in having these approved. Its leadership waged a vigorous campaign to overcome these constraints, and TVA is now allowed to borrow funds for power facilities projects, and is financed by a combination of borrowing, appropriations, and earnings. (Power operating expenses are paid by revenues generated from the same activity.)

During the past 65 years, the TVA has accumulated and used over $10 billion for power projects, and over $2 billion for other projects. Starting in 1961, the TVA was required to make a return on investment (in power facilities) to the U.S. Treasury. By the end of 1979 over $1.7 billion had been returned. Finally, the TVA makes returns to the states in its area as well. Each year, 5 percent of gross power revenues for the preceding year are divided among the states. The TVA also returns to the states and counties funds equaling the property taxes that prevailed before the TVA moved in.·

The Political Balance Sheet

Since its inception, the TVA's constitutionality has been questioned, its production and sales practices challenged, and its tax-exempt status criticized. Even as late as the 1980s, the TVA was in the national spotlight because of quality control and safety problems in several of its plants. Nuclear plants provided almost a quarter of its 1996 electrical output, but they have been expensive, and several units have been either sidelined or never commissioned.

Despite all the controversy, the TVA has achieved its major goals. The TVA manages almost 1,000 kilometers (over 600 miles) of the Tennessee River system. It opened up much of the system for navigation. Since its completion of a system of dams along the Tennessee River, there has been no major flood damage in the area. TVA projects have played a key role in the development of what was previously an economically depressed area. TVA power was important for wartime production of aluminum metal for airplane production, and TVA still operates hydro plants for TAPOCO (a subsidiary of Alcoa), and for the U.S. Corps of Engineers (on the Cumberland River system).

To what extent has the TVA been a political, social, economic, and engineering success? The political controversy associated with the TVA reflects deeply held views about proper roles for government and private enterprise, so no real agreement seems likely on that point. The TVA has undoubtedly helped to improve local economic, social, and recreational conditions. It has provided an effective flood control system, large amounts of cheap electricity, and over 1,000 kilometers of navigable waterways. Moreover, the TVA's efforts have improved local agricultural practices by their support for reforestation and soil conservation and air and water pollution control.

Although there is clear evidence of improved local conditions attributable directly to TVA activity, the TVA did not, unfortunately, stimulate strong *continued* economic growth in the area. TVA has also had costly problems with its nuclear power units. In the end it has not been accepted either as a model for regional development, or as a benchmark for other utility companies, as was once intended. So—what do you think about the TVA? Was it a good idea? Should the federal government be so involved at the state and local level? What do you see as the central issues?

Source

Based on work by Seema R. Patel, Stanford University, 1997.

References

Various Encyclopedias
TVA World Wide Web site, particularly the TVA Information Statement for the 1996 financial year, at: <www.tva.gov/finance/home.htm>.

5

A Perspective on Economics

The ideas of economists ... both when they are right and when they are wrong, are more powerful than is commonly understood. Indeed the world is ruled by little else. Practical men, who believe themselves to be quite exempt from any intellectual influence, are usually the slaves of some defunct economist.

— *John Maynard Keynes (1883–1946) British economist*

his chapter deals with "wealth" and provides a context for Chapters 6 to 9. We use "wealth" throughout this book to mean sustainable material improvement in the fullest sense, not simply the accumulation of assets.

Economic theories and policies have a powerful role in decision making at all levels and a major impact on the work of professional engineers; economics has enormous social and political prestige. Yet this prestige may not be wholly deserved. The discipline of economics claims to give insight into the operation of the economy and to offer a rational basis for its analysis. However, engineers need to be aware of the values underlying economic ideas and of some significant shortcomings in present theory and practice. They should also be able to assess the likely effects of economic policies on their own work.

Many key tenets of economics cannot be readily tested. When the proverbial bridge falls down, the structural principles applied in its design and construction are exposed to immediate public scrutiny. There is no such decisive testing of the ideas of economists; no laboratory experiment to isolate significant causal influences. Applied techniques of statistical inference have been developed since the 1930s (econometrics) with the ambition of putting economic theories on a firm quantitative footing. However, early optimism regarding such techniques has faded— they are neither robust nor discriminating enough for decisive testing of hypotheses. As we suggested in Chapter 2, this is not really surprising, given economists' use of very simple system representations for complex systems. Consequently, values and methodological preferences continue to influence reigning economic ideas significantly.

Economics and economists can have major effects, both for good and ill, on the functioning of the global and U.S. economies and on individual enterprises. Engineers need to understand economics well enough to

---

Key Concepts and Issues

Wealth

We might imagine that the concept of wealth is readily understood. However, it has proved extremely elusive. We use wealth in this book to mean sustainable material (and cultural) improvement, not simply the accumulation of assets.

Economics has generally failed to acknowledge the social character of the gratification derived from the possession of material objects. People who have done so, such as Thorstein Veblen (who invented the term "conspicuous consumption"), have been marginalized. Economists have also overemphasized the notion of wealth as individual property, neglecting the public dimension of individual well-being. This dimension was enshrined in the term "commonwealth," and is reflected today in what economists call "public goods" (defense, parks, etc.) and the welfare state.

In practice, much of economics is now associated with a common-sense version of wealth, based on an "adding-up" approach to all goods and services. The common denominator is money, an imperfect but functional measure. This approach to national wealth is embodied in a country's national accounts (see Table 5.1 later in this chapter), and is addressed by much public policy supporting economic "growth." Some of its limitations are raised in the box on Development in Chapter 8.

Conspicuous consumption and manipulation of the reported values of assets by "paper entrepreneurs" tend to obscure both the processes and the so-

challenge the prevailing assumptions, if necessary, and make positive inputs to policy debates.

THE ORIGINS OF ECONOMICS: BUILDING A NATIONAL ECONOMY ON STATE AND MARKET

Economics as a recognizable body of thought started in western Europe, roughly in the seventeenth century. There was a natural interest in issues that arose with the consolidation of the market economy and the reduced influence of the Church in social and political life. The state was an integral element in the evolving economic system and thus figured in the speculations of political economists.

cial importance of material wealth creation. Associated with this is a widespread concern that much of what is called wealth creation merely involves the transfer of value from one group or location to another.

The key social role engineers play is in creating value via new and improved products, processes, and systems. Innovation is central to creating value. The issues of wealth and wealth creation are therefore of fundamental practical and ethical importance to engineers. The engineering profession needs a clear understanding of these concepts, both for its own purposes and to encourage broader community recognition of the profession's role in wealth creation.

The engineering profession has focused on meeting needs and responding to opportunities, often with great success, and resulting in "wealth creation." The profession has not, however, generally addressed the broad social and political issues associated with both the types of jobs created and the distribution of the resulting wealth. The design parameters for engineering work have also generally neglected these questions. Historically, technological change has commonly increased economic inequality among those affected.

The creation of material wealth has taken a heavy toll on the earth's resources. Given suitable design criteria, it is possible for value-adding to actually reduce resource consumption: high-efficiency lighting is one example. We need to increase the total value of the world's assets in a resource-sustainable manner. We also need to distribute access to them more equitably.

The rise of the market economy went hand in hand with the rise of the nation state in western Europe. This period, from roughly the late sixteenth century to the early nineteenth century, goes under the name of mercantilism. Global businesses developed, such as the (British) East India Company and its continental counterparts, supported by nation states and commonly involving the expansion of trade though colonization.

The issue of stability of the (national) economy was of long-term significance to political economists. The political structure was increasingly based on the nation state, which was in perennial conflict with other states. Economic activity was local, regional, national, and international, with varying regulatory structures regarding production and exchange. How could it be ensured that economic activity contributed to the stability of the nation, and that political imperatives (especially war) contributed to economic development?

Two key issues are representative of the theoretical concerns of the political economists. First, how did the distribution of national wealth affect stability and growth? Second, should trade in goods and the associated movement in bullion be regulated in any way? Behind the first issue was the question of what the elite classes (and the monarch) did with their wealth. Should they be abstemious or profligate (in current language, savers or consumers)? And which elite class in particular deserved support as the best embodiment of the "national interest"? Much of the debate was bound up with whether it was the land-owning gentry or the merchants and manufacturers whose interests came closest to representing the broader national interest.

Despite the lack of definitive answers to these questions, by the early eighteenth century Britain had the most successful combination of attitudes and institutions. It is noteworthy that British economic supremacy was developed before Britain moved to embrace free trade; that move, in the nineteenth century, was driven more by strategic conceptions of British self-interest than by conviction of the intellectual merits of the free trade argument.

Adam Smith (1723–1790) wrote his famous book *Wealth of Nations* in the mid-eighteenth century. Smith marks the transition between mercantilist thinkers and what came to be called "classical economics." Smith drew substantially on the French "Physiocrat" school, which was probably the real precursor of economics as a distinguishable discipline. The Physiocrats conceived of economic activity systemically (albeit overemphasizing the role of agriculture). Smith borrowed this system's perspective. He criticized the inefficiencies of the mercantilist system, but he was not a defender of unrestrained free markets. Smith's values were complex, but they were all rooted in the eighteenth century. Smith criticized the British state as corrupt and inefficient (which it continued to be), but he also claimed that business competition would not be conducive to the public good unless constrained by a substantial dose of moral virtue inculcated into its participants.

Smith was neither a democrat nor egalitarian; he was in many respects old-fashioned. He believed in a plutocratic nondemocratic state, ruled by the wealthy few. He hoped the masses would benefit from the "trickle down" from economic development, but he believed that they should be socialized into accepting their subordinate lot. Moreover, he was influenced by the prevailing philosophy of natural law that if the natural world has a built-in tendency to order, so too should the social world. His argument for *laissez-faire* was thus in part dependent on wishful thinking and (influenced by the Physiocrats) on the persistence of a world that continued to be dominated by the "traditional" agrarian and mercantile (trading) sectors. Smith did not foresee the major changes in capitalism that would come when it took on an industrial form and became more large scale and urban, with subsequent pressures for state action.

We have devoted some attention to Adam Smith to clarify some misconceptions and make some useful generalizations about economic thought and its evolution. Although Smith's analysis was impressive for the breadth and closeness of his reasoning, as with economic thought before and since, it was not value free. Smith's arguments were deeply influenced by the social philosophy he had absorbed. In addition, Smith was writing in the mid-eighteenth century, and his work could hardly be expected to be of comparable relevance one or two hundred years later. However, Smith's analysis was exemplary in that it was concerned with the "big picture" and formed part of a seamless body of analysis of society; there was no rigid compartmentalization of intellectual disciplines, as exists today.

The character of economic analysis was to change soon after Smith, with the development of a body of economic thought that came to be called "classical economics." The classical economics school was essentially an Anglo-French phenomenon, linked to such thinkers as Adam Smith, David Ricardo (1772–1823), Thomas Malthus (1766–1834), Jean-Baptiste Say (1767–1832), Nassau Senior (1790–1864), and John Stuart Mill (1806–1873). An inevitable trade-off occurred: economic thought became more systematized, but it also gradually became more detached from a broader social analysis. The means to systemization were two-fold. The first was the assumption of "economic man": man was seen as autonomous, rational, calculating, selfish, and hedonistic. The second was a variety of assumptions regarding economic, social, and technical structures that completed or "closed" the theoretical system. The classical economists hoped to achieve both "rigor" and "relevance" simultaneously; the economic dimension was claimed to be growing in importance (so "economic man" was seen to be a reasonable assumption), and the delimiting structural assumptions were presumed to be empirically valid at the time, at least in Britain.

The emphasis on the "big picture" and on some relevant empirical data remained, but the imposition of conceptual order on a complex world was of dominant importance to the economists. This priority can be appreciated when one recognizes that the assumptions were seen as vehicles for introducing the analytical prowess of classical mechanics (from natural philosophy, as that branch of physics was then called) into the study of human behavior. The ambition of economists, which remains deeply entrenched to this day, was to establish social laws comparable in form to laws in the natural sciences. Scientific principles were taken to be introduced by the vehicle of "economic man": the individual as the basic unit of society (the atom), driven by the pursuit of maximum profit/happiness (a dynamic). The conservation principle is introduced by the assumptions that generate systemic closure.

Classical economics thus relied on notions that the economic system was both essentially simple in its organizing principles and ultimately stable. One example of the structural assumptions was that of the determination of the wages of the laboring classes. It was assumed that wages would be automatically regulated in some way; first by a "subsistence" theory of wages (population growth of the working classes would persistently drive wages to a subsistence minimum), and later by the assumption of a fixed fund of capital to pay wages, called the wages fund. However, the prospect of a "law" of wages was increasingly threatened by the possibility that wages could be influenced by government action, by popular pressure, or by trade unions. The possibility that a key price in the system (wages) might not be determined by some simple natural rule drove economists to despair.

The refusal to confront these complexities meant that classical economic theory did not keep pace with economic and political developments. The relevance of economic theory in Britain declined considerably over the course of the nineteenth century. The issue of systemic stability was gradually subsumed within a single assumption called Say's Law, that "supply creates its own demand." Under this assumption, gluts of commodities would be impossible. Production would be taken up by consumption; savings would be invested. The inadequacy of Say's Law was demonstrated by perennial economic crises after 1875, but it was not seriously attacked within the economics discipline until the 1930s.

If there was any continuous thread of economic debate from the early days, it concerned monetary subjects. Mercantilist writers were preoccupied with ensuring an adequate supply of money to finance trade and industry (within national boundaries). As credit instruments gradually came to supplement precious metals, the key issue concerned the creation and maintenance of a stable currency. In Britain, the Bank of England was established in the late seventeenth century to facilitate monetary control. By the nineteenth century, monetary debates hinged on such matters as should the Bank of England monopolize the issue of paper banknotes in order to keep their total linked strictly to gold backing? Should gold provide the hard basis for a means of exchange that would have global legitimacy, or should gold and silver?

Britain tied its domestic currency to gold. Because Britain was the most powerful economy, many countries deemed it necessary to copy Britain in order to become a major economic power. Ultimately, a "gold standard" was established by the 1870s. The linkage of key currencies to gold (with complementary support from silver, used for example by France and India) was used to regulate international trade and capital flows until World War I. The gold standard benefited Britain particularly because London was the center for financial mediation. Indeed, Britain's income surplus increasingly came less from manufactured exports and more from income from financial mediation and overseas investment.

Belatedly, the troubled state of the British economy after World War I and the trauma of the Great Depression of the 1930s forced a chink in the complacency of English economic theory. Through it emerged modern macroeconomics. However, before we move on to modern macroeconomics we need to examine the development of microeconomics, which has since come to dominate economic theory.

An Analytical Distinction: Macroeconomics and Microeconomics

Economics is now conventionally divided into two distinct levels of analysis, the macro and the micro. The broad level is captured by macroeconomics, which gives the overview, the "big picture" of national and global economies. Macroeconomics deals with aggregates such as national output and investment levels, unemployment and inflation, and the price and quantity of credit. It does not show how individual enterprises or sections of the economy operate or are evolving. The study of the economy at this more specific level is formally captured by microeconomics.

There is some utility in this distinction for purposes of conceptual specialization. However, the assumption of a tidy distinction between macroeconomics and microeconomics is a fiction, and has been detrimental to better understanding of the relationship between the two. Macroeconomic policy necessarily has an uneven impact at the micro level. This uneven impact is rarely acknowledged in the textbooks or among the policy makers, particularly in the English-speaking world. For example, high interest rates hit those sections of the community with atypical debt demands, in particular, farmers and home mortgagees. In countries such as Japan and Germany, policies that in the U.S. are presumed to have only an aggregate effect (e.g., monetary and tax policies) have been used deliberately for their discriminatory impact in order to favor some sections of the community over others.

Although the clear differentiation between macroeconomics and microeconomics is a fiction, the distinction will be used here to explain the preoccupations of English-language economists.

MICROECONOMICS

Microeconomics is the name given to the study of individuals as economic decision makers within the context of an economic system characterized simply as a "market economy."

Microeconomics has been developed since the last quarter of the nineteenth century, and is the central focus of the neoclassical school, which dates from around the 1870s. It is linked to such people as the Englishmen William Jevons (1835–1882) and Alfred Marshall (1842–1924), Europeans Leon Walras (1834–1910) and Carl Menger (1840–1921), and the American J. B. Clark (1847–1938). The contributions of this cluster of individuals were gradually simplified and cemented by disciples into a set of principles to be henceforth associated with a "school" of thought. To this school is attributed a rigorous theory of relative prices and of the price mechanism and, more generally, a comprehensive theory of the capitalist system as a universalized market economy.

Several assumptions are implicit, primarily that the market mechanism is all-pervasive and is automatic (hence impersonal) in its operations. In other words, the "market" is the system. This intellectual structure has its positive and negative dimensions; it has produced powerful and elegant theories of economic behavior; but, for both understanding and policy applications, the utility of such theories has been persistently questioned. Economic theorizing increasingly took place through intellectual exchange between professional academics within universities and became divorced from practice; separate disciplines were constructed with their own sets of rules. "Political economy" became "economics." Neoclassical economics replaced classical economics. Ironically, classical economics was losing respect because of its emphasis of rigor over relevance, yet the rise of neoclassical economics actually reinforced this emphasis. This highly formalized theory of the market economy has come to dominate all English-speaking economics. Whatever the intentions of its founders, the neoclassical approach has come to serve two functions, one analytical and the other ideological, which have combined to inhibit an adequate understanding of economic processes.

▪ The Analytical Role

The first key function of the neoclassical system is analytical. The neoclassical approach to human behavior involves reducing it to a problem of logic. Neoclassical theory is essentially logico-deductive theory, in that it sets up a logical framework from which it deduces conclusions. To achieve this, the new school took over the concept of "economic man" from classical economics (economic man is the force that drives the system to optimal outcomes). In addition, the structural/environmental assumptions of classical theory were dramatically simplified.

The "big picture" was replaced by the assumption of a universal market mechanism. The context usually assumed is that of the more specific "purely competitive" market in which all trade takes place between large numbers of economic agents. No single agent has any economic power but has to take the price structure as given. This is a favorite assumption because it satisfies analytical and ideological requirements simultaneously. It is mathematically tractable, because the economic agent's environment is given rather than contingent. It is also ideologically acceptable, because economic power is eliminated from the system. Logical deductions are then made about behavior, given the assumed context. The structure dictates optimum outcomes; attention then focuses on the character of these optimum ("equilibrium") outcomes.

Within this strict analytical framework, some useful propositions have been generated. Neoclassical theory has made a useful contribution to analyzing the character of consumer demand in a narrowly economic dimension. This has been done by dissecting the impact of price changes on commodity demand into a "substitution" effect and an "income" effect, and considering their different implications. These principles can be applied in analyzing effects of changes in the tax structure, for example, on consumer demand. (One product is a substitute for another if it enjoys increased demand when the other's price rises and the consumer's income is also raised just enough to compensate for the associated drop in living standards.)

The "price elasticity of demand" is a useful tool, a means of conceptualizing differential demand responses to price increases. For example, what will be the expected demand response for the basic family automobile if the price is increased by $1,000? If demand falls off dramatically, demand is said to be elastic; if demand is sustained, demand is said to be inelastic. The strength of product demand is important in corporate pricing strategy. There are regular price hikes on milk and bread, for example, because the industries judge (correctly) that the demand for them is reasonably price-inelastic. The strength of demand is also important for policy makers. High sales taxes can be levied on alcohol and tobacco products and petrol, because these products are price-inelastic.

In general, however, inventing subject matter that is analytically manageable has taken precedence over dealing with subject matter of historical and political importance. Classical and neoclassical economics both display an admiration for the presumed elegance of the natural sciences and of mathematical techniques. When the human subject matter of economics refused to bend to simple propositions, economists created an artificial world (or "system representation") to which their simple propositions could be applied. The rise of neoclassical economics was integrally linked to the application of the mathematical calculus (and later matrix algebra) to human behavior, a problem of optimization within a known and fixed environment. Since then the demands of mathematical analysis have dictated the directions of (micro)economic research. The prestige of neoclassical economics has been overwhelmingly linked to its sacrificing of substance for form.

■ The Ideological Role

The neoclassical school fulfills an ideological role in the defense of capitalism. However, it does this indirectly rather than directly. It bypasses the problems of explaining the "market" system in all its complexity and the ethical dilemmas put-forth in real-world economies. For example, contemporary market systems are dominated by very large private corporations. These corporations might produce "the goods," but they also possess substantial economic power and political influence. Neoclassical economics deals with this practical and ethical dilemma by ignoring it, treating the existence of market power as an anomaly.

An ideological ingredient was perhaps more deliberate when neoclassical economics was first developed. Economic thought had long sought a universal and basic substance called "value" that would provide the ultimate link between objects of production and exchange and that would be the ultimate objective source of observable prices. The classical school's concept of value was essentially linked to (ultimate) labor inputs. This approach left complex analytical problems unresolved (how to treat different degrees of capital intensity in production?) It also had unpleasant ideological connotations that were being appropriated by socialist thinkers: if the product was produced by labor, why didn't laborers deserve all the product?

Neoclassical economists resolved the theoretical and ideological problems of classical economics by pragmatically relocating the origins of "value." On the cost side, they generalized all inputs as "factors of production." Thus, landowners, creditors, machinery, labor, management, and so on, were all to be treated similarly. Each of these factor inputs was in principle rewarded justly, according to its respective contribution. This approach eliminated the unique contribution of labor in the material transformation that takes place in commodity production. It also generated a theory of income distribution (the marginal productivity theory) that has since then been used to justify real-world income distributions.

The neoclassicals also added and emphasized a new demand side where value was dependent on the (subjective) utility of the commodity for the consumer. This conceptual reorientation undoubtedly captured an important and neglected dimension in understanding market exchange. However, this dimension has tended to be inflated into an over-arching world-view. The neoclassicals considered the essential driving force of the economic system to be the subjective material interests of the individual consumer. Integral to this world-view are psychological assumptions of utilitarian and liberal philosophy: the autonomous, self-interested, rational individual as the center of society. Those who postulated these assumptions believed them to be innate in human nature and thus universally applicable to analyzing human action across the ages.

In general, the two pillars of microeconomics are its emphasis on analytical elegance and its indirect ideological support of the capitalist economic system. Any attempt at providing an adequate explanation of the structural workings of capitalist economies, or an ethical treatment of their successes and failings (as Adam Smith did in the eighteenth century), has been abandoned. In particular, the neoclassical analysis of production and of the business enterprise in general has been overly mechanical. The nature of production and of the business enterprise (the social institution in which production takes place) is of vital interest to engineers. Unfortunately, one has to recognize economists' contribution to these subjects as counterproductive. It is worth devoting some space to the analysis of the business enterprise.

Beyond Microeconomics

▮ The Complex Business Enterprise

In economics textbooks, the firm is treated essentially as a logic machine, processing inputs into outputs in the most efficient manner, in both technical and economic terms. This theoretical firm responds unerringly to the profit imperative, hounded constantly by the market mechanism to improve or be eliminated.

As a first approximation, this treatment may not be so bad. But it is strictly a first approximation. The single most important factor in any business environment is uncertainty. The typical business is subject not merely to the law of the jungle, but to a jungle without laws. Much business behavior can be understood as a pragmatic attempt to reduce uncertainty by evolving gradually with known inputs, known processes, known products, and known customers.

The strongest forms of this strategy involve tying up these processes legally through horizontal and vertical integration and enlisting government support against remaining uncertainties. Horizontal integration involves eliminating competition between firms in the same industry by a reduction in their numbers (takeovers, mergers) or by formal restrictive arrangements between legally separate firms (trusts, cartels). Vertical integration involves the bringing together, within one corporate entity, of supplying and customer companies, so that ultimately one spans the range of activities from production of raw materials through to marketing the resulting consumer products (an example is described in Case Study 7.4, p. 354). Originally nonrelated-party market transactions are thus internalized within larger business enterprises. Of course, all competition and uncertainty can never be eradicated. However, by the above means they can be significantly reduced for long periods. Moreover, "super" profits earned during periods of market control can be used as a cushion to carry companies through lean times.

Economists have studied practices of market domination within an applied subdiscipline called *industry economics* (Scherer and Ross 1990).

However, these studies have been inhibited by the unsatisfactory theory at the core of the discipline. There is inadequate treatment of the strategic processes and structural pressures businesses face.

The internal bureaucratic environment of business has also been neglected. The more that impersonal transactions are internalized within the firm, the more complex the business enterprise becomes. The evolution of business in the last one hundred years has been largely the evolution of private bureaucracies. Managerial strategy has been in large part the evolution of techniques to make unwieldy institutions workable (cf. Chandler 1977; Womack et al. 1990). Business enterprises, especially major corporations, have become complex social institutions. Engineers discover this when they go to work for them, especially if they find themselves as their CEOs!

Economists have belatedly acknowledged these changing structures (e.g., Williamson 1975). To understand business bureaucracies and their evolution, the contributions of economists have to be augmented by those of historians of technology, sociologists, and management theorists. Mainstream microeconomics takes no account of the role or impact of specific technologies and their linkages with specific labor skills, or of the more intangible culture of the company.

There is also the fundamentally important issue of technological change. What have been the key changes over the last few hundred years and how do we account for them? Smart business managers appreciate that one cannot rely indefinitely on forms of "negative" market control (takeovers, government support, etc.) More positive forms of market control must be pursued, especially through technological advancement. Karl Marx (1818–1883) saw the importance of this mechanism as a dynamic force in capitalist development and, more profoundly, general social change. Subsequently, the conservative Austrian (later Harvard Professor), Joseph Schumpeter (1883–1950), emphasized this dynamic edge to business strategy, calling it "the process of creative destruction." Historical explanation of technical change has mostly been left to a specialist group of economic historians such as Nathan Rosenberg (Rosenberg 1982). Many of these issues are dealt with in the next chapter.

One other aspect of the internal environment of the company deserves attention. It concerns what one might call the "culture" of the business firm. This vitally affects the efficiency and profitability of the firm, but has not been given proper attention by economists. Sociologists have given it some emphasis; it is also dealt with by management theorists.

The culture of the company affects even the goals themselves. Such culture is reflected in relationships between various corporate divisions (production, finance, accounts, sales) and is reflected in which group wields effective central command. Engineers will recognize these features in the perennial conflict between engineering values and finance/accounting ones. Strong national cultures can have an influence on the balance between such values in nationally based firms. For example, manufacturing companies are much more likely to have engineers or scientists as senior managers in Germany and the United States than in Britain or Australia.

▌ The Control of Labor: A Special Problem

Wage labor is a unique ingredient in the business enterprise, requiring particular attention. Neoclassical economists downplayed the special nature of wage labor, treating all inputs similarly as inanimate factors of production. By default, a separate discipline (industrial relations) has grown up to cater to the peculiar conditions faced in the employment of wage labor. In the past, engineering education has tended to neglect the human element in production, but at work engineers neglect it at their peril.

The structure of capitalist production is inherently nondemocratic, notwithstanding the myths of the democratic Western market economy. Unless this issue is faced squarely, the difficulties of the workplace can never be understood. Hierarchy and subordination are essential features of the capitalist workplace. These features were originally formally codified by appropriating "master and servant" legislation, and are now embodied in the concept of "managerial prerogatives." Early large commercial enterprises (the railways and post office) borrowed their labor practices from the armed forces. The textile industries borrowed their labor practices from the patriarchal family. From both precedents the culture of "obedience to authority" was appropriated. Because many wage and salaried workers find workplace hierarchy irksome, considerable energy must be devoted to coping with lack of cooperation. Modern trends towards fewer layers of management and more effective consultation with employees partly address these issues, as discussed in the section on Lean Production in Chapter 7 (p. 380).

Historically, a variety of mechanisms have been used to offset resistance. They highlight the underlying problems of ensuring "efficient" capitalist production. They also show how essential ingredients of capitalist production have evolved qualitatively over time.

Two main approaches have been either to repress dissent or to channel and absorb it: the "stick" and the "carrot." Each approach has involved various techniques. Repressive strategies have depended upon violence and draconian laws, both very important in American and Japanese history. Accommodative strategies have included monetary payoffs and welfare schemes. In general, repressive strategies have over time given way to accommodative strategies, linked to an historically evolving balance of power. This balance of forces differs across countries. Currently, the ascendant libertarian ("free market") political environment and localized depressed economic environments have generated a more repressive balance of forces. This is recognizable in attempts to impose rather than negotiate changes in working conditions, often accompanied by vigorous media campaigns against particular groups of workers. The U.S. presents a complex picture; its highly differentiated labor market means that accommodative and repressive measures are applied at the same time to different segments of the labor force.

Sophisticated means of dealing with labor have moved beyond the interpersonal to more structured and impersonal methods of labor control. In this category are forms of technical control, especially control by the rhythm of the machine. Henry Ford's assembly line was a major development in this direction. Another related (though distinct) development is the manipulation of tasks. One famous variant of this, discussed in Chapter 2, was called *scientific management* by its proponents and Taylorism by its critics. It aimed at separation of mental and manual labor, and minute division of tasks within the latter.

These approaches increased labor productivity, but historically they were also designed explicitly to minimize labor's control of the job. Engineers were central to this process of technical control and transformation of work. In the late nineteenth and early twentieth centuries, industry leaders found engineers useful precisely because of the latter's role in facilitating these processes.

With the change from mass production to flexible manufacturing, managerial hierarchies have to some extent been softened. In the 1990s, the greater importance of product development and product quality require the increasing involvement of production workers in defining tasks as well as carrying them out. However, the extent of these changes should not be overemphasized; typically only the most skilled workers are beneficiaries of these changes.

These comments on wage labor, the nature of work and of technical change, highlight the fact that the production process is not purely a technical issue. Labor discipline is a vital element in the choice of technique. The choice of technique is a social as well as a technical matter.

Engineers and Economists

Engineers and economists could profit from exposure to each other's expertise and perspective. They have long ago gone their separate ways, but there have been instances of useful interaction.

The Contribution of Engineers to Economic Theory

This is one area of interesting overlap (we are unaware of any reverse contribution of economists to engineering theory). Perhaps the most significant individual is A. W. H. Phillips (1914–1975). Phillips, an engineer by training, became an academic economist. From a macroeconomic orientation, he tried to apply his background in systems theory, viewing the economy as a closed loop control system. Efforts along these lines had not proved particularly successful because of the complexity of the subject matter. However, Phillips took a more empirical approach, studying the correlation between the unemployment rate and the rate of change of money wages. His work (that implied a trade-off between unemployment and inflation) became generalized as the "Phillip's Curve." It has served as a key (if crude) guide for policy makers in the post-1950 period.

The contribution of engineers to economic theory was more a phenomenon of the nineteenth century, before disciplines became established and segmented. At the École des Ponts et Chaussées in Paris, individuals such as Jules Dupuit (1804–1866) fulfilled their engineering function in the development of "bridges and roadways," but at the same time inquired into their economic benefits and appropriate pricing for their use. They were pioneers in what later came to be called "public economics."

Other engineers contributed to what came to be called microeconomics: consumer utility, monopoly pricing, transport pricing, and so on. Some of this was a by-product of the engineer's professional responsibilities—for example, Dionysius Lardner (1793–1859) was a railroad expert (Ekelund and Hooks 1973)—and some of it was the product of idle speculation. Certainly, the engineer's analytical equipment (deep familiarity with mathematical techniques, especially the relatively new field of the calculus) provided a natural basis for the economic theory that they produced. Their work was inadequately acknowledged by economists, partly because they were engineers, and partly because they were contributing to forging a "paradigm" (microeconomics) that developed a critical mass of adherents only in the twentieth century.

Project Evaluation

Another arena in which the interests of engineers and economists (and accountants) have overlapped is project evaluation, or the effective use of capital (sometimes called "engineering economy"). It has generated a journal, *The Engineering Economist*, and textbooks (cf. Gönen 1990). Its audience is the many engineers who become managers. Engineers played the key role in the development of engineering economy, addressing the need to estimate pricing for one-off jobs, to value long-life assets, and to allocate costs. What started as a pragmatic accompaniment to the producer's technical skills fed into the attempts of people (later to be called "accountants" and "economists") to formalize these processes. The techniques that were developed include cost accountancy in general and, more specifically, cost-benefit analysis and discounted cash flow analysis, used for estimating the present value of a flow of values across different time periods. More sophisticated tools, such as linear programming and critical path analysis, have been developed to facilitate efficient resource allocation for routine applications.

Economists have contributed to developing these techniques, which provide inputs to engineering decision making and project management. However, it would be naive to expect simple mechanical solutions to complex problems that involve a high degree of uncertainty and the need for judgment. A permanent tension exists between the hope for mechanical solutions and the refusal of the subject matter to be reduced to a problem of logic. Seeing economics from this engineering economy perspective, engineers may have tended to assume that it consists simply of a tool kit of objective techniques that can be used for problem solving and decision making. However, engineering economics is marginal to microeconomics, which is essentially an intellectual system with a theoretical and philosophical focus.

Production

Economics has consistently avoided getting its hands "dirty." It has been unable to confront the physical character of the production process. The economists' theory of production is highly abstract, essentially part of price theory; it relates the relative prices of inputs with the proportionality of their employment (e.g., if labor prices are lower, proportionally more labor will be used). Production itself is treated as a black box.

(continued on next page)

■ The Variety of Market Economies

The fact that social elements (the internal and external environments of the company) can differ across capitalist countries is of vital significance. Differences in business culture are merely one reflection of a broader phenomenon. Economic activity can never be detached from its social context, and is mediated by cultural conventions embodied in institutional structure (cf. Thurow 1993). "Market" economies are not all the same. What is ultimately significant is not the existence of a pervasive market mechanism, but the character of the institutions within which the market operates. The label of "the market" is used glibly in English-speaking countries without an appreciation of its complexity and variety. Economists pay attention to quantitative differences in economic variables, such as the savings propensity of the population or the share of government

Institutionalist economist and social critic, proposed a theory that society developed in permanent tension between two "instincts," the first creative, the second predatory. Veblen applied this theory to the American business world at the turn of the century. He lamented that industry (embodying the spirit of workmanship) was being dominated by business (profit on any terms). The traditional craftsman was dying off; who would replace him? Veblen's hopes were placed in the rising class of engineers, "[whose] class consciousness has taken the immediate form of a growing sense of waste and confusion in the management of industry." Some engineers, especially a radical element within the American Society of Mechanical Engineers, agreed with Veblen (Layton 1971).

In so far as they shared a world view, these engineers were captured by the idea of applying rational scientific and technical approaches not merely to industry, but to broader social and political processes; there were thought to be technocratic solutions to social problems. However, Veblen overstated the role of engineers in the production process itself. He also overstated the degree of dissent, and misunderstood the source of the rebel engineers' dissent, which was centered on the drive for professional status against the increasing reality of subordination as employees. The philosophical push led to the "Technocracy" movement after World War I, but petered out after a decade of self-serving publicity. More sober movements in the name of "efficiency" had a broader impact, not least through the person of Herbert Hoover. By this stage, Veblen had become embittered and new generations of economists and engineers developed along divergent paths.

spending in the gross national product. These are useful indicators of underlying national differences, but the qualitative aspects are equally important. The differences in institutional structures and cultures in different capitalist countries are typically ignored by economists, for whom a market economy is a market economy—period. The Key Concepts and Issues box below, The Social Fabric of the Market, illustrates these cultural differences.

Each country possesses a different culture and a different cluster of institutions within whose possibilities and constraints both the private sector and governments work. Key institutional blocks are located in the producing sectors, in finance, labor organizations, bureaucracy, science, and in education and training. The character of and relationship between these key blocks is critical. The structure of domestic family life is also important, reflecting the role of women in the broader economy. Any

The Social Fabric of the Market

The market mechanism is commonly seen as the key institution underpinning Western economic society. More extreme approaches present the "free market" as the means by which efficiency and material abundance are generated. These notions have also been presented as ways of alleviating the poverty of Third World countries, and for the reconstruction of ex-communist countries.

There is little understanding, especially by economists, that the market does not exist by itself. The "market" in all capitalist economies has always depended on a range of supportive social structures for its effective operation.

In Germany, in matters such as the inhibition of excess capacity or the pursuit of technical advance, policy development for key companies and industries is overseen by equity holding investment banks. In Japan, companies and industries are sheltered by membership in a *keiretsu* family of integrated firms, typically with bureaucratic mediation. For the Chinese community throughout South East Asia, common ethnicity and extended family networks often provide the social cement that constrains the endemic uncertainty of commerce (see Case Study 7.3). Lacking more benevolent means of coordination, Americans have underpinned their free market economy with a military industrial complex (see Case Study 5.1 at the end of this chapter). Australia has used tariffs, subsidies, and public infrastructure provision. The varying institutions developed to support the market thus give capitalism a qualitatively different character across national boundaries, with different implications both for international competitiveness and for social equity.

The most important supportive structure is the state itself, or rather the system of nation states. All capitalist countries have, without exception, developed with the help of state support. State responsibilities have actually increased pragmatically in the last one hundred years. This accretion of responsibilities is not an external "interference" in normal market processes, but constitutes an integral element of the political compromises necessary to build a workable market economy. In short, the pure market is a pure fiction.

particular country's economic performance is partly a product of the way these institutions work and interact. Wage labor is treated differently across countries; the structure and culture of management differ and the institutions of economic policy differ. In general, what we would call "the market" differs across countries because of the intangible elements of national culture. For example, Taiwan has staffed its key economic development bureaucracy with engineers rather than economists (Wade 1990, Ch. 7). German financial institutions are much more conservative than their counterparts in English-speaking countries, and have been less speculative in their recent dealings. This is not a matter of legislation or regulation but of internalized culture.

One of the great cultural divides in "market" economies is between English-speaking and non-English-speaking countries. The English-speaking world has been profoundly influenced by the philosophical culture of liberalism, a culture that has been built into the behavior and interaction of institutions. Liberalism is a complex and subtle philosophy, often simplistically equated with support for "economic freedom" or "free markets." This is simplistic because economic freedom is always constrained, and markets are never free. For our purposes, the essential legacy of liberalism is a detached, arm's-length relationship between institutions. This is reflected in our attitude towards "competition." Competitors are not supposed to have anything to do with each other, a mental set enforced by antitrust legislation. Yet in Europe and Asia, cartel arrangements between firms doing comparable business have been tolerated, even supported, as a means of stabilizing the business environment. Businesses in the English-speaking world have more typically attempted to stabilize their environment through mergers and takeovers. Firms originating in various countries will thus expect to compete globally according to different rules.

VARIETIES OF OPINION IN ECONOMICS

So far, we have presented economics as if it were an essentially monolithic discipline. Neoclassical economics does dominate the discipline, to the extent that the conceptual orientation of that school dictates the discipline's self-perceptions of what economics is about. However, there has always been dissent. Dissent exists and thrives because, as noted earlier, there is no decisive test of the truth content or successful policy application of an economic proposition.

Some of the dissent exists within and on the margins of the orthodox neoclassical school; other dissent has been sufficiently distinct and concentrated in its adherents as to constitute discernible alternative schools of thought (cf. Barber 1991). The most successful school of dissent has been Keynesianism (treated below). Another significant dissenting school is that of Institutionalism. As Institutionalism has been centered on the United States, it will be treated in a separate section on the American economic tradition. Other significant schools of dissent are those of Marxism, Feminism, and Green economics.

■ Marxism, Feminism, and Green Economics

Many social scientists and economists, finding elements of the workings of capitalist economies ethically distasteful, have sought alternative conceptual approaches to interpret the workings of capitalism. There have been numerous political and ethical critiques of capitalism, but many of those seeking an economic critique have depended on Marxist economics because it promised a rigorous theory of the capitalist economy and its development. Marxism is based on the work of Karl Marx (1818–1883) and Friedrich Engels (1820–1895). Marxist economics is actually an offshoot of respectable English classical economics, borrowing from it two key conceptual ingredients: the class basis of economic society and the ultimate origins of economic value in labor inputs. Marx's dominant ambition was to demonstrate that a system of selective property ownership and free labor exchange would be innately exploitative. He also sought to understand the imperatives of profit in a dynamic context, predicting technical transformation, the concentration of capital, perennial economic crises, and painful social dislocation for those involved in "backward" production techniques.

For ideological reasons, the direct influence of Marxism within the economics discipline has been limited. Although major economists in the eighteenth and nineteenth centuries analyzed their society in class terms, doing the same today would threaten an economist's professional reputation. This is ironic, in that the language of class appears daily in the newspapers, not least in the perennial conflict between employers (capital) and employees (labor) over wages and conditions. Interest in Marxism has surged whenever capitalist economies have experienced major economic difficulties, as during the 1930s and the 1970s. The decay of centrally planned Eastern bloc economies has led to a renewed condemnation of and lack of confidence in socialist thought and politics. Marxist economics, however, stands or falls on other criteria, as it offers an analysis of capitalist society rather than a blueprint for the future. Theoretically, Marxist economics is hampered by its conceptual dependence on the labor theory of value, which is not readily linked to the complexities of modern production processes. The practical usefulness of Marxist eco-

Two Giants of Economics

Karl Marx (1818–1883), has been described as the last of the classical economists. He is also a major figure in philosophy and sociology. Born in Trier in the western part of Germany, he studied at Bonn and Berlin, becoming increasingly active in the revolutionary struggles that climaxed in the Revolutions of 1848. His best-known work is the 1848 *Manifesto of the Communist Party*, written with his lifelong friend, collaborator, and financial supporter, Friedrich Engels. Marx's major contribution to economics, *Das Kapital* (*Capital*, in its English translation), is a detailed analysis of the operation of the capitalism of his day, an analysis that he saw as essential preparation for revolutionary change to a more humane social and economic system. Volume I was published in 1867, and Volumes II and III were completed by Engels after Marx's death and published in 1885 and 1894. Something of a recluse, Marx lived in poverty for much of his life, deliberately refusing to let capitalist society turn him into a money-making machine.

John Maynard Keynes (1883–1946) was born in the year Marx died. The father of modern macroeconomics, his efforts were aimed at reforming capitalism to ensure its survival. His great contribution to economic theory, *The General Theory of Employment, Interest and Money*, was published in 1936. Keynes recognized the social costs of speculation. In a comment still relevant today he noted: "When the capital development of a country becomes a by-product of the activities of a casino, the job is likely to be ill done." Keynes had none of Marx's reluctance to make money. His speculation on the foreign currency and commodity markets made him a personal fortune of more than £2 million, and increased one small fund at his Cambridge College from £30,000 to £380,000.

Source: Based on Wonnacott, P. and Wonnacott, R. 1979, *Economics*, McGraw-Hill, New York.

nomics has been limited by the absence of a statistical accounting framework that would give its conceptual categories a substantive reference point. However, Marxist economics has exerted a substantial indirect influence. In the third world context, it has tempered the confidence of first world development economists, leading to theories of "underdevelopment" and "dependency" (discussed in Chapter 8) in which first world influences are claimed to hinder rather than help third world development. Radical economists have also contributed to an institutionally and empirically based reinterpretation of the economic development of key first world countries and their role in the global economy. Radical economic analyses can be found in journals such as the U.S.-based *Review of Radical Political Economics* and the U.K.-based *Capital and Class.*

A feminist school of economics has arisen recently, generating substantial criticisms of economic orthodoxy. Feminist economists have looked systematically at a broad range of institutions and their economic linkages. They have been particularly concerned with the family and the household as important economic institutions, as well as their interdependence and changing relations with the business and government sectors. Feminists have provided insights on the evolution of the sphere of production in the different sites from household to factory/mine/office, and the evolving economic functions of the household (production giving way to consumption and reproduction of the labor force). They have also highlighted the extent to which all economic institutions (including business enterprises) are dominated by hierarchies based on gender (see Chapter 3), implying that even so-called "economic" institutions are driven by multiple objectives. Feminist perspectives are represented, for example, in the U.S.-based journal *Feminist Economics.*

Green economics has arisen because of what its proponents see as economists' neglect of the environmental dimension (see Chapter 9). The early classical economists had a subtle appreciation of the distinction between renewable and nonrenewable inputs. That distinction disappeared with the rise of the neoclassical school, which treats all inputs on a common conceptual basis. The resource and environmental degradation aspects of production were sidelined with the development of national accounting systems (treated below) that ignored the costs of resource depletion and pollution in production processes. Mainstream economists have fought back by claiming that green economists overstate the degree of environmental degradation and that environmental problems (e.g., depletion of fossil fuels) will ultimately be solved by the workings of the profit motive. Green economists' attempts to incorporate environmental concerns explicitly into the economic calculus are reflected in the U.S.-based journals *Journal of Ecological Economics* and *Capital, Nature, Socialism* (cf. Diesendorf and Hamilton 1997).

■ The Cumulative Causation School

There is a loose grouping of individuals who can be distinguished by their adherence to a theory of "circular and cumulative causation." This approach might find sympathy amongst those with an engineering training. People who have contributed to this approach include Adam Smith himself, the American Allyn Young (1876–1929), the Swede Gunnar Myrdal (1898–1990), and the Anglo-Hungarian Nicholas Kaldor (1908–1988).

The essence of the theory is that economic activity can only be understood as an organic and dynamic process; dynamic because the system is seen as continuing to evolve qualitatively, and organic because the components of the whole are seen as inevitably inter-related. The inter-relation is not merely in an impersonal detached sense, as implied in "general equilibrium" theory, a macroeconomic version of neoclassical economics. The key difference is the concern with the degree of symbiosis and coherence between the component parts. The current buzzword to describe this phenomenon is *synergy*. Do the parts come together productively or unproductively? With the economic system in inevitable movement, this school claims a cumulative impact, a feedback effect, so that systemic stability is less likely than either integrative growth or disintegration. Some members of the school emphasize that deliberate political action can help to generate institutional coherence that is conducive to integrative growth.

This group has had most recognition for its analysis of underdevelopment in the third world (see also Chapter 8). The approach can, however, be applied more broadly. Indeed, some proponents claim that the first world, the Western industrialized countries, demonstrates the plausibility of the theory, in that its emphasis on manufacturing industry, and the capital goods sector in particular, produces an integrating effect that gives the first world its essential character. Manufacturing industry generates cost economies that are not strictly a product of scale but of a continuing division of labor through process specialization. These cost economies are not a product of a specific company but of the sector as a whole.

Some proponents see the integrative process as one that involves more than technical factors and includes political and social ones as well. Integrative growth generates a more democratic political system, higher educational standards, and so on. In this view, integrative growth is not merely a process of material enhancement, but is also a civilizing influence. Gunnar Myrdal put such ideas into practice as an active social democratic member of parliament in Sweden, and as a key UN administrator in third world programs in the post-World War II period.

Some dissenting schools have obvious ideological dimensions that limit their influence. However, the cumulative causation approach does not fall into that category. This approach has probably been neglected for methodological reasons; its conceptual orientation cannot be captured by the deterministic analytical structures preferred by economists.

■ The American Economic Tradition

By the mid-nineteenth century, the most influential school of economics was classical economics, essentially an Anglo-French product. American economic thought was sparse, not least because the few tertiary education colleges emphasized the classics, theology, and moral philosophy (Barber 1988, Ch. 1). Classical economics was imported from England and France (southerners had greater tolerance of French thinkers), though typically with pragmatic modifications for a better local "fit." Regardless of the modifications, American would-be economists absorbed the ethical and political underpinnings of classical economics, namely that economic imperatives dictated *laissez-faire* on social issues (nothing could be done) and free trade in economic relations. These precepts were seen as consistent with both natural law and God's will. Representative of this tradition was Yale's William Graham Sumner (1840–1910).

An alternative indigenous tradition had developed at the birth of the Republic, mostly within politics. This approach saw national economic development, including a strong indigenous manufacturing sector that would support and reinforce a healthy agriculture, as the major objective. This was to be achieved by the assertive support of the national government (provision of infrastructure, use of the protective tariff, etc.). James Madison, John Adams, and Alexander Hamilton were key figures (though they fought over details). These individuals were essentially appropriating the "mercantilist" tradition that had favored Britain in its relation with its colonies, as a formula to be used for the development of American greatness: if it had worked for Britain, why not for the United States? (Williams 1966). This general approach was later reproduced and publicized by Henry Carey (1793–1879), whose views found an academic platform at the new Wharton School of the University of Pennsylvania. This approach also founded a political platform with successive Republican administrations after the Civil War (Barber 1988, Ch. 9).

After the Civil War, economic development proceeded apace. There was a substantial push for education of a practical nature. Although it only passed through Congress because the antagonistic southern states were absent, the 1862 Morrill Act embodied this spirit, generating federal funds for institutions offering training in "agriculture, the mechanical arts and the applied sciences." During this period the training of engineers expanded dramatically. Rapid development, however, brought economic (crises), physical (land use), and social (class conflict) problems. In spite of a influential sentiment of *laissez-faire* and self-help, government activity expanded substantially towards the end of the century (as it did in all countries at this time). Often the fight was not over whether to have government involved in dealing with a problem but over what level of government would be responsible.

In this context, a new grouping of "institutionalist" economists arose to try to interpret these issues and to offer solutions. There was a commitment to the notion that economic activity was part of a broader social system that should be not only driven by "free enterprise" but also channeled by social action for the broader public good. There was also a novel commitment to empirical study and the systematic collection of information as a basis for informed analysis and purposive action. An academic tradition in Germany (the Historical School), where many Americans had studied, provided precedents for these views. Institutionalism became the dominant school of economics in the United States, effectively, until World War II; it still flourishes, albeit on the margins of the discipline.

Institutionalists vary in their emphasis and approach (their preoccupations are reflected in their U.S.-based journal, *Journal of Economic Issues*). Unlike the neoclassical school, there is no tight body of theory to which individual members might subscribe. One of the best-known modern members is J. K. Galbraith (b. 1908) (cf. Galbraith 1975). Analytically, institutionalists pay attention to economic structure, and they have paid particular attention to the evolution of the large corporation and its economic and ethical implications. There are individual economists in other countries with similar conceptual and philosophical orientations.

Ideologically, most institutionalists are "middle of the road." They are anxious to steer a political middle course between a *laissez-faire*, dog-eat-dog world on the right and socialism on the left. Their general aim has been to reform or regulate social and economic institutions within capitalism in order to make it a system that is not merely productive but also ethically just (and hence politically viable).

A representative of the American institutionalist tradition is John Commons (1862–1945). Commons emphasized the subtle cultural conventions that bound together seemingly individualized economic activity, and was active as an adviser and drafter of reforming legislation. He was a major contributor to the Wisconsin progressive movement that was exemplary in building supportive structures for "small" farmers and business.

Another important figure is Wesley Mitchell (1874–1948). Perhaps the most significant empirical economist in American history, Mitchell was a pioneer in attempting to understand and control the business cycle. He was instrumental in the establishment of the National Bureau of Economic Research.

For institutionalists, the "positive" forces are those contributing to the "productive" side of economic activity: tradespeople, engineers, or scientists. The negative forces inhibiting the activities of the productive forces typically include financial speculators or "paper entrepreneurs." This dichotomy has been a persistent theme in reformist economics. A notable representative of this tradition was the Norwegian-American Thorstein Veblen (1857–1929), writing in the age of the "robber barons" of capitalism (cf. Josephson 1962/1934).

Institutionalists have generally looked favorably on technology, seeing technological "advance" as the underlying basis for expansion in material affluence and beyond that, of culture. Engineers could find some kindred spirits within this institutionalist tradition. Nevertheless, the institutionalist tradition has probably been too accepting of the vision of a technological imperative as the driving force of civilization, assuming that new technology can readily be harnessed to serve the public good.

Modern Macroeconomics

Macroeconomists have exhibited a greater degree of flexibility than microeconomists. Macroeconomics has always been tied pragmatically to policy considerations. Moreover, macroeconomics is not tied as closely to logico-deductive analysis as is microeconomics and therefore suffers fewer analytical constraints. The Great Depression of the 1930s was responsible for renewed emphasis on macroeconomic theorizing.

The depression brought many responses: pragmatic (the United States), repressive (Germany), or enlightened and far-sighted (Sweden). Britain's major contribution was not in the realm of useful policies but of ideas. The Englishman John Maynard Keynes (see Key Concepts and Issues box on Keynesianism) constructed a theory that challenged the dominance of neoclassical microeconomics (Stewart 1986). Keynesianism was couched at the level of aggregates (especially those of the national economy). Keynes modified the narrow macroeconomic inheritance based on Say's Law (which effectively implied that long-term, large-scale unemployment was impossible).

Sympathizers found a significant place for Keynesian ideas in post-1945 textbooks and in the minds of policy makers. The theory was neither coherent nor revolutionary (although many Americans thought otherwise). However, it was compatible with ideological and political conditions in English-speaking countries while not being too radical in its demands on governments.

As with microeconomics, outsiders often assume that macroeconomics offers a toolkit of ideas and techniques that transcend ideological and intellectual differences and are available to be generally applied. There are in fact some handy techniques, the use of which has become widely acceptable, but these are intellectual artifacts with built-in biases. Moreover, all systems of macroeconomic theory are simplified, stylized representations of a complex reality. As the complex institutional character of capitalism changes, theories tend to ossify and outlast their usefulness. One always needs to take economic theories with a grain of salt. Keynes, for example, built his own system on the basis of the specifics of British capitalism in the 1920s. Even so, his theoretical explanations of Britain's problems were overly simplistic and downplayed the long-term qualitative transformation of the British economy. Keynes' ideas were simplified even further in the textbook versions that became dominant after the 1950s. Keynesianism is best known as a form of liberal reformist economics (comparable to institutionalism), concerned to tinker with the system in order to save it from both *laissez-faire* proponents and socialists.

Since the 1960s approaches that have arisen in dissent from Keynesianism have been more consistent with *laissez-faire* ideology and with neoclassical microeconomics. Yet even these approaches have been based on a "systemic" approach that owes much to Keynesian macroeconomics. They each include an intellectual system with a semblance of coherence, a more concrete accounting and statistical edifice to make it operational, and an institutional establishment to carry out economic policy.

KEYNESIAN THEORY

The intellectual essence of Keynesianism is that economic instability, observed since the rise of commerce, is not just a monetary phenomenon. The adjustment process itself operates predominantly through quantities ("real" variables) rather than through prices ("monetary" variables). The level of (national) output is the key mechanism of adjustment of the system, in turn conceived of as the equilibration of the flows out of the system (savings) and the flows into the system (real investment). The increasing significance placed on systemic adjustment of quantitative changes is linked to historical institutional changes in the nature of capitalism. Some important changes are the increasing significance of industry over agriculture and the growth of large corporations and unions, reducing wage and price volatility.

The two main elements of simple textbook Keynesianism will be briefly discussed here: the theory (see Key Concepts and Issues box on Keynesianism) and the implied policy tools.

Keynesianism

For Keynes, effective aggregate demand (total expenditure) is the driving force of the system. The components of aggregate demand are consumption expenditure, investment expenditure, government expenditure, and net expenditure from overseas (exports less imports).

Total expenditure generates aggregate national output. Recent levels of these descriptors of the U.S. economy are given in Table 5.1 (see following section on GNP and GDP).

What determines the components of total expenditure? In the simplest versions:

♦ Consumption is dependent on the level of national income. The level of consumption is therefore derived endogenously (within the economic system). Nevertheless, consumption is quantitatively the most significant component of total expenditure (see Table 5.1).

♦ Investment expenditure is determined exogenously (outside the system). Investment expenditure is dependent on expectations of future profitability, which are fickle. Investment is thus the most volatile component of total expenditure and the key source of instability in capitalism.

♦ Government expenditure is determined exogenously, by political considerations.

♦ Imports are dependent on national income.

The Keynesian emphasis is upon policy tools that are presumed to work only at the aggregate level, and are directed to stabilizing the aggregate level of economic activity. There are four major sets of instruments associated with fiscal policy, monetary policy, wages policy, and exchange rate policy. Because of various institutional limitations regarding the last two, macroeconomic policy is essentially delivered through fiscal and monetary instruments.

Fiscal policy is mostly effected through government expenditures and revenues. The key instrument is the annual budget. Budgets can be either generous or austere, which have, respectively, expansionary or restrictive impacts on the national economy. Budgets are also the means by which the allocative and redistributional priorities of the national government

◆ Exports are considered exogenous. In more complicated models, exports are considered dependent on relative prices in the U.S. compared with those in the rest of the world, in turn dependent on the exchange rate, relative labor costs, and inflation rates.

The discretionary character of government spending is crucial, as it can be used to stabilize the system. This is of fundamental importance in recessions and depressions because the profit motive that drives the private sector inhibits any automatic self-regenerating mechanism from operating speedily. The problems of uncontrolled capitalism are multilayered. Investment expenditure is volatile and over a number of successive accounting periods any disturbance to the system is amplified by multiple (though declining) adjustments through a feedback mechanism called the "multiplier" effect. The effects of variations in investment expenditure on national income and living standards are even more pronounced.

In the Keynesian model, money wages are inflexible downwards. Keynes still believed that wage levels were determined ultimately by objective market forces. The process of adjustment was not, however, through the medium of money wages (now politically difficult), but through adjustments in the level of output and employment—a disruptive process. Keynes also understood the average level of money wages as underpinning the cost structure, which also made the general price level *sticky* (resistant to change), especially downwards.

---

are established and reconciled. This distributive feature tends to be downplayed by economists.

Monetary policy is effected through influencing the quantity and price of credit available from financial institutions. A range of instruments is used, the balance of which has changed over time. Purchases and sales of government securities (open market operations) have been used to influence the general level of liquidity. Financial deregulation since the early 1980s has severely curtailed the range of instruments at the disposal of the authorities. In general, the dominant monetary instrument in the United States is the Federal Reserve's discount rate. The intention is that movements in the discount rate feed through to other rates and ultimately influence the general cost of credit.

Wages policy cannot be derived directly from Keynesian theory. In practice, a policy rule of thumb has been established: keeping average wages growth below the growth in average productivity. However, this kind of overarching wages policy can only be implemented if there exists centralized institutional machinery with legal and/or moral authority. No such machinery exists in the United States.

Exchange rate policy can also be used as a stabilization device, adjusting rates to offset instability resulting from the net effect of international activity. In practice, such ready manipulation is not always feasible. Exchange rates are now determined in the open market, in which central banks are only one player; but exchange rate policy was always problematic because any movement has multiple effects that simultaneously benefit and harm various interest groups.

■ The Development of National Accounting Systems

One other development of major consequence occurred in time for the post-1945 boom. This was the invention of national accounting techniques. Political arithmetic arose during the seventeenth century, but progress was slow. Even by the mid-nineteenth century, policy makers were still dependent on statistics from a handful of prices and goods. American institutionalist economists were most active in developing new statistical techniques to study movements in trade and production. The late nineteenth century saw perennial economic crises with major social disruptions: they needed to be better understood. War was an important motivator—the U.S. National Bureau of Economic Research was begun after World War I, providing the first institutional base for studies in national accounting systems. The Russians under Soviet rule soon followed. The British were slower in this field, but caught up during World War II, forced on by the planning needs of a national emergency. These conceptual developments occurred independently of the work of Keynes.

■ Gross National Product and Gross Domestic Product

By 1945, the analytical vision of Keynesianism and the work of the statistical researchers had meshed to generate the conceptual models and associated statistical techniques that now dominate the way in which the West concretely understands the performance of its national economies. This period saw the introduction of the concept of gross national product/gross domestic product (GNP/GDP) as the key indicators by which to measure the performance of the national economy. GDP is the sum of the values of all the marketed goods and services, plus government expenditure, over a specified period of time, usually a year. It is "gross" because it does not allow for depreciation of production plant and equipment. It also does not include intermediate products. GNP is equal to GDP plus the net income earned by domestic residents from abroad, less domestically earned income belonging to foreigners abroad. It is a measure of the resources available to the country and its citizens. The difference between GDP and GNP is usually relatively small. GNP, and GNP per capita (GNP/p.c.) are key indicators by which national economies are compared internationally, especially in the World Bank statistics.

The data generated by these indicators are not empirically "true" facts, but artifacts. Thus, as discussed in Chapter 8, the very concepts of "advanced" or first world countries and "undeveloped" or third world countries are defined in terms of their placing on the GNP/p.c. list. GNP figures are averages, giving no indications of the distribution of income. Many economic processes that are important but hard to quantify are also excluded from these measures, including household production, polluting activities, and the consumption of nonrenewable resources.

National accounts are constructed as double-entry accounting tables, with expenditure and its components on one side, and income and its components on the other (Table 5.1). Although the two sides are considered theoretically identical, in practice the summation of income and expenditure components provide alternative estimates of GNP. Strictly, there is no

TABLE 5.1

Gross Domestic Product: U.S. 1997 $ Billion (Current Dollars)					
Expenditure side			**Income Side**		
Consumption		5485.8	Employee compensation		4703.6
Investment		1242.5	Wages & salaries	3878.6	
income in stocks	68.4		Supplements	825.0	
Government		1452.7	Proprietors' income		544.5
federal	523.8		Rental income		147.9
defense	350.3				
state & local	928.9		Corporate profits		805.0
			Net interest		448.7
Exports		957.1	[National income		6649.7]
minus			Indirect taxes less subsidies		593.3
Imports		−1058.1	Depreciation		867.9
			[Gross national product		8060.1]
			Net factor income		
			payable overseas		19.8
Gross domestic product		8079.9	Gross domestic product		8079.9

Source: Survey of Current Business, June 1998.

concept of causation implied in the accounts. De facto, however, there is the implication of a Keynesian-type causation, that aggregate demand (the expenditure side) is the driving force behind the level of GNP.

A Key Macroeconomic Concern: Inflation and its Causes

Discussion of inflation will be used to distinguish between the approaches taken by different schools of macroeconomic thought. Economists do not agree on explanations of or solutions for inflation. Keynes' own theory of inflation is obscure, and Keynesian economists developed their own versions. Some crude empirical correlations had been made between unemployment levels and the rate of change in wages. Empirical correlations were also noted between unemployment levels and the rate of change in prices. This unemployment/inflation relationship was dubbed the Phillips Curve, and has been the basis for a key theory of inflation and of policy approaches to it.

Simply put, inflation is interpreted as a product of excess aggregate demand. The causal mechanism at work has never been made clear, but the process appears to operate through the degree of tightness of the labor market. The higher the effective demand, the lower the unemployment rate, the greater the pressure on wages, and thence on inflation. Notions of "demand-pull" inflation and of "cost-push" inflation have often been used and contrasted, but in this Phillips Curve process these expressions effectively amount to the same thing: the way to control inflation is to weaken wage-push pressures. This may be achieved either through self-conscious wage restraint by workers and unions or by the deliberate creation of unemployment. Although wage pressures are, in formal terms, merely symptomatic of inflationary pressures, they can and have been taken as the independent cause of inflation. Inflation can then be attributed simply to worker selfishness.

In the mid-1970s, monetarism arose as an orthodox alternative to Keynesianism. According to the simple Keynesian model, inflation should be reduced as unemployment increases. This did not happen after the late 1960s (Figure 5.1). Monetarism brought an alternative explanation of inflation that inflation is due to growth of the money supply in excess of the needs of trade. This is an excess demand theory of inflation but one purely of monetary determination. The source of excess demand is excess purchasing power. In this view, unemployment cannot ultimately be influenced by government fiscal policy. There is a natural level of unemployment dictated by labor market conditions. Fiscal policy designed to

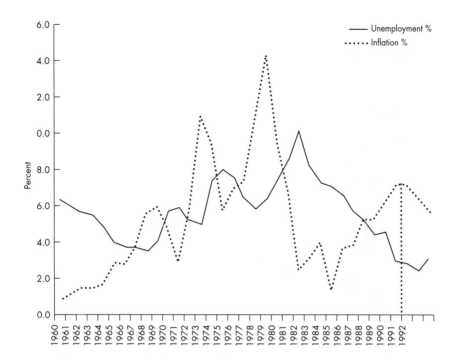

Figure 5.1 Unemployment and Inflation Rates—U.S.

reduce unemployment will only generate inflationary tendencies through the creation of excess purchasing power. Unemployment should thus be left to find its own level. Inflation can be cured only by basing money supply growth purely on the expected growth in real output. Inflation is thus due to government irresponsibility and/or incompetence. Get the money supply right and there will never be (long-term) inflation.

A third key analysis of inflation is that of the institutionalists. They put inflation down to the private possession of market power, in particular the power of large corporations to set prices (administered pricing). According to institutionalists, administered pricing inflation became a key feature of the post-1945 boom period, with large corporations acquiring market power and increasing prices during recessions in order to maintain profit flow. Coupled with the possibility of well-established union power within large organizations, one gets an institutionalized mechanism for an inflationary process from within the "real" economy. Governments are often forced to ratify this discretionary process—unless they allow faster money-supply growth they face higher unemployment: that is, money supply growth might accompany inflation, but is ultimately a symptom rather than a cause. This institutionalist theory of inflation can be generalized in the concept of inflation as a social phenomenon, essentially the product of conflict over limited resources. Inflation is one particular means of redistributing income, wealth, and command over resources. An important example of heightened social conflict over resources resulted from the American government's attempt to finance increasing Vietnam War expenditures in the 1960s without resorting to unpopular taxation increases. The inflationary effects of "printing dollars" fed right through the global economy. Economists have neglected the social origins of these inflationary pressures. Inflation is likely to be lower both in economies and at times in which resources are being used productively, because the material claims of different groups can be met from real growth.

There are also other immediate causes of inflation. One is the price-push of goods whose quantity cannot readily be increased. Housing stock in desirable locations is a crucial example. The increasing cost of accommodation feeds on into other prices through various social claims on income. Economists have not given adequate attention to these sources of inflation.

The institutionalist theory of inflation has never found favor with orthodox macroeconomists. Monetarism itself has fallen out of favor because its policy rule is too simplistic and ultimately destructive. A new approach has arisen since the 1980s that now dominates macroeconomic policy thinking, at least in the English-speaking world. This approach is extremely pragmatic, has little theoretical underpinning, and has no label. Its dominant focus is the level of inflation (following monetarism), and it is centered on the use of monetary instruments to facilitate the maximum feasible rate of economic growth compatible with keeping inflation under control. In practice, the level of unemployment is taken as a proxy for growth and inflation trade-offs. The catch-phrase for this orientation is the "nonaccelerating inflation rate of unemployment" or NAIRU. The significance of this focus can be readily ascertained from the public pronouncements of the current Chairman of the Board of Governors of the Federal Reserve System, Alan Greenspan.

The monetary authorities are now the dominant institutions of macroeconomic policy. In practice, most governments complement monetary policy by attempting to inhibit wage growth directly. Right-of-center governments have preferred to inhibit wage-earner rights, including by imposing restrictions on union legitimacy and power. Left-of-center governments have preferred to encourage voluntary cooperation from wage earners and their unions for wage restraint.

The subject of inflation has been used above to illustrate the interaction between economic theory and actual economic processes. We need to broaden this comparison of theory and history by examining the public policy process in its historical context, and in conditions peculiar to the United States. This provides the necessary backdrop to the current economic environment, and highlights the issues that have to be addressed for economic policies that are both appropriate and feasible.

Economic Theory and Public Policy

■ The Post-1945 Boom in Theory and Practice

The post-World War II boom is probably the most favorable period in capitalist history. It is therefore important to examine its character and its origins.

The global depression of the 1930s was ended by militarization and global war. The subsequent boom was created by a complex set of historically specific forces. From an international perspective, at least three elements were fundamental:

1. the construction of an international economic order, based on a new set of institutions and relationships and derived from the wealth and power of the United States, a structure blessed with the label *Pax Americana* (the American peace);

2. wide-ranging interventionist activities by national governments to support industrial development;

3. a "favorable" environment for profitable capital accumulation: elements inherited from war (forced savings in some of the victor countries, and the destruction of plant and infrastructure and the displacement of personnel in Europe), and elements generated from Cold War politics (a repressive political environment that inhibited labor militancy).

In 1945, the United States had 50% of global industrial production and 75% of the world's gold stocks. It was in a position to dictate the general terms of the post-war order, and it did. The United States became the global hegemonic power, playing the same role that Britain had assumed in the nineteenth century (Brett 1985). The U.S. dollar became the key international currency. The formal basis for the dollar's credibility was that it was tied to gold, at $35 to the ounce; however, the fundamental basis for the dollar's acceptability was the size and strength of the U.S. economy itself. Countries such as Great Britain and Australia (with left-leaning governments) hoped that the newly created International Monetary Fund and the World Bank would facilitate full-employment policies in the first world and generous lending for development in the third world. However, the United States dictated that the charters were established on relatively parsimonious terms; lending policies were ultimately dictated by political considerations. The United States filled the void itself by expanding international liquidity with a massive three-pronged expenditure outflow: loans and grants, especially through the Marshall Plan; military expenditure; and direct foreign investment. These expenditures benefited European interests and U.S. interests simultaneously (for example, Marshall Plan grants were conditional on recipients purchasing U.S. materials). The outbreak of the Korean War in 1951 ensured that comparable assistance flowed to (non-Communist) Asia, especially Japan. The other key international institution was the General Agreement on Tariffs and Trade (GATT), belatedly established in 1948. GATT provided the ground rules for freeing up international trade. Under GATT, international trade was greatly expanded, albeit on terms that suited the United States and industrialized Europe. Tariffs on manufactures and equipment were gradually reduced, but these countries retained protection against third world exports, notably agriculture and textiles.

In the United States itself, a new right-of-center coalition between business and government underpinned sustained economic growth (Kolko 1984, Ch. 9). On a scale and breadth never previously experienced in the United States, the federal government subsidized the development of a range of industries. The center of gravity of this underpinning was defense preparedness. Industries (and associated firms) that benefited directly included aerospace and electronics; sectors that benefited indirectly included nuclear power, shipping, steel, mining and agriculture (preparedness in materials and supplies), construction and educational institutions. Defense spending constituted, on average, well over 8 percent of GNP during the 1950s and 1960s, providing a "floor" to the economy. Employment in defense industries constituted about 16 percent of employment in all durable goods manufacturing through the 1960s (compared with 3 percent in 1939). Engineers, of course, were substantial beneficiaries from this development. (Case Study 5.1 at the end of this chapter describes the rise of the military-industrial complex and discusses its implications.)

A substantial "bottom-up" development occurred that complemented military-supported activity. A dramatic rise in consumer spending (including durables) fostered a substantial consumer goods sector. This spending was initially fueled by the release of war-time forced savings and reinforced by the development of consumer credit finance. The social basis for this expansion was the phenomenon of family formation, and the spatial embodiment was "suburbanization." The symbols of this dimension were the mortgage-financed home, durable household goods (whitegoods and furnishings), and the automobile. The technical foundation lay in the extension of the principles of mass-production, which cheapened the cost of production and the price of a wide range of commodities. The quality of work life was drab for production workers, but the pay packets expanded to assist affordability of the "household package." This set of complementary elements has come to be known as "Fordism," named after Henry Ford, whose decisive contribution to assembly-line production was discussed in Chapter 2. Affluence was still denied to a significant percentage of the American population, but a larger percentage than before now enjoyed a "middle class" existence.

The long boom was thus a product of complex and historically specific forces. In general, economists have not provided us with an adequate explanation of the most successful economic period in twentieth century history. There has been a preference for fairly simple explanations, based on theoretical preconceptions. Orthodox economists see the boom as substantially dependent on the presumed rise of global free trade during the period. More typically, the boom was believed by many to be the result of Keynesian ideas. This belief involves three ingredients: (1) the acceptance of Keynesian principles by government officials; (2) the development of appropriate instruments (purely at the macroeconomic level) to put these principles into effect; and (3) a change of moral stance, embodied in a political commitment to full employment. This belief thus involved both technical and ideological optimism. The ideological optimism implied that governments would commit themselves to a full-employment agenda. The technical optimism implied that governments could develop and apply policies (assisted by new national accounting data) that would eradicate economic depressions and mitigate the business cycle indefinitely. The hope was that the capitalist system could turn on (near) full employment at will by policy manipulation purely at the macroeconomic level, without interfering with the imperatives of the private sector in the structure of the economy. Although there is an element of truth in this picture, it does not represent the real policy significance of the postwar experience.

More flexible macroeconomic policy instruments have certainly been developed, including budgetary manipulation and monetary manipulation of credit and interest rates. However, three significant qualifications are necessary. First, macroeconomic policy was of only marginal significance in the creation and sustaining of the boom, as discussed above. Second, the instruments were too imperfect to be a means of eradicating the business cycle. Third, at least in the United States, the moral stance did not change at all. The Employment Act of 1945 was emasculated of any commitment to full employment. Macroeconomic policy kept on being used for old purposes (especially during the Eisenhower years): to repress rather than expand economic activity; that is, budget balancing and restricting inflation were considered more important than generating employment.

Even when macroeconomic policy instruments have been responsible for the expansion of activity, this has not been motivated by egalitarian purposes. Kennedy-era tax cuts went to business; Johnson's Great Society program was pragmatically motivated to head off racial civil war; Reagan-era tax cuts went to the well-off, and were essentially mechanisms for wealth redistribution. Perversely, military spending *was* motivated partly by employment-generating ambitions; for that reason, the postwar boom in the United States has been pragmatically labeled a period of "military Keynesianism." However, economists had little to do with this expansionary mechanism.

Economists of all persuasions have substantially over-stated the degree of influence of economic theories on government action and on the economy in general. In general, one has to be cautious about accepting that academically generated conceptual tools have ready applicability, and that their meaning and use is invariant across time. By concentrating on theory, economists have missed two key lessons of the post-1945 boom.

The first lesson is that successful government involvement in favorable periods of expansion in material affluence is not just a matter of getting the aggregate settings right. In fact, that is rarely achieved. Effective government involvement required and still continues to require structural intervention into the details of the economy. The second lesson is that one cannot create sustained growth at the flick of a wrist, because growth depends upon complex and historically specific circumstances. Capitalist profitability is not just a matter of a well-oiled and liberated market mechanism; it is a matter of constructing and reconstructing a supportive institutional foundation, which includes appropriate technologies, business organizational structures, and even the broader political and ideological settings. Some economic historians have claimed that material affluence proceeds in regular long cycles of boom and bust, "Kondratieff" cycles of approximately fifty years in duration. Certainly there have been long booms and long busts in the last two hundred years, but the notion of a regular long cycle is too simplistic.

There are, however, some consistent ingredients in the favorable periods. For better or worse, material development appears to depend on a global politico-economic hierarchy, enforcing stability at the global level. The two most pronounced booms in capitalist history have been the twenty-five-year periods after 1850 and 1950. They were distinguished by the super-power dominance of Britain in the first period and the United States in the second. Another key feature of long booms is that of qualitative technological or infrastructural change, with dramatic implications for the whole structure of production. This is an issue of substantial relevance to engineers. Railways were vital to the mid-nineteenth century boom, providing multiplier effects back through their impact on the metal industries, and forward through their integration of markets. Factors such as the gradual diffusion of steam power, the spread of electricity, and oil replacing coal have also been important. World War II and the Cold War brought about major developments in a cluster of technologies: electronics, advanced materials, nuclear power, and so on. In other words, capitalist development has been crucially dependent upon major qualitative transformations of the techniques and organization of production, and the organization of social and political life.

∎ The End of the Postwar Boom: New Theories, New Policies

The combination of specific developments favorable to the post-1945 boom had fallen apart by the mid-1970s. During the boom era, as noted, economists had a stylized and simplistic interpretation of the boom and appropriate economic policy. The faltering of the boom led to a revision of economic theory, but the tendency remained to interpret developments simplistically and to offer simple solutions to complex problems. The deterioration of boom conditions in the late 1960s provided the opportunity for orthodox economists to question the validity of Keynesian ideas and associated policies of assertive macroeconomic demand management. Economic downturn and the 1970s oil price hikes permitted the return to influence of more orthodox economic ideas.

As noted above, macroeconomic theory evolved from Keynesianism to Monetarism, and later to the "NAIRU" anti-inflation model. A more orthodox macroeconomics was complemented by a growing emphasis on a purist application of microeconomic theory to policy measures at the structural level. The structures of contemporary economies were said to be impeded by market distortions, requiring a freeing up of the market mechanism from "over-regulation" and a liberation of the capital markets from the "crowding out" of government borrowing. In the United States, this push was labeled the "supply-side" revolution, and found special favor with the Reagan administration. In other countries (especially Great Britain) this push was associated with the privatization of publicly owned enterprises. Many essentially capitalist countries have historically used public ownership for the delivery of broad aims from key industries and infrastructure. The United States, however, coupled private ownership with substantial regulation of key services in a model that dated from the late nineteenth century. Ironically, the preference for private ownership of key industries meant that the United States had more regulatory structures on industry than other countries.

The supply-side push led to the dismantling of some regulatory structures, most notably in telecommunications and air travel. Environmental regulations were also a target in this new age; such regulations had been built up dramatically from the 1960s, but they were particularly irksome to some businesses. However, recent developments require close examination and interpretation; there has been no sweeping move from "big government, high regulation" to "small government, low regulation." The main thrust of the Reagan era, for example, was not a dramatic reduction in the size of government but a reorientation of its priorities. Welfare spending was significantly reduced and military spending significantly expanded.

Economic theory and economic history diverged early with the rise of neoclassical economics in the late nineteenth century. Economic theory developed a highly simplified model of capitalism as a pure market economy. The developers of the theory appreciated the highly abstract nature of their model, but later generations of economists have tended to assume that it mirrors the actual historical process. Any deviation between theory and the real-world is now presumed to be a fault of the real world. There has been a tendency for economists to try to solve real-world problems by direct application of their simple models. They have sought stabilizing mechanisms for socioeconomic problems within quite mechanically determined processes. Problems with the United States economy have often been taken as due to distortions in the workings of the market mechanism. The impersonal pricing process is taken as the motive force of the theoretical market. The whole of contemporary mainstream opinion can be summarized thus: clear away impediments to the free market and let price flexibility do its job in allocating resources efficiently. Mainstream economists miss the fact that the tasks of resource allocation and providing stability have always depended on a broad range of socially and politically constructed institutions. We need to look at specific developments within the United States to appreciate the special characteristics of the U.S. economy and the forces that shape it.

THE U.S. ECONOMY AND THE GLOBAL ECONOMY

The U.S. economy is a very significant component of the global economy and U.S. policy makers have a significant influence on economic developments elsewhere. Although international trade has accounted for a smaller percentage of GNP than in other countries, the U.S. economy is best understood in its global context.

The degree of American economic superiority in 1945 could not last. For the United States to play the role of global power involved special conditions and, more fundamentally, inconsistent requirements (Brett 1985, Ch. 5). As long as the U.S. dollar played the role of key global currency, global reconstruction required that the availability of dollars outside the United States be sustained. This required the United States to persistently run a deficit on its balance of payments, which it did from 1950. This was manageable as long as the United States was an industrial powerhouse compared with the rest of the world. But other countries were rapidly reconstructing, and with U.S. assistance. Indeed, U.S. political objectives of containing the communist bloc led to the further undermining of its competitive position: the United States gave privileged access to its huge domestic market to the exports of its political allies and supported the formation of the European Economic Community. By the mid-1960s, even though the domestic economy was booming, the situation had become untenable. Attempts to find additional sources of international credit were inadequate. The fundamental inconsistency was that international liquidity depended unduly on the dollar, which required the U.S. balance of payments to be in *deficit*, but the credibility of the dollar depended upon the U.S. balance ultimately returning to *surplus*.

Ultimately, it was the "conservative" President Richard Nixon who adopted a radical (if temporary) solution to the problem. Nixon removed the convertibility of dollar into gold, devalued the dollar, and introduced import controls. The dollar soon devalued further on the open market, enhancing the competitiveness of American exports. Nixon enhanced agricultural subsidies and made a rapprochement with the Soviet Union and China (especially with an eye on their rural imports); in effect, initiating a rural-sector-led recovery. Countries that held dollars helped pay for the U.S. reconstruction. Gradually the dollar has been joined by other strong currencies in creating a "basket" of currencies that facilitates the financing of international trade.

Nixon dealt with only one part of the problem. The United States was confronting both higher productivity production and lower wage costs in a broader range of products from other countries. This was especially reflected in key industries such as steel, automobiles, and consumer electronics. For steel, the U.S. producers ("Big Steel") persisted in expanding in obsolescent "open hearth" technology, delaying the introduction of superior technology (basic oxygen and electric furnaces and continuous-process mills). For automobiles, U.S. producers depended upon large-throughput single-model technology, discussed in more detail in Chapter 2. Both steel and automobiles were representative of mass production labor relations: blue collar workers who were relatively well-paid but were alienated and were not consulted on potential productivity improvements. Various protective barriers were erected; for example, "voluntary export restraints" were imposed on Japanese producers. With government support, U.S. producers in some of these industries ultimately responded to these challenges, but it took them until the 1990s. Before coming to the U.S. economic resurgence, it is informative to examine the evolution of the "balance on current account" and the peculiarities of the U.S. trade structure. These dimensions give us a "window" into the entire U.S. economy, helping us to understand its unique characteristics and its recent problems.

▮ The U.S. Trade Structure

The "balance of payments" of a country is a set of accounts that summarizes the economic relations for a particular year between the residents of that country (and its governments) and the residents of all other countries. The balance of payments comprises two accounts: the balance on current account and the balance on capital account, now collected for each country on a uniform basis. The balance on current account summarizes the transactions for that year, separated into four categories (see Table 5.2). The balance on capital account summarizes the changes in the levels of assets held and liabilities incurred by residents/governments of one country in real form or in securities of other countries. The two accounts form the two sides of a "double-entry" table constituting the balance of payments, with the net figure from each account offsetting the other (in practice, usually with a "statistical discrepancy"). Changes are sourced in both directions; decisions to trade internationally are made autonomously, and decisions to purchase assets are made autonomously. A surplus on the current account (e.g., from an increase in net exports) is registered as an increase in assets in the capital account. The returns on assets held overseas and shown in the capital account are registered as income in the current account. The U.S. government and private corporations engaged in a spree of spending, giving, lending, and buying after the late 1940s, some of which "returned" in the form of surpluses on trade. However, by the late 1970s, trade itself was hemorrhaging into persistent deficits and the balance on current account deserves close examination.

Table 5.2 shows the *net* level for each category for selected years over the last twenty-five years: merchandise trade, services trade, income, and transfers. For example, in 1975–1976, Americans spent $9.5 billion more on merchandise imports than they received on merchandise exports (the total figures, not shown, were $124.3 billion in imports and $114.7 billion in exports). Table 5.2 shows that the net balance on *merchandise trade* has deteriorated dramatically in the last twenty years. There have been periods of improvement (the late 1970s and especially the late 1980s) but these periods involved a decline in the rate of growth of GNP. That is, economic growth has been associated with rising deficits on merchandise trade. U.S. officials have for some time been concerned with the bilateral (one-on-one) overall deficit in trade with Japan, in particular. In 1995, the bilateral deficit with Japan was approximately $60 billion and approximately $34 billion with China. These compare with a total net deficit on merchandise trade of $172 billion.

By contrast, U.S. residents typically generate a net surplus on *services trade* (Hollywood movies, Disneyland, education fees, consulting services, etc.). U.S. residents used to earn a net surplus on *income* from overseas

TABLE 5.2

Balance on Current Account (Selected Years, $ Billion)*						
Year	1976	1981	1986	1991	1995	1997
Merchandise trade	−9.5	−28.0	−145.1	−74.1	−172.0	−196.0
Services trade	2.7	12.4	7.0	45.2	68.2	90.0
Net income	16.0	32.4	12.1	14.8	−9.1	−9.5
Current transfers	−5.2	−11.9	−24.5	4.8	−35.1	−39.9
Balance	3.8	4.8	−150.5	−9.3	−148.2	−155.4

*Net levels: credit − debit.
Source: *Balance of Payments Statistics Yearbook*, 1998, International Monetary Fund.

investments, but net income turned from a surplus into a deficit in the 1990s. In short, the deficit on merchandise trade is driving an overall deficit balance on current account. The net revenue from the other categories either do not offset or add to the net deficit on merchandise trade (Figure 5.2 shows 1997 values). (Note that these figures are somewhat misleading, in that a U.S.-owned subsidiary producing in and exporting from another country to the United States will contribute to the merchandise trade deficit. The multinational character of many large companies generates an anomaly. In principle, earnings from such activity ought to be registered in the "income" category; however, the deteriorating income balance suggests it is unlikely that registration of the income of U.S. multinational corporations is reliable. This constitutes a divergence between the interests of U.S. shareholders and those of "the United States" as a national entity. Regardless of these anomalies, the balance on current account is taken as an indication of the "health" of the U.S. economy by financial authorities and by those whose liabilities are designated in U.S. dollars.)

Some inclusions and features of the four components of the current account are as follows:

I. *Merchandise trade:* exports/imports of commodities: aircraft; computers; soybeans; and so on. The merchandise net deficit

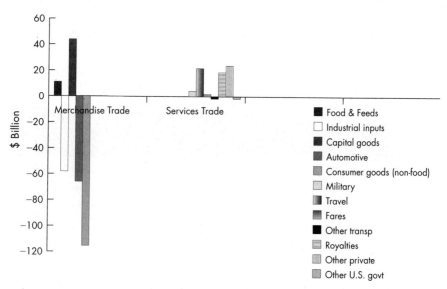

Figure 5.2 Net Balance of U.S. Merchandise and Services Trade, 1997

Chapter 5 • A Perspective on Economics

started to blow out in the early 1980s, was pegged back in the early 1990s, and blew out from 1992–1993 onwards

2. *Services trade:* includes shipping revenues/costs; commercial and consultancy services; cultural trade; tourism; and so on.

3. *Net income:* includes unrepatriated profits, dividends, and interest payable on equity investment to head office of private companies and financial institutions; and interest payable by governments and public enterprises on overseas loans.

4. *Current transfers:* includes foreign aid; family transfers; and so on.

It is also instructive to look "inside" the merchandise trade and services trade categories of the balance on current account. Again, total levels of exports and imports for each subcategory are excluded, but the net levels highlight the respective contributions of the subcategories to the balances on merchandise and services trade. Table 5.3a indicates the United States has net trade surpluses in both rural products and capital goods. This is an unusual phenomenon, although it is shared with some western European counties. A surplus in capital-goods trade is the mark of an advanced industrialized country. A surplus in rural goods reflects a fertile land mass, a productive rural sector, and an important political constituency. Rural exports are indirectly and directly subsidized.

U.S. domestic production is near comprehensive, but Table 5.3a highlights the global linkages of a resource-intensive industrial growth machine. The United States now has a major dependence on foreign industrial inputs. In addition, in the last ten years there has been a

TABLE 5.3a

Composition of Merchandise Trade (Selected Years, $ Billion)*						
Year	1980	1985	1988	1991	1993	1997
Food and feeds	+17.7	+2.7	+8.9	+9.6	+12.8	+11.8
Industrial inputs	−60.4	−52.8	−32.7	−23.2	−40.5	−59.0
Capital goods	+44.7	+18.0	+16.9	+45.7	+29.8	+41.1
Automotive	−10.8	−40.0	−54.5	−45.7	−50.0	−66.7
Consumer goods (nonfood)	−16.5	−51.7	−69.4	−60.9	−79.3	−115.6
Balance	−25.5	−122.2	−127.0	−74.1	−132.6	−198.0

* Net levels: credit − debit.
Source: Statistical Abstract of the United States, 1998.

substantial rise in foreign penetration of the U.S. automobile market and a substantial rise in consumer goods sources overseas (including from U.S. subsidiaries). The bilateral deficits with Japan and China, mentioned above, are relevant in this regard. In 1995, for example, the United States experienced a $27.2 billion deficit ($31.3 billion minus $4.1 billion) with Japan in automobiles; and the United States experienced a $21.8 billion deficit ($22 billion minus $0.2 billion) with China in clothing, footwear, and miscellaneous manufactures. The deficits highlight why U.S. officials have long claimed that Japan inhibits access to significant U.S. products, and why China has recently been subject to lobbying from the United States on fair labor practices and so on.

Table 5.3a highlights the fact that the substantial merchandise trade deficit is systemic and cannot readily be reduced. Table 5.3b shows the breakdown of services trade by key categories. This table shows that exports of a range of services are increasingly important to the U.S. economy. Travel (tourism) revenue is moving in favor of the United States; so is airline income. Surprisingly, net revenue from goods transport is in deficit (this might reflect the fact that many shipping companies are registered in exotic countries such as Liberia). The substantial growth in earnings from royalties and license fees reflects the strength of the United States in both entertainment and technology development. The substantial growth in the "other private" category reflects the strength of the United States in business services and in education. These developments

TABLE 5.3b

Composition of Services Trade (Selected Years, $ Billion)*						
Year	1980	1985	1988	1991	1993	1997
Military	−1.9	−4.4	−6.4	−5.3	0.5	6.8
Travel	0.2	−6.7	−2.7	12.1	17.2	22.0
Passenger fares	1.0	−2.0	1.3	5.9	5.3	2.7
Other transport	−0.2	−1.0	−1.2	−1.9	−2.6	−2.0
Royalties	6.4	5.6	9.5	14.1	15.7	19.3
Other private	3.4	9.8	12.2	20.3	23.1	24.3
Other U.S. Govt.	−0.8	−0.9	−1.2	−1.4	−1.4	−2.0
Balance	6.1	0.3	11.6	44.7	57.8	87.7

* Net levels: credit − debit. Two categories here, expenditures relating to military transactions and miscellaneous government services, are not subject to commercial considerations.

show why U.S. trade officials now devote much lobbying to what they regard as trade impediments to the penetration of these services into other markets.

∎ U.S. Variations on the Capitalist Theme

This chapter has already shown that market economies vary in their character. The United States is, of course, in the English-language camp, and firmly rooted in a liberalist heritage. The conventional wisdom has it that liberalism (meaning "economic freedom" rather than "left-wing politics") is intimately related to healthy market economies. The reality is that the institutional fragmentation generated by liberalism has both advantages and disadvantages for healthy market economies. Advantages lie in the inhibition to special deals that are conducive to inefficiency and corruption. Disadvantages lie in the stark necessity for institutional coordination in generating economic dynamism, under perennial pressure from a competition for advantage that involves not merely business enterprises but governments at every level.

The response within the United States (as within other English-language countries) has been to break the strict principles of liberalism. Structures for industry assistance have been constructed pragmatically. The farming sector has been a long-time beneficiary of such pragmatism. Perhaps the most important reflection of this "rule breaking" is in the phenomenon of the military-industrial complex, detailed in Case Study 5.1 at the end of this chapter. This complex is also a reflection of another peculiar characteristic of the United States: that it is a global hegemonic power, in which political, military, and economic power are inseparably intertwined. Industrial development has been a vehicle for the enhancement of military capacity, but military capacity has also been a vehicle for the pursuit of industrial development. Certain industries have benefited from this linkage, in particular aerospace and electronics. Indeed, the export capacity of the United States came to depend substantially on military-related industries. It has been estimated that defense-oriented industries—aerospace, ordnance (military aircraft, ships, combat vehicles, supplies), and particular electronics—constituted over 20 percent of all the U.S. exports in early 1950s, and 12 to 15 percent during the 1960s. Defense-oriented exports thus contributed substantially to assisting the U.S. economy to achieve a balance of trade surplus during the 1950s and 1960s.

However, much defense-oriented research and development is not readily amenable to civilian application. The ambiguity and tension is reflected in the development of semiconductor technology. The military bureaucracy was fundamentally responsible for initiating and underpinning the development of computer chip technology in the United States in the 1950s. Production of integrated circuits for defense purposes was still over 50 percent of all production of integrated circuits in 1966, just before total production "took off." On the other hand, military imperatives increasingly pushed priorities that ran counter to commercial imperatives (for example, the conditions under which circuitry was expected to operate), and the military bureaucracy resisted initiatives to counter the competitive onslaught of Japanese semiconductors firms in the 1970s.

Apart from the lack of congruence between military and commercial priorities in the most favored industries, broader industrial and social priorities were not addressed. Commercial technology was being neglected; for example, by the mid-1960s, the merchant fleet and metal-working machinery was outdated. With the largest number of research scientists and engineers in the world, trade in machine tools went into deficit.

By the 1980s, the United States was faced with two long-term changes of substantial importance. It now shared industrial preeminence with a number of other countries, whose firms were not merely competing with U.S. exports but penetrating the U.S. market itself. Moreover, the Soviet bloc was collapsing, and the Cold War bogey as an inducement to the perennial reinforcement of military priorities was losing credibility. These "adverse" developments had to be surmounted.

RECENT DEVELOPMENTS AND DIRECTIONS FOR CHANGE

In principle, the U.S. economy ought to have been well-suited for a resilient accommodation to challenges. However, there were subtle elements constraining the pace and character of change. The liberalist culture of the U.S. influenced the unspoken rules by which change was supposed to take place. If an industry was going under, no matter what its importance (steel, automobiles?) and it couldn't get its own act together, then there wasn't much one could do about it. It deserved to fail, and those were the rules of the game. There was also the entrenched "cowboy" culture in American business, embodied in perennial takeover waves in which efficiency is less a motive (and even less an outcome) than empire building or asset stripping. The institutionalist economists have been criticizing this phenomenon for almost one hundred years. Its most recent instance was in the late 1980s. Finally, the forces ensconced in the "military-industrial complex" were resistant to a loss of influence; new enemies were needed (such as North Korea and Iraq) to replace the Russians.

As in all English-speaking market economies, reputable economists were disdainful of any positive initiatives of a more collective or public nature (usually labeled "industry policy") to facilitate viable reconstruction. Reputable economists in general also denied that there was anything to be learned from overseas experience. Indeed, the most publicized response to developing problems was the preference not for more "intervention," but less intervention—the aim being a "freer" economy. This was the "supply-side" revolution, and it is not accidental that this influential viewpoint was substantially the product of well-funded, right-of-center think-tanks (such as the American Enterprise Institute, the Hoover Institution, and the Heritage Foundation). As a consequence, supply-side ideas had substantial impact on policies, reflected in the deregulation push, noted above. The experience of an industry sector as far away as Australia (see Case Study 5.2 at the end of the chapter) illustrates the pervasiveness of such theories.

However, there were other important and divergent responses. The U.S. federal government cranked up the imperial machine and attempted, in effect, to throw its weight around. U.S. agencies attempted to leverage their institutionalized international power to influence outcomes more favorable to American business. This occurred through the full range of options: unilateral, bilateral, multilateral, and regional, even though such options are supposed to be mutually incompatible. Unilaterally, the United States pragmatically imposed a range of import restrictions to manage the deteriorating trade deficit. The United States aggressively pursued bilateral negotiations with Japan in particular, harassing the Japanese government over such issues as market access, the low value of the yen, continued American subsidization of Japanese security, and, more generally, the whole character of Japanese economic policy making. The United States has pursued, through the multilateral structures of the new World Trade Organization (WTO), market-opening and market-securing measures in what are perceived as its current areas of "comparative advantage"—capital investment, services trade, intellectual property, and so on—while continuing to benefit from subsidized farm exports. Finally, the United States has pursued greater regional economic integration with its neighbors Canada and Mexico through the North American Free Trade Agreement (NAFTA) as a regional defense mechanism against European and Asian economic power. These measures (with the exception of the continuing fight with the Japanese) have partly contributed to American economic resurgence.

There has also been much activity that runs counter to the rules of liberalism. American business firms defied the economists and acknowledged that they might learn more productive procedures from overseas firms. For example, Ford and GM belatedly recognized that they might learn something from the way in which their rivals (Toyota or Mercedes Benz) went about their business in matters such as internal culture, relations with suppliers, and relations with government institutions. American firms were also following new international corporate strategic practices and forming joint ventures with other companies (especially with foreign companies), that is, cooperating with some companies in order to compete more aggressively with others.

An institution with its origins in the American Revolution illustrates the pragmatism that has always characterized U.S. government policy; this is the U.S. Army Corps of Engineers, discussed in Case Study 4.2 (see p. 180). In practice, governments at all levels construct industry and other policies to support business and employment within their constituency.

State governments have always been active in forming regional groupings and developing pragmatic measures to draw industry within their territory (tax concessions and so on), but recently more sophistication has been applied to the task. In the mid-1980s, a number of eastern and north-eastern states (Michigan is representative) initiated manufacturing modernization programs. These were especially directed at small and medium-size enterprises. They transcend earlier hopes for technology as a quick-fix solution, to concentrate on the less tangible elements of organizational culture within the firm and organizational networks outside the firm (Best and Forrant 1996). Again, these programs were inspired by structures in use in Europe and Asia.

At the federal level, Congress passed the Omnibus Trade and Competition Act of 1988 giving the Department of Commerce a new role in assertive industrial policy. Some programs complement state government programs on a broader scale; for example, the creation of Manufacturing Technology Centers, which aim to facilitate advice and information flow, and to improve cost effectiveness by utilizing local cooperative support networks (local governments, educational institutions, etc.). Symbolic of recent experimentation is the establishment of a joint venture called Sematech, involving all the large semiconductor firms *and* the federal government as joint shareholders. This form of collaboration is unprecedented in American history, but it was seen as necessary by the participants in order to continue to compete with the Japanese. Adoption of this uncharacteristic structure was facilitated by the fact that Sematech support comes from the federal defense budget.

The U.S. economy has recently bounced back and essentially since 1995 has been defying expectations in terms of sustained economic growth. The trade balance remains in substantial deficit, but inflation is low and official unemployment is low (under 3 percent) by the standards of the entire postwar period (about 4 percent). By the late 1990s, U.S. commentators had become so confident in the U.S. economy that they could look upon overseas differences in institutions and culture as aberrations, the ideal now being the U.S. "model." The 1997 crisis in many Asian countries merely reinforced this self-confidence. The popular explanation is that the U.S. economy's success is due to its embodying the "free-market" model, whereas economies in both Europe and Asia are hampered by restrictions on market freedom. Overseas differences in institutional structure and policy priorities are tacitly acknowledged, but seen as market distortions. However, this is a dangerously misleading explanation of the peculiarities of the U.S. economy itself, as we have tried to show. The mid-to late 1990s vibrancy of the U.S. economy was undoubtedly a reality, as reflected in the macroeconomic indicators. Yet the sources of that vibrancy are complex: a combination of market deregulation, refashioning of business culture, government support at both the domestic and international levels, and so on. An adequate understanding of the U.S. economy and its successes and failures can only flow from an adequate examination of the historical and empirical details.

Discussion Questions

1. Outline the essential arguments of and differences between the major schools of macroeconomics.
2. What are the essential characteristics of the neoclassical school's model of microeconomics, and what strengths and weaknesses follow?
3. Discuss the idea that "the economy can be considered to be a machine that can be adjusted to give required outputs or results."
4. Speculate on the nature of the intellectual "mindset" of engineers and economists and what differences and similarities exist in their respective orientations.
5. Speculate as to what factors have been important in influencing the rise to prominence of particular schools of thought in economics?
6. Outline your understanding of the key sources of favorable conditions constituting the "long boom" in the United States in the post-1945 period.
7. Discuss the stated aims and assumptions of the 1980s proposals for economic "deregulation" and the outcome of following such a policy prescription.
8. What is your understanding of the term "the military-industrial complex"? What impact, if any, has it exerted on the structure of the U.S. economy?
9. What are some of the challenges in facilitating an institutional structure conducive to industrial dynamism in the U.S. economy?

FURTHER READING

Gönen, T. 1990, *Engineering Economy for Engineering Managers*, John Wiley & Co., NY. A good example of the range of texts aimed at supporting decision making on engineering investment.

Markusen, A., Yudken, J. 1992, *Dismantling the Cold War Economy*, Basic Books, NY. Raises important issues about future directions for technological development.

Rosenberg, N. 1982, *Inside the Black Box: Technology and Economics*, Cambridge University Press, Cambridge. All of Rosenberg's work on the economic aspects of technological development is well worth reading.

REFERENCES

Babbage, C. 1963/1832, *On the Economy of Machinery and Manufactures*, Augustus Kelley, NY.

Barber, W., Ed. 1988, *Breaking the Academic Mould: Economists and American Higher Learning in the Nineteenth Century*, Wesleyan University Press, Middletown.

———, 1991, *A History of Economic Thought*, Penguin, Harmondsworth.

Best, M., Forrant, R. 1996, Creating Industrial Capacity: Pentagon-Led versus Production-Led Industrial Policies, in J. Michie, J. Grieve Smith (Eds.), *Creating Industrial Capacity*, Oxford University Press, NY.

Brett, E. A. 1985, *The World Economy Since the War: The Politics of Uneven Development*, Macmillan, London.

Chandler, A. 1977, *The Visible Hand*, Belknap Press, Cambridge.

Diesendorf, M., Hamilton, C. 1997, *Human Ecology, Human Economy*, Allen and Unwin, St. Leonards, NSW Australia.

Ekelund, R. B., Hooks, D. L. 1973, Ellet, Dupuit and Lardner: On Nineteenth Century Engineer and Economic Analysis, *Nebraska Journal of Economics and Business*, 12, p. 3.

Galbraith, J. K. 1975, *Economics and the Public Purpose*, Penguin, Harmondsworth.

Gönen, T. 1990, *Engineering Economy for Engineering Managers*, John Wiley & Co., NY.

Josephson, M., 1962/1934, *The Robber Barons: The Great American Capitalists, 1861–1901*, Harvest, NY.

Kolko, G. 1984, *Main Currents in Modern American History*, Pantheon, NY.

Layton, E. T. 1971, *The Revolt of the Engineers: Social Responsibility and the American Engineering Profession*, Case Western Reserve University Press, Cleveland.

Livingstone, I., Assuncao, L. M. 1993, Engineering vs. Economics in Water Development: Dam Construction and Drought in North-East Brazil, *Journal of Agricultural Economics*, 44:1.

Rosenberg, N. 1975, Problems in the economist's conceptualization of technological innovation, *History of Political Economy*, 7:4.

—, 1982, *Inside the Black Box: Technology and Economics*, Cambridge University Press, Cambridge.

Scherer, F., Ross, D. 1990, *Industrial Market Structure and Economic Performance*, 3rd Ed., Houghton Mifflin, Boston.

Stewart, M. 1986, *Keynes and After*, Penguin, Harmondsworth.

Thurow, L. 1993, *Head to Head*, Morrow, NY.

Wade, R. 1990, *Governing the Market: Economic Theory and the Role of Government in East Asian Industrialization*, Princeton University Press, Princeton.

Williams, W. A. 1966, *The Contours of American History*, Quadrangle, Chicago.

Williamson, O. W. 1975, *Markets and Hierarchies*, Free Press, NY.

Womack, J. P., Jones, D. T., Roos, D. 1990, *The Machine that Changed the World: The Story of Lean Production*, Macmillan, NY.

THE MILITARY-INDUSTRIAL COMPLEX

The term *military-industrial complex* appeared in President Eisenhower's farewell speech in 1960. Since then the term has been an integral part of the language in which the priorities of post-1945 U.S. development are discussed. Appelbaum and Chambliss (1997) define the military-industrial complex (MIC) as "institutions connected by a common interest in weapons or other defense spending." It includes a set of interlocking institutions (elements of the military, the defense bureaucracy, private firms, educational and professional institutions) and associated imperatives. These imperatives essentially involve the commitment of a significant percentage of national resources to military preparedness during peacetime in the interests of "national security," with significant implications for training, research, industrial development, and ultimately lifestyle.

The American MIC acquired the beginnings of its modern form and scale in the late 1940s, but its roots go further back. Warfare was central to the formation of the modern nation state, and has remained one of its major characteristics. In the twentieth century, with its two world wars and hundreds of other conflicts, the scale of slaughter escalated, and nearly 100 million people have lost their lives. The stark reality is that warfare requires resources, and resources require industrial capacity. Moreover, technological advantage in war has been a strategic imperative from very early days. Military considerations have for centuries, certainly since the days of European mercantilism, been a significant spur for industrial development.

The U.S. Civil War was probably the first modern war of any scale. It saw the first appearance of iron ships and, less significantly, rockets, submarines, and torpedoes. More importantly, the demands of servicing 3 million personnel with effective transportation and communication stimulated inventiveness in key processes (railways, bridges, and the telegraph) and standardization of production (the rifle, armor plating, screw propulsion, and the steam turbine), providing lessons in metal working and mass production for the burgeoning industrial age.

Some individuals, especially engineers, carried the lessons of war over to industry. However, the first self-conscious linkage between industry and military preparedness came later in the nineteenth century. Before the 1880s, most military hardware in the United States was procured from government arsenals and yards. The "New Navy" program of the mid-1880s had two parallel ambitions: to modernize the navy from wood/sail to steel/steam-powered and to build an indigenous ship-building capacity. Almost immediately, a "revolving door" was established, through which personnel moved between the navy and private ship builders. Lobby groups were established to push for permanent "preparedness." Federal support grew for related research and development, explicitly copying Germany. In the early part of the century, the federal government moved to support the sale of naval vessels in Asia and Latin America, justified on the grounds of maintaining the industrial base in shipbuilding. The new linkage reflected a dramatic change in the nature of war; there was a reduced dependence on personnel and an increased dependence on machinery. More than shipping was involved, and backward linkages were forged. The Navy pressed for modernization of steel technology, which in turn facilitated developments in machine tools, electrical machinery, and automobiles, for example.

The first World War brought private firms and individuals directly into the war machine, most centrally into the War Industries Board (WIB). Symbolic of this direct input was DuPont, with substantial responsibility for munitions supply. During the war, there were major inefficiencies in supply operations. Part of the problem was within the War Department itself, which was both unwilling and unable to cooperate with WIB—unable because as many as eight supply bureaus purchased independently and competitively for the army. Moreover, the WIB was organized along commodity lines whereas the War Department's supply network was structured by function. Belated collaboration occurred towards the end of the war, providing an early framework for recognition of mutual interests at the highest level and the seed for a future military-industrial bureaucracy.

What happened after the war was more important. The entire interwar period was spent working out the elements of a workable structure. Basic issues had to be confronted: how much planning for a future war was necessary, how was such planning compatible with American political traditions, and who would be in charge? Congress set the ball rolling with the 1920 National Defense Act, placing supply under a (civilian) assistant secretary of war. The military, however, had to be dragged out of its traditional notion that strategy dictated supply considerations rather than the two being complementary. By the 1930s, the need for comprehensive industrial mobilization was being accepted. The early years of Roosevelt's New Deal administration (from 1932) saw the influence of planners, but of a "progressive" variety. The Senate's Nye Committee recommended controls on the prerogatives of private property during wartime, but it was out of step with developments.

The year 1937 was a turning point. War preparation began in earnest (long before Pearl Harbor!) and the War Resources Board (WRB) was established with senior military and industrial figures. The War Production Board was created in 1942 on the ashes of the WRB. By then big business had regained the legitimacy and influence it had lost during the early New Deal, and mobilization occurred with minimal restraints on business prerogatives. The difference from World War I was that this slow process of forging close collaboration ensured that the institutional structure would survive and thrive in peacetime conditions. This administrative structure was complemented by the forging of a bond between science/engineering and industry. Representative of this link was the 1940 National Defense Research Council (NDRC). The NDRC channeled funds especially into electronics and communications, providing the model for the postwar subsidization of research and development (R&D). The most significant outcome was the Manhattan Project, progenitor of the atomic bomb.

If the Navy was the driving force in the 1880s, it was the Air Force's turn after 1945. The 1947 restructuring of the military created the Department of Defense and raised the Air Force to be co-equal with the Army and Navy. Judged in terms of claim on resources and influence on strategy, the Air Force became preeminent. Air power was seen as an alternative to an army dominance that threatened the militarization of society. Paradoxically, it also promised cheap defense in a period of budgetary contradictions. Air power quickly became an expensive option, for two reasons: (1) the rapid loss of the U.S. monopoly on nuclear weapons, generating the "arms race," and (2) the need to generate a "hi-tech" industrial base centered on an advanced aircraft industry. Strategic considerations came to dominate budgetary considerations, rather than vice versa. Expenditures escalated for a third reason: the

(continued on next page)

competition for resources between the ser-vices, indicating that the MIC is not a mono-lithic entity. The Navy created the Office of Naval Research in 1946 to preempt Air-Force dominance. At an average of roughly $180 billion (in 1982 dollars) expenditure per annum since 1951, defense expenditure by the end of the century would approach $9 trillion (1982 dollars).

Navy planning in the late nineteenth cen-tury dictated a deep involvement in the de-velopment and substainability of the relevant industrial segments. Air-Force dominance dic-tated the same on a larger scale. New Deal pro-gressive values were becoming a casualty of the Cold War, and the war-time planning ap-paratus in all areas other than defense was being dismantled. Ironically, the Pentagon was left as the only force with a brief for orderly development of American society! Although individuals formed families, spent their sav-ings, borrowed more, and acquired their share of consumer durables that generated sub-stantial private-sector profits, it was the MIC that set the broader priorities. In other coun-tries, the priorities of industrial development were set explicitly and self-consciously; in the United States, industrial policy was deter-mined under the auspices of the MIC.

The MIC has channeled industrial prior-ities: the locus of industrial support is on aerospace, communications, and electronics (ACE). The MIC has channeled both acade-mic and applied research priorities, through both direct and indirect (the National Sci-ence Foundation, etc.) funding of research. The elite of the scientific/technical wings of American academia (most notably MIT and Stanford) have been nurtured and socialized by the MIC, becoming an adjunct part of it themselves. The MIC has channeled educa-tional and training priorities, influencing the vision of key professions, notably engineer-ing. The post-Sputnik National Defense

Education Act subsidized training in appro-priate engineering and science subdisciplines. The MIC has even exerted a profound influ-ence on the spatial location of industrial de-velopment. MIC spending moved the center of gravity from the previous industrial heart-land centered on the midwest to favored re-gions called the "gunbelt": the south-west, the pacific coast, the north-east, and outliers such as Virginia.

In the early 1990s, it was estimated that about one half of all R&D in the United States was funded by the federal government. About two-thirds of federal R&D spending and about one-third of all R&D spending is motivated by defense priorities. In 1988 total U.S. defense R&D was $40.1 billion, com-pared with $1.1 billion in Germany and $0.4 billion in Japan. The ACE complex benefits disproportionately. Engineering is a major beneficiary. Of a total support from the De-partment of Defense for university research of $445 million in 1978 and $785 in 1988, the engineering discipline received $152 million (34 percent) and $321 million (41 percent), respectively. In 1988, the Department of De-fense provided 44 percent of all federal fund-ing to university research in engineering. The priorities of R&D spending in turn dictate the location of those who perform R&D. Some 15 to 20 percent of all employed sci-entists and engineers in the United States are employed in defense-related work.

Defense spending has varied considerably since the war, increasing significantly during the Korean and Vietnam wars. However, the MIC has remained deeply entrenched. It is not monolithic, as reflected in shifting feder-al budgetary priorities in the mid-1960s. Yet the strong revival of the MIC's budgetary for-tunes in the late Carter, Reagan, and Bush Presidencies is testament to its resilience, even in the face of disappearing "enemies." The hold of the MIC on the priorities of industrial policy has been dented a little, but remains firm. In the face of the declining competitiveness of American industry, interpreted by the military as a loss of a critical mass of domestic suppliers for military purposes, the Pentagon has facilitated the development of a "dual-use" strategy, assisting (through the post-Sputnik Defense Advanced Research Projects Agency) the same industries as before, but also including advanced materials, machine tools, optics, and so on. Radical though this development might seem, it is not very dramatic. The range of products/-processes assisted remains restricted; military imperatives still dominate; and hard-liners in the military think the diluted priorities have already gone too far towards civilian ends. A nonmilitary industry policy rooted in commercial imperatives is still in the utopian category.

However, since the early 1990s there have been atypical and increasing pressures on the military budget, which was previously generally untouchable. Some members of the military (Colin Powell) have publicly acknowledged the implications of the decay of the Soviet Union. There has been a growing constituency of business people, administrators, and academics concerned about the loss of American competitiveness in key industries. Finally, and most important, both the federal budget deficit and overseas debt blew out dramatically in the 1980s. The MIC is facing a fiscal crisis. The cut in the defense allocation in the 1991 budget, with the Defense Department accepting further cuts, symbolized this new state of mind. Yet change has occurred as a snail's pace. Base closures were some of the first outcomes of the cuts, hardly getting at the root of the problem. Many bills put into Congress to facilitate "economic conversion" have failed to pass. For lack of federal direction, legislatures in vulnerable states

(Arizona, Texas, California) are generating conversion packages, albeit of a small-scale nature. In short, the MIC is facing considerable scrutiny, but its influence on federal government priorities will continue in the near future.

The phenomenon of a MIC is not the sole possession of the United States. The United Kingdom and France have them also, for example. Of course, the other major representation was in the former Soviet Union. As in the United States, the Soviet MIC exerted enormous influence on government priorities. Indeed, it has been rumored that U.S. foreign policy deliberately escalated the arms race in order to cause so much imbalance in Soviet society that it would collapse. Collapse it did and, although elements of the Soviet MIC survive in the Russian confederation, it is merely a shadow of its past strength. Indeed, western groups that recognize the tenuous economic situation in eastern Europe are trying to promote stability by assisting in the dismantling of the former Soviet MIC. By contrast, the U.S. MIC remains robust, but the defense of its prerogatives will be very much harder as its reason for being disappears and its domestic industrial support base declines.

References

Appelbaum, R. P., Chambliss, W. J. 1997, *Sociology*, 2nd Ed., Addison Wesley Longman, NY.

Hooks, G. 1990, The Rise of the Pentagon and U.S. State Building: The Defense Program as Industrial Policy, *American Journal of Sociology*, 96,2.

Markusen, A., et. al. 1991, *The Rise of the Gunbelt: the Military Remapping of Industrial America*, Oxford University Press, NY.

Markusen, A., Yudken, J. 1992, *Dismantling the Cold War Economy*, Basic Books, NY.

MacCorquodale, P. L., et al. 1993, *Engineers and Economic Conversion: From the Military to the Marketplace*, Springer-Verlag, NY.

AUSTRALIAN HEAVY ENGINEERING AND THE NORTH WEST SHELF

The dominance of an orthodox mindset in Australian economic and political thinking has made it much harder for heavy industry there to win construction contracts for the offshore gas industry, particularly on the significant North West Shelf development off the coast of Western Australia. Because this corporate mentality is likely to prevail in future developments in the region, including in the massive reserves of the Timor Sea, this case study is of wider interest. It shows some of the linkages between prevailing economic theories, the history and prospects of manufacturing, and the operation of political processes. It captures the dilemmas facing the Australian economy (and, similarly, the economies of many other nations) and its industrial viability.

The market in heavy engineering products operates through an uncertain and expensive tendering process. Scale, demand instability, and product complexity combine to produce a more demanding environment than appears in textbook theories of the market mechanism. The heavy engineering industry has been crucial for Australian economic development, but its long-term future is uncertain. The late 1970s resources boom created great expectations and then destroyed them. One-fifth of the workforce, about 25,000 employees, lost their jobs in the early 1980s.

In that context, the North West Shelf gas project was very important. It has been the biggest private resource development in Australian history and the second stage involved considerable spin-offs. The contract chain descends through project management, design, fabrication, manufacture, and installation. This cluster effect involves a potential qualitative transformation of technology, skills, and institutional innovation, with implications for sustainable long-term industrial activity. The involvement of local industry in this process ought therefore to have been a matter of national interest.

The first stage of the Shelf development in the early 1980s did provide such contract opportunities. Between the first stage (North Rankin) and the second (Goodwyn), the operating company, Woodside Petroleum, adopted a more "hard-nosed" attitude to tender negotiations. Its signals have since been ambiguous, keeping its suitors on tenterhooks. At issue is who decides "local content," and what criteria apply. In this development, Woodside (now majority owned by Shell) appears to have been less than enthusiastic about Australian content. In submissions to government inquiries, Woodside has stated its strong opposition to local content guidelines.

This attitude is nurtured by the "market forces" orientation of the political-economic culture in Australia, embodied in a hands-off approach by government bureaucrats. Another dimension is a marked "colonial cringe." Australian heavy engineers are said to lack a track record in large-scale projects. Some problems with local industry's contribution to the first stage fueled this opinion.

However, the performance of overseas manufacturers has also been uneven. For the Goodwyn stage, a Korean company (subsidized by the Korean government) was contracted for the high-tech processing modules but botched the job, and the modules had to be rebuilt in a Western Australian yard. The "jacket" contract was let to an Indonesian-based British company, but it also botched the job and the work had to be repaired by others. Both contracts were costly exercises. In short, although Asian-based contractors may have cheap labor, quality and productivity cannot be guaranteed.

Track record is only part of the story. Local content rules threaten established client-supplier relationships in the international oil and gas industry. Such relationships can move from being a rational risk-reducing mechanism into a nonmarket patronage network.

Tendering processes in such an uncertain and unsupportive environment are inevitably costly, pragmatic, and messy. For example, one Australian company spent a small fortune developing a tender that was never likely to be taken seriously. A successful local bidder was forced by its own cost concessions to reopen the bidding process for its subcontractors; it had to set up its own welding shop by default, and ultimate returns were negligible.

By international standards, the joint venture partners in the North West Shelf have been allowed a high level of discretion in the tendering process. In no other country is the environment pertaining to offshore development so permissive. At the tough end of the regulatory spectrum are countries such as Brazil, Malaysia, and Indonesia, where local content rules are detailed and discriminatory. In countries with heavy engineering capability (Japan and Korea), public-private links ensure government production and export assistance. Even at the soft end, including Britain, Norway, Canada, and New Zealand, regulations have aimed at equal access for local suppliers to the tendering maze, regulating the adequacy and

transparency of the project time scale, specification standards, and so on.

Regulatory structures are not viewed sympathetically by civil servants in Australia. Yet, given official concern about the persistent deficits on Australian current account (cf. the U.S. position in Tables 5.2 and 5.3), it is remarkable that the possibility of some causal relation between Australian content and such deficits has not been recognized.

The case has been put quite forcefully. A 1989 federal government report contrasted the strong policy structures in place overseas with the bureaucratic failings in Australia: absence of strategic vision, lack of departmental coordination, and inadequate information bases. It emphasized the importance of using such projects as a long-term foundation for an indigenous capacity in design engineering and project management. Local industry has continued to push for more coherent alternatives, similar to those embodied in the British and Norwegian regulatory structures. These formally embody a "full and fair opportunity" for domestic industry to compete. Such structures are actually compatible with the orthodox economic perspective of facilitating market processes. Regulations covering the North Sea were passed during the Callaghan Labour years in Britain, yet continued during the conservative Thatcher period. The impact of such regulations on the British off-shore supply industry has been most significant.

In Australia, in spite of such overseas precedents and recommendations at home, an interdepartmental committee dominated by orthodox economists recommended inaction. Not only was it apparently ignorant of overseas practices, but it was also

(continued on next page)

reluctant to propose effective policy initiatives because they might offend multinational oil and gas companies. In mid-1990 a political deal was arranged, belatedly and pragmatically, to enhance Australian content in the second stage. It included the state and federal governments and the relevant unions. The federal government offered duty-free entry of component inputs (estimated to be worth at least $20 million); the West Australian state government offered payroll tax concessions; the unions offered industrial harmony, and the local contracting companies discounted their bids. As a consequence, local companies got some contracting crumbs from Woodside.

The national Labor government in 1990–1991 reiterated its commitment to industry self-regulation, claiming that a broader commitment to a level playing field philosophy prevented the establishment of local content rules. Even more pressure was placed on local industry in 1991 by the government offering duty-free entry to equipment deemed more advanced than that available in Australia. This was accompanied by an overall decline in tariff protection for heavy engineering.

The North West Shelf project has generated public benefits in the form of royalties and tax revenues; it has also generated some work for the domestic heavy engineering industry and contributed to the establishment of a skills base, such as in geomechanics technology. Yet there has been no government concern for the net impact on the Australian economy or, in particular, on the balance on current account. The project will export substantial volumes of liquefied natural gas, contributing to the merchandise trade balance. However, there has been no concern for the negative side of the balance sheet: the cost of imported equipment, the interest costs of imported capital, and the royalties on imported technology (coupled with the possibilities of discretionary transfer pricing of these costs). There is no awareness that a heavy engineering industry sustained on domestic projects can enhance its capacity as an export-earner.

The fast track to attaining cooperation of project operators is to make acceptance of such goals a prerequisite for the granting of licenses for resources development. However, offshore waters are the constitutional domain of national government, which in Australia has consistently been indifferent to assertive action. Given this indifference, the state government has tried to go it alone with "local content" guidelines, but has had meager success to date. Attention has now moved to the Timor Sea, in which the Northern Territory government (with even less leverage than Western Australia) has a specific interest. Woodside has proposed another major project (the $900 million Laminaria development), but little of its infrastructure will be sourced locally.

Before the national Labor government lost office in 1996, the matter was again referred to a government committee. It issued its report belatedly in 1998. In the face of aggressive submissions from industry, ill-informed submissions from the relevant bureaucracy, and (by now) a compliant national government, the report was remarkably critical of the status quo. However, it was ignored by the media and, to date, by the (now) conservative government.

A major concern illustrated in this case study is the unsatisfactory relationship between a multinational project operator and local industry. The overall picture is a pessimistic one. The vital importance of a more enlightened economic policy structure, one that could encourage the viability and dynamism of local industry through its integration into large projects involving global capital and technology, is a lesson not yet learned in Australia.

References

Australia, House of Representatives Standing Committee on Industry, Science and Technology 1989, *The North-West Shelf: A Sea of Lost Opportunities?* Canberra, November.
— 1998, *A Sea of Indifference: Australian Industry Participation in the North West Shelf Project*, Canberra, March.
Jones, E. 1997, Slim pickings on the heavy side, *Australian Financial Review*, 1 September.

6

Innovation

However far modern science and technics have fallen short of their inherent possibilities, they have taught mankind at least one lesson: Nothing is impossible.

— *Lewis Mumford (1895–1990) U.S. social philosopher*

In this chapter we distinguish between *"invention"* and *"innovation,"* terms that are used to describe key aspects of the technology formation process. We discuss the design, research, and development activities linking them, and look at some of the social constraints on innovation. Innovation is essentially a creative process; we explore some of its attributes and associated activities, which are complex and inherently difficult to analyze satisfactorily. The creative and intuitive leaps required for good design are central to innovation, but are all too easily overlooked. We look at the ways inventors, innovators, designers, and researchers work and the sorts of organizational structures they need.

While addressing the dynamics of innovation it is important to consider how innovation is embedded in the social and economic structures of society. Choices that are made by engineers and others regarding the needs that innovations should address and the ways innovation should be carried out have a major effect on the quality of life in the societies in which these innovations are applied.

The timing of decisions is vitally important to the success of the innovation process. This raises questions about the extent economic activity is cyclical. If it is, do particular types of innovation tend to cluster at different stages in an economic cycle?

Finally we discuss the social context of innovation and some of the ways it is constrained. We consider ways in which professional engineers may incur liability as a result of their work, and discuss the increasing importance of legal constraints on product innovation.

Key Concepts and Issues

The Innovation Process

In this book we use the term *innovation* for the process of developing new ideas and inventions to the point at which they are ready for commercial or public sector use. The purpose of innovation is the creation of new technology.

The innovation process is driven by new ways of doing things, known as "technology-push," and by the identification of needs, known as "demand-pull." Innovation is the basis of a dynamic engineering profession; its outcomes include new and improved products, processes, and systems.

New technology is essential to economic development, technological progress, and the advancement of science; it plays a principal role in the creation of real wealth. Economists, sociologists, and technology analysts have developed many theories as to how new technology is created. In these theories, innovation usually plays the central role.

The word innovation is derived from the Latin *innovare*, to renew or alter. The term is thus used both for the process of deploying the product of invention and for the successful result.

The four main elements of the innovation process are: invention, innovation, design, and research and development.

To some extent invention is always happening, but there are times that are more propitious for it than others. In the past, apart from Schumpeter (1939) and Rosenberg (1976), economists generally do not seem to have known what to make of invention. They have usually assumed that it is exogenous (originates outside the economic system) and therefore, by implication, does not need to be paid for. In recent years some economists have begun to treat invention as endogenous, a legitimate part of their system. Coombs et al. (1987) give examples of this. Data from Mensch (1979) lends support to the idea of some coupling between invention and economic activity, and there is now increasing acceptance of this.

Design is central to innovation, and to engineering generally, but is seldom mentioned. Engineers who actually create products, processes, and systems need to manage the innovation process much more deliberately.

In many areas of technology, research and development (R&D) run broadly in parallel with continuing innovation, sometimes leading, sometimes following.

INVENTION

The term *invention* is commonly reserved for an original concept or discovery. It consists of "the first idea, sketch or contrivance of a new product, process or system, which may or may not be patented" (Freeman et al. 1982). The education and training engineers receive should prepare them to be inventors and designers. Most engineers are employed by a company or corporation that pays them to invent products, processes, or systems, and the engineers' employers generally retain the ownership of the inventions. Case Study 6.3 highlights one outcome of employee dissatisfaction with this arrangement. Some organizations in innovation-intensive areas are starting to share the ownership of inventions with their employees. Engineers and others continually come up with new ideas and new ways of solving problems. Whether or not they should really be called inventions is probably a matter of their degree of novelty. Are these "inventions" solutions that would be obvious to anyone familiar with the "state-of-the-art" in the technical area involved, or do they demonstrate a special level of creativity?

For clarity in this discussion it is important to distinguish between a new idea, or invention, and an innovation, which is the result of turning the invention into a new product, process, or system that is part of the ordinary commercial or social activity of a country. The importance of distinguishing between invention and innovation is recognized by sociologists (Hill 1988) and economists. Invention is an essential part, but only a part, of the commercial development process.

Some people argue that the complexity and scale of technology are now such that the individual inventor no longer exists, or even argue that everything there was to be invented has already been invented. Others argue, equally forcefully, that invention can only be performed by gifted individuals (Jewkes et al. 1969). As popular television programs have demonstrated, individual inventors do still exist, not all of them with a formal technological education, although education would often be helpful in turning their particular invention into a successful product.

Is invention really a lottery, a flash of divine inspiration, or can we actually devise organizational structures to assist invention, such as design teams? The answer is important. We really do need a Thomas Kuhn to help us to understand the paradigms of engineering and its associated management functions (Kuhn 1962).

The following propositions can be used to sum up invention:

- ◆ invention cannot simply be "ordered" to happen; it depends too much upon the intuitive flash of insight;

- invention can be encouraged by a sympathetic social ethos and an awareness by the most creative people of the fields that are most in need of new discoveries;
- the social and economic context in which the inventor is embedded focus the process of invention on changing the particular world in which the inventor lives;
- invention is not always dependent upon advanced technological training, but is more likely to emerge from such a background;
- invention is only weakly dependent upon the particular phase of the economic cycle, but encouragement is likely to be greater and the invention more likely to be exploited if times are favorable.

The last point is a vital one because it affects whether the invention obtains the backing that will take it through the early research and development stages that are essential to commercial success. For this to happen there must be awareness of a potential demand, and resources must be made available. A former research director of a major British chemical company, ICI, discussing the prospects for a novel product, has said "I would not like to make an invention now. You wouldn't get the money [to develop it]. You have to make an invention at the right part of the investment cycle." (Birchall 1973). Later in the chapter we will look in some detail at the effects of trends and cycles.

Wiener (1993) usefully distinguished between the early part of the process of invention, a "pre-industrial" stage, when an idea for a product was conceived (for example the idea of recording and playing back sound), and the later part of the process, when actual products were being defined that could put the idea into practice. The idea could antedate its industrial use by several decades. The later stages of formulating and realizing a product are more closely tied to the economic cycle and have a tendency to be carried out simultaneously by individuals in different countries.

Lawrence Hargrave (1850–1915), the great Australian aeronautical inventor and experimenter, illustrated the lack of interest in protecting and commercializing his work that has cost his nation the economic benefit of much original effort. Hargrave's most significant work was probably his 1889 development of the radial rotary airscrew engine, which played a major part in European aviation. His curved airfoil surfaces were used in the wings of the Wright brothers successful airplane. The Wright brothers carefully protected their work, but Hargrave rejected the idea of protecting his rights in his inventions. He was unable to secure financial or engineering backing for his work. The fact that none of his Australian audience believed in the possibility of powered flight would not have helped (Moyal 1976).

What Are Inventors Like?

Is the individual inventor typically an eccentric, a "mad scientist," a genius, a villain, or a hero? Essential personal characteristics of successful inventors seem to include:

- intelligence (but by no means a genius IQ);
- burning curiosity;
- a high degree of autonomy and self-sufficiency;
- a tendency to introversion, but open to stimulation;
- resilient and nonconformist;
- risk-taking; hard working;
- intuitive and creative;
- not afraid of apparent contradictions; and
- not predisposed to reach a premature conclusion.

Individual inventors tend to be loners, who pay scant regard to established conventions and existing authorities. It may be precisely these perceived negative attributes that make a person an inventor in the first place. If society does not produce and tolerate people who are prepared to question accepted wisdom and authority, then it will produce no inventions.

One of the great inventors and innovators was Thomas Alva Edison (1847–1931), who made improvements to the telegraph and telephone, invented the phonograph, an incandescent lamp, and a dynamo to run it. He also founded the laboratory that later became the research headquarters of the General Electric Company. The contrast between the working styles of Edison and his sometime employee, Nikola Tesla (1856–1943), the inventor of the electric induction motor, is discussed in Case Study 6.1.

Not all experiences with the patent system have been unsatisfactory. As we discuss in Chapter 8, global engineering demands global protection of intellectual property. One prolific inventor, who founded an international firm supplying equipment to the mining industry, said:

> It would be difficult to over-emphasize the role of the patent, trademark, design and copyright systems, administered by governments on a world-wide basis, in protecting the market for a product resulting from years of research, development, prototype and field testing and the gaining of user acceptance together with the very large investment entailed before any return is achieved (Warman, 1989).

EDISON AND TESLA: CONTRASTING STYLES OF INNOVATION

Thomas Edison and Nikola Tesla are the two men given credit for bringing electricity to the world in a practical sense. Their very different approaches to innovation provide a useful lens for looking at development of new technologies.

Edison succeeded by tireless tinkering. Once he got an idea for a project he wouldn't let it go. His method of innovating is often called a "trial-and-error" method. He employed an array of machinists in his laboratory, so that at any moment he could ask for a part or a model to be crafted to try his ideas for a new design. Often this was a sketch on a piece of paper, with a note like, "John Ott—make this" (Millard 1990, p. 35). Probably one of his longest on-going projects was the search for a better light filament. His list of ideas in this area was endless; he tried every material he could get his hands on, including a wide variety of fruits. He sent explorers around the world, at great expense, to find a certain variety of bamboo so that he could see if it would make a good filament. This continual process of trial and error is how Edison liked to work, and he designed his West Orange laboratory around it.

Edison's laboratory had many of the characteristics of the old guild system. Edison handed on tasks and projects to his many employees and he, the master craftsman, worked on his own projects, as well as helping those working for him. More than anything else, Edison stressed the importance of craftsmanship. If a worker made shoddy products, he was fired. Communication and learning from others were high-ly valued. The entire lab was built with free and easy communication in mind; the experimental rooms were grouped together and Edison encouraged his workers to share their answers. When one worker came up with a particularly good or ingenious solution to a problem, he showed it to the others so they could learn from it. Beyond his abilities as an inventor, Edison was also an agile businessman and this greatly aided his climb to the top. He showed great ability in attaining financial backing, profiting from his inventions, and saving money everywhere possible. He realized that selling off the patents for inventions was rarely profitable for the inventor, so he founded businesses to apply his inventions, especially the incandescent lamp. He set up everything: central power stations, ore mining, wire laying, and installation companies. His business skills took him beyond being simply "the wizard of Menlo Park"; they made him a dominant force in the new technological era.

In total contrast to Edison, Tesla was hopeless at business and was more an eccentric genius than an experimenter. Tesla scorned Edison's trial and error method because of the time and resources it wasted. Tesla claimed that he "never made a mistake." (Hunt and Draper 1964, p. 129) There were two reasons for this. Tesla was well educated, so he could replace much guesswork with solid scientific reasoning. In addition, he had an amazingly vivid imagination, so vivid that when talking about an object he could imagine it tangibly in front of him. This mental ability allowed him to envisage a device and then tinker with it mentally. He visualized it so sharply that he could see where the machine would have problems. He would then work out how to fix these problems, and make the changes in his mental picture of his machine, perfecting it until he could find no more problems with it. Only then would he have this mental image turned into reality, and invariably it worked as envisaged. What Edison did in the laboratory, Tesla did in his mind.

Tesla's laboratories were quite different from Edison's. Tesla had few workers, and he required that they be highly skilled, even able to work without drawings. He expected his workers to be able, at least in part, to visualize a problem and solve it mentally the way he did. Unfortunately, the media and society generally were unsympathetic to Tesla because of his ethnic background and eccentricities. Tesla came from eastern Europe, so he was seen as mysterious, coming from a land of "wandering gypsy bands with their gaudy wagons, smoldering campfires, furious dances, and the plaintive strains of violins." (Hunt and Draper 1964, p. 183)

Furthermore, as is not uncommon with genius, Tesla's personality was full of eccentricities. One of the most apparent was his acute germ phobia. He insisted on the most rigorous standards of cleanliness and sterilization. He also had almost no social life; while very young, he had decided not to marry or engage in any sort of relationship with a female, and he avoided the distracting presence of females in his life until the day he died. Tesla was plagued with misunderstanding throughout his career. When journalists could not understand his advanced theories, they made up their own interpretations that threatened to destroy his reputation as a rational scientist.

Tesla and Edison show two rather extreme styles in innovation. Tesla's approach is unrealistic on a large scale because it requires inherent genius in the innovator. On the other hand, Edison's approach didn't leave room for Tesla to exercise his genius while he worked for Edison in the latter half of 1884. Edison was economically and personally committed to direct current (DC) circuitry, so Tesla's ideas about an alternating current (AC) motor were not given much attention by Edison. Livor, a new business associate, went on Tesla's behalf to talk to Bachelor, one of Edison's closest partners, about getting a raise, but Bachelor flatly denied him saying "No, the woods are full of men like Tesla." (Seifer 1996, p. 38)

Tesla's true potential was not seen by Edison and his group. Tesla left Edison after he was refused a completion agreement of $50,000, which Tesla said he had been promised but had failed to get in writing—an example of Edison's occasionally questionable approach to business and Tesla's rather trusting approach. Tesla's value to Edison was minimized because Edison's system didn't allow Tesla to contribute his genius in areas that were outside of Edison's goals and ideas.

Edison's guild-style laboratory had exemplary aspects in that it insisted on precision quality and it left room for tinkering and working on individual projects so that people could try out their ideas, but the workers were largely unskilled. Edison himself disliked mathematics and theory. Tesla commented, "I was almost a sorry witness of his [Edison's] doings, knowing that just a little theory and calculation would have saved him 90 percent of the labor … the truly prodigious amount of his

(continued on next page)

actual accomplishments is little short of a miracle." (Seifer 1996, p. 34) Obviously, Edison's style of innovation had room for improvement. The year before he died, Edison complained: "I have made very little profit from my inventions ... I have taken 1,180 patents, up to date ... these patents have cost me more than they have returned me in royalties ... We have a miserable system in the United States for protecting inventions from infringement." (Jewkes et al. 1958, p. 112)

These different styles can help us explore our own approaches to research and innovation today. Do we allow people to use laboratories to work on their own innovations and projects? Do our methods allow open communication between machinists and experimenters so that each has the opportunity to learn from the other's solutions? Are our systems open enough to recognize and make full use of genius like Tesla's when it comes along?

Source

Based on work by Daniel Skarbeck, Stanford University 1997.

References

Hunt, I., Draper, W. 1964, *Lightning in His Hand: The Life Story of Nikola Tesla*, Sage Books, Denver.

Jewkes, J., Sawers, D., Stillerman, R. 1958, *The Sources of Invention*, Macmillan, London.

Millard, A. 1990, *Edison and the Business of Innovation*, Johns Hopkins University Press, Maryland.

Seifer, M. 1996, *Wizard: The Life and Times of Nikola Tesla*, Carol Publishing Group, NJ.

Air Commodore Sir Frank Whittle (1907–1996) patented the gas turbine engine in 1930. He complained in 1981: "It has been the custom of certain individuals to treat me as a gifted amateur, inventor, etc., and talk of taking my child and sending it away to school, to say that I have no production experience, etc., and, I believe, to represent me as a somewhat difficult and temperamental individual" (Whittle 1981).

The Irish dramatist George Bernard Shaw (1856–1950) summed up the essential characteristics of the inventor in his *Maxims for Revolutionists* (Shaw 1903): "The reasonable man adapts himself to the world: the unreasonable one persists in trying to adapt the world to himself. Therefore all progress depends on the unreasonable man." The temperament that makes people good inventors would appear to make it difficult to assimilate them into industrial and production organizations, let alone the commercial world. Are these habitual questioners of authority really fit to manage projects? Individual inventors commonly need to look for someone else to fund, produce, and market their inventions; this is the important role of the product champion, who will carry the invention through the innovation process and drive it to commercial success. We need to give much more attention to the development of astute product champions.

Invention and Innovation

Intellectual Property Rights: Who Owns an Invention?

Invention seems essentially something an individual does. One model of the process of invention is that it takes place in a brief, earth-shattering moment, as a blinding flash of inspiration. Ownership of the invention is then, *prima facie*, with the individual inventor. But how adequate is this model? Even if correct, it says nothing about the physical and mental preparation of the inventor that leads up to it. In organizations set up to support and encourage invention and innovation, is it meaningful (or even possible) to identify which member of the team is the inventor?

Intellectual property rights can be defined and protected in a variety of ways (see Chapter 8):

- novel ideas can be patented: in return for making the idea public, the inventor is granted control over its use for some limited period;
- designs and trademarks can be registered;
- copyright exists on drawings, film, audio, and videotape.

The extent of protection varies (in some countries parodies are breaches of copyright).

Why Don't All Inventions Become Innovations?

What may initially seem an obvious solution to some actual or potential need is rarely easy to realize in practice. Carrying through the innovation process requires much more resources than invention. Most inventions never become commercially effective realities because:

- the idea may not be good enough;
- not enough resources are available to develop the idea;
- well-established comparable or competing solutions already exist.

How Does An Invention Become An Innovation?

Development of the three main types of innovation—product, process, and system—probably demands progressively larger levels of resources. The relative complexity of the prospective innovation is also important. Innovation of a product typically requires organization of design and development, production, and marketing. This may require more resources than an individual can command. The general acceptance in the nineteenth century of the limited-liability company as a means of pooling resources was a major advance in this area.

Don Frey (1991) vividly describes the frustration of failing to sell potentially successful ideas to the people within his organization who could have brought them to realization. He also describes the challenges and joys he experienced as a product champion. Despite Henry Ford II's grudging approval, and with a lead time of only eighteen months, his team turned the dull Ford Falcon into the once-in-a-lifetime success of the Mustang, which sold more than 400,000 in the first year. His later achievements for Ford included the introduction of disk brakes and radial tires. When Frey moved to Bell and Howell in the early 1970s, he had to look beyond short-term financial criteria and consider where the company business as a whole needed to go, so as to survive the transition to the video-digital age. The success of Bell and Howell's videocassette recorder and CD-ROM projects would not have been possible without his long-term support. One of the reasons for Silicon Valley's success (Case Studies 2.4, on p. 98, and 6.3, later this chapter) has been the presence of a group of venture capitalists who could, as a condition for providing finance, induce inventive groups of technologists to work with people with managerial and financial skills.

Most inventions languish unused and unsung; for a variety of reasons they never make it through the innovation process. Based on detailed study of a number of inventions and innovations, workers at the Open University developed a prescription (see box, Ten Steps to Innovation) for successfully negotiating the path from invention to innovation.

INNOVATION

Innovation is the process of developing the original concept or invention through to the economic deployment of the resulting product or process. Technological innovation has been defined by the Organization for Economic Cooperation and Development (OECD) as "comprising those technical, industrial, commercial and other steps that lead to the successful marketing of manufactured products and/or to the commercial use of technically new processes or equipment." Effective development of policies to encourage change requires, as a first step, a clear appreciation of the processes involved. This is one reason why a weak and vague definition of innovation in terms such as "the introduction of change in order to improve performance," is so unhelpful.

It is useful to distinguish between innovation and entrepreneurship. According to Drucker (1985), the term entrepreneur was coined by J. B. Say around 1800 to describe a person "who shifts economic resources out of

Ten Steps To Innovation

1. Identify a need.
2. Produce an original and creative solution or find a new way to use an existing one; you now have an invention!
3. Check for originality (patent and literature searches).
4. Don't upset the apple cart (like trying to alter the way a keyboard is arranged).
5. Build a working model (both to check out the idea and to demonstrate it to others).
6. Learn the patent system and protect the invention (not too early or too late).
7. Decide how to produce it; be realistic about demand and costs.
8. Sell yourself with the invention.
9. Find yourself a product champion (inventors doesn't usually have the skills, authority, or capital to make their invention a success).
10. Persevere; successful innovation requires exceptional commitment and effort.

After: Elliott 1986.

an area of lower and into an area of higher productivity and greater yield." The principal ends and means of entrepreneurship are therefore economic; it need not involve technology at all. On the other hand, innovation is essentially about generation of new technology.

The most common model of how products or processes are created is known as the "linear model of innovation." This model treats the innovation process as a sequence from basic research, through applied research, to experimental development, and thence to deployment. The model seems to have been first used by Dr. Vannevar Bush in a 1945 report to President Truman. It sought to define ways to apply in peacetime the approaches used to achieve wartime technological successes, such as the development of the atomic bomb. Bush stated:

> Basic (pure) research leads to new knowledge. It provides scientific capital. It creates the fund from which the practical applications of knowledge must be drawn. New products and new processes are founded on new principles and new conceptions that in turn, are painstakingly developed by research in the purest realms of science (Bush 1960).

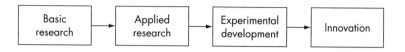

Simple Linear Model of Innovation

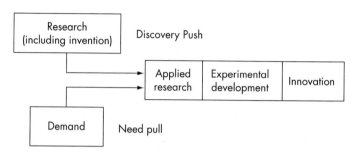

Revised Linear Model of Innovation

Figure 6.1   Simple and Revised Linear Models of Innovation

The linear model of innovation has been described by Ronayne (1984). It was further refined, as indicated in Figure 6.1, by developing the technology-push and demand-pull models. Technology-push is said to occur when a discovery or invention provides the stimulus to initiate the innovation process (e.g., the laser, which became the basis for compact disc data-storage technology). Demand-pull is when management or customer need is the stimulus. The pickup truck or coupe-utility, usually abbreviated in Australia to utility or "ute," is a classic example of demand-pull. In the 1930s a farmer wrote to Ford Australia saying that he could not afford both a truck and a car. He asked them to make a vehicle with a coupe front to take the family to church on Sundays. The back was to be a roadster-utility box for taking the pigs to town on Mondays. Ford's chief body engineer, Lewis Brandt, drew up the idea. It looked so good that Ford went on to successful production, and the rest of the industry soon followed.

Despite the attractive simplicity of linear models, innovation inherently involves many complex and sophisticated interactions between different groups of people. Any linear representation must therefore be recognized as a rather gross oversimplification.

Kline's more sophisticated "chain-linked" model, Figure 6.2, reflects his multidisciplinary approach to innovation. It draws on the thinking

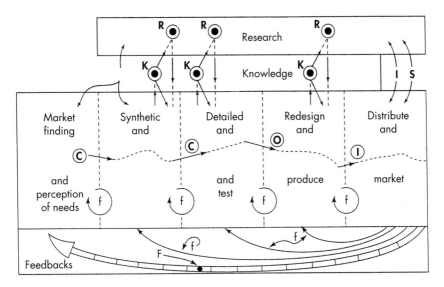

Figure 6.2 The Chain-Linked Model of Innovation. Key: f, F = Feedback information loops; C-C-O-I = Central chain of innovation; I = Instruments for research and industrial use (two-way flow); S = Support of long-range research by industrial corporations; C = Enabling of designs by science and questions arising in science from problems in design; K, R = Links to existing knowledge and research output.

of a number of disciplines, each with an incomplete model of innovation. This model places science alongside the innovation process, with accumulated knowledge acting as an intermediary that is called upon in stages. Research is only invoked when the required information is not otherwise forthcoming. Kline's model suggests a rather weak coupling between research and innovation. It challenges conventional wisdom about the purposes and organization of research, particularly applied research. It has gained increasingly wide acceptance, despite the fact that many countries and companies would be unhappy with the idea that it really described how they operate (Kline 1995, p. 184).

Jonas (1979) has argued that the hallmark of modern progress over the last two centuries has been a kind of dance between science and technology, with each feeding on the other: "Intellectual curiosity is seconded by interminably self-renewing practical aim." This implies a much more complex and dynamic web of interactions than would be allowed by a linear model of innovation. It also reminds us that science has gained at least as much from technology (including engineering) as technology has from science.

A useful distinction may also be drawn between product innovation and process innovation. Product innovation includes both the progressive improvement of an existing product and the introduction of a new product to the market. The latter case involves not only the risk of problems in the manufacturing process, but also the risk that the product may not sell. Product innovation may therefore be less controllable than process innovation; it is a higher risk operation, but with potentially higher returns.

Process innovation is the introduction of a new or improved process, often for manufacturing a particular product. Although process improvement is often of a cost- or labor-saving nature, whenever it is relatively expensive to implement it tends to be a long-term strategy.

Much process innovation has an incremental rather than a step-like character. This is central to approaches such as the Toyota production system, discussed in more detail in Chapter 7, which seeks to remove all storage of intermediate products from the production process. Any limitations or bottlenecks are then immediately apparent and have to be addressed. Because process innovation usually takes place within an organization, the risks and returns are substantially under management control.

The most comprehensive approach to innovation is system innovation, which involves developing all the elements needed to carry out a production or other function. A traditional example of system innovation is the turnkey project, where all the purchaser has to do is turn a key and the system is expected to work according to specification. In a computing environment this could involve a complete new assemblage of personnel (with all necessary training), procedures, hardware (equipment), software (computer operating systems and applications), and all necessary documentation. With quality certification becoming increasingly the norm, the fundamental importance of adequate documentation can hardly be overstated. The sheer physical quantity of documentation on paper has become overwhelming. Even for the older conventional submarines still in service, the documentation can weigh twenty tons. Innovation in recording, storage, and presentation of data has become a major challenge, and the documentation for new submarines is only manageable because it is handled electronically.

DESIGN

The stimulus to innovation may come from invention (possibly arising through government or corporate basic research), or market demand or, more probably, an interaction of both; but the link from the stimulus to the end product will involve design. As with innovation, the term *design* is commonly used to describe both the process and the outcome. Social, cultural, and other perspectives are involved in framing design problems and shaping the choices made in the design process. These perspectives make design a social, value-laden process.

As Alexander (1964) and Cross (1989) point out, people have always designed things. In traditional societies and many craft activities, designing is not really separated from making. However, with the introduction of new technologies, and the advent of mass production, designing and making have been separated. Under these circumstances, the design process must be essentially complete before production can start. Design includes all the creative, analytical, and documentation activity that needs to be carried out to define a product, process, or system so that it can be made. Design is thus the process of converting an idea into a form, of changing existing conditions into preferred ones, of putting onto paper or into a computer the specific features of a real product, process, or system. Design is a creative and disciplined problem-solving process that involves defining the problem as well as resolving it. Using the principles and methodologies of art and science, technical information and imagination are used to define a new structure, machine, process, or system that will perform a desired function with the maximum economy and efficiency. Because the outcomes of design are normally intended to be used by people, the design process needs to include input from likely users in as direct a way as possible. An example is computer software design, where users are now recognized as an essential part of the design team.

The importance of the design process is often overlooked by nonengineers. The concept of *synthetic design* includes invention and preliminary and analytic design. Preliminary design is the creation of a new arrangement of a product or system created within (or just beyond?) the current state-of-the-art. It is not yet a detailed physical design but a general layout, with only enough detail and calculation to confirm the viability of the concept. Analytic design is the use of mathematical or other models to arrive at optimal or near-optimal designs under known constraints. This process has been radically advanced by the advent of computer simulations.

Design is a central engineering activity, accomplished despite huge gaps in knowledge; it ranges from the development of computer software or consumer goods to the planning of construction work or the organization of a factory or production run. Little wonder that the hallmarks of a good engineer are creativity and ingenuity; Regrettably the engineering curriculum does not always aid the development of these qualities. About halfway through most undergraduate courses, engineering students are commonly struck by a sudden and distressing disjunction. Previously the emphasis was on engineering mathematics and science. They were dealing with problems where all the necessary information was available and there was one, and only one, correct answer. Suddenly they meet design, where all the necessary information is available only for trivial problems. Even worse, there is a virtually unlimited number of solutions, all of them compromises, and all to a greater or lesser extent wrong! Something is too heavy, too costly, unreliable, and so on. Scientific knowledge is an indispensable analytical tool for the engineer and often provides a basis for new developments. However, as we discussed in systems terms in Chapter 2, our analysis is never an analysis of the real problem in all its complexity, but of more or less simplified models.

Koen argued that the design process involves so much iteration and feedback that any model of it should be a guideline rather than a rule. Such a model is a morphology, a set of heuristics useful for solving an engineering problem. Koen summarized the engineering method as being the use of heuristics, including all the types indicated in the box opposite, with the purpose of causing the best change in a poorly understood situation within the available resources. This is essentially similar to the "soft systems" approaches discussed in Chapter 2.

■ The Design Process

Any good designer has a distinctive approach, and there are probably as many models of the design process as there are writers on design. We like the model given in Figure 6.3 (see p. 292). The steps are shown in sequence, but each step sheds light on the previous ones and may require going back through one or more of them. Thorough understanding and documentation of the design decision-making process has become even more important with the introduction of strict legal liability for products, discussed later in this chapter, and of the quality processes described in Chapter 7. The accompanying Key Concepts and Issues box (Steps in the Design Process) gives an outline of the design process. Design according to this model is still essentially an iterative procedure.

Case Study 6.4 at the end of the chapter shows how the design process may be divided up in practice. J. J. C. Bradfield carried out the first four steps, from reviewing the transport needs of Sydney through to specifying the preferred alternatives for a bridge across the harbor. Ralph Freeman was responsible for the second part of the design process—analyzing the alternatives in detail and documenting the preferred solution for Dorman, Long to implement.

The Heuristic

The heuristic approach is one commonly used approach by engineers to analyzing real problems and finding acceptable solutions to them. It is an approach, based on experience, that may solve a particular kind of problem, but offers no guarantee of success. The most commonly encountered form of the heuristic is the rule of thumb.

A detailed and stimulating review of the central place of the heuristic in modern engineering practice was given by Koen (1984). He described four characteristic features:

1. it does not guarantee a solution;
2. it may contradict other heuristics;
3. it reduces the search time for solving a problem; and
4. its acceptance depends on the immediate context instead of on an absolute standard.

Koen categorized engineering heuristics into five groups:

1. simple rules of thumb and orders of magnitude;
2. factors of safety;
3. determinants of engineers' attitudes to their work, such as: always give an answer, the best you can with the resources available;
4. ways to keep risk within acceptable bounds by making small changes in the state-of-the-art;
5. rules of thumb in resource allocation: resources need to be allocated as long as the cost of not knowing exceeds the cost of finding out. (©1984 by the American Society for Engineering Education)

Heuristics resemble the fuzzy logic increasingly used in expert systems. Heuristics are more art than science. They would seem to fit comfortably with the models of learning and knowledge proposed by Kolb (1984).

Both the review of the problem area and the choice of the specific need that is to be addressed are relatively subjective processes. They set the design agenda and belong in a broadly political and commercial strategic domain. Engineers should be encouraged to be much more involved in this key part of the design process. This is the point at which broad issues such as ecological sustainability of design outcomes are most

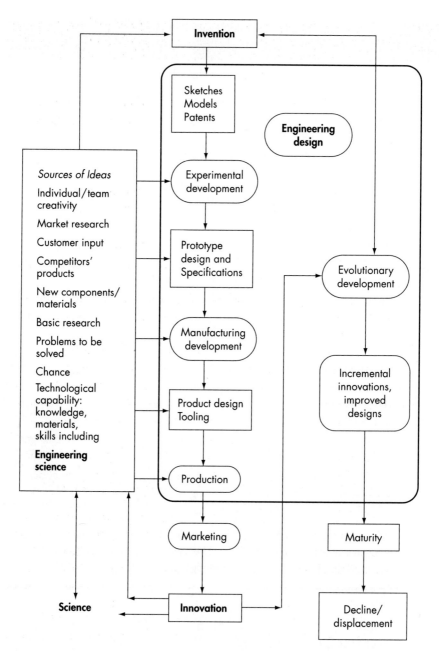

Figure 6.3 A Model of the Design Process. *Source:* Adapted from Roy and Cross 1983

Steps in the Design Process

1. Review the problem area and select the need that is to be addressed.
2. Define the requirements that a solution must meet—prepare a *functional specification*.
3. Synthesize a range of alternative solutions.
4. Model the alternatives.
5. Analyze the alternatives in terms of the functional specification.
6. Select, document, and communicate the preferred solution.
7. Implement.
8. Repeat, as often as necessary, some or all of the above steps.

effectively addressed. It is also where basic ethical choices are made about professional priorities, including what problems and issues will and will not be addressed.

Hudspith (1996) describes the skills discussed here as "the art of inquiry." This includes formulating good questions and seeking out good answers. Good questions are those that look for: (1) underlying causes or key features of an issue; (2) understanding of the origins of a situation; and (3) linkages between problems or issues. Finding good answers calls for research skills, including a willingness to go out into the community and find things out at first hand. Hudspith argues that both types of skill need to be learned. His innovative Engineering and Society program at McMaster University in Canada, described in Case Study 3.4 (p. 154), is designed to teach both types of skills. An important benefit from such a program is that its graduates are effectively prepared for working in interdisciplinary design teams. They are also likely to appreciate the need for their designs to be compatible with the context in which they will be applied. This awareness is an aspect of what Vandenburg (1996) describes as "preventive engineering," an approach intended to prevent or minimize the negative effects of any technological change.

The preparation of the functional specification, spelling out the need that is to be addressed, is a surprisingly challenging part of the process. No possibility should be ruled out before it has been considered thoroughly. The difficulty is in defining the requirements in sufficient detail so that costs and benefits can be assessed and alternative proposed solutions compared without unnecessarily constraining the solution.

The most creative part of the technical process is probably synthesizing a range of solutions. It requires immersion in the problem, an open mind, and a broad experience of possible directions in which a solution can reasonably be sought. Techniques such as brainstorming are useful. Creativity must not be compromised by moving too quickly to the subsequent analytical stage.

For a complex artifact, the design process may be divided into investigation of the system as a whole and of its various components. In a large engineering firm the organizational structure will reflect this specialization. An aircraft engine company will have, for example, separate sections for compressors, turbines, combustors, and nozzles. The component design process is further subdivided into theoretical and experimental modeling or rig-testing. Although usually regarded as part of the design process, these activities tend to be the ones most closely associated with research and development. Getting these separate groups to work together effectively is essential. As the auto industry example in Chapter 7 shows, it may require an organizational structure that rewards professionals as project team members rather than as technical specialists.

Checklists can help to ensure that no significant parameters are neglected. The potential users, usage patterns and environment, manufacture, serviceability, and sale or distribution of the project, process, or system all need to be analyzed. Techniques such as trade-off tables are available to help choose between alternatives that meet the essential requirements of the functional specification.

Once a clear decision is made, the preferred solution must be optimized. A design audit is also needed to check performance parameters and factors such as safety, maintainability, and reliability. The selected design must be documented thoroughly so that it can be communicated clearly and unambiguously. When the design audit is complete it is possible to proceed to building prototypes, testing, production, and marketing. In general the cycle is a continuing one and once the product reaches the market the design cycle starts again, if this has not already happened. After a solution has been chosen and implemented, management systems need to ensure that feedback is obtained on performance in practice, both to make continuing improvements to the design and for liability reasons.

A rational model of the design process is shown in the accompanying Key Concepts and Issues box on the design process. In practice design is rarely as orderly or coherent a process as the model suggests. Despite its limitations, this model is extremely useful, especially for beginners. Management of the design process is discussed in Chapter 7; increasingly complex expectations for new products, processes, and systems virtually demand design by teams rather than individuals. Diversity in the design team is important for the variety of perspectives it brings. With globally distributed teams, early face-to-face contact is an important part of team building. Initial team-building exercises can help to build the mutual understanding, confidence, and trust that are essential for effective operation. It is not simply a matter of throwing a diverse group together and leaving them to it. Prompt and constructive resolution of tensions in the team is essential to reap the creative benefits of diversity; conflict resolution may require skilled group management.

■ The Importance of Good Design

It is hard to overstate the importance of engineering design to the success of manufacturing industry, both now and in the future. During the design stage of the development of a product, the limits on cost, quality, performance, and technological content are essentially determined. Although the costs of this early stage of product development are perhaps only 10 percent of overall costs, this stage of the process determines between 75 and 90 percent of the total costs (White et al. 1990). Unless good decisions are made at the design stage, the manufactured product has little or no future. High-quality design is also a means whereby firms that are not at the leading edge of technology can develop attractive and reliable products that are competitive with any in the world.

The new focus of engineering is on designing, manufacturing, and marketing products and services that meet the specific wants of specific consumers. Along with this, manufacturing must rapidly become flexible enough to meet rapidly changing individual demands. The ideal production batch size may now be as small as one. The new systems that are being developed to provide this flexibility seem to offer, as an exciting side effect, the possibility of providing much more satisfying work.

The choice of the country whose codes, standards and conditions of contract are used for international project work can be important, because it tends to advantage that country's suppliers of goods and services. It can assist and encourage the participation of local suppliers, manufacturers, and contractors, leading to: better service and maintenance, more competitive local industry, lower balance of payments problems, technological spin-off, and more employment and skills transfer.

Moves towards sustainability suggest future directions for technology generally and for professional engineering activity in particular. In what contexts will ecologically sustainable technology have to operate? Will sweeping social and economic changes be essential prerequisites for long-term solutions to the problems we currently face? What pressures will be needed to bring on these changes? There may not be simple quantitative answers to these questions, but they must not be ignored.

RESEARCH AND DEVELOPMENT

Following (or sometimes preceding) invention, and supporting the innovation process, are the activities known as research and development (R&D), a continuous range of activities from the most academic of university research to the field testing of new hardware. The Key Concepts and Issues box classifies different stages of R&D, based on the work of Hitch and McKean (1960) and Zuckerman (1961).

The principal qualities required of a good research worker are curiosity, creativity, initiative, and analytical ability. In development work the rather idiosyncratic qualities of curiosity and creativity may be augmented or even replaced by the ability to work well in a team. Basic research is usually relatively cheap; development is expensive. Because much more money is spent on development than on research, and because it involves teamwork rather than individual brilliance, its management function is correspondingly more developed. This is not to say that management skills are not required for research. Applied research costs money and is usually performed by a team under contract, with the normal requirements for proposals, estimates, budgetary control, progress reporting, contract review, and so on. These requirements, however, are less exhaustive for research and, as compared with development, are usually interpreted more liberally.

The management of development is an area of critical importance for the company and the nation. Development costs are high, and dependent production costs are even higher. The loss of a project manager is generally reckoned to delay a major project by a year. This is a serious problem, because delays cost money and cause political pressures to build up.

Management techniques and style for R&D need to be rather different from those for, say, marketing or military operations. As Hitch and McKean (1960) point out, the armed services have all too frequently tried to command research workers to invent new weapons to specification, just as they would command an infantry platoon to march by the right flank. The results can be disastrous, especially when applied in a university situation. We look in the next chapter at the changing organizational approaches in engineering, and many of the lessons there apply to R&D

The Process of Research and Development

1. *Pure basic research:* Research pursued by scientists to satisfy their own curiosity and to advance knowledge. This includes specialist work at the frontiers of knowledge and reviews of broader fields. An important characteristic of the work is that it should be original and publishable. Basic research is especially, but not exclusively, the province of universities. Universities are long-term institutions, and a university that values its international standing will value these characteristics highly.

2. *Directed basic research:* Sometimes called strategic research, this is basic research chosen to be undertaken in a field of potential value and importance. The pursuit of economic, social, and technological goals reveals fields in which further scientific knowledge is needed. Although the work should have the same characteristics and quality as basic research, because it is goal-oriented rather than curiosity-driven, some direction of the work is appropriate. This work may take place in a university, a research institute, or in industry.

3. *Project applied research:* The objective here is the attainment of a defined practical goal. The goal may be a product or a new process. Any suitably equipped institution may undertake applied research, but security classification or commercial considerations may inhibit publication, making the work less attractive to some of the best researchers. Because of its larger scale, project applied research is usually performed in groups.

4. *Operational applied research:* This is the improvement of existing processes or operations. It relates closely to process innovation or to maintenance engineering and gives a systematic basis for operations management. A typical example would be the work of research laboratories that provide operational support to the defense forces.

5. *Component development:* The simultaneous development of the various subsystems of a product or process. Economic pressures dictate that all components be developed simultaneously where feasible. Once the satisfactory performance of the separate components has been demonstrated they may be combined into what is, hopefully, a working unit. Aircraft, cars, ships, trains, and their engines are all produced in this way.

6. *Systems development:* The development of complete systems, including reviewing the interaction of the various components and all testing and field evaluation.

management. Creativity and skillful probing are generally more effective research attributes than orderly scheduling to a clearly defined objective. When employed in development, a rigid approach is likely to result in a clock-watching response from the individual contractor. If the contract stipulates acquisition of thirty-six data points, the contractor will take thirty-six data points, no more and no less, no matter how interesting or uninteresting the test proves to be. This is not the contractor's fault, merely a sad recognition of commercial reality.

The question of contract research, especially for the military, raises important issues that have started to be addressed in studies of the politics and sociology of science. What sort of social and physical environments have to be established to encourage and support the creativity of original thinkers? How might such people react to the sorts of constraints imposed by a world of security clearances and loyalty checks?

An equally difficult problem is that even when new weapons (or other innovations) are developed, unless they are compatible with the existing culture they may be sabotaged. Fallows (1985) illustrates this point with an analysis of the fate of the M-16 rifle. Studies of actual combat in World War II had shown that nearly 80 percent of combat soldiers did not fire their weapons during battles. Firing an ordinary rifle seemed rather futile: its user had no real control over the nearby area. However, it was found that soldiers with automatic weapons, which could hose down an area, did fire, and so did those close to them. The conclusion reached was that every infantryman needed to be provided with a low-recoil automatic weapon. A lightweight rifle was developed that fired a very light, high-velocity bullet, so that three times as many rounds of ammunition could be carried as compared with the standard army rifle. The problem was that this approach conflicted with the prevailing ordnance corps culture. The army weapons establishment believed that a rifle should be designed for aimed fire and accuracy over long distances. This very senior group detested the new weapon that was being imposed on them from outside. They sabotaged it, requiring a number of arbitrary and inappropriate modifications, including the use of an unsuitable propellant that caused fouling and jamming in action. This sabotage caused unnecessarily increased casualty rates among American infantrymen in the Vietnam War. A congressional investigating committee delivered scathing criticism of army behavior, but its report was quickly buried.

There is a world of difference between bureaucratically induced one-dimensional activity and the gadfly brilliance of the basic researchers surveyed by Kuhn (1962). In Britain, Darlington (1948) prepared the way by diagnosing a hardening of the arteries of scientific journals, research groups, and learned societies. Instead of a Ministry of Science, he prescribed a Ministry of Disturbance, to undermine complacency. In the words of Hinshelwood (1965):

> If we wish to plan research, we can only do so by assembling a community of people of varied and complementary talents, operating with strategic flexibility and in an atmosphere of curiosity in which the members know and understand what the others are talking about and respect their leaders.

Wiener (1993) complained about the inflexibility of highly structured and organized research where the roles of individual scientists were assigned in advance. He likened this arrangement to a bulky and ponderous machine that was not easily redirected to meet "the needs of the world of ideas." He added that:

> New ideas are conceived in the intellects of individual scientists, and they are particularly likely to originate where there are many well-trained intellects, and above all where intellect is valued.

An example of institutionalized megadollar science is the current work on nuclear fusion. When progress is good, the flow of R&D will be continuous, with ideas moving from inception as basic research, through applied research, to development. Although sequentiality is not essential, it is certainly useful to consider fundamental research and applied research as different phases of the R&D process. Management and institutional barriers between two phases will tend to inhibit the flow. By the same token, however, these barriers could be skillfully exploited by a management seeking to impose broader goals such as cost or arms control. These are the locations at which the major program reviews should take place. Effective management control requires review processes along the way; incipient innovation may flounder at these review processes for technical or financial reasons. If an idea survives the tests in the earlier research phases, it emerges into the usually costly later phases of development. Striking the best balance between the different facets of R&D is difficult and is a problem that needs to be addressed in both national and global contexts.

Quinn (1980) criticizes "the gimmickry of simplistic formal planning and Management by Objectives approaches," and advocates "logical incrementalism" as an interesting alternative that could overcome these difficulties. Benveniste (1987) reminds us that the management of professionals, particularly in relatively open-ended activities such as R&D, needs to be different from traditional approaches to the management of a blue collar work force. He stresses the need to distinguish in the management of professionals between the organization's stated goals and its actual operational goals. Operational, or real, goals include many transactions that cannot be explained in terms of the rhetoric of the organization's stated goals.

Failure by any country to take R&D seriously would show a short-term outlook that is a recipe for technological sterility. Similar traps that countries need to avoid are excessive reliance on strategic planning and other mechanistic approaches. These techniques typically lack the dynamism and adaptability needed to address a rapidly changing technological and social environment.

∎ Research and Development Spending

The United States accounts for roughly 44 percent of the industrial world's R&D investment total, easily leading all other countries in terms of such support. The National Science Foundation projected total U.S. investment in R&D to reach $205.7 billion in 1997, divided as shown in Figure 6.4.

In inflation-adjusted terms, R&D increased by 3.8 percent in 1997, 3.2 percent in 1996, and 5.9 percent in 1995. Development funding of new and improved products, processes, and services totals over 60 percent of all R&D funding. In real terms it has actually been falling slowly since 1990 (largely a result of post-Cold-War curtailment of defense spending).

Before 1970, Federal agencies supplied slightly more than half of all dollars spent on R&D in the United States. By 1980, industry had become the leading source. The most recent data show industry providing about 60 percent of the R&D funds in the United States, whereas the federal

Research and Development in America, 1997

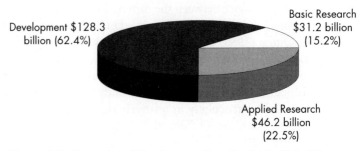

Figure 6.4 Research and Development Spending in the United States, 1997. *Source:* National Science Foundation.

TABLE 6.1

Engineering R&D Expenditure at American Universities and Colleges*			
	1981	1990	1995
Current dollars	$967	$2656	$3545
Constant 1987 dollars	$1243	$2371	$2776

*Dollars are in millions.
Source: U.S. Bureau of the Census, Statistical Abstract of the United States 1997, p. 606.

government provides 35 percent. Universities, colleges, and nonprofit organizations provide most of the remaining 5 percent. Industry, the federal government, and academia are major *performers* of R&D, responsible for 70 percent, 10 percent, and 13 percent, respectively, of national R&D spending. Almost all development spending is by industrial firms (Office of Science and Technology Policy 1995a, b, c, d).

Engineering R&D expenditure at universities and colleges in America has increased in recent years, as shown in Table 6.1.

How does the U.S. R&D picture compare with other countries? The largest seven R&D-performing countries are the United States, Japan, Italy, Canada, Germany, France, and the United Kingdom. Industry/business was the leading R&D performer in each, and the majority of industry's R&D performance was funded by industry itself, followed by government funding. Government share of funding for industry R&D performance ranged from as little as 1 percent in Japan to about 20 percent in the United States and France. Figure 6.5 shows this expenditure for 1996-97, expressed as a percentage of GDP.

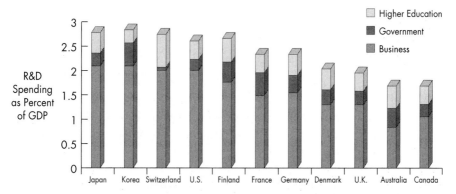

Figure 6.5 National Research and Development Spending as Percentage of GDP, 1996–1997. *Source:* Australian Bureau of Statistics, 1999, p. 599.

Total R&D expenditures generally stagnated or declined in each of the largest R&D-performing countries in the early 1990s. Geopolitical changes have resulted in cutbacks in government support for defense-related R&D, particularly in the United States and France. Additionally, for most of the seven top R&D-spending countries, economic recessions and general budgetary constraints had the effect of trimming both industrial and government sources of R&D support.

THE EFFECTS OF TRENDS AND CYCLES

One interesting hypothesis is that there are long-term patterns of economic activity and that particular types of innovative activity are advantaged at different parts of the pattern. This hypothesis does seem to fit much of the available evidence. If a parameter such as GNP per capita, national energy requirement, or research funding is plotted over time, certain underlying tendencies become apparent. If the parameter is plotted on a logarithmic scale, a straight line plot may result. This would indicate that there has been an exponential growth pattern.

The concept of exponential growth as a secular trend (a consistent variation from a reference level) is useful. It traditionally provided an effective guide for decision making for the planners charged with the provision of electricity supply. This guidance was important because there was usually a ten-year lead time between the decision to build an additional plant and its commissioning. A study undertaken for the Club of Rome on *The Limits to Growth* (Meadows et al. 1972) performed a valuable service by arguing that, for an Earth of finite size with finite resources, there were absolute limits to growth. Most of the scenarios it examined predicted an eventual catastrophic termination to the exponential growth curve. However, the system representation it used was greatly oversimplified, and the detailed models have not stood up well to criticism. The continuing importance of this work was that it prompted a deeper analysis of the sustainability of continuing economic growth.

The economic historian looks not only at the secular trend of economic growth, but also at singular and periodic events, and random behaviors that disrupt the curve. Although the long-term trends may be clear, the short-term behavior of an economy is as unpredictable as the weather. There will be fluctuations that can only be classified as random. Statistical information on economic performance will reflect all sorts of external factors, including political developments and even the weather itself, which has a strong effect on productivity in outdoor work. Recent developments in nonlinear dynamics have demonstrated that behavior that may appear to be random is not necessarily so; the theory suggests that deterministic behavior patterns may be discernible within so-called chaos. An excellent introduction to this field has been given by Gleick (1987). Although truly random behavior is neither deterministic nor predictable, statistical analysis can be employed to classify the data and give guidance on the anticipated range.

▌ Individual Product Life: The Sigmoid Curve

How do individual products perform? Our investigation should proceed beyond the smooth curve of a broad indicator, such as GNP, and focus on the components that make it up. Each of these may have a characteristic lifetime. Aggregate growth curves that are exponential in type may actually be the summation of many separate sigmoid (S-shaped) curves, where each curve represents the sales of an individual product. The model assumes that each particular product starts out on an exponential growth path, reaches the limits of its useful life, tapers off, and is eventually superseded by a "superior" product. The spread of steamships and railroads (Hart 1949), the mechanization of farming, and the introduction of radio and television all followed this trend. Another example is the improvement of power station efficiency discussed in Chapter 9. The example of Figure 6.6 is taken from Grilliches (1957) and records the introduction of hybrid corn into the United States.

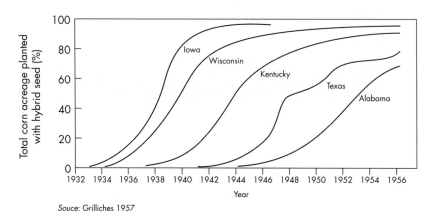

Souce: Grilliches 1957

Figure 6.6 Rate of Diffusion of Hybrid Corn in the U.S. *Source:* Z. Grilliches 1957, "Hybrid corn: An explanation in the economies of technical change," *Econometrica.*

The definition of the sigmoid curve is:

$$y = 1/\left(1 + e^{-x}\right)$$

where

$$x = k\left(t - t_0\right)$$

The elapsed time is $\left(t - t_0\right)$ and k is a constant that describes the rate of diffusion of the product.

The curve is characterized by an exponential growth followed by an asymptotic leveling off to some upper limit, the phenomenon of saturation. In this sort of model it should be clear that the sigmoid curve is in itself a major simplification, and the determination of the diffusion rate constant k is an essentially empirical process. Economists are beginning to integrate such models of diffusion with the more sophisticated elements of microeconomic thinking.

▪ Cyclical Behavior

Although scattered data on prices are available from millennia ago, the systematic recording of such data developed over the last five hundred years. For sustained analysis the information is only satisfactory over the last two hundred years. Figure 6.7 presents data on British and American wholesale prices from 1800 to 1950. The prices represent an aggregate for wheat, cotton, wool, coal, beef, timber, iron, and sugar. Each data point represents an average over a period of ten years. Wholesale prices give an indication of the ratio of monetary to real income. A high price index indicates a high level of demand-pull, and a possible surplus (profit) for investment.

A spectrum analysis of such data over the last two hundred years indicates three predominant wavelengths, where the fundamental wavelength is of approximately fifty-three years. This is termed the Kondratieff Cycle. There are also two significant shorter wavelengths that become familiar to anyone who invests in the stock exchange or worries about the likely selling price of their house.

Although work at Harvard confirmed a fifty-four-year cycle in British wheat prices since the thirteenth century (Bannock et al. 1987), only a few cycles have been recorded for most data, and there is a range of possible explanations for long-cycle behavior. For the economy as a whole, it is difficult to decide whether a genuinely periodic phenomenon has been observed or whether each peak had an autonomous cause. It has been suggested that all the peaks coincide with wartime activity, that gold

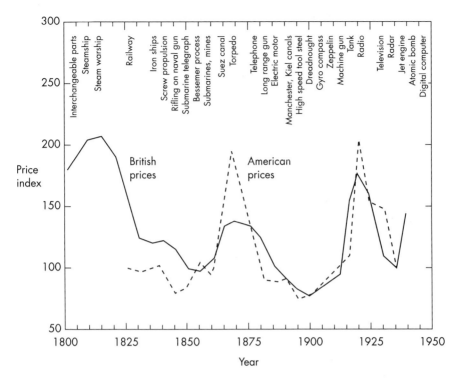

Figure 6.7 Long Cycle Activity and some Major Innovations

was discovered just prior to each expansion, and even that the British Empire expanded in a similar cycle. Nevertheless, the trends observed from 1950 to the present have accorded remarkably well with expected long-cycle behavior.

Long-cycle theory seems plausible as an historic phenomenon. It could, in principle, be used to forecast the long-term behavior of the economy or even important political and social trends. If the mechanism is periodic, then the economy last peaked around 1966. We would then be in an economic trough where innovative activity should nevertheless be strong. It would be some years, however, before its benefits for the economy would be seen (Mensch 1979).

Mensch claimed to observe long-cycle behavior not only in economic activity and innovation but also in invention, although for the latter the peaks were more diffused. On average the economic peak occurs about twenty-eight years after the innovation peak, which itself occurs about twenty-seven years after the corresponding peak in inventive activity.

It can be argued from a reading of history that most inventions occurred in times of prosperity that afforded a few the freedom to invent. If this is correct, then a possible explanation for long cycles is the average two-generation gestation period between a major invention and the associated economic return, which may in turn favor further invention. In a real sense the inventor is working for the benefit of his or her grandchildren; this gives poignancy to the old saw "rags to riches and back again in three generations."

■ Clustering of Innovations

If an argument for long-cycle behavior is to be credible, it must demonstrate the mechanism and reasons for the periodicity. Schumpeter (1939) based his interpretation of the role of innovation on the theory of cyclical behavior. He rigorously excluded inventions that he considered to be exogenous to (not accounted for by the model of) the business cycle. He therefore placed great emphasis on the role of the entrepreneur. Schumpeter found that the introduction of innovations was bunched into periods of rapid expansion, when cash-rich investors are like Christmas shoppers and the risks of a new product failing to sell are smallest. By the peak of the cycle the risk of failure has increased. Case Study 2.4 (see p. 98) suggests that the availability of essential supporting services in specific locations may also encourage particular sorts of innovation.

It is possible to examine the history of innovations to see when they were taken up for production. At the same time economic parameters may be plotted in an attempt to detect any correlation. It could be argued that the immediate perceived need of society provides the demand-pull for the innovation.

Schumpeter recognized the limitations of his hypothesis: "If innovation and qualitative change are the fundamental elements in economic development then no amount of quantitative analysis can reveal the truly significant pattern." His contribution was not, therefore, as deterministic as it superficially appeared. A later example of the sort of pattern he seemed to have in mind started with the inventions of the jet engine around 1930 and of the atomic bomb, ballistic missile, and digital computer during World War II. By the early 1950s, the economic cycle was poised for sustained expansion, and these inventions were deployed as full economic innovations. Each product innovation started its own new sigmoid curve, and superposition of these and others contributed to the long-cycle activity. This is similar to the view ascribed by Williams (1982) to those historians who treat the long-cycle upswings as successive industrial revolutions.

Advances in technology result from a number of dramatic precursor events and a far greater number of relatively pedestrian accomplishments. The direction of growth could be predicted from the former; the rate of growth is dictated by the latter. One of the causes of the eventual tapering off of exponential growth in a specific technology is the cumulative effect of serious competition in that area. This competition is likely to stimulate innovation in production processes aimed at increasing productivity and maintaining profit margins. As a technology comes of age it tends to become specialized, over-organized, and unprofitable.

ESSENTIAL INTERACTIONS AND DRIVING FORCES

As the above descriptions of the design process and the R&D process indicate, there are marked similarities between the two. Each starts with a review, which may be on quite a broad front, proceeds to work through the clarifying definition phases, then on to the analysis and testing of alternatives, and culminates in implementation. Is this coincidental or could we be looking at two facets of the same process?

The best way to address this question is probably to ask: What is actually done in industry? Do we have different and parallel design and development efforts? Do we have an activity that is called research? The answers appear to vary from company to company and from product to product. The culture in some innovation-based organizations has been such that research is esteemed more highly than in a university; staff are encouraged to do basic research, to publish results, and to attend conferences. The availability of tax concessions for research is an incentive to classify a wide range of design and investigatory activities as research. To name just a few, the research laboratories of IBM and Bell have been justly famous, although shake-outs in American industry in the early 1990s trimmed IBM's spending rather severely. In other companies the opposite is the case. If research is done, it is hidden in the budget and called development testing or some other euphemism. The organizational culture determines what is respectable and therefore determines how activities are described and funded.

In mechanical engineering the distinction between design and development is a little more clear cut, if only because design tends to take place at the desk, computer terminal, or drawing board, and development tends to take place on the test bed. Some companies deliberately foster this difference. Others regard it as a gap that should be bridged, for example, by having the same engineers responsible for both design and development aspects of a project. This gives a healthy vertical integration of experience for an engineer who works on a given product from inception through to manufacture. The use of computer workstations for computer aided design (CAD) and for data acquisition, and the extensive use of simulation software are tending to blur the distinction. Even so, it is clear that at a certain point design stops and development takes over. Although the links between them are strong it is nevertheless meaningful to discuss design and development as two separate and distinct processes.

Design and development interact in an iterative manner, often according to an explicitly stated cycle. A U.S. example of this is the annual model changes of a particular car, such as the Chevrolet Corvette. The Corvette, first sold in 1953, was intended to test the market for an American-made sports car that could compete with British cars such as the MG and Triumph. Typically, automotive engineers work on the designs of cars well ahead of their intended market release, and each more complex design takes longer to develop. Therefore, the development cycle is one of decreasing frequency. Chevrolet is no exception. In 1953, Chevrolet engineers and stylers were working on the 1955 Corvette. In 1959, the all-new 1963 Corvette was already on the drawing boards. The latest Corvette, the 345 horsepower (257 kilowatt) C5 model, launched in 1997, had its genesis in 1988, taking a full nine years to bring to market.

Various inputs, including previous development programs, new legislation, customer surveys, and others, will foster identified and desirable design changes. The design process will have been through all of the steps described above and will, at the modeling stage, interact with research to launch a new development phase. As the new model is developed there is close interaction with the design office to enable the final design for that year's model to be "cast in concrete" by the deadline.

If design and development are not different aspects of the same process they might perhaps be viewed as a double helix of two closely coupled spirals, much like DNA molecules. They must proceed according to the same wavelength and be able to interact closely. Design could be expected to precede development but be closely related to it.

The earlier phases of the development cycle, associated with invention and research, will tend to be technology-push phases. The inventor has an idea and is in the difficult situation of having to sell it to a sponsor and obtain backing. In these early days management and financial institutions need to be extremely sensitive and perceptive to decide which inventions can eventually succeed and which are more likely to be a waste of resources. Case Study 6.2 on the Gossamer Condor and Albatross illustrates a very direct demand-pull: prize money of £150,000 sterling, and international fame for the winning team.

THE SOCIAL CONTEXT FOR INNOVATION

The innovation process is affected by a range of practical, social, and ethical constraints, as well as the cost and the associated difficulty of finding funding. Merson (1989) described how, in the fourteenth century, China led the world in technology. The Chinese rulers, under pressure from barbarian invaders, then took a deliberate decision to discourage exploration and contact with other cultures in order to foster internal stability. The price of the resulting harmony was minimal technological progress. The Europeans overtook the Chinese, especially, but not only, in military technology. This reminds us that the social context is important for innovation. It may also suggest either that a certain degree of openness, of ferment of ideas, may be important for innovation.

The world wars in the twentieth century were each a major stimulus to the development of technology. It seems an unfortunate fact that many rapid and important technological developments, such as the birth of the nuclear industry, have been in response to the pressures of actual or threatened war. Society and technology interact strongly, and such interactions are complex and not necessarily benign. This is one reason why technological innovation is such an inadequate and unsatisfactory measure of social progress.

The importance of the social context in promoting or constraining innovation has been discussed in detail in this chapter. We discuss social issues in the context of technology transfer in Chapter 8.

■ The Organizational Context

As an innovation proceeds through its development phases, increasing resources of capital and human power are required. Provision of capital requires surplus productivity and an organization capable of correctly channeling available funds. The role of the project manager is especially vital.

Because the potential innovation can be killed easily and immediately at this expensive stage it is important that a demand-pull should emerge. The most important function of technological marketing is to be aware

GOSSAMER CONDOR AND ALBATROSS

At 8:40 A.M. on June 12, 1979, Bryan Allen landed the Gossamer Albatross on the beach at Cap Gris Nez, France after pedaling the human-powered aircraft for 2 hours and 49 minutes across the English Channel. Allen won the Royal Aeronautical Society's second Kremer Prize and closed a remarkable chapter in the history of aviation (Grosser 1981).

A British industrialist, Henry Kremer, had offered prizes of £50,000 and £100,000 sterling, respectively, for the first human-powered flights:

◆ around a figure-of-eight circuit encompassing two markers half a mile apart; and

◆ across the English Channel from England to France.

The Kremer contest recaptured the excitement of aviation's early days. Sheer flying performance was the challenge, and normal commercial, military, or even safety considerations were virtually absent. In Los Angeles, Dr. Paul MacCready took up this challenge. He recruited a brilliant team that was ideal for the task. MacCready was determined to win both prizes. As is commonly the case with novel design problems, achieving optimum results required winning the goodwill and support of potential collaborators. For the figure-of-eight course

the Gossamer Condor was designed, and for the channel crossing the aircraft designed was the Gossamer Albatross. The optimum human power source was found to be a cyclist driving the propeller.

After surveying the available information on human power output achievable as a function of time, MacCready realized, in a flash of inspiration while on vacation, that there were important advantages in machines that could fly very slowly, thus demanding low-drive power that could be sustained over relatively long periods. There were many associated technical benefits. The undercarriage could simply be two flimsy plastic wheels from a toy fire engine. The huge wing area and low cruising speed (only 13 kph) meant that starting and landing could take place at a fast walking pace, greatly facilitating manual assistance from the ground. The structure was to be of weight-saving, wire-braced construction—basically a canard pusher with the pilot suspended on wires beneath the wing. (A canard configuration has the wing at the rear and a smaller airfoil at the front as a stabilizer. A pusher uses a propeller at the rear pushing it along.)

The aerodynamics, materials, and control aspects of such a design all needed careful consideration and extensive development. In the original Condor design, hang-glider

of the market and the technological possibilities, and to marry the two. In most technically advanced countries, innovation also occurs, at least in the later development stages, within an organizational context, usually a large company. This fact in itself generates expectations of the innovation. There are promotion prospects and even jobs on the line. The organization and its employees have strong vested interests. The innovation, which has been pushed from below by the researchers and pulled from

construction predominated. The wings were of a sail-like construction with a single mylar surface. The pilot was open to the elements. By the second (successful) version of the Condor this had evolved to a wing with a sophisticated airfoil shape and an enclosed cockpit for the pilot, which also served as an aerodynamic keel for increased stability.

Good aerodynamics is obviously essential in aircraft design. The success of these aircraft depended on the wing designs of Dr. Peter Lissaman. The original twenty-seven-meter span, hang-glider-type wing had a circular leading edge spar, a mylar sail shaped by a few ribs, and a taut trailing edge wire. The flow over this wing separated badly at the leading edge at angles of attack other than the design value. The wing also had virtually no turning capability. In response to these difficulties, Lissaman used an airfoil design program to produce a new family of wing sections. These were to be true airfoils with low drag over a wide range of attack angles, high lift-to-drag ratio, and a nearly flat pressure (under) surface to enable the mylar to retain the desired profile. In conjunction with the new family of airfoil sections, a tapered span was to be used with an increased length of 29.5 meters (Burke 1980).

The secret to the turning capability required for the first Kremer prize effort was suitable aerodynamics and good control. These MacCready achieved (after much consideration and "trial and error") by ap-propriate reverse twist on the inner portions of the wing, thus giving a more favorable angle of attack. A control lever was used to alter the wing twist in flight, making the required 180-degree turns possible.

The structural designers' motto was: "If it doesn't break, it's too heavy." A wide variety of materials was used and much experimentation took place. For the Albatross, carbon-epoxy composites were used in compression members and kevlar was used in tension members. Wing ribs were cut from styrofoam sheet capped with carbon-epoxy strips. Tensilized mylar gave an excellent wing surface. Without the cyclist, the total mass was only twenty-five kilograms.

The four-meter diameter propellers also needed special attention; for the Condor, a constant chord was used, but for the Albatross, with its requirement for efficient cruise operation, a special propeller with an unusual highly tapered plan was designed by a team at the Massachusetts Institute of Technology.

Source

Based on personal communication from Peter Lissaman, University of Southern California, and:

Burke, J. D. 1980 The Gossamer Condor and the Gossamer Albatross: A Case Study, In *Aircraft Design*, AIAA Professional Study Series, Report No AV-R-80/540.

Grosser, M. 1981, *Gossamer Odyssey*, Dover, NY.

above by the management, acquires a certain momentum in terms of the mass of personnel and hardware involved and the pace at which it is moving. For a given company or national organization, this is not the only innovation: others are coming along at the same time. If the planners and product champions are effective, some of these innovations will be linked into a dynamic and synergistic system. Case Study 6.3 describes the social dynamics of some organizations driven by these sorts of goals.

FAIRCHILD AND THE FAIRCHILDREN

"Fairchild" and "Fairchildren" are magical names in the semiconductor industry. Fairchild Semiconductor was a startup company formed by eight young engineers from Shockley Semiconductor Laboratories. The Fairchildren were companies that, in turn, spun off from Fairchild in the 1960s. The chain is in fact even longer; Shockley itself was formed by engineers from Bell Labs. Subsequently, new companies such as Zilog and Sequent Computer Systems have sprung from Fairchildren such as Intel. A somewhat ironic pattern seems to have emerged that is possibly inherent in the nature of innovation itself: startup companies are most likely to spring from other startups.

Is there a common thread that makes the proliferation of startups more likely in certain engineering and/or business climates? Why is the draw of a startup so great for an engineer? Is it a problem that switching companies has become so attractive for employees? What happens to the parent firm?

The first spin-off from Fairchild was Rheem Semiconductor, an unsuccessful venture founded by Fairchild's former general manager in 1959. In 1961, Signetics Corporation was founded by four Fairchild employees, and later that year four of the founding engineers of Fairchild left and created Amelco. Several employees of Fairchild founded General Microelectronics in 1963, subsequently becoming millionaires. Other Fairchildren include Molectro, Intel, Advanced Micro Devices, Four Phase, Qualidyne, Computer Micro Technology, Advanced Memory Systems, Precision Monolithics, and LSI Logic. One engineer stated: "It got to the point where people were practically driving trucks over to Fairchild Semiconductor and loading up with employees." (Hanson 1982, p. 115)

Why did so many employees leave Fairchild? It had been the first company to shape Silicon Valley into a place of innovation, so it may appear strange that its talented engineers would leave it to pursue their own innovations. Working conditions at Fairchild were not bad. In fact, the only serious problem was the restriction that Fairchild Camera and Instrument, the parent company, placed on Fairchild Semiconductor, which was that semiconductor profits were to be spread throughout the company. However, east-coast management styles were very different and more strict, and the engineers may have felt they did not have as free a rein as they wanted.

One inducement was the ready availability of venture capital in Silicon Valley. Venture capitalists liked Fairchild employees, who knew all about the newest technology. Robert Noyce commented: "There was this beautiful new horizon with all sorts of money available to go ahead and try whatever you wanted to try." (Hanson 1982, p. 114). Gilbert Masters, a Fairchild employee in 1967 and 1968, explains that engineers at Fairchild were bored, emphasizing that Fairchild was "a good place to work, but there was the lure of the industry and new ideas: people knew they were taking a risk and they wanted a challenge."

The pattern of "chronic entrepreneurship," or startups from startups, undoubtedly pushed technology forward in Silicon Valley. When an environment is created that is conducive to innovation and discovery, it becomes difficult to keep employees from being enticed away by other, more attractive job opportunities. The bright young engineers not only wanted to use their talents to create new technology, but they also knew that there could be a huge payback for their hard work. By the era of Fairchild Semiconductor, the boring, tie- or lab-coat-wearing engineer had become a risk taker, no longer committed to playing it safe. Engineers acknowledged the challenge and excitement in the industry and many decided to go for it. "It was the beginning of an entrepreneurial flourish in this valley," Robert Noyce remarked. "Suddenly it became apparent to people like myself, who had always assumed they would be working for a salary for the rest of their lives, that they could get some equity in a startup company. That was a great revelation—and a great motivation too." (Hanson 1982, p. 91) The desire of individuals to leave the security of a company and try their luck in an expanding field is still a widespread phenomenon, and engineering firms presently have to provide big incentives to keep their brightest engineers.

The fact that firms such as Shockley, Fairchild, and Intel have had trouble keeping their most innovative employees at first appears to be a negative phenomenon, but it may actually be something positive—both for the general public and the workers themselves. Engineers moving between companies broadened their experience and benefited from their job move, with big pay increases, as well as excitement. Furthermore, there are few drawbacks to changing jobs in Silicon Valley because most of the people have worked with or for everyone else. One engineer explained, "It wasn't that big a catastrophe to quit your job on Friday and have another job on Monday, and this was just as true for company executives. You didn't necessarily even have to tell your wife ... You didn't have to sell your house, and your kids didn't have to change schools." (Hanson 1982, p. 113) The consumer potentially benefits as well, because engineers spread their talents and ideas around, helping to create new technology.

When the "would-be" Fairchildren contemplated leaving Fairchild, they were aware of the possibility of failure, but they had many factors on their side, including a huge market, plenty of money, and amazing talent. Although businesses in the fast-moving Silicon Valley cannot remain static, the cycle of startups from startups may now be slowing as the industry matures. Fairchild was a huge company at the dawn of an industry, and its own founders became Fairchildren themselves when innovation beckoned them. Startup companies today have bigger problems finding capital and a niche; moreover, current industry leaders such as Intel have the means to fund their employees and provide motivating research and design products. As the new "best and brightest of the semiconductor industry" show us where Silicon Valley is headed, we will find out if the current giants can create a varied enough environment to slow the model of Fairchild and the Fairchildren, or if they too will eventually find themselves gutted to create yet more startups.

Source

Based on work by Wendy Marinaccio, Stanford University 1997.

References

Hanson, D. 1982, *The New Alchemists*. Little, Brown, Canada.
Masters, G. 1997, Personal interview. 10 Nov.
Sutherland, E. 1997, "Technological Trends." 1995. Online. Internet. 27 Oct.

Organizations and Innovation

How Big Do Innovative Organizations Need To Be?

There are roles for both small and large companies in innovation. Small organizations can respond faster to changing conditions. For example, they played a major role in developing high-efficiency lighting devices (Davis 1993). Where the minimum economic scale of activity is very large, the organizations involved also tend to be large. The advantages of reducing commercial uncertainty by internalizing activities were discussed in the previous chapter.

In many countries there seem to be advantages for companies to be either small or large, but not middle-sized. The institutional structure of many financial systems does not encourage the creation of medium-sized companies, which are widely seen as being captives of the organizations that finance them. Medium-sized companies are neither as flexible as small companies nor as stable and able to finance and defend themselves as large ones.

How Do Organizations Involved in Innovation Ensure That They Have a Future?

Organizations may be set up initially to market a single product. To be successful they will then probably need to provide a range of technical and other support, including accessories, detailed operating instructions, and training packages. Even so, individual products have limited commercial lives. Once the market for the original product is saturated, there need to be other products that build on the reputation and support the marketing of the original product. One way to address this problem is to support and encourage other inventors.

Lotus was perhaps the most high-profile example of the single-product software company. Although its personal computer spreadsheet, Lotus 1-2-3, led the market for many years, it was increasingly challenged by comparable products that were supported by compatible word processing and other applications. By 1993 these products were being bundled together and marketed as complete office software systems. Lotus survived by broadening its product range to match.

Structuring Organizations To Develop and Capture New Ideas

Who should own and benefit from new ideas? How can an innovation-based company ensure its survival? Ownership of intellectual property is important both for individual enterprises and for nations. Noble (1977) described the way the patent system developed in the United States. It was initially intended to reward and protect the lone inventor. However, defense of patents became increasingly complex and costly. Even the inventor of the first triode valve, Lee de Forest (1873–1961), was eventually bankrupted by the cost of patent litigation.

During the late nineteenth and early twentieth century, the patent system increasingly became a means to corporate rather than individual ends, particularly in the science-based industries. The present situation was the logical and perhaps inevitable outcome. Professionals are now specifically employed to invent and innovate. Their employment contracts have commonly included assignment of all patent rights to the employer (although more inclusive management approaches are leading to changes to this approach). Some outstanding examples of organizations established to carry out invention and innovation include:

Menlo Park (1876)

Thomas Alva Edison's laboratory, where, with the help of several very talented associates he invented the phonograph in 1877 and demonstrated the carbon filament electric lamp in 1879. One of Edison's associates at Menlo Park, Nikola Tesla, subsequently invented the alternating current (AC) induction motor (see Case Study 6.1 on p. 280).

Bell Labs (formal title AT&T Bell Laboratories, Inc). (1925)

This is one of the world's most prestigious research facilities. It was incorporated in 1925 as Bell Telephone Laboratories, Inc., but can be traced back to 1883, when the American Telephone and Telegraph's Mechanical Department was set up. The two principal engineering divisions in 1925 were Apparatus Development and Systems Development, and the labs did early work in the discipline that emerged after World War II as systems engineering. Bell staff have produced thousands of scientific and engineering inventions and innovations, including the first synchronous-sound motion-picture system in 1926 and the transistor, for which Bell researchers John Bardeen, Walter H. Brattain, and William B. Shockley were awarded the 1956 Nobel Prize for Physics.

National Aeronautics and Space Administration (NASA) (1958)

An American government organization that has developed many innovations, NASA started out as the National Advisory Committee on Aeronautics (NACA), which was chartered in 1915 and was operational from 1917 to 1958. Near the end of World War II, NACA started studies on supersonic flight using a series of specially constructed research aircraft.

The first of NACA's rocket-powered research aircraft, the X-1, was a bullet-shaped airplane built by the Bell Aircraft Company. On October 14, 1947, with U.S. Air Force Captain Charles "Chuck" Yeager as pilot, the aircraft flew faster than the speed of sound for the first time. Yeager ignited the four-chambered XLR-11 rocket engines after being air-launched from under the

(continued on next page)

Organizations and Innovation *(continued)*

bomb bay of a B-29 at 21,000 feet (6.4 kilometers). The 6,000-pound (26.7 kilonewton) thrust ethyl alcohol/liquid oxygen burning rockets pushed him up to a speed of 700 miles per hour (1,125 kilometers per hour) in level flight. Continuing research in the program resulted in notable increases in knowledge about the dynamics of high-speed manned flight in winged aircraft.

In 1958, President Eisenhower signed the National Aeronautics and Space Act, which officially formed NASA from NACA. Near the end of NASA's X-plane program, three X-15 aircraft made a total of 199 flights. Flight maximums of 354,2000 feet (108 kilometers) in altitude and 4,520 miles per hour (7,270 kilometers per hour) in speed were reached. The final X-15 was manufactured by North American Aviation, now known as Rockwell International Corporation.

NASA went on to land men on the moon, develop the space shuttle, and put the Hubble Space Telescope into orbit. Numerous technological advances resulted from these and many other NASA projects. NASA established its Commercial Development and Technology Transfer Program to help transfer new technologies developed in the course of NASA activities to the public and private sector. NASA innovations have found application in everything from health care to integrated circuits.

▮ Mobilizing Resources for Development

A major problem with innovations, particularly when they reach the expensive development stages, is the need for a substantial supply of patient capital, capital on which there is not a pressure for short-term returns. One of the reasons for the importance of large, commonly multinational, corporations in the innovation process has been that they can generate funds internally for long-term developments and so, despite the short-term vagaries of the stock and bond markets, can support a continuous stream of innovations at various stages of development.

Shareholders, particularly in English-speaking capitalist countries such as the United Kingdom, the United States, and Australia have recently demanded short-term returns. There is a conflict between these expectations and the relatively long lead times required for technological development. The principal justifications for government intervention have often been related to this conflict. When the Thatcher government in Britain attempted to privatize the nuclear power industry, it discovered, to its cost, that the half-lives of fission products were not particularly consistent with the dividend return expectations of the stock market!

▮ The Political Context

One country that has generally managed these conflicting objectives effectively has been Japan, where market share has ranked with profit as a corporate objective. Cooperation between public and private sectors has also been the predominant mode of technological enterprise. Johnson (1992) argued that one of the most important elements in public-private sector cooperation in Japan was government assistance with financing, commercialization, and sale of products. There was a national commitment to the development of industry; the Japanese system of collecting private savings through the postal system and concentrating them in government accounts was an important mechanism for generating investment funds for innovation.

Johnson suggested that the Japanese form of government-business relationship was not unique; it was just that they had worked harder at it than most. Much the same economic relationship existed in the United States between the large firms producing military equipment (that often led the way in the deployment of innovations), the armed forces, and the associated bureaucracy, where it was commonly referred to by the political epithet of "military-industrial complex" (see Case Study 5.1 on p. 266). Along with "multinational company," the term itself is often seen to have worrying or distasteful connotations. This is partly because the very size and power of the organizations involved raises concerns about their controllability. The focus on military rather than civil applications of technology also occasions concern. It is a significant comment on the effectiveness of the MIC that aviation, space vehicles, and atomic energy have all been sectors in which the United States is preeminent.

At the heart of these huge and complex organizations are networks of ordinary people. Many are highly qualified and many are also ambitious—many wish to preserve their job security, even if they are prepared to risk several job moves in the course of their careers. Most probably see their work as a question of survival. Moving simultaneously through the networks are numbers of innovations, often intricately linked, that are fast becoming products. For many employees, the future of their jobs is inextricably bound up with these. Some may be potentially beneficial to society and others potentially harmful, raising important ethical questions for the engineering profession, but each has a momentum of its own. In the world of the theater we have long been accustomed to the theatrical imperative that says "the show must go on." This is far more true of a major engineering project, which may give employment to thousands, and could constitute an essential link in a highly sophisticated national defense network: the "technological imperative."

With the end of the Cold War we have seen that even the "technological imperative," of arms development to "keep ahead of the enemy," can be overridden by political and economic imperatives. For example, contracts for military aircraft have been canceled. In both the United States and the nations of the former U.S.S.R., efforts are being made to redirect organizations that were formerly engaged on military research and development to working for peaceful purposes. Human and economic resources that would formerly have gone to military projects are being redeployed to work towards goals such as ecologically sustainable development. This could prove to be a most interesting and exciting challenge (Markusen and Yudken 1992). It also raises important questions about the role of government, particularly as to how governments can most effectively encourage innovations that meet social aims.

The international experience with domestic appliances shows that well-targeted government incentives can spur significant improvement in efficiency. For example, U.S. initiatives setting performance criteria for washing machines and refrigerators have led to the development of significantly more efficient products. At the other extreme, protection of an inefficient local industry, such as has happened in Britain, has led to the situation where British refrigerators are not competitive on the export market. Indeed, the most efficient British refrigerators are still less efficient than the least efficient German units (Herring 1994). The lesson here is that setting challenging but realistic performance targets, and requiring manufacturers to provide clear and adequate information on actual equipment performance, can promote more sustainable design. It also indicates that in our increasingly global economy ill-advised support for the makers of inefficient products may eventually destroy whole industries.

Consumer Rights and Protection: Product Liability

We have argued that a major purpose of engineering is the creation of new technology: that is, innovation. We see engineering as an imprecise, creative activity based on an heuristic approach, although owing much to science. A quite different view was taken, however, by two judges in the United States, who concluded that engineering was an exact science, and proceeded to specify damages accordingly for an engineering failure. Their view has gained wide legal acceptance. In some areas of engineering activity, it has completely stifled the development of new products. Even engineering design textbooks have been affected, with design data being omitted or heavily qualified to protect their producers from litigation. One of the implications of the approach presented here is that we need to be much more open about the possibilities of failures occurring and about the acceptance of risk. One publication on risk management argues that in order to contain legal liability problems we will need to change our rhetoric from "making decisions and providing answers" to "facilitating decisions and offering options" (Miller 1988).

As early as the 1930s there were public expressions of concern at the behavior of large corporations, whose thrust for increased market share and control was not seen to be in the public interest. These concerns resurfaced slowly after World War II. In 1962 President Kennedy declared that consumers were entitled to: "the right to safety, the right to be informed, the right to choose and the right to be heard" (Hearne 1979, p. 7).

A dramatic example of the need to restrain U.S. corporations was the conspiracy by a motor vehicle/tire/petrol cartel headed by General Motors, Firestone, and Standard Oil (Esso) to develop the market for their products by destroying public transport systems. They bought up the electrically driven trolley networks in forty-five cities and replaced them with diesel buses, then ran the services down, forcing commuters into buying their own cars and also increasing tire and gasoline sales. The importance of the automobile and its associated industries for the economic health of the United States led the chairman of General Motors, Charles Erwin Wilson, to assert in 1953 while this conspiracy was in full swing that "for years I thought what was good for the country was good for General Motors, and vice versa. The difference did not exist" (Sheehan 1988).

Los Angeles exemplifies the results of this process; there, most journeys must be made by car. The air quality is often poor, causing illness and masking the intrinsic beauty of the landscape. Many Asian cities, most notably Bangkok, have even higher levels of congestion and pollution. In the early 1960s, Mumford (1961, p. 510) summed up the absurdity of relying on motor cars for inner city transport:

> *two-thirds* of central Los Angeles are occupied by streets, freeways, parking facilities, garages ... The last stage of the process already beckons truly progressive minds—to evict the remaining inhabitants and turn the entire area over to automatically propelled vehicles, completely emancipated from any rational human purpose.

In fact, Los Angeles has been trying for some decades to install mass-transit systems, without much success. The geology of the area does not help. Experience after the Northridge earthquake was that people used mass-transit while they had to, but that mass-transit systems did not retain them as customers once they could use their cars again. For those with access to them, cars are certainly perceived as secure and comforting—effectively an extension of one's own space. As Case Study 2.1 (p. 62) demonstrates, an issue that urgently needs to be addressed is the proper comparative costing of road and other options.

Ralph Nader (1973) was probably the first to successfully challenge the worship of the automobile. His book *Unsafe at Any Speed* dramatically exposed some of the shortcomings of the American auto industry. The title referred to the Chevrolet Corvair, which had such poor suspension geometry that it could roll over if it hit a bump going round a moderate curve at sixty kilometers per hour. The irony of the Corvair was that its roll-over problem could be corrected simply by fitting the stabilizer bar that was available from third parties, offered as a "performance" accessory. One of Nader's concerns was the dominance of styling and sales over engineering and safety considerations. Another was the complete failure of industry self-regulation to provide any effective public accountability. Nader's book still makes compelling reading.

A more recent problem was with the Ford Pinto. Ford testing showed that in a rear-end collision, the fuel tank could be punctured. The cost of fitting a protective device to prevent this would be around $11 per vehicle. Ford decided against the protective device because its cost would be significantly higher than the likely claims on the company for compensation for the deaths and injuries they expected to result from the unsatisfactory design (Harris et al. 1997, p. 44). It is little wonder that quality of life rather than commodity ownership is becoming a major issue in affluent countries.

▪ Is Someone Always at Fault?

Partly as a reaction to the motor industry's insensitivity to safety issues, the attitude arose, particularly in the United States, that the hazards, injuries, and deaths on the roads were *all* caused by the manufacturers—though perhaps aided and abetted by the road makers. Issues such as the mismatch in a collision between a car and a semi-trailer, or the inherent hazard of traveling at high speed on a narrow path with a fallible and not always sober controller, were conveniently neglected. So were the social structures that put people into this situation instead of into public transport that, technically at least, is inherently much safer. The attitude is remarkably similar to the view held in some tribal societies that all injury and illness are caused by the ill-will of some other person. Its ultimate extension is that "life should be safe," and that all injuries (if not ill-health) are therefore the result of negligence by someone and should be compensated. This approach may reflect people's feelings that they have lost control of their lives; instead control has been passed over to the marketers and lawyers, who should in return ensure that life is safe.

The problem has been compounded by exaggerated claims for the safety, reliability, and social benefits of all sorts of extremely complex and demanding technologies, including those used in the nuclear industry. Such claims have fed the unfortunate notion, mentioned above, that engineering is an exact science. One result of these developments has been rapidly increasing government regulation. (The somewhat uncertain nature of engineering is discussed in some detail in Chapter 10.)

An implication of the approach presented here is that we need to be much more open about the possibilities of failures occurring and about the acceptance of risk (Petroski 1985; Wynne 1988). In the United States the cost of litigation has become a major constraint on the introduction of new products and processes. It is widely considered to have destroyed the light-aircraft industry, and the expected cost of litigation is said to account for half the price of a motorcycle helmet.

▌ Legislation Affecting Innovation

The law provides a framework for private economic activity. One consequence of this is that private marketing decisions are to some extent subject to public scrutiny. It is beyond the scope of this book to give an overview of the law and the legal system, but we do need to consider how the law affects product commercialization.

In most countries, there is a plethora of state and federal legislation affecting business. Three main purposes of such laws are to protect:

1. companies from each other, particularly by prevention of unfair competition;

2. consumers from unfair business practices, such as misleading advertising or deceptive packaging; and

3. the broad interests of society against unrestrained business behavior, including requiring proper environmental and occupational health and safety safeguards.

The major federal legislation affecting business in the United States is the 1914 Federal Trade Commission Act, which created the Federal Trade Commission (FTC). The FTC Act made it illegal for persons, partnerships, or corporations to use unfair methods of competition and unfair or deceptive acts or practices in or affecting commerce. The problem for the courts has been defining the term "unfair." The act also made it unlawful for any person, partnership, or corporation to disseminate, or cause to be disseminated, any false advertisement. Again, it has been left to the courts to define "false."

The 1938 Wheeler-Lea Amendment to the FTC Act increased the FTC power to control unfair or deceptive acts or practices, including false advertising. The 1975 Magnussion-Moss Warranty and FTC Improvements Act authorized the FTC to set requirements for written warranties and service contracts, and recognized consumer class actions for breach of warranties to consumers. The FTC Act Amendment Act of 1994 further strengthened the original FTC Act.

One result of the changing consumer climate has been an increased recognition of the need to protect consumers from products that can injure them. In the United States, millions of people are injured by defective products every year, including some 150,000 disabled and 30,000 killed (Nagarajan 1986). It is important to compensate people who have been injured and, retrospectively, to punish manufacturers who produce dangerous products, and distributors who sell them.

Compensation and punishment do not, however, directly address the central issue, which is to stop unsafe goods from coming on to the market. Design and manufacturing need to focus on producing safer products. Insurers could play a key role by requiring appropriate quality processes to be in place before they provide product liability insurance. When liability is being considered, the product is taken to include the packaging, labeling, and associated warnings. Design is also broadly interpreted, taking in the whole of the process from conceptualization through material selection, manufacture, testing, evaluation, and review of safety. Design also includes checking product performance, reviewing complaints, and correcting any problems that may emerge.

Standards bodies such as the International Organization for Standardization (ISO), the American National Standards Institute (ANSI), the Canadian Standards Association (CSA), the British Standards Institution, the Japan Industrial Standards Committee, and the Standards Association of Australia (SAA) have promoted uniform approaches to product design and assessment, but such standards are only mandatory when specified by legislation. The ISO is a nongovernmental organization that was established in 1947 for the purpose of developing worldwide standards, improving international communication and collaboration, and promoting the smooth and equitable growth of international trade. Used widely throughout government and industry, ANSI standards cover a variety of items, from standards for architectural products and consumer goods to nuclear safety standards. CSA standards are used extensively throughout Canada, as well as by firms exporting to Canada. Other developed countries such as Germany have comparable Deutsche Industrie Normen (DIN) provisions. Design standards are also produced by some professional bodies, for example, the American Society of Mechanical Engineers (ASME) standards for boiler and pressure vessel design are used internationally, particularly in critical areas such as nuclear engineering.

■ From Negligence To Strict Liability

There are two main legal approaches to product liability that focus explicitly on manufacturers and, implicitly, on designers. These approaches are based, respectively, on negligence and strict liability.

Under the concept of negligence, the injured party had firstly to show that there was a defect in the product and that this defect caused him or her injury or loss. Secondly, the injured party had to show that the defect resulted from the negligence of the designer or the manufacturer, in that they failed to exercise a reasonable duty of care in the design or manufacture of the product. As a first step, this duty of care would have included ensuring that the product met the relevant statutory requirements for the various locations where it was to be sold and the conditions in which it was to operate. Once these requirements had been addressed, the duty to design a safe product included consideration of all the likely ways in which the product could be used. There was a requirement for consideration of both the seriousness of the likely risk and the probability of its occurring. The rather more demanding concept of strict liability, which now applies in most developed countries, is outlined in the box.

Engineers need to be aware of their legal responsibilities. Vaughn and Borgman (1999) is a widely used text that provides a good general introduction to the law for engineers. A comment from the web site of Kiesel, Boucher and Larson (1999), a law firm specializing in the product liability area, reflects the public perception:

> Since ... the 1960s we have come a long way. However, despite the numerous and proud victories that have been achieved in product liability, product injury and death in America remain a national disgrace. Every year thousands continue to be killed and millions continue to be injured by products. Every year, the cost of accidental injury and death in America exceeds the national

Strict Liability

Under strict liability laws, the injured party has to prove that the injury or loss was caused by the product and that it did not arise solely from unreasonable use (or misuse) of the product. The onus of proof is then on the manufacturer to show that it is not liable. Defenses could be that:

♦ The product was not defective. A product is only defective if it fails to provide the degree of safety that people generally are entitled to expect, taking into account all the circumstances.

♦ The defect that caused the damage probably did not exist when the product was put into circulation, or came into being after that time.

♦ The manufacturer or producer did not put the product into circulation.

♦ The product was only defective because it had to be in order to meet the requirements of a mandatory standard.

♦ The state of scientific and technical knowledge at the time the product was put into circulation was such as not to enable the existence of the defect to be discovered.

Quality systems, being introduced for other reasons, are well suited for ensuring that products comply with government regulations. Kannegieter (1992) mentioned additional precautions to minimize exposure to legal action:

Comprehensive and accurate records should be kept, covering what the engineer took into account at the time of providing a design or service, what reasoning process was followed and what conclusion was arrived at. This can be used ... [to prove that] ... the engineer acted reasonably, taking account of what knowledge was available at the time ... On the product side it is particularly good to keep a record of who has received the goods ... [for] ... product recall if necessary.

defense budget. And, as the complexity of our society and the mass marketing of products continue to increase, so do the social costs resulting from such defective products as asbestos, the Dalkon Shield, the Ford Pinto, L-Tryptophan and Thalidomide to name just a few. These stories and others should stand to remind us that we still have a long way to go (Stewart 1999).

■ Increased Emphasis On Safety

Engineers need to review existing and proposed products to ensure that they are "safe," bearing in mind the levels of education and skill of the likely users, the environments in which they will be used, possibilities for misuse, and changing community expectations. Where hazards cannot easily or inexpensively be removed by good design, the inherent risks must be communicated clearly to those within the manufacturing organization responsible for preparing product warnings and instructions, which have become increasingly important in minimizing liability.

Although it is clearly essential for designers and manufacturers to make their products as safe as practicable, nothing can be foolproof. Fools are so ingenious! The main defense seems likely to be that a product was safe and must therefore have been misused to cause injury. The requirement to prove this could be a major cost and a serious constraint on product development. In fairness, Kannegieter (1992) quoted a review of the effects of the change to strict liability in Britain and Germany, which concluded that there were unlikely to be widespread implications for consumers, manufacturers, or insurers.

Perhaps the major social danger of American litigation over product liability is that manufacturers may withdraw good products that meet real needs because the costs of defending claims against them are too high. Public liability claims, or the difficulties and costs of obtaining insurance cover against them, can also pose a serious threat to public facilities such as swimming pools and children's playgrounds. Although we need to continue to strive towards the ideal of "safe" products, the engineering profession must also help raise the level of public understanding as to the impossibility of eliminating risk from real life.

CONCLUSION

The consequences of legislative constraints on products are not necessarily negative. Tough, well-directed legislation can encourage innovation in directions that meet social aims, including promoting the more sustainable approaches discussed in detail in Chapter 9. Such legislative requirements can play roles analogous to those of informed, tough, and demanding clients, stimulating developments that can be internationally competitive. For this sort of benefit to be realized, however, legislation and policy making need to be based on coherent and well-informed technology policies. Our discussion of the design process in this chapter shows why such policies need to spell out broad targets (in design process terms, these could be described as *needs*, even as *functional specifications*), but not constrain how these targets are to be achieved.

The underlying philosophical approach for much of this chapter has been to focus on the technological issues and on how technological change shapes society. This type of approach, which can appear to imply that the advance of technology is inevitable, has been described as technological determinism. The historical evidence is clear: there really is nothing inevitable about the development of specific products, or the advance of technology generally.

One issue is that, once developed, particular technologies acquire something like "momentum," in that they represent significant investments in doing particular things in particular ways. Obvious examples include the arrangement of the typewriter keyboard, or the side of the road on which we drive, rather arbitrary, but costly to change. A more fundamental issue is that the values implicit in the corporate design context may constrain the range of acceptable solutions. We will explore this issue in later chapters.

Discussion Questions

1. What are your views on long wave theory and its implications?
2. Could Watt have succeeded without Boulton?
3. Why did printing (or gunpowder) have such important consequences in Europe, but not in China, where it was invented?
4. Discuss Whitehead's claim that: "the greatest invention in the nineteenth century was the invention of the method of inventions."
5. What are the current implications for the United States of the clustering of innovations?
6. How confident can we be of identifying those technologies that might be important to the United States in a "fifth Kondratieff wave"?
7. Do most countries currently provide sufficient encouragement and support to inventors and inventions? Is this important and can it be improved?
8. What are the main barriers to successful innovation in the United States?
9. What are the biggest opportunities in innovation in the United States?
10. The linear model of innovation is primitive but does include technology and research. What model might replace this and still do justice to technological issues?
11. Is R&D directly related to the design cycle or does "knowledge" act as an intermediary, as in Kline's model?
12. You may consider the argument advanced by Melman (1974) contradictory to a determinist approach. If so, what argument do you consider to be more plausible, and why?
13. Does technology have a dynamic of its own?
14. Government spending, particularly military spending, can play an important role in the development of particular technologies. Discuss some of the problems involved in commercializing one such technology (e.g., nuclear power).

15. Are political stability and technological innovation related? How?
16. Most large cities have problems, but Los Angeles has more than its share. To what extent are the problems a result of the technologies deployed, a product of the society, or an outcome of natural disasters?
17. Does litigation threaten innovation in the United States? How could the legal system support innovation?

FURTHER READING

Alexander, C. et al. 1979. *The Timeless Way of Building*; 1977 *A Pattern Language: Towns, Buildings, Construction*; 1975, *The Oregon Experiment*, a trilogy, Oxford University Press, NY. Classic analyses of human cultural values and needs and the design process.

Vaughn, R. C., Borgman, S. R., 1999, *Legal Aspects of Engineering*, 6th Ed., Kendall/Hunt Publishing, Dubuque, Iowa. This text is widely used in U.S. engineering schools.

REFERENCES

Alexander, C., 1964, *Notes on the synthesis of form*, Harvard University Press, Cambridge, Mass.

Australian Bureau of Statistics 1999, *Year Book Australia 1999*, Commonwealth of Australia, Canberra.

Bannock, G., Baxter, R. E., Davis, E. 1987, *The Penguin Dictionary of Economics*, Penguin, London.

Benveniste, G. 1987, *Professionalizing the Organization*, Jossey-Bass, San Francisco.

Birchall, D. 1973, Address to British Association.

Bush, V. 1960, *Science—The Endless Frontier*, National Science Foundation, Washington.

Coombs, R., Saviotti, P., Walsh, V. 1987, *Economics and Technological Change*, Macmillan Education, Basingstoke.

Cross, N. 1989, *Engineering Design Methods*, John Wiley, Chichester.

Darlington, C. D. 1948, *The Conflict of Science and Society*, London.

Davis, F. 1993, Engines of Energy Innovation: Smaller Manufacturers of Efficient Lighting Products, *Energy*, 18 (2), February, pp. 185-190.

Drucker, P. F. 1985, *Innovation and Entrepreneurship*, Harper & Row, NY.

Elliott, D. 1986, *Design Criteria and Innovation: Some Conclusions*, T-362, Open University Press, Milton Keynes.

Fallows, J. 1985, The American Army and the M-16 Rifle, In: MacKenzie, D., Wijcman, J., *The Social Shaping of Technology*, Open University Press, Milton Keynes.

Freeman, C., Clark, J., Soete, L. 1982, *Unemployment and Technical Innovation*, Frances Pinter, London.

Frey, D. 1991, My Life as Product Champion, *Harvard Business Review*, September-October.

Gleick, J. 1987, *Chaos*, Viking, NY.

Grilliches, Z. 1957, Hybrid Corn: An Exploration in the Economics of Technical Change, *Econometrica*, p. 502.

Harris, C. E., Pritchard, M. S., Rabins, M. J., Harris, C. E., Jr. 1997, *Practising Ethical Engineering*, IEEE, Washington, D.C.

Hart, H. 1949, *Technology and International Relations*, Chicago University Press, Chicago.

Hearne, J. J. 1979, *Marketing for Managers*, Edward Arnold, Melbourne.

Herring, Horace 1994, Is Britain a Third World Country: The Case of German Refrigerators, *Energy Policy*, Vol. 22 pp. 779-787.

Hill, S. 1988, *The Tragedy of Technology*, Pluto, London.

Hinshelwood, C. 1965, Science and the scientists, *Nature* 207 (1055).

Hitch, C. H., McKean, R. N. 1960, *The Economics of Defense in the Nuclear Age*, Harvard University Press, Cambridge, Mass.

Hudspith, R. C. 1996, The Art of Inquiry and Its Relevance to Preventive Engineering, In: *Engineering Education in Transition: Proceedings of the 10th Canadian Conference on Engineering Education*, pp. 315-322, Queen's University, Kingston.

Jewkes, J., Sawers, D., Stillerman, R. 1958, 1969, *The Sources of Invention*, Macmillan, London.

Johnson, C. 1992, MITI and the Rise of Japan, In: Carroll, J., Manne, R., *Shutdown*, Text Publishing, Melbourne, pp. 65-76.

Jonas, H. 1979, *Towards a Philosophy of Technology*, Hastings Center Report, February.

Kannegieter, T. 1992, Advice on Precautions Regarding New Class Action Law, *Engineers Australia*, 3, April, pp. 29-30.

Kline, S. J. 1995, *Conceptual Foundations for Multi-disciplinary Thinking*, Stanford University Press, Stanford, CA.

Koen, B. V. 1984, Toward a definition of the engineering method, *Engineering Education*, December, pp. 150-155.

Kolb, D. A. 1984, *Experiential Learning: Experience as the Source of Learning and Development*, Prentice-Hall, Englewood Cliffs, NJ.

Kuhn, T. S. 1962, *The Structure of Scientific Revolutions*, Chicago University Press, Chicago.

Markusen, A., Yudken, J. 1992, *Dismantling the Cold War Economy*, Basic Books, New York.

Meadows, D. H. et al. 1972, *The Limits to Growth*, Potomac, Washington DC.

Melman, S. 1974, The Myth of Autonomous Technology, In: Cross, N., Elliott, D., Roy, R. Eds., *Man-made Futures*, Hutchinson, London.

Mensch, G. 1979, *Stalemates in Technology: Innovations Overcome the Depression*, Ballinger, Cambridge, Mass.

Merson, J. 1989, *Roads to Xanadu: East and West in the Making of the Modern World*, Child and Associates and ABC Enterprises, Crows Nest, Australia. (Published in the U.S. as *The Genius That Was China*)

Miller, P. O. 1988, Beliefs, *Proceedings, IEAust Bicentennial Engineering Conference*, Sydney.

Moyal, A. M. 1976, *Scientists in Nineteenth Century Australia: A Documentary History*, Cassell, Melbourne.

Mumford, L. 1961, *The City in History: Its Origins, Its Transformations, and Its Prospects*, Harcourt Brace Jovanovich, NY.

Nader, R. 1973, *Unsafe at Any Speed: The Designed-in Dangers of the American Automobile*, Bantam, London.

Nagarajan, R. 1986, Legal Considerations in the Design Process Now and in the Future, *Conference on Teaching Engineering Designers for the 21st Century*, 5-6 February, University of New South Wales, Sydney.

Noble, D. F. 1977, *America by Design: Science, Technology and the Rise of Corporate Capitalism*, Oxford University Press, Oxford.

Petroski, H. 1985, *To Engineer is Human: The Role of Failure in Successful Design*, Vintage Books, NY.

Office of Science and Technology Policy.

———, 1995a. *National Patterns of R&D Resources: 1994, Final Report*. NSF 95-304, Arlington, VA.

———, 1995b. *Research and Development in Industry 1992*. NSF 95-324, Arlington, VA.

———, 1995c. *Academic Science/Engineering: R&D Expenditures, Fiscal Year 1993*. NSF 95-332, Arlington, VA.

———, 1995d. *Federal Funds for Research and Development: Fiscal Years 1993, 1994, and 1995*. NSF 95-334, Arlington, VA.

Quinn, J. B. 1980, *Strategies for Change*, Richard D. Irwin, Homewood, Illinois.

Ronayne, J. 1984, *Science in Government*, Edward Arnold, Caulfield East.

Rosenberg, N. 1976, *Perspectives on Technology*, Cambridge University Press, Cambridge.

Roy, R., Cross, N. 1983, Bicycles: Invention and Innovation, *Course Notes for T263*, Units 5-7, Open University, Milton Keynes.

Schumpeter, J. A. 1939, *Business Cycles*, McGraw-Hill, NY.

Shaw, G. B. 1903 (reprinted 1987), Maxims for Revolutionists: Reason, in *Man and Superman*, Viking, London.

Sheehan, P. 1988, The GM Conspiracy: How a Greedy Corporation Ate Its Own Market, *Sydney Morning Herald*, 21 May, p. 72.

Stewart, L. S. 1999, *Final Argument: Communicating the Philosophy and Concepts of Strict Liability*, on the World Wide Web at: <http://www.kiesellaw.com/lawyers/strick.html>

Vandenburg, W. H. 1996, The Concept of Preventive Engineering and its Potential Contribution to Canada's Viability, In: *Engineering Education in Transition: Proceedings of the 10th Canadian Conference on Engineering Education*, Queen's University, Kingston.

Vaughn, R. C., Borgman, S. R. 1999, *Legal Aspects of Engineering*, 6th Ed., Kendall/Hunt Publishing, Dubuque, Iowa.

Warman, C. 1989, *Peter Nicol Russell Memorial Address*, IEAust, Canberra, Australia.

White, J. D., Apple, N., Haselgrove, G., Dowe, R., Clyde, D. H. 1990, *Creating Wealth Through Manufacturing—The Role of Government in Developing an Internationally Competitive Design and Engineering Capability in Australia*, Canberra.

Whittle, F. 1981, *Jet*, Pergamon, Oxford.

Wiener, N. 1993, *Invention: The Care and Feeding of Ideas*, MIT Press, Cambridge, MA.

Williams, B. 1982, *Living with Technology*, 1982 Boyer Lectures, ABC, Crows Nest, Sydney.

Wynne, B. 1988, Unruly Technology: Practical Rules, Impractical Discourses and Public Understanding, *Social Studies of Science*, 18, pp. 147-167.

Zuckerman, S. 1961, *The Management and Control of R&D*, HMSO, London.

SYDNEY HARBOUR BRIDGE DESIGN

The question: "What is design?" can be illustrated by debates over the Sydney Harbour Bridge, opened in 1932. At the time, the public was bemused by the heated argument between two very eminent engineers as to who had really designed it. The argument finally came down to whose name would be immortalized on the memorial bronze on the bridge itself!

The two contenders for the honor were:

Dr. J. J. C. Bradfield, Chief Engineer of the State Department of Works

In 1909 J. J. C. Bradfield was working in Sydney with the State Department of Works when the government set up an investigation of the city's transport problems. The possibility of using a tunnel under Sydney Harbour to connect the city with North Sydney was suggested. Bradfield opposed tunneling—he favored a bridge, spanning a distance of almost 500 meters; and he designed two alternative cantilever bridges for the crossing. In 1912 Bradfield was appointed Chief Engineer of the Metropolitan Railway Construction Branch of the Department. A bridge over Sydney Harbor became part of his overall design for an effective transport system for Sydney. The government sent him overseas to look at the latest trends in railway and bridge design. On his return, his paper on "The Transit Problems of Greater Sydney" laid out the core of his vision: suburban electrification, a city underground railway, and a bridge to link the city with the northern suburbs.

In 1922, Parliament approved the Sydney Harbour Bridge Bill. Earlier in the year Bradfield had again been overseas looking at bridge designs, and on seeing the Hell Gate arch bridge in New York, realized that an arch bridge over Sydney Harbour would be cheaper than a cantilever design. The final tender for the construction of the Sydney Harbour Bridge called for either a cantilever or arch bridge, in accordance with official designs provided by Bradfield. Bradfield's designs were general designs and estimates, capable of variation until the contract was finally let.

Mr. Ralph Freeman, Consulting Engineer, Dorman, Long and Company

In 1921 Ralph Freeman was a senior partner in the British engineering firm Douglas Fox and Partners (in 1938 it became Freeman, Fox and Partners). He had assisted in the design of the arch bridge over the Zambezi River at Victoria Falls. This bridge, completed in 1903, was built by the Cleveland Bridge Company.

In 1922, Cleveland Bridge hired Freeman as a consulting engineer to prepare their tender for the Sydney Harbor Bridge. From the first, Freeman favored an arch design, and work was well under way on the tender when the managing director of Cleveland Bridge Company died, and the company decided not to go ahead with their tender. They allowed Freeman to transfer to another firm. So it was that Freeman and Dorman, Long and Company came together and eventually won the contract to construct the Sydney Harbor Bridge.

So Who Really Designed the Sydney Harbour Bridge?

During Freeman's first visit to Sydney in 1926, he and Bradfield got on well together. The controversy started in 1928, when the

English journal *Engineering* published two articles written by staff members from material supplied by Bradfield. In the first, Bradfield was given credit for the design and Freeman was named as the consulting engineer to the contractor. Freeman tried unsuccessfully to stop publication of the second article. When Freeman visited Sydney again in 1929, the argument came into the open when *The Sydney Morning Herald* published three articles by Freeman, under the title "Harbour Bridge. Designer's Story. Engineering Romance": In the first article Freeman stated:

> The word "design" applied to a bridge has a clear meaning. It means the selection of quality of the steel to be used, and the preparation of the calculations and drawings required by the builder in order to build the bridge ... Subject to the limitations of the specification as to span, headway above the harbour, position and breadth of railway and roadway tracks and the load to be carried, the design of the Sydney Harbour Bridge was made by me.

In the last of his articles Freeman mentioned in passing the role of Bradfield:

> To him also is due the credit for the conception of the bridge project and the decision as to the type of bridge to be built.

The following day, the same newspaper carried a story that the Minister for Works and Railways was calling for a report from Dr. Bradfield on the subject. The Minister confused the matter even further by commenting that he would describe the bridge as "a Bradfield—Dorman, Long design." The public debate was launched. Bradfield's report to the Minister was not slow in coming:

Mr. Freeman's claim that he "designed" the bridge is based on a definition of the word "design" as meaning the selection of the quality of steel to be used and the preparation of the calculations and drawings required. These matters are not the "design" of the structure. A great part of the design is done before these matters came up for decision, when the determining features such as location, grades, span length, number of spars, approaches, width and arrangement of deck or decks, depth of trusses, lateral and transverse bracing, suitability of foundations, feasibility of erection, etc., are considered. The design of the structure depends upon these major considerations, and after these have been settled, the detail drawings can be commenced.

To this Freeman replied:

> The selection of the steel is the basis from which any design must be evolved. Until the selection is made no design can be commenced. The completed bridges from the same span, height, loading, and type produced from different kinds of steel would vary greatly from one another in appearance, weight and cost.

Much of the argument revolved around Clause 20 of the tender, which stated:

> The contractor must satisfy himself as to the sufficiency and suitability of the design, plans and specifications upon which the bridge is to be built, as the contractor will be ... required to guarantee the satisfactory erection and completion of the bridge ... it is to be expressly understood that he undertakes the entire responsibility not only for the materials and construction of the bridge,

(continued on next page)

but also for the design, calculations, specifications and plans furnished to or by him, for the quantities submitted by him in his tender as accurately representing the amount of work involved and for the sufficiency of the bridge for the loads therein specified.

During his travels, Bradfield had become aware of the need for a contract that gave the builder overall responsibility, and he probably drafted the contract. To him it was obvious that:

the meaning of the clause is that the contractors must assume responsibility for the design, not as the designers, but as the builders of the bridge.

The debate in the local press continued. Besides disagreements over official clauses in the tender, there were arguments over the technical specifications. For Bradfield:

The specification did much more than set out the span, headway, position and breadth of the railway and roadway tracks and the loads to be carried, as stated by Mr. Freeman; it completely governed the calculations and details of the structure, design, workmanship, model tests, fabrication and erection, materials of construction, including even the composition of the paint ... Tenders had to

determine the weight of steel themselves in accordance with the class of steel they produced.

However, Freeman considered the specifications far from adequate!

A compromise had to be reached when Dorman, Long threatened to sue the state government if it erected a plaque on the bridge naming Bradfield as the designer. Sir Arthur Dorman, a friend of both Freeman and Bradfield, was responsible for the final wording. The plaque spells out the division of responsibilities between Bradfield, who carried out the conceptual design, in the course of which he made initial calculations, and Freeman, who did the detailed design.

For successful innovation both areas are critically important. As each man was no doubt aware, any significant design involves a degree of iteration and concept refinement. Bradfield's wisdom in clearly assigning responsibility for the detailed design to the constructor is vividly demonstrated in Case Study 7.5 (see p. 392), which tells a very different story about Australian bridges.

Reference

Raxworthy, R. 1989, *The Unreasonable Man: The Life, and Works of J.J.C. Bradfield*, Hale & Iremonger, Sydney.

7

Management in Engineering

Getting results through people is a skill that cannot be learned in the classroom.

— J. Paul Getty (1892–1976) Oil millionaire, arts patron

Management is essential to the operation of any complex enterprise. It involves mobilizing and directing substantial human and physical resources. Most professional engineers spend much of their working lives on tasks they identify as "management and supervision."

In addition to these general management responsibilities, engineers also manage technical resources. The specific requirements for engineering management capability depend on the technical content and character of the resources to be managed. In essentially traditional production facilities, such as machine shops, the focus is on accuracy and reliability. In continuous production plants the emphasis moves to process efficiency and plant availability. In industries such as software production where the product is increasingly abstract, critical issues include interpreting and recovering from equipment crashes. The mix of appropriate managerial skills will vary according to the organizational context.

Rather than try to cover every area of management in which engineers are likely to be involved, in this chapter we will focus on general engineering management skills and competencies. We will then look at some recent international changes in manufacturing to explore the sorts of changes that are happening in management thinking and practice.

Modern Organizations

The leading organizations, including engineering organizations, have changed radically in the last decades of the twentieth century. There are many views on what further changes will be needed if organizations are to continue to be successful going into the twenty-first century. Although there is no consensus on what will be the best approaches for the future, the limitations of traditional models are generally acknowledged. "Fast Capitalism," the ideology that underpins the major current business organizational approaches, is described and reviewed by Gee et al. (1996).

The classical analysis of the functions of management drew heavily on nineteenth century military organizational concepts (e.g., Fayol 1949). The focus was on the organization and its priorities, with industrial discipline, imposed from above, as a significant theme. Even the terminology was strongly hierarchical. As is immediately apparent from every-day experience, and from any modern text on organizations and organizational methods, both the focus of management activities and the context within which they take place are changing.

There are new paradigms for management. The mission of the organization must give clear and realistic guidance as to what its core activities and values are. Networking, both inside and outside the organization, permits increasing group capability, while tending to reduce the number of levels of hierarchy. Partly as a result, management structures are becoming flatter. There is a major emphasis on improving communication. This will be increasingly important because change, and effective change management, are the new order of the day. Human resources departments already have major ongoing responsibility for changing organizational culture and this seems likely to continue, indeed to accelerate. Peters (1994) argues that "change" is too slow and conservative a word to describe the pace and scope of the transformations required for organizations to be successful in today's climate. He suggests that "abandonment" and "revolution" better capture the required spirit.

A focus on the requirements of the customer or client is seen as driving decision making for every person in the organization. Who is my customer (or client)? The answer now is—everyone inside or outside the firm for whom I do something in the course of my work. Marketing is a central organizational concern. Marketing is much more than selling. Ultimately, it is about aligning customer wants and organization capability. It will increasingly drive the reshaping of organizations, even engineering-intensive ones. Beder (1998) reminds us that in organizations dominated by engineers, there has been a tendency to define problems in a narrow technical way, with the emphasis on the efficiency and effectiveness of carrying out a particular task (such as sewage "disposal" or electricity generation) at minimum cost, regardless of the associated environmental and other impacts. A socially responsible marketing focus encourages much broader definition of the problems to be solved and much wider consultation with those likely to be affected. Marketing of broadly based engineering services, as well as goods, has taken on new importance in our "information age." It is no longer enough for an organization to make a product or offer a service and hope it will satisfy its clients. Potential clients need to be part of the development process, and marketing is now a essential responsibility of engineering managers.

What sorts of people will thrive in the very different organizations of the future? In the twentieth century, the key thing was to have a job. In the twenty-first century, the essential skill will be to remain employable. Individuals must take personal responsibility for their own careers. For engineers this includes their continuing education and professional development. This is not a new idea: professionals have always had an ethical responsibility to remain current in their technical area. What is new is the expectation of also having to learn how to handle new organizational approaches and broader social and other responsibilities.

Engineering management requires both leadership and organizational skills, in addition to technical competence. Kotter (1990) has usefully argued that leadership is required to introduce or to cope with change, whereas management is required to handle complexity within and around organizations. The engineer, concerned with innovation and with organizations, therefore needs to understand something of both leadership and management.

Leadership is essential for successful organizational change, but true "change agents" are not always appreciated. The engineer who discovers, invents, or introduces new technology is an innovator, but is often (mistakenly) also taken automatically to be an agent of change. In fact, the new technology may be used in the same old ways, while little else changes. In management terms, change is about new ways of *doing* things (possibly new things, possibly the same old things). Effective change management changes the culture of the organization (and it may include introducing new technologies). Senge (1990) suggests that successful organizations will increasingly be organized as critical learning systems, consciously open to questioning everything, including the contexts in which learning must occur. Within such organizations there will be both formal and informal chains of command, and engineers (and others) need to be aware of how to work effectively through both types so that they can survive and the organization can thrive.

Leadership and Management

The distinction between leadership and management, once rather blurred, is becoming much clearer. They are increasingly recognized as different roles for different purposes. Leadership is associated with the transformation of current practices; management is associated with the ongoing routine transactions of the organization and maintenance of a specified level of performance. In theory the same individual might perform both roles. In practice the different tasks associated with the two roles require different personal characteristics, and it is unusual for one person to have talent in both areas.

The word *leadership* is derived from the Anglo-Saxon verb *laeden* meaning to travel or go. The leader was the person who showed the way, whether as guide, steersman, or navigator (Adair 1989). Leadership encompasses direction-setting and inspirational and motivational capacities, but most of all the vision of what needs to be done and what needs to be changed, and the empowering of those in the organization to make the changes happen (De-Pree 1992).

The word *management* comes from the Italian word *managgiare*, which meant to train horses in a riding school. Management is usually described as achieving objectives through people. Typically, objectives are set by the owners or directors of the operation, and are usually stated as maximizing competitiveness in the market, or cost-effectiveness. The exclusive right of managers to set objectives is usually assumed, as is their right to manage the other people in the operation so as to achieve these objectives. These characteristics reflect basic values implicit in "liberal[ist] democratic" capitalist society, with its explicit protection of private ownership and control of capital. Such characteristics are not inevitable and have been successfully challenged, but they do set the context in which other models and approaches have to establish themselves.

In most of the current literature, the word *management* is used as a generic term to describe the broad area that includes both leadership and management. Management is thus used in two senses: that of an overall area that includes the questions of innovation and leadership, and in the more particular sense of coping with the complexities of organization.

The traditional management operations—planning, organizing, commanding, coordinating and controlling—as listed by Fayol (1949), plus some additional functions such as forecasting, characterized a number of early approaches. Typically, descriptions of management functions were narrow, static, prescriptive, and authoritarian. They may still have some relevance in traditional industrial and workshop situations where the focus is, for example, on reliable and accurate production of machined parts. However, they are no longer useful approaches where the production process is essentially continuous. They are even less adequate for dynamic modern technological environments such as software development, described by Clegg (1998) as "postmodern" production processes, where the focus is on presenting increasingly abstract information.

What is modern engineering management really like? Mintzberg (1975), having surveyed the actual work routines of managers, argued for a broader and less structured view of management. Tom Peters is a leading management thinker and proponent of organizational change. He argues that hierarchical business structures are going into the shredder, and are increasingly being replaced by flexible, fast-responding, *ad hoc* project teams (Peters 1992).

Performance criteria for companies and the managers within them vary widely. What pressures do the stock market and shareholders put on short- and long-term growth and performance? How can organizations continue to improve their performance? What should the ideal performance criteria be? What is the relative importance of short-term profitability versus long-term market share? What responsibility can (and should) managers take for providing stable and continuing employment? The accepted answers to these questions vary between industries and national cultures and vary over time. Benchmarking—systematically making annual comparisons with other broadly similar organizations—provides an objective basis for developing answers to these questions.

Case Study 7.2 gives some interesting insights into some of the management issues involved in a large engineering project that draws on engineering management expertise from very different cultures.

Much of the engineering enterprise of the future seems likely to be made up of self-organizing, self-motivating, dynamically changing units, able to respond quickly and effectively to changing demands. All those involved in such an organization will need to be committed to its goals and to continuing to learn throughout their working lives. Product data will be held in a single computerized database and will be available to every workstation over broad-band communications networks. Although the head office and the major ownership of an enterprise are likely to continue to be located in a single country, the operational elements will be distributed globally so as to optimize marketing effectiveness and customer response times.

BILL GATES, MICROSOFT, AND THE INFORMATION AGE

In March 1986 Microsoft sold its first shares to the public for $21 per share. The initial public offering raised $61 million. At thirty-one years old, Bill Gates became the world's youngest billionaire.

And today, while dealing with lawsuits from competitors and federal and state governments, Microsoft and Bill Gates continue to prosper. Microsoft is now the world's leading provider of software for personal computers, employing more than 20,000 people in some fifty-six countries. And Bill Gates is now the richest man in the world.

What accounts for Microsoft's continued success? Initially, its fortune was derived from the innovation of others who went before. Bill Gates and his colleagues turned the innovations of people such as Gary Kildall, Tim Patterson, and the engineers, scientists, and programmers at places such as Xerox's Palo Alto Research Center (PARC) and molded them into commercially successful products. Of course, the birth and growth of the information industry has carried Microsoft along with it. And to some degree, Bill and his company were in the right place at the right time.

To remain competitive, Microsoft recognizes that it must hire and retain software developers who possess the skills necessary for the effective implementation of increasingly complex products. In the early years of software development, programmers tended to work alone. However, as programs grew in size and complexity, teamwork and cooperation became of critical importance. Creativity and the ability to think outside the envelope have always been a requirement. Therefore, when Microsoft interviews job candidates, it expects see an openness to questions, an eagerness to puzzle through technical questions, and a willingness to think aloud. Interviewers tend to present relatively complex problems, with nonobvious solutions, to the candidates. Of course, Microsoft seeks good problem-solving skills and a demonstrated ability to work as a member of a team.

When creating large, complex programs (such as Windows 2000), the main issues for software development include functionality, usability, accuracy, reliability, efficiency, recovery, and diagnosis. Programs must provide the functions users require, and must offer a good user interface. The code must be accurate (correct), reliable (dependable), and efficient (fast). Programs should be able to recover from errors when they do occur, without crashing the computer. Ideally, in a "physician heal thyself" fashion, the software should be able to diagnose its own problems and avoid them in the future. The skills and capabilities of professionals at Microsoft and other software companies must include the ability to deal with all of these issues, while easily fitting into the Microsoft's team-based structure.

References

Gates, B. 1995, *The Road Ahead*, Viking Penguin.

Cringely, R. X. 1992, *Accidental Empires*, Harper Business, NY.

Wallace, J. 1993, *Hard Drive: Bill Gates and the Making of the Microsoft Empire*, HarperCollins, NY.

<http://ei.cs.vt.edu/[sim]history/Gates.Mirick.html>

<http://www.si.edu/resource/tours/comphist/gates.htm>

How To Succeed in (Changing) Business

Since the early 1990s, lifetime employment has been under challenge. Drucker and Nakauchi (1997) discuss this development in the context of how the individual can continue to progress professionally and personally. They stress that knowledge workers (including engineers) will need to be able to anticipate, create, and manage change. Change will increasingly mean:

- greater social mobility, especially through education;
- the shift in employment emphasis from skill, which changes relatively slowly, to knowledge, with knowledge workers needing to go back to "school" every three or four years so as not to become obsolescent; and
- the need to learn to "reinvent" and revitalize ourselves many times over during our working lives so as to remain interested and effective, because lifespans have become longer; people are healthier and able to be productive for a much longer period.

Drucker spells out the importance of:

- high personal goals;
- continuing self-improvement;
- learning how to broaden one's knowledge in a systematic way;
- review and reflection so as to be able to decide on personal priorities; and
- continually asking oneself what one must do to be effective in one's *present* position.

Changing responsibilities require us to concentrate on the effectiveness of what we are doing *now*. A useful way is to write down the anticipated results when we make a key decision. We are then in a position to learn from the whole process when we routinely review the actual results nine to twelve months later. The habit of self-reflection has on-going spin-offs. Most people fail in new positions (including promotions) because they continue to do much the same things in much the same way as in their previous situation—a recipe for disaster!

Encouraging all these learning strategies is a central responsibility of management. Complacency about previous successes is a major threat, in that it can destroy the ability to learn how to meet new challenges. To take advantage of new opportunities, Drucker argues that organized innovation needs to be built into systems, incorporated into personnel policies within the organization by incentive, encouragement and reward structures. The leaders in any organization should systematically view each change in internal or external circumstances as a potential opportunity.

HONG KONG'S NEW AIRPORT

During its fifty-year history, the approach into Hong Kong's Kai Tak Airport has been one of the most dramatic in the world, thrilling millions of airline passengers (not to mention providing several serious incidents). Problems associated with its operation were the impetus for numerous improvements in international safety standards, including the mandatory provision of second crews. The average inbound flight time is more than six hours, and the approach is over one of the world's most densely populated neighborhoods. As the 1980s drew to a close, the need to replace Kai Tak was clear. In addition to safety concerns, analysts predicted serious economic losses for Hong Kong if Kai Tak had to operate at or above its designed capacity for extended periods. A new airport was essential to reaffirming Hong Kong's position as the crossroads of Asia.

The problem was that any civil projects lasting through the 1997 hand-over period required the approval of Chinese, British, and Hong Kong governments, with conflicting interests looming large. Amazingly (and very encouragingly), the three sides worked out the basics of the $26-billion project in 1991.

The project was overseen by the New Airport Projects Coordination Office (NAPCO). As well as the new airport, nearly four times as large as Kai Tak and with twice the passenger capacity, works included development of a new (ninth) town and building a 1.6 kilometer long suspension bridge, a 2.4-kilometer tunnel and a mass-transit railway.

NAPCO's task was to organize contractors for the over 200 component projects of the Airport Core Program (ACP) while ensuring uniform adherence to an overall strategy. Problems successfully resolved by NAPCO during ACP construction included: political budget disputes, conflicts between contractors, and unforeseen technical circumstances.

Political wrangling between Britain and China meant that contractors had to start work even before the final design decisions had been made. NAPCO's approach was to delegate responsibility at the project level to the contractors themselves, increasing their self-reliance and thereby reducing NAPCO's own risk. As long as they stayed within pre-approved scope and budget requirements, contractors were permitted to do whatever was necessary to complete the task at hand. One of the contractors in charge of the terminal project assembled the 1.2-kilometer-long steel roof in modular units overseas because it found Hong Kong's labor supply inadequate. Problems that could not be resolved at the project level were reviewed and evaluated by NAPCO with regard to the

BROAD KNOWLEDGE AND MANAGEMENT SKILLS FOR PROFESSIONAL ENGINEERS

It may be useful to think of the need for general management knowledge and skills in terms of three levels: the relatively new graduate, the experienced engineer, and the senior manager, who may be more involved with

overall agenda. Most of these problems resulted in relatively minor scope/budget modifications.

The key to success of the ACP was probably the "level playing field" approach to development. NAPCO maintained the basic regulatory structure while fostering an open and competitive bidding environment. The ACP invited bids worldwide, achieving both low cost and high quality. By 1995, 157 major contracts had been awarded to firms from seventeen countries, led by Japan. Low bidders won projects, and qualified firms were employed, regardless of nationality. A Dutch company might work alongside a South Korean company under the supervision of a team of American consultants. Because of the size of the ACP, firms were willing to make great efforts to get and keep projects. Any single project could be carried out by one of a large number of firms. Competition was keen, and the ACP required most participating contractors to have International Standards Organization (ISO) 9000 quality certification, so only very well-prepared firms won bids. An important additional safeguard was ensuring that firms with political ties to NAPCO did not get preferential treatment in the bidding process. In short, the ACP appears to have provided an outstanding example of market-driven allocation of resources (with the resources being high-quality engineering services).

Opponents of multinational engineering projects argue that scenarios such as this raise problems of standardization, and that differences in approach in different countries are simply too large to reconcile. Perhaps partly because of the sheer size of the project, the ACP had surprisingly few problems of this nature. In the conceptual-design stage the engineers paid close attention to the contingency requirements of individual projects and groups of projects alike. They made sure that it was the responsibility of contractors to identify and document complications and changes in their needs well in advance or risk losing the contract. Where problems emerged, slight modifications to overall strategy were usually enough to resolve them.

The ACP certainly had its share of critics, but it is hard to dispute the claim that it ran a very successful public works project. It was the largest construction project ever, yet it proceeded on time, under budget, and without major headaches. The new airport at Chek Lap Kok took over from Kai Tak in July 1998.

Source

Based on work by Sean Eric Smith, Stanford University, 1997.

References included a number of magazine articles and the Hong Kong airport web site at <www.hkairport.com>.

general, rather than technical, management issues. When engineers are promoted through successive levels of responsibility within an organization (or move to other firms), they use different mixes of management skills. At higher levels of management, their focus moves to the organization as a whole; strategic issues and planning for the future become more important.

What knowledge, skills, and personal qualities do professional engineers need in order to be most effective in engineering enterprises? The model of the design process in the previous chapter requires rather more than a good technical grounding, which many students may assume to be sufficient. What do engineers—particularly new graduates, who are commonly (but not always) employees in relatively large organizations—do at work? What sorts of skills will they use as they take on more managerial functions in the course of their professional careers?

We need to avoid a narrowly instrumental approach to defining the skills required. Clegg (1998) argues that because organizations need to be able to learn to do new things and to do old things in new ways, they always require more skills than they are using at any one time. Creativity, enthusiasm, the desire to explore and to satisfy one's curiosity are all essential aspects of successful engineering practice; case studies throughout this book demonstrate its almost infinite variety. They also show that professional engineers are generally fortunate (and unusual), in that they like their work. They find it challenging and stimulating, particularly in such areas as design and R&D—so much so that enthusiasm for their technical work can be a problem for engineers moving into management. They often find it difficult to let go of their technical tasks and move to the more people-oriented focus required for their new roles.

Engineers must be able to recognize that the technical aspects are only part of any problem, and therefore only part of an acceptable solution. To explore and ultimately resolve the nontechnical aspects of problems, engineers must be able to work well with nonengineering players and value their contributions. For instance, optimal resolution of aesthetic, environmental, commercial, and other issues requires these factors to be taken into account as the design is proceeding, not to be accommodated by bolt-on afterthoughts.

Management is not something that only managers do. Everyone exercises some managerial skills in their private lives. People manage their own time and resources, often in situations with conflicting goals and very busy personal schedules. They manage housework and a family budget, organize projects such as gardening and home or car repairs, and still manage social and community activities. Parents manage their children and orchestrate their numerous activities, sometimes while also caring for aged or infirm family members.

Can we encourage people to apply these same time-management and organizational skills to their work? Everyone in an organization is involved in some degree of management. Even if they are not yet in "management positions," engineers should recognize that the management skills of people at all levels of an organization contribute to building positive and constructive relationships within the organization, and they should be valued and developed. As organizational structures become flatter and less authoritarian, these basic management skills are increasingly important.

There is no doubt that engineering managers are part of "management," but is "engineering management" actually engineering? Can the skills and functions called "management" be separated from the specific engineering processes or activities being managed? This issue is part of the wider question of how engineering should be defined. In many countries in the late 1980s and early 1990s, there was significant "de-engineering" of organizations responsible for the infrastructure and operation of public works and transport. The question then arose: To what extent is engineering knowledge and skill still needed for the effective management of these organizations? In such situations, is the management of engineering still an integral part of engineering? Of course, not all organizations have changed at the same pace over the last few decades. At the same time, management thinking has changed. For instance, "re-engineering" (a potentially confusing term) is now widely used by management consultants to describe the process of getting a firm to focus on what its basic aims are, and then changing its operating practices to achieve those aims most effectively. Will "de-engineered" agencies eventually come full circle?

With flatter organizational structures there are fewer middle managers. This imposes greater demands for management skills on all those in the layers that remain. Peters (1994, p. 162–163) describes what one major engineering organization, ABB Power Transformers, a company with annual revenues of $1 billion, did to ensure that the resources of all its twenty-five highly autonomous operations in seventeen countries would be available to support any individual unit. Incentives for sharing knowledge in a timely, even pro-active way, were critical: one step taken by ABB (and other such firms) was to ensure that knowledge-sharing is explicitly expected and rewarded. Similar strategies are relevant to in-house R&D departments, where individuals take responsibility for keeping up with advances in particular areas of technology and for disseminating relevant information to those on the production side.

■ Cognitive Areas in Engineering Practice

As we indicated in the Introduction to this book, expectations of engineers and of engineering education have been changing rapidly over the last decade. Clearly, there is more to being an effective engineer than mathematics and engineering science. This may come as something of a shock to engineering undergraduates, selected for entry to their courses on the basis of their high performance in mathematics and the physical sciences. Competence in engineering science is certainly needed in order to cope with both coursework and subsequent practice as an engineer. Expertise and a high, even outstanding, level of individual technical competence may be the basis for winning contracts. However, a critical part of engineering practice is applying this competence to generating marketable outcomes.

A traditional academic focus on engineering science has tended to squeeze engineering undergraduate courses into a narrow mold and to promote a rather limited view of engineering itself. One result can be a tendency for graduate engineers to focus only on research, design, and development and to lack awareness of the importance of an effective professional engineering contribution to marketing and production at individual, enterprise, and national levels.

What sorts of academic disciplines are important to engineering? Figure 7.1 is a model of the cognitive continuum relevant to engineering practice (adapted from Parkin 1994). It indicates where management fits into engineering and shows the major cognitive areas underpinning engineering practice, ranked from the most analytical at the top to the most intuitive at the bottom. Parkin suggests that engineering practice starts with the engineering science elements necessary for the basic design of products and processes. He then includes, as engineering management, all the other functions intrinsic to the practice of the profession. Parkin divides the other disciplines, the nonengineering science areas of knowledge and skill, many of which are treated in this book, into:

- organizational disciplines, which look inwards and are relevant to improving the competitiveness of the enterprise; and

- social disciplines, which look outwards to address the place of the enterprise in society. These disciplines are important to understanding the opportunities and constraints of the social and business context in which the enterprise operates.

Inevitably there is some overlap, and some disciplines would fit into both groupings. In his definition of design, Parkin focuses on the engineering science elements. As the Key Concepts and Issues box on Design (p. 293) shows , we believe this focus to be much too limited. All engineers need broad competencies such as team working skills. Specific competency requirements depend largely on two factors: the particular technical area of work and the stage the engineer has reached in his or her career.

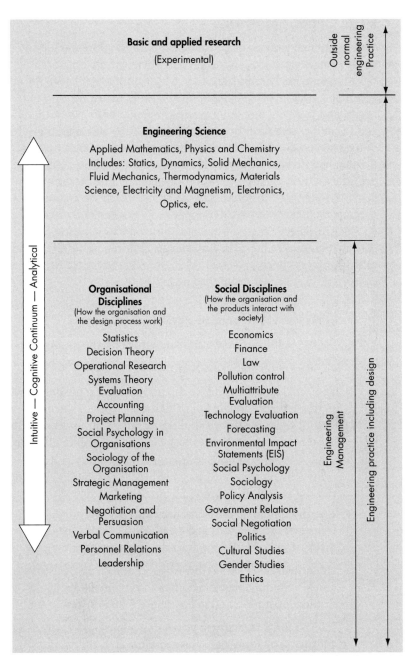

Figure 7.1 A Cognitive Model of Engineering Practice. *Source:* Adapted with permission of the American Society of Civil Engineers from Parkin, J. V. 1994, "Judgmental model of engineering management," *Journal of Management in Engineering*, 10 (1), ASCE, Engineering Management Division, January–February.

■ The Graduate Profile

Undergraduate courses obviously need both to address the immediate needs of new graduates and to prepare them for lifelong learning. The most fundamentally important areas for all graduates relate to the generic competencies involved in learning new skills, matching skills to project needs, and carrying out skills inventories and audits. Sound judgment and the ability to deliver quality outputs on time with a clear awareness of the real costs of the hours of work involved are probably characteristics that have to be developed by learning in the workplace, although appropriate educational preparation can help. All of these complement general design and problem-solving skills, which we assume all engineering graduates will have developed.

The ABET 2000 Criteria listed in the Introduction are intended to ensure that graduates have the knowledge, abilities, and understandings called for by modern professional engineering practice. Nonspecialist competencies specified or implied include:

> oral and written English communication; computer use; design; engineering management; and occupational health and safety awareness.

Additional key skills, awareness, and knowledge for the future seem likely to include:

- ◆ Risk and liability management: understanding the organizations in which professionals work and the societies in which the products of their efforts are applied, as well as an understanding of the ethical and legal issues that can lead to liability problems.

- ◆ Management and leadership: thinking critically and possessing team skills, with an awareness of processes as well as results and the ability to lead and monitor projects.

- ◆ Interdisciplinary team skills: working effectively in diverse teams using systems methods and creative/holistic approaches, such as mechatronics for new product development.

- ◆ Documentation skills: recording engineering decisions of all kinds, including preparing operations manuals for quality certification, operation and maintenance manuals, and so on.

- ◆ International awareness: possessing a familiarity with overseas cultures, languages and skills requirements, and markets.

- ◆ Ethical awareness: dealing with values questions in choice and implementation of technologies, including economic, environmental, and conservation issues.

- Social awareness: possessing good social skills in order to interact appropriately with all fellow workers, whatever their level in the corporate structure, sex, race, and so on.

- Marketing: possessing a basic understanding of how products and services are marketed, and in particular how they are marketed by one's employer.

- Finance and accounting: understanding company balance sheets and financial and income statements, leading on to financial analysis and engineering economy.

- Law and regulation: dealing with the many government bureaucracies and regulations, contracts, intellectual property, industrial relations, occupational safety and health, and the engineer's legal liability.

Many of these issues are already being addressed in engineering courses. English language competency is extremely important, because engineers can spend up to 90 percent of their time writing, depending on type of employment, and increasing with seniority. Most of this effort goes into writing memos, letters, and project documentation. Beer et al. (1997) stresses the importance of writing proper memos and letters, pointing out how to start them, how to refer to prior correspondence, how to get to the point quickly, how to close, how to know when you're done. Other subjects include the importance of correct spelling and grammar and the use of efficient wording to create concise documents. Effective oral and written communication is fundamental for documentation, a centrally important but seriously undervalued area. The engineer's responsibilities in these areas have increased sharply with the general adoption of the quality standards discussed later in this chapter.

■ Continuing Education for Management

The importance of continuing professional education is increasingly apparent. As graduates take on more management responsibility, they need to learn new skills. Some engineers acquire MBA degrees and then tend to move away from engineering towards more general management. A wide range of postgraduate material is now available with a focus on engineering as well as general management, and increasingly in distance-education format (although this mode may not be ideal for developing interpersonal skills). Models for the future may include degrees that draw equally on general management and engineering management material.

One of the most difficult aspects of solving engineering management problems is to extract the key issues from their context and formulate relevant questions clearly and accurately. Partly for this reason, case studies play a major part in management teaching. Unfortunately, although anecdotes abound, good in-depth case studies are less readily available for engineering. The case studies in this chapter need to be supplemented by a range of others. Harvard and other leading business schools have developed business case studies; there are also increasing numbers of engineering and management ethics case studies, such as those included in Harris et al. (1995). Case studies are important in clarifying ethical and other value-intensive issues and in preparing engineers to play leading roles in developing organizations with a long-term perspective.

Diversity in Engineering Management

As we discussed in Chapter 3, diversity is a critical social and professional issue. To achieve the full benefits of a range of different approaches and different management styles, and to capitalize on the potential for cross-fertilization of ideas, we need to look both inward and outward. We should be open to change, to different ways of reaching our goals, and receptive to other models that challenge traditional stereotypes.

We have already urged the desirability of engineering becoming more representative of the population as a whole. There are several distinct problems here: gender and minority issues are different, and different problems arise at every stage, from freshman intakes to engineering courses, through the ranks of professional employment, to promotion and management positions. Not only should the profession attract at least three times as many women as it does at present, an important challenge is to ensure that these students are retained within the profession when they graduate, partly so that they may in due course move into engineering management positions. Concerted action, not just pious platitudes, will almost certainly be required; progress by way of slow incremental increases in the broadly representative ranks of engineering managers will not achieve the desired ends nearly fast enough. We need to remind ourselves very firmly that supporting and encouraging the professional development of colleagues and subordinates are specifically included in modern professional codes of ethics. Positive role models, formal mentor programs, greater sensitivity to the real, practical barriers that still exist (for instance, in combining parental with professional responsibilities) and the more subtly inimical workplace culture, and our *determination* to eradicate these adverse factors; these all need to contribute to strategies to gain greater diversity in engineering management.

In a global engineering context we also need to recognize that the internal environment of business enterprises can vary greatly, calling for very different types of expertise and managerial strategies. Case Studies 7.3 and 7.4 illustrate, in very different contexts, some of the cluster of issues that arise once we start examining the case for greater diversity.

There are advantages for everyone, not just "minority groups," from broader, more "human" and inclusive personnel policies and everyday work practices; and the key role of management in setting the agenda is very clear. In his early days at Microsoft, Bill Gates was known for his ability to be a six-hour turnaround man—that is, from the time he left work on one day until the time he arrived on the next, only six hours had passed. Employees, particularly those in management positions, were expected to put in similar sorts of hours. A former vice-president of Microsoft, Daniel Petre, has recently written very revealingly about the work ethic at corporate headquarters in Seattle and his eventual rejection of what he saw as an inhuman regimen (Petre 1998).

It is not just women who want to combine motherhood with a career who find the conflicting demands of work and family almost impossible to reconcile. Bright, caring men who want to succeed but also want to be good fathers are also discriminated against by a culture that accepts as the norm early morning, late night, and weekend attendance at the workplace, the need to take work home regularly, to be away from home frequently on business trips, and to be "on call" routinely, not just for the occasional emergency. Gee et al. (1996) highlight the extent to which the modern management approaches advocated by Peters (1994) and others break down the barriers between work and home and take over previously personal space. The value of family and community life is diminished, and ultimately the managers themselves, and through them the organizations they work for, also suffer. Unaccommodating, basically negative attitudes to competing responsibilities or interests outside of work inevitably translate into a "skewed" employee profile. Real diversity is one of the first casualties, in spite of the claims of "official" company policies.

What sorts of problems can female engineers expect to encounter as they move into engineering management roles? The situation of women in general management gives some guidance to the likely difficulties. With equal employment opportunity legislation in place, the barriers to women in management are now less visible, although there are still concerns about a "glass ceiling," through which women can see, but not rise.

A 1997 survey of the Fortune 500 companies by Catalyst for Women Inc., found that women hold only 51 of the 1,728 highest positions in

LILLIAN MOLLER GILBRETH AND THE PSYCHOLOGY OF MANAGEMENT

Today it seems that an engineer is all but required to sacrifice personal life for career. In a society where 40 hours per week is the norm for most jobs, 60, 70, even 80-hour work weeks are common for engineers. In addition to the hours spent in the office, work is often taken home. These working conditions cause undue stress, may decrease productivity, and often prevent an engineer from participating in a meaningful way in social activities. One wonders what Lillian Gilbreth, often called "the first lady of engineering" due to her extensive work in industrial engineering, might think of this situation (Society of Women Engineers 1978).

Lillian Moller Gilbreth (1878–1972), along with her husband, Frank Bunker Gilbreth, was a pioneer in the field of industrial management. Before they met, Frank Gilbreth had begun to study motion and how to modify motions in order to increase productivity and reduce fatigue. Lillian had already completed a Master's degree in English. Lillian became interested in his work and joined him in it after they were married in 1904. After graduate studies in Psychology at U.C. Berkeley, Gilbreth obtained her Ph.D. in Psychology from Brown in 1915. Her thesis was *The Psychology of Management*. Lillian used her knowledge of behavioral science to expand the scope of scientific management from the purely physiological to include psychological aspects as well. In the process, she profoundly changed the workplace for many Americans, even today. Frank concentrated primarily on the physical aspects of work such as lifting, walking, and bending, but Lillian's focus was on workers' health and well-being,

which she believed would bring greater yields and profits for the company. In the early 1900s this was a revolutionary concept.

Her humane approach emphasized the individuality of workers. The concept that each employee had different motivations, methods of learning new ideas and processes, and goals had the potential to completely recast the relationship between management and employees. She pushed for psychological considerations to be incorporated into time and motion analysis: giving employees incentives to work efficiently and diligently, providing them with portable skill sets for multiple jobs and industries, and furnishing them with a good work environment, complete with improved lighting, rest breaks, and ergonomic equipment. Gilbreth applied her understanding of the significance of human factors to improving the processes of engineering design and manufacturing. For working women, Gilbreth advocated a system where managerial tasks became more specialized and subdivided, so that women would be evaluated by more than one male manager. This afforded women a safety buffer against harassment by any one man in a dominant position (Trescott 1983).

For twenty years, the Gilbreths distinguished themselves in the fledgling field of scientific management, seeking the "one best way" of doing work. Their intent was not simply generating larger profit for shop owners by making workers more productive. They always maintained that workers should share in any greater profit that resulted from the increase in productivity. They were also concerned with increasing quality of life;

they were convinced that refining the motions used to do a job would not only make the worker more productive but it would also reduce fatigue and therefore increase enjoyment of life. "Motion study and all its refinements and techniques were attempts to create rhythmic work methods by means of which human beings might find their own greatest number of happiness moments" (Yost 1949, p. 250). In *Fatigue Study*, the Gilbreths wrote that fatigue elimination can only be successful if "happiness minutes" are increased. If workers really are happier, "then your fatigue eliminating work has been worth while in the highest sense of the term, no matter what the financial outcome" (Gilbreth and Gilbreth 1916, p. 150).

Until Frank's death in 1924, Lillian worked primarily in the background, although she was always present and her husband considered her an equal partner in both his work and his life. After Frank's death Lillian found herself with eleven children to raise and little savings. She decided to continue the work they had begun, but the engineering world was not yet ready for a female engineer—even most existing clients did not renew their consulting contracts with Gilbreth, Inc. In spite of this, Dr. Gilbreth was in great demand as a speaker and she traveled throughout the world explaining motion study and the techniques she and Frank had devised. At the same time, she remained involved in her children's lives, ensuring, among other things, that all eleven completed college.

Consulting jobs were rare, but she continued lecturing and became affiliated with a number of colleges. She was asked to open a management school to train managers in applying motion-study techniques to improve productivity. Although industry leaders could not imagine inviting a woman to their factories to tell them how to improve productivity, they were quite willing to send their managers to her for individual training. Dr. Gilbreth operated a small school for several years until colleges began offering management courses. She went on to design the first "efficiency kitchen," the legacy of which is seen in many homes today. She also did ground-breaking work with disabled homemakers, enabling them to do more with less fatigue. The list of Lillian Gilbreth's achievements is long, but the theme is always the same: enabling people to do more while improving their quality of life and increasing opportunities for "happiness minutes."

Dr. Gilbreth continued working, traveling, and lecturing, for forty-four years. She earned several awards for her work in engineering management. Following the Gilbreth Centennial meeting held in 1968 by the American Society of Mechanical Engineers to observe Frank's 100th birthday, she finally retired. She was 90 years old. During her years of travel and work, Lillian always made time for family and friends. Her son, Frank, commented that "... if anyone keeps track of how many Happiness Minutes each of us tries to create in her lifetime, she must have a high score." (Gilbreth 1970, p. 254) In many ways, she was the epitome of applied motion study—she had a highly productive life and career, while at the same time she created "happiness minutes" for herself and those she loved.

So what would Lillian Gilbreth think of the work styles and lifestyles of engineers today? Likely, she would find much room for improvement. When an engineer, male or female, must spend unreasonably long

(continued on next page)

hours at work, he or she misses many of the possible "happiness minutes" in life. Stress and fatigue levels are very high, which may result in diminished productivity. Engineers must often miss social activities, an important part of life that they must learn to live without—to their own and their families' detriment. I believe that Dr. Lillian Gilbreth would determine that there are better and more productive ways for engineers to work.

Source

Contributed by Michelle L. Toich and including work by Amy Chang, Stanford University, 1997

References

Gilbreth, F. B., Gilbreth, L. M. 1916, *Fatigue Study*, reprinted in 1973 by Hive Publishing Co., Easton, PA.

Gilbreth, F. B., Jr. 1970, *Time Out for Happiness*, Crowell Co., NY.

Society of Women Engineers 1978, Lillian Moller Gilbreth: Remarkable first lady of engineering, *Society of Women Engineers Newsletter 25 (Nov–Dec 1978)*: pp. 1–2.

Trescott, M. M. 1983, Lillian Moller Gilbreth and the Founding of Modern Industrial Engineering, *Machina Ex Dea*, Pergamon Press, NY.

Yost, E. 1949, *Frank and Lillian Gilbreth: Partners for Life*, Rutgers UP, New Brunswick.

these companies. In the same firms, corporate officer positions held by women rose by only one-tenth from the previous year, to 10.6 percent. At 5 percent, or twenty-three of these companies, women filled one quarter or more of the corporate officer positions. The highest figure was 44 percent at Reebok International. Companies with two or more women who are top earners include H.F. Ahmanson (a consumer and small-business banking and related financial services provider), Avon Products, Estee Lauder, and Nordstrom.

The survey also showed that only 20 percent of female corporate executives were in revenue-generating line positions, compared with 41 percent of male corporate executives. This is a problem because career path studies show that line positions are critical for advancing to the highest levels. A 1997 report from the Geneva-based International Labor Organization (ILO), showed that U.S. women managers only earned 68 percent as much as their male counterparts. In Canada women held over 20 percent of managerial positions, up from 4 percent twenty years ago. In both Japan and the United States, certain jobs were so segregated that nearly half of the women professionals worked in just two occupations, nursing and teaching. In its 1996 survey of 300 companies in Britain, the ILO found that just **3 percent** of board members were women. One effect of discrimination in the workplace and the resulting frustration has been more women leaving to set up their own very successful businesses (San Francisco Chronicle, Dec. 12, 1997, p. E4).

▪ Working in Teams

Diversity can be achieved by different means. One modern slant on the traditional management tasks listed above is that they will increasingly happen within multiskilled project teams. Many problems are too complex for any one individual to handle effectively. This is part of the rationale for greater emphasis on teams and team skills. Engineers can expect to spend more time working in teams of people drawn from diverse professional backgrounds. The importance of team work (and the impossibility, now, of engineers trying to "do it all themselves") is highlighted by the fact that no program of which the authors are aware, including two five-year "Engineering and Society" and "Engineering and Management" degrees at McMaster University in Canada (Case Study 3.4, p. 154), offers all the engineering practice areas shown in Figure 7.1 in a single program!

In the future much engineering will be done by teams that draw together the necessary range of expertise (see Figure 7.1) from a variety of engineering specializations. Engineers will also work with other professionals: social and environmental scientists, economists, and lawyers. Mutual understanding and respect is essential for this to be effective. Above all, the management content of their courses must help engineers learn to develop the common language necessary for working effectively in these groups and benefiting fully from the skills and knowledge base of other members. Broadly based design teams facilitate some of the informal learning that takes place in the workplace and is essential to effective organizations.

It is not a simple matter for engineers to learn to work in diverse teams, but there are some guidelines on how to proceed successfully. Wilde (1997) points out that team selection and team building are important and complex issues. Diversity is important for both creativity and effectiveness. In this context, diversity includes culture, gender, and personality type. In its very successful engineering design program, Stanford University has regularly used a modified Myers-Briggs style of personality testing to identify personality diversity as an input to team formation. Students are encouraged to form diverse design teams by project rules that give them a wider choice of project as a result. The members of these teams need to develop high-level information, communication, and conflict resolution skills. Conflict resolution is not about ignoring or avoiding problems. It involves bringing out and clarifying areas of difference or conflict, and then working on them to resolve them promptly and constructively.

FORMOSA PLASTICS GROUP AND THE "WANG DYNASTY"

How different from the typical modern western model can the management of a complex business enterprise be, and still succeed? This case describes the leadership within the vertically integrated, family-based group of companies built up since 1954 by a Taiwanese businessman, Yung-Ching Wang. By 1997, his Formosa Plastics Group (FPG) was valued at $14 billion. It was the largest nongovernment company in Taiwan, and the top polyvinyl chloride (PVC) producer in the world.

With three wives, two sons, and eight daughters, Y. C. Wang has been able to have each branch of his conglomerate under the management of a family member. This model seems characteristic of Chinese-culture business structures, in which there is an expectation that important decisions will be taken by a member of the family rather than by outsiders (regardless of the level of expertise or seniority of the "outsider" concerned). This pattern is not in fact restricted to Chinese culture, but seems typical of family-run businesses worldwide. The management challenge for such enterprises is to bring in outside expertise when necessary and then to accept it and act on it. The "outsiders," who commonly include engineers, may find this environment very frustrating. Their problem may simply come down to: how to be accepted as one of the family? "Marrying in" is one traditional way of accelerating the process, but is not always an acceptable option!

Born in 1917 in Taipei, Y. C. Wang's first businesses were a rice mill and then timber and plywood companies. To make the newly developed plastic resins, Y. C. Wang created Formosa Plastics in 1954, buying the technology from the Japanese and using money borrowed from the U.S. AID Mission to China. No other Taiwanese company was then willing to invest in plastics resin manufacture. Four years later Y. C. Wang founded Nan Ya Plastics to process the raw PVC produced by Formosa Plastics into products such as pipes and toys. Suddenly Wang was his own best customer (a key ingredient for successful vertical integration). In 1965 he established Formosa Chemicals and Fiber—another move towards greater integration, because rayon was needed to produce rayon backing for PVC imitation leather. By 1997 various companies had branched out from the three major subsidiaries of FPG. There were also fourteen plants in the United States under the name Formosa Plastics Corporation USA (FPC-USA). By the end of 1998 FPG planned to complete construction of Taiwan's sixth naphtha cracking plant to produce ethylene, a major ingredient in the PVC process. The new plant would cut production costs and reduce dependence on chemical companies in Japan and in Korea.

Y. C. Wang worked at preparing his family members for business leadership roles. With a poor educational background himself, he made sure that all his children were well-educated. All of them studied abroad, but they also felt the impact of their father's traditional beliefs. Growing up in poverty taught Y. C. Wang to make the most of everything. Despite the family wealth, lessons of frugality were emphasized throughout his children's lives. Cher, one of his daughters, believes that her father tried particularly to teach them the value of ordinary, everyday things. This emphasis on frugality has clearly worked to Y. C. Wang's advantage in his business enterprises—he has concentrated on reducing production costs and being more efficient.

Vertical integration also became ingrained in the minds of his children. Their combination of international mindsets and traditional attitudes is seen as what makes the Wangs successful today. Vertical integration has in fact been central to the way Y. C. Wang built his empire. The expansion of the three subsidiaries, and especially Nan Ya Plastics, into electronics, demonstrates the continuing innovation and competitive edge necessary to survive in today's markets.

This success story is not without its blemishes, however. A family-managed firm is not necessarily a good neighbor. There have been serious environmental problems associated with the group's production processes, and FPG has been responsible for many environmental law violations, especially in the past decade. This has hindered FPG expansion. In Wallace, Louisiana, the local community, supported by civil rights groups, environmentalists, and preservationists, stopped FPC-USA from building a wood pulp and rayon plant. Already overrun with chemical and oil plants, Wallace is in a poor African-American area known locally as "Cancer Alley," and the proposal to locate the plant in Wallace was widely regarded as environmental racism.

Between the mid-1980s and 1997, the group's FPC-USA ethylene plant at Lavaca Bay in Texas was fined by the Texas Water Commission, the Texas Air Control Board, and the Occupational Safety and Health Administration for violations relating to toxic wastewater contamination, leakage of pollutants, and other environmental hazards. In 1991 the EPA fined FPC-USA $3.37 million for hazardous waste violations—the largest fine EPA has ever administered.

FPG has learned to comply with environmental standards while still producing cheaply. In 1993 Taiwan's EPA named one of FPG's plants as one of the ten firms who has successfully attained economic growth simultaneously with controlled pollution. There are some continuing environmental problems, but FPG remains successful, partly because of the entrepreneurial skills of the family. FPG access to funding comes from its own revenues, profits from other affiliated family companies, selling of FPG bonds, and loans by government-run banks. Associated family businesses have the same funding strategy, since the banks trust Y. C. Wang's guarantees.

A critical issue for a family-run conglomerate such as Wang's is leadership succession. Y. C. Wang's brother, Y. T. Wang, is the president of FPC, and his nephew, William, runs Formosa Chemicals & Fiber. Two daughters, Charlene and Cher, are the presidents of First International Computer, Inc. (electronics) and Everex Systems Inc. (computers), respectively. One of Y. C. Wang's sons, Winston, former vice-president of Nan Ya Plastics, instigated successful expansion of the family business into electronics, producing chemicals for the electronics industry, particularly for semiconductors. By the late 1990s, Y. C. Wang was looking to retire and Winston was expected to step into the leadership. However, a commitment to family values

(continued on next page)

may have put at least a temporary stop to this. Winston's refusal to end a rather visible extra-marital affair has resulted in his being forced out of the entire business by his father. It will be interesting to see how the ongoing leadership of the group is eventually resolved.

Source

Based on work by Beatrice Bi-Ann Lee, Stanford University, 1997.

References

Relevant World Wide Web sites, selected issues of: Asiaweek, Chemical Week, China Economic News Service, Environmental Action Magazine, Oil and Gas Journal, New York Times, and Trade Winds.

MANAGING RISK

Risk management is an important technique that can help engineers meet their ethical responsibility to the community. However, it should be recognized that, in part, it utilizes subjective, value-laden approaches that need to be opened up for consideration, not suppressed as if they were simply methodological issues.

Engineers often have a rather narrow perception of the aims of technology; for example, engineers often think of the aim as being to supply electricity at minimum cost without regard to environmental or other issues. Problems with electricity supply in newly privatized systems in Australia and New Zealand in recent years have highlighted public concerns with a "risk management" approach to infrastructure provision. Briefly, the central issue is a change in the attitude of electricity suppliers to the risks inherent in their operation. In the past they had been "risk-averse," that is, they had avoided the likelihood of supply failure by providing significant overcapacity and redundancy in the supply systems. This allowed for plant failures and for "accidents," such as the severing of critical power supply cables by ditch-digging machinery carrying out other tasks (which apparently happened in one case). However, part of the rationale for "corporatization" and "privatization" of electricity provision has been the argument that levels of overcapacity and redundancy were excessive, unwarranted, and ultimately uneconomic. Electricity was thought to be able to be supplied with "acceptable" reliability by systematically analyzing the risks and managing them more efficiently. The risk analysis in these cases seems demonstrably inadequate.

Risk Management

The process involves the following steps:

- ◆ Establish the strategic, organizational, and risk-management context in which the remainder of the process will take place, and establish how tolerable risk criteria will be agreed.
- ◆ Identify the hazards.
- ◆ Analyze each source of risk, and combine these analyses to estimate an aggregate level of risk.
- ◆ Assess the estimated risk against the tolerability criteria and prioritize risks for treatment.
- ◆ Treat the risks: implement a risk-management plan.
- ◆ Monitor and review: regularly evaluate the performance of the risk management plan.

Risk management has been successfully applied to individual projects and activities, but is likely to fail when important hazards are not identified. It is not an "objective" technique and its application involves political and moral choices, which should be subject to public scrutiny.

There is no empirical evidence that it is well suited to the management of global change issues, such as sustainability. The aggregate of many environmental decisions, each based on risk management, may not ensure sustainability.

Source: Institution of Engineers Australia 1997, Section 3, p. 4.

Beder (1998, Ch. 12) takes this discussion further, pointing out that engineers can only say something such as: "On the basis of the assumptions we made, and the limited applicability of the models we used, our assessment is that the project will meet acceptable risk criteria." She goes on to discuss the difference in the social models used by proponents and opponents of technology. Proponents tend to assume that organizations and equipment always function perfectly, with complete communication of relevant information and in accordance with established rules. This is clearly not true, as both our own common sense and disasters such as Chernobyl remind us. "Watchdogs" such as Greenpeace focus on the uncertainties and the deviations from ideal technical performance. Communities affected by proposals tend to focus on the social dimension, looking at the past performance of the organizations involved and their openness and general demeanor.

Wynne (1989) has shown that in practice the public do make rational judgments on risk, based on how risk is controlled. The example of the Ford Pinto (discussed in Chapter 6 in the section Consumer Rights and Protection: Product Liability, p. 320) illustrates the importance of a proper and open process. In this case Ford decided not to protect the fuel tank from penetration in the case of a serious rear impact, on the basis of an assessment that the cost of protection would be greater than the cost of the losses (including compensation for deaths and injuries) from vehicles catching fire. Subsequent experience showed that this was not a socially acceptable decision (Harris et al. 1997).

PROJECT PLANNING

Project planning is used by all modern managers and in all areas of engineering. It has become so entrenched in management that projects are unlikely to proceed without a critical path analysis by which managers can assess the initial proposal and then monitor progress to see that milestones are being achieved. Project planning is the process of deciding what is to be done, how, when, where, and by whom. It also involves developing the structure (time framework, availability of physical and human resources, and so on) for the project.

A range of important and useful techniques has been developed to support planning exercises. However, difficulties are likely to emerge in technology-intensive organizations when there is an undue emphasis on short-term performance at the expense of continuing to develop the technologies that will underpin future profitability and market share. The bottom line, this year's profit, may be a necessary criterion for the success of an enterprise, but it is not a sufficient one. The fact that public companies in the United States are so driven by the next quarter's results gives rise to real concern for their long-term performance.

Projects vary greatly in scale and complexity. In construction activity, the need is usually to decide on the elements of the construction process and arrange them most effectively. The process has been described to one of the authors as "throwing a huge amount of money into a big hole as quickly as possible." The engineering management challenge is to do it in a planned and monitored way. In this sort of project, a work breakdown study (WBS) is first carried out so as to divide the work to be done into predictable, independent tasks or activities.

The simplest type of planning diagram is probably a Gantt bar chart, which shows time periods as columns and activities as rows. It is quite adequate for a straightforward activity such as a project being carried out by one student, as shown in Figure 7.2. At a more sophisticated level, it can also be used to allocate resources (e.g., the number of people committed to a particular activity) and as a visual check on progress. In this example, progress at the review date, the end of week 9, has been demonstrated by thickening the lines to show the proportion of completed activities. In this case, the commissioning of the test rig is shown as having been completed a week ahead of schedule, whereas the writing up is a week behind.

When a project involves a number of interdependent activities, they can be arranged into a network, using arrowed lines to show the order in which they need to be carried out (hence its other name, an arrow diagram). Based on past experience, estimates can be made of the time each activity will take. The longest path from start to finish through the network is then the shortest time in which the project can be carried out. This path is known as the critical path, and this approach is called the *critical path method*. The arrow diagram can be constructed either with activities on the nodes (AON) or with activities on the arrows or arcs

Project Name: **Typical Student Project**

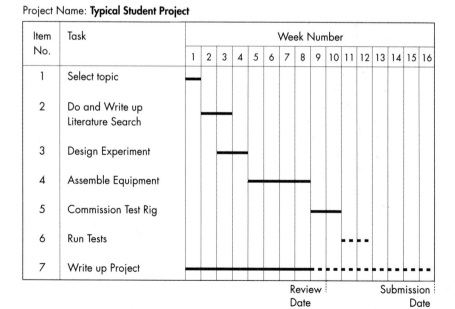

Figure 7.2 Gantt Bar Chart for a Student Project

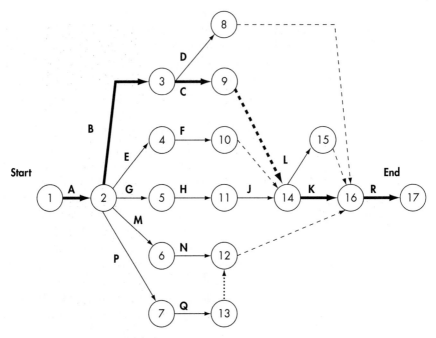

Figure 7.3 Critical Path Chart for a Village Micro-Hydro Project.
Source: Courtesy of APACE (1992).

(AOA). We show AOA in Figure 7.3. This is a planning diagram for the technical activities involved in the building and commissioning of a microhydroelectric plant to provide lighting and some industrial power in a village in the Solomon Islands. Background to the project is given in Case Study 8.1 (p. 406). (Incidentally, the penstock is the piping that carries water under pressure from the dam down to the turbine.)

The input of the recipients to project choice, direction, and design was essential. In fact, for the village to have the level of commitment to the project necessary for the system to continue to work effectively after installation, there has normally been a lead-up period of at least two years before the first sod was turned. During this period, members of the village community organized themselves into committees to plan and administer the system and to raise funds for its installation. A preliminary visit to the site was made by some of the technical people involved. A work-breakdown study process was used to break the project into smaller parts (also known as work packages) and to determine the activities that must immediately precede each package. The time to carry out each package was estimated on the basis of previous experience. Dummy arrows

TABLE 7.1

Activity	Coords	Immediate Predecessor/s	Description	Time (days)
			Microhydro-Installation Project Activities	
A	(1,2)	—	Final survey of sites and routes	6
B	(2,3)	A	Prepare dam site	10
C	(3,9)	B	Build forebay of dam, settling areas, strainers	20
D	(3,8)	B	Build dam	24
E	(2,4)	A	Prepare penstock route	10
F	(4,10)	E	Run penstock	18
G	(2,5)	A	Prepare turbine house site	3
H	(5,11)	G	Build turbine house	6
J	(11,14)	H	Install turbine	10
K	(14,16)	J	Wire up generator control unit	4
L	(14,15)	C, F, J	Connect ends of penstocks	1
M	(2,6)	A	Install house and industrial wiring	24
N	(6,12)	M	Wire up village distribution boards	6
P	(2,7)	A	Prepare route: generator to village	6
Q	(7,13)	P	Run power transmission line	24
R	(16,17)	D, K, L, N, Q	Commission and optimize system	8

were used to show logical connections rather than tasks. The critical path here was 1-2-3-9-14-16-17, with a duration of forty-eight working days. The separate activities shown on Figure 7.3 are listed in Table 7.1. All the essential technical skills needed to be transferred to the villagers for them to be able to keep the system working effectively, so the activities shown included training by doing.

The critical path method is characterized as activity-oriented. The time to complete a project can be reduced by speeding up one or more of the activities on the critical path (A, B, C, K, or R in this case), although this normally implies some additional cost. Activities that are not on the critical path have some float: that is, they do not need to be started immediately upon completion of the preceding activity.

A variation on this approach is the PERT (Program Evaluation and Review Technique) approach, created in the late 1950s to plan and accelerate the development of the American Polaris ballistic missile. It helps the planner to deal with uncertainties. Each activity on the PERT network is allocated three times for completion: most optimistic; most probable; most pessimistic. A specific relationship is normally assumed between the three, allowing calculation of the probable project completion time. Because of the focus on outcomes, PERT is described as event-oriented (Babcock 1991). A range of personal computer programs of varying complexity (and cost) is now available for producing these diagrams.

QUALITY AND MANAGEMENT

Quality has become a fashionable management term and is now used in a variety of ways. In its traditional and still most widely recognized sense it means a product or service of distinction. This usage certainly applies in most engineering research. There are also important fields of engineering activity and products (such as luxury cars) for which the traditional meaning is still the most relevant. In the management context, however, there are new usages and new shades of meaning. One approach to teasing out the meaning of quality is to think about the issue in terms of "soft" versus "hard" quality, a distinction explained in the Key Concepts and Issues box.

Before employees can take responsibility for quality they need appropriate training: workers and their unions should ideally be part of the team for effective system development. Worker participation in goal development is particularly important with new information-based technologies, for which worker commitment and sense of responsibility are critical. Although the very idea may be disquieting to many professional engineers, there is evidence that modern unions can be important in upgrading the work process (Kaminski et al. 1996). Unions can play an important role in setting the direction and form of technological change, particularly as the focus in production moves towards quality and sustainability.

■ Documentation

In engineering generally, proper documentation is recognized as essential to quality and indeed to corporate survival. Computer aids to writing, drafting, and design can help make the integrity, coherence, and security of the resulting documentation a major aid to the success of the modern engineering enterprise. As Case Study 7.5 at the end of this chapter shows, contracts for engineering work need to specify responsibility for information handling and transmission. The process of preparing operating manuals can show up limitations in both policies and processes and be an incentive to resolve them.

Documentation, described by McGregor (1997) as both process and product, is also central to protection against product liability claims. Formal processes that ensure quality within the organization and warrant it to customers have to be integrated into the daily life of the enterprise. Complete commitment at all levels of management is essential for these processes to be effective.

After exhaustive public consultation, the ISO 9000 series quality management standards have been adopted around the world. This series of standards lists the quality system requirements that need to be addressed, aimed at ensuring consistent quality processes. They are complementary rather than alternative to specified technical requirements for the products or services covered.

In practice, ISO 9000 requires two levels of quality documentation within an enterprise. The first is a relatively brief public policy statement, signed by the chief executive, that spells out the commitment of the enterprise to quality and explains the significance for the organization of each paragraph of the relevant ISO standard.

A second, more detailed document (or series of documents, such as manuals) spells out for every activity and for each section of the company involved in handling it approved procedures and processes. This second document typically incorporates much of the technical expertise of the firm (and the accumulated skill and tacit process knowledge of the workers) and is confidential. As Case Study 7.2 (on p. 340) indicates, quality certification is increasingly a prerequisite for tendering for work in both the public and private sectors.

In parallel with the ISO 9000 series, the ISO 14000 standards deal with environmental management. Both sets of standards call for the same systematic analysis and documentation processes. The initial driving force for companies to comply with environmental standards is often legislative requirements, but a major survey in Europe in 1996 suggested that companies also find that the review process helps increase staff concern for environmental issues. Involvement in the documentation process can raise awareness and perform a useful training function.

Formal auditing of organizational performance in quality, environmental management, and other areas, such as occupational health and

Soft Quality

James Adams suggests that we think about "soft" or subjective quality in terms of a palette of characteristics, some or all of which we use to distinguish "good products" from "bad products." These characteristics include:

- symbolism—visionary technology (the world's "highest"; "fastest"; "most advanced" ...);
- elegance and sophistication—possibly useless and inefficient (cuff links?);
- emotional content—associated with high fashion or personal commitment;
- craftsmanship—the feel of leather and wood, solidity;
- consonance with global constraints—environmental friendliness; and
- human fit—ergonomics.

(Adams 1997, Personal communication. The relevant course Adams offers at Stanford University is now simply called "Good Products, Bad Products.")

It is easy to recognize how, in marketing, some or all of these characteristics are used to denote quality. They can give products the right sort of "feel"; for example, telephone handsets commonly have some weight added to give the right feeling of robustness. These characteristics are clearly relevant to successful product design.

safety (OH&S), is increasingly common. Computer packages have been developed to assist with auditing, and to document the outcomes in a consistent and accessible way.

SOME LEGAL ISSUES FOR ENGINEERS

▍Intellectual Property

In our information society, intellectual property is central to the success of all modern, technologically oriented companies. It goes almost without saying that a high priority should be given to proper recording of,

Hard Quality

Engineers also deal with "hard" quality through objective measures signifying conformance to an objective standard of fitness for purpose (for example, in terms of defects per vehicle). Arising initially in manufacturing engineering, but now finding currency in many other areas (including higher education), these measures are the basis of quality control and quality assurance. Although widely accepted, it should be noted that they can involve a leveling down process that is inimical to traditional concepts of quality. This is not inevitable, however, and in fact the outcome of these approaches can be products or approaches that much better meet the requirements of the users.

Total Quality Management

The two central concepts of total quality management (TQM) are those of customer service and individual employee responsibility for quality. Customers include internal departments as well as external clients. All areas need to work to improve understanding of what their customers want and need so as to improve the service they provide and to develop clear performance targets. TQM emphasizes that effective management must go beyond a narrow focus on worker shortcomings. The approach recognizes that up to 85 percent of quality problems are in the design of the system and can therefore only be addressed by management (Sprouster 1987). Lean manufacturing (discussed below) illustrates many of the issues.

and security for, the knowledge on which an organization's survival is based. Regrettably, this is not always the case. Clear and well-developed corporate policies on intellectual property are essential. Engineers can make a significant contribution to the formulation and ongoing implementation of these policies. They need to understand their organization's portfolio of intellectual property and how it can be protected. This includes maintaining control over new products and novel features of existing products. The organization has a reasonable expectation that it will reap the rewards for original research, design, and development work by selling its products at satisfactory margins and getting adequate returns on its investment. This may require preventing theft and use by other organizations. Typically, the identifiable intellectual property of an engineering organization will include some or all of:

- patents;
- trademarks;
- registered designs;
- confidential information; and
- trade secrets and know-how in design, manufacturing, and materials science.

One way of estimating the value of intellectual property is to subtract the book value of tangible assets and assumed values of goodwill and reputation from the total value of the company. The difference gives an indication of the value of the organization's intellectual property. For software and information/knowledge-based companies, this figure can approach 100 percent of the company's value. For manufacturing organizations developing leading-edge technologies ("technology-leaders") the value of intellectual property can easily be half the total value of the company. This may seem surprisingly high, but it is important to recognize that the organization's principal tool for generating products is its intellectual property. This is increasingly relevant in today's global marketplace where, in principle, products can be made anywhere around the globe, although there are practical limitations, particularly for complex manufacturing technologies and products. Case Study 8.6 (see p. 442) on IBM's choice of a location for manufacturing parts for data storage units illustrates some of the issues involved.

■ Patents and Other Protective Measures

The choice of ways to protect innovations depends on the specific features requiring protection. The patent system is intended to protect the interests of the originator of a novel material, process, or product. In return for making details of the novelty public, the patent provides the devisor with a monopoly on its use for a limited period. Of course, once an idea is in the public domain, efforts can be made to get around the constraints on its use.

There are other limitations on the patent system. Not all ideas are new, and not all new ideas result in products that are patentable. Even where a patent is granted, it may be challenged in court, and effective patent defense requires deep pockets. Prolific inventors, from Edison to the present day, have complained about the excessive cost of defending their patents.

Patents only apply in the countries where they are applied for and granted. Not all countries have the same attitude towards intellectual property rights, as tensions in the 1990s between the United States and China demonstrated. Despite its limitations, the patent system is a primary method of protecting intellectual property. Arthur Bishop is the Australian inventor of a widely used power steering system for automobiles. Bishop argues that effective protection may require developing not only advanced and patentable products, but also advanced and preferably patentable machines and technologies for producing them. This nested, two-level protection increases the cost (and difficulty) of market entry for potential competitors, and reduces the price at which they must be able to produce.

One of the most common ways of appropriating intellectual property in engineering products is to "reverse engineer" them. This typically involves purchasing an item and trying to work out how it has been made, including taking detailed measurements of it. Some engineering items, such as pump impellers or liners, may be very difficult to measure, partly because of their complexity, and partly because the finishing process commonly removes reference and set-out points used in their original manufacture. Establishing manufacturing tolerances is even more difficult, involving measuring and analyzing a number of samples of the same item. Despite the difficulties, a serious replicator has to establish item tolerances. A major market for replicated parts is the wearing elements of equipment, including pump liners. Purchasers expect alternative replacement liners to fit properly into pump casings.

Copyrights on original drawings made within or for an organization can provide a significant degree of protection, as they highlight the questionable legality of reverse engineering. Recreation of a drawing that is essentially identical to that used by the original manufacturer may be held by the courts to be an infringement of that manufacturer's copyright. Not surprisingly, stealing the original drawings is an approach that has been used to short-cut the process. An organization that stole original drawings but claimed to have reverse-engineered their product was caught when the imitation was found to have features that had been on the original drawing but were not incorporated in the item as actually manufactured.

Where a design feature gives a product a uniquely distinctive visual appearance, it may be possible to register the design. This form of protection should probably receive more attention.

With many sophisticated modern materials, for example synthetic elastomers, examination of the end product may not give a useful indication of the ingredients or the production process. Here the most effective protection may be obtained by treating the formula as "confidential information" or as a "trade secret." This requires careful protection within the organization and only revealing it outside the organization on the basis of strict confidentiality agreements. The formula for Coca Cola™ concentrate is one very visible example of a trade secret.

■ Security Procedures

What steps should organizations take to conserve and protect intellectual property systematically? Some of the most important and widely used actions include the following:

1. All staff involved in technology generation and change need to sign confidentiality agreements. These agreements normally include provisions to assign to the organization inventions that staff make in the course of their duties. In some countries copyright in the work of employees is automatically assigned to the employer unless specified otherwise in the terms of employment. Otherwise, provision for such assignment needs to be made.

2. All subcontractors, consultants, trainers, and so on need to sign similar agreements before they are exposed to confidential corporate information. Minimizing the use of outside staff may be prudent, particularly in more sensitive areas.

3. Confidentiality agreements are important in reminding employees and associates of their obligation to keep the company's intellectual property secure. They can also discourage departing employees from taking secrets away with them. Lawsuits in the U.S. microprocessor industry indicate that poaching of employees in order to steal intellectual property is an ongoing activity.

4. A common provision in employment contracts for technical professionals is that they will not take employment with a direct competitor for a certain period after they leave the firm. There are obvious difficulties with enforcing such contracts, which are clearly a restraint on the individual's right to trade. Enforcement may require showing both that the degree of restraint was reasonable, and that the payment the employees received while in employment took suitable account of the personal cost of such restraint.

5. Other procedures may include: strict security procedures for visitors; strict drawing office controls on and records of copying and removing drawings; shredders readily available for destroying confidential material immediately it is no longer required; strict protocols on computer aided design systems, with different levels of access and weekly password changes; ensuring that drawings

have appropriate confidentiality and copyright warning notices; and ensuring that all staff are aware of the value of the intellectual property and the need for vigilance in protecting it.

Once a problem is detected, it is essential to act promptly and decisively. Any delay is likely to increase the resulting losses and weaken the case for redress. The starting point for this action is for the relevant technical managers to retain the best possible local lawyers with experience in intellectual property law. Close collaboration between legal and technical people is essential—the lawyers need to learn about the technology and appreciate the engineering involved, while the company's technical people need to come to appreciate how the law is applied. Thorough preparation, including comprehensive documentation, is essential for success. After all, the stakes can be high. Multimillion-dollar judgments have been given against manufacturing organizations found to have infringed copyright in the process of copying products (Wightley 1997).

■ Other Legal Issues for Engineers

As citizens, engineers encounter the same sorts of legal issues as others. Our focus here is on legal issues that are likely to concern engineers in their professional lives. Vaughn and Borgman (1999) offers a useful general introduction to the law for engineers.

Legal issues can become very complex, particularly in the context of global engineering practice. Aside from laws that regulate the professional conduct of engineers, in most countries there is a plethora of legislation affecting business that is also likely to impinge on their work. Three main purposes of such laws are to protect:

- ◆ companies from each other, particularly by prevention of unfair competition;
- ◆ consumers from unfair business practices, such as misleading advertising or deceptive packaging; and
- ◆ the broad interests of society against unrestrained business behavior, including requiring proper environmental and occupational health and safety safeguards.

Peters (1992) argues that outsourcing and networking are becoming increasingly important ways of building organizational capability. In this context, an organization's reputation for integrity will be fundamental to its continued operation. Regrettably, not all organizations conduct themselves in ways consistent with professional expectations and attitudes.

Ignorance of the law is rarely an excuse for breaking it. Engineers have a responsibility to inform themselves about what the law expects of them professionally. In England and in former British colonies a system of common law is the norm. Broadly, this means that elected legislators in parliaments or congresses enact laws, and then judges interpret them. However, one of the complexities associated with the global practice of engineering is that the legal systems in different countries vary in form and detail.

Although the ideal is that the law will always be applied fairly and impartially, lawyers and law enforcement officers, like engineers, are fallible human beings. In many parts of the world, including Eastern Europe and the former Soviet Union, expectations of law and government are also changing rapidly. The 1990s was marked by sudden and rapid change as these countries moved from centrally controlled to market economies. Similar uncertainties exist in some third world nations. It would be unwise, and almost certainly incorrect, to assume that technical professionals there will approach negotiations about contracts, for example, with the same understandings as Western engineers. At the same time, the system of civil laws to enforce such contracts is still weak. The best protection against default is that contracts are fair, achievable, and clearly recognized by both parties as being in their best interests.

How should negotiations be carried out under such circumstances? For successful collaboration, especially long-term collaboration, the essential, underlying approach should be the same as anywhere else in the world. There needs to be a strong ongoing effort to ensure an open flow of information, particularly when the circumstances of the parties change, as they will over time.

For professional engineers there are three legal and professional areas of responsibility of particular importance:

1. their work must be done competently;
2. the products, processes, and systems with whose design they are associated must conform to contractual specifications and be as safe as practicable, given the state of the art at the time; and
3. the work situations they design or over which they have control must conform to local legal standards and should be as safe as practicable, given the constraints of current technologies and knowledge.

The first of these points is discussed in Chapter 11 in the context of ethics and professionalism. The second usually arises in the context of product liability, and is discussed in terms of its effects on innovation in Chapter 6. The third point reflects the fact that engineers are often responsible for the safety of the operation of factories or other facilities. In the United States, the Occupational Safety and Health Administration (OSHA), established in 1970, has a major role in work safety that was discussed in Chapter 3.

Responsibility in many professional situations is vicarious, that is, the engineer only represents the organization in a formal sense. However, in some countries and jurisdictions, engineers may have personal responsibility for product liability, occupational safety and health matters, and so on. Verdicts of negligence may then result in fines levied against the individual as well as the organization and, in extreme cases, in prison terms for individuals. The same personal liability may apply to company executives in cases where companies engage in price-fixing agreements or other activities that are explicitly forbidden by the law of the land.

A broad problem associated with liability issues, dealt with in detail elsewhere in this book, is that engineering is not a science, or even an applied science. Engineering is a pragmatic, design-based activity that draws to a greater or less degree on science and mathematics. Regrettably there is widespread lack of understanding in the legal profession, including the judiciary, of this characteristic of engineering activity. Even more regrettably, this misunderstanding has occasionally extended to sections of the engineering profession. One result has been quite unrealistic expectations of the degree of certainty with which engineers can predict the outcomes of practical situations, even perhaps a touch of *hubris*, fatal overconfidence, in the analysis of the safety and reliability of engineering systems.

MARKETING

A strong focus on client wants and needs and marketing strategies is vital for the viability of all engineering organizations. Marketing also plays a major role in helping enterprises to adapt to changing demands and circumstances. We noted earlier in this chapter the role of marketing in matching the capabilities of buyer and seller. Marketing is now central to modern Western societies, which commonly take it as an article of faith that the market mechanism is the best way of allowing all levels of the economy to find their optimum levels of activity. Modern industrial economies are based on commodities, goods, and services produced for sale in the market to consumers. Marketing them is much more than just selling. It includes a whole process, from discovering a market opportunity through to timely and cost-effective delivery of the goods or services to satisfy it. The process of supplying the market became much more sophisticated during the twentieth century. In affluent countries, the focus moved from a concentration on products and promotional or selling effort, to creating and then meeting consumer wants, as discussed in Chapter 3.

Marketing is relevant to selling services as well as products. Good marketing should advantage both parties to the exchange, although it may skew the direction in which choices are made. In selling, the focus is on the seller's needs, whereas in the definitions of marketing given above, the focus is on the buyer's needs. Although this approach may suit the consumer goods area, a broader and perhaps more productive definition

cent of the students voted to pay the extortionist so that their bid would be considered. How might you have voted? Harris et al. (1995) offer a very helpful introduction to thinking about moral issues in engineering.

Transparency International (TI) is an organization that was set up specifically to discourage corruption and assist with eliminating it. One of TI's projects is a world wide web site with a wide variety of information, including a rather distressing "league table" of countries in terms of their levels of corruption. One of TI's strategies is to support the establishment of what it describes as *islands of integrity*, which can exist even within regions or industries in which corruption is endemic. This can be done by, for example, spelling out to all those involved that particular attention will be paid to this project, and that companies or organizations found to be transgressing will be publicly exposed, and then banned from competing for contracts in the country or region for a period of years.

This approach has had some significant successes. On one large power-station contract in a Latin American country in the mid-1990s, the savings were estimated at $300 million. This level of saving is important because it means that repayments of infrastructure loans can be made from the income from power sales, rather than becoming an additional debt burden on an already poor nation and people.

of marketing for our purposes would be: "identifying and matching the needs of the potential purchaser with those of the intending seller."

The more traditional approach to market analysis was the internally focused one of forecasting. Although this was acceptable in relatively static or stable situations, it was not very good at detecting impending changes in patterns of demand, and it tended to assume that the organization could impose what it wanted to produce on the market, rather than expecting to adjust its product to suit market trends. This is still essentially the position of the automobile maker in China described in Case Study 4.4 (see p. 192), but it is no longer the case in the West. Continuing consultation with current or potential clients, and listening to both their words and their silences, is much more appropriate to the marketing approach.

For a long-term view, the usual approach is to set up a range of scenarios. These are outlines of the general situation that assume particular consumption or other patterns to dominate, and attempts to project what

Marketing

Two useful definitions of marketing are:

1. *"the process of identifying and satisfying a client's needs at an acceptable level of profit"* (Rosier and Whitmore 1988, p. 1); and
2. *"human activity directed at satisfying needs and wants through exchange processes"* (Kotler et al. 1989, p. 4).

These definitions highlight the need to focus on likely consumer interest, rather than on items that the technical staff want to develop because they are at the forefront of technology. The importance of good design and flexible manufacturing to this approach is obvious.

Needs include "physiological needs for food, clothing, warmth, and safety; social needs for belonging, influence, and affection; and individual needs for knowledge and self-expression." These needs are inherent in human nature.

Wants are the form that culture and individual personality give to basic needs. Producers act to build the desires or wants for their products, promoting them as the way to satisfy a particular need.

Exchange takes place between two parties, each having something of value to the other. Exchange is commonly and conveniently of goods or services

---

their influence will be on the issues in which we are interested. For a central electricity supplier, for example, with a lead time of six or seven years to bring a new coal-fired power plant on line, there are now major uncertainties. Scenarios might range from ones in which extreme measures were taken to minimize use of central generation of electricity from coal, through a "business as usual" scenario, to one in which major efforts were being made to replace gasoline- and diesel-fueled cars by electric vehicles and electrical-powered public transport. These three scenarios should give a reasonable indication of the likely level of uncertainty, allowing strategies to be adopted that minimize expenditure while providing reliable power supply. One reason for working out scenarios is to allow for political developments such as the Kyoto Protocol on climate change, which may sharply change the context for decision making. Options for dealing with uncertainties include early retirement of a less-efficient plant or upgrading it and keeping it in service longer than originally intended. Consideration also needs to be given to alternatives such as intervening

for money. In the international context, particularly with developing or formerly socialist countries, it may also be through direct barter, or more sophisticated counter-trade arrangements.

An important insight for sellers is the distinction between wants and needs. A manufacturer may think that a consumer needs drill bits, but they are actually wants. Needs felt by the consumers are, for example, to make their homes more attractive. To do this they might like to hang pictures, for which they need to drill holes in walls and put plugs in them for hooks for hanging the pictures. The needs are thus not the drills, but the use made of the holes the drill bits make. Focusing on products, rather than the needs they meet, has been called "marketing myopia". Levitt (1960) who coined the phrase, applied it to the railway operators in the United States who failed to remember that there were other ways to meet consumer needs for transport, and were devastated by the rise of private motoring.

In consumer goods, much attention is paid to developing brand images and targeting products at emotional needs, suggesting that using the product will make you attractive, sophisticated, and/or loved. As one marketing manager put it: "My company makes cosmetics, and sells hope." (Hearne 1982, p. 2). A cynical view of this whole process is that needs are just wants backed by purchasing power.

on the demand side to encourage more efficient energy use, for example, by supporting energy audits and implementing their recommendations.

■ Developing A Marketing Strategy

A marketing strategy must be based on an awareness of the broad aims, capabilities, and past performance of the people in the organization; on sensitivity to the needs of the potential clients; and on awareness of the competition. This is summarized in Figure 7.4. It is important to recognize that although developing a marketing strategy needs to be a continuing process in any organization, it is not a substitute for action.

Although there are many examples of highly profitable opportunistic organizations, a successful enterprise usually needs clear goals and objectives that differentiate it from its competitors and locate it clearly in its market. The SWOT analysis of an organization—reviewing its strengths, weaknesses, opportunities, and threats—helps focus systematically on its

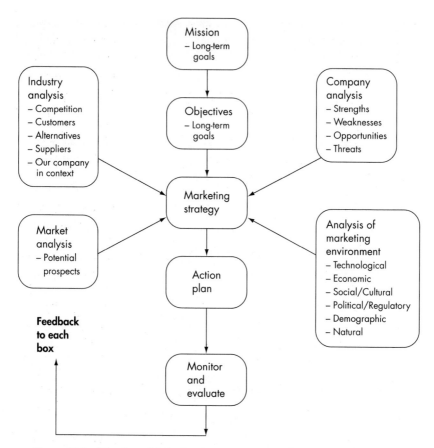

Figure 7.4 Planning a Marketing Strategy. *Source:* After Rosier and Whitmore 1988.

resources and competition. Analysis of the industry, the market, and the operating environment are important in formulating a marketing strategy and an implementation plan. Prompt feedback of information on performance and problems is important for a range of reasons, including minimizing liability.

▌ Ethical Issues and Marketing

There are obvious ethical concerns with approaches to marketing that do not address the global need to move towards sustainability. The whole relationship between marketing and consumption, particularly of non-renewable resources, will be increasingly important for engineers. Because innovation is essentially an iterative process, it can be expected (particularly in products that are still developing) that earlier models will become obsolete, possibly quite quickly. The instant obsolescence of computers is only one important example. But what is a reasonable time scale? Do product cycles allow proper engineering development to take place? The differences between European, North American, and Japanese cars and motorcycles, for example, raise some interesting questions about obsolescence. As is discussed in Case Study 8.4 (see p. 418), short model lives can create serious problems for the countries to which they are exported, particularly countries with relatively small markets.

Questions of psychological and technical obsolescence raise fundamental ethical issues. One outcome of consumerism is that more goods and services are consumed than are needed to provide a comfortable standard of living. Yet the concept of limiting economic growth poses a major problem to the marketer, who is: "locked into a pragmatic world in which he has to sell to survive and sell more to succeed" (Hearne 1982, p. 6). Hearne suggests that: "the principle that the consumer's avaricious pursuit of his own needs benefits society as a whole has been deeply embedded in Western philosophy and is a cornerstone of the capitalist system." The principles of liberalism (the belief in the free market and unfettered competition) and the liberal[ist] view of human nature and how society should operate, enshrine a belief in the unlimited capacity of the individual to consume. This is one important view of what constitutes freedom. Under capitalism, it is widely seen as a basic force driving the development of the economy.

As we discussed in Chapter 5, the ideal of a free market is not even a good first approximation to reality in many industries. Huge capital and other resources are often needed to be competitive, and are particularly needed to lead the development of new products. Markets thus tend to be dominated by a few major suppliers. Government limits on the number of commercial television channels in some countries create a near-monopoly situation, making each channel much more valuable than it would otherwise be. In these circumstances, regulation to ensure good corporate citizenship is probably inevitable.

The most recent trends in marketing are towards selling "quality of life" (satisfaction, or perhaps "soft quality," both difficult to quantify) rather than "standard of living" (rate of consumption). Examples might include an emphasis on miniaturization and elegance, rather than power (although the 1990s love affair with the four-wheel-drive and the light truck as a domestic vehicle reflects a different imagery). Questions of equity for people with less purchasing power, particularly in poor countries, do not generally seem to be addressed at all.

Marketing: The Example of Levi Strauss and Company

The fluctuating fortunes of jeans-maker Levi Strauss (LS&Co.) dramatically illustrates the magnitude of market changes. Jeans started off in California in the nineteenth century as practical, hard-wearing work clothes. In the 1960s, they were taken up by the generation born after World War II as a protest against the neat, conformist values of their parents. Market growth continued through the 1960s and 1970s, and in the late 1970s Levi had an annual earnings increase of 37 percent.

However, the market peaked in 1981. It was only possible to increase sales by diversifying styles and materials, and Levi's profitability was based on standardized products and economies of scale. In any case, parents were by then wearing jeans, so they were no longer anti-establishment, and falling birth rates had reduced the relative size of the teenage market. A progression through flares and designer jeans to stone-wash and acid-wash jeans, with a maximum life of perhaps three months, saw their rugged, western image fade: fashion turned a durable item into a transient one. Ironically, the major firms that had driven the change were not structured to profit from it, and seem to have been surprised by the changes in the market (Kotler et al. 1989, pp. 371–372). In 1985, a takeover saw Levi Strauss become a private company again.

In 1986 Levi's introduced a new brand line, "Dockers," aimed at broadening its market coverage. Since then, Levi's has energetically promoted a more casual business lifestyle in support of its range of products. In 1994, taking advantage of more flexible approaches to production and its significant level of

Engineers are important in all this economic activity and, as managers, are expected to participate in the marketing strategy of their organization, whether it be a private commercial firm selling products or a government authority selling services to the public. Engineers need to be aware of the associated social, political, economic, environmental, and ethical issues. These include questions of long-term sustainability and a proper balance between the resources consumed in production, and the durability and utility of the resulting products.

MODERN PRODUCTION SYSTEMS

As was discussed in Chapter 2, "scientific management" increased the division of labor in the factory. Control over the organization and planning of work was taken away from the shop floor and from the people

domestic production, Levi's introduced its "Personal Pair" program. For a premium price, this allowed a purchaser to be fitted in the store for a made-to-order pair of jeans, delivered three days later.

In 1997 Levi Strauss & Co. announced that it would close some facilities in 1998 "in response to a long-standing need to bring the company's manufacturing capacity in line with its U.S. production needs." The company said that its excess capacity was fueled by increased production efficiencies, a changing apparel market, and a shift in the company's brand management strategy. Gordon Shank, president of Levi Strauss explained: "to maintain the leadership position of our Levi's® brand, we are focusing our efforts on enhancing the brand... rather than pursuing volume for volume's sake."

The eleven plants to close employed 6,395 workers, 34 percent of LS&Co.'s total manufacturing workforce in the United States and Canada. LS&Co. would retain one of the largest apparel manufacturing bases in North America. "Our owned and operated facilities in the United States currently provide us with a competitive advantage," said Shank. "They offer speed to market, flexibility and mass customization capabilities." At the beginning of 1998, Levi Strauss & Co. was still the world's largest brand-name apparel manufacturer, with 1996 sales of $7.1 billion. It owned and operated thirty-two apparel manufacturing plants and finishing centers in the United States and five in Canada. The company manufactured and marketed branded jeans, casual sportswear, and dress pants under the Levi's®, Dockers®, and Slates® brands, and employed approximately 37,500 people worldwide.

actually carrying out the tasks. This became the basis for mass production of standard items. It tremendously increased the productivity of the manual work involved, but at the cost of sharply reducing the likelihood of job satisfaction. To what extent is this approach still effective in today's flexible production of high-quality products?

There is little point in making products that no one wants. In the era of mass production, it was assumed that problems could be solved by selling a bit harder. A modern approach is to bring the intended customers into the decision-making process, which requires much greater flexibility in the manufacturing process. Human skills and capabilities need to be drawn on effectively to meet this new challenge.

Even outside manufacturing, it is clear that maximizing the benefits from computer-based equipment requires broadly trained staff who understand the way the technology operates, and are committed to making it work effectively. In banking, for example, the integrity of the data entered is of critical importance. The systems used are extremely vulnerable to error, so clerical staff need to feel responsibility for results. The systems appear increasingly abstract to the operators, so intellectual mastery of them is essential. Tasks become increasingly interdependent because there is no longer a processing chain in which errors can be picked up by staff further down the line (elaborated in Thompson 1967). Effective operation of these computer-based systems requires workers who are involved with their work, who have become part of a team during the process of being consulted, and who are involved in the process of selecting and implementing the new technology (Mathews 1989, p. 211).

▌ Lean Production

One approach to manufacturing that emphasizes worker skill and involvement was the development of what was known initially as the "Toyota production system" (see box), and later, as it was adopted more widely, "lean production." The productivity increases flowing from this series of cumulative changes to Ford and Taylor's model of mass production are truly startling. Womack et al. (1990) found that assembly time at lean plants was down by from one-third to almost one half. Defect rates were around one-third those for traditional plants, as good as and often better than the performance of luxury makers.

The lean production system is not restricted to Japan. Performance at the New United Motor Manufacturing Inc., (NUMMI) assembly plant in Fremont, California, compares well with Toyota in Japan although the distance from parts suppliers increases the average parts supply on hand from the two hours common in Japan to two days. It is still a far cry from the two weeks common in mass-production plants.

Lean production is not a bed of roses for the workers involved. It puts much of the responsibility for the successful operation of the plant and for product quality on the employees on the production line. However, a key feature is that it starts to treat workers as intelligent human beings. Of course other factors affect the figures, but one indication of the social impact is that absenteeism, which runs as high as 25 percent in some European mass production firms, is 5 percent or less in lean-production plants in Japan (Womack 1990, p. 89).

This has been a very brief overview of a complex and interconnected set of changes in production; we have not described associated changes in product design or marketing. However, the inherent flexibility and responsiveness of this approach, and the associated determination to deal with conflicts promptly and resolve problems as they arise, are great advantages in both these areas. The flexibility allows a marketing-oriented responsiveness to customer demands, rather than a focus on price and the hard sell.

For careers in automobile engineering, one implication of this newer approach is that the focus moves from advancement through a technical specialty to advancement as part of a series of successful design teams. In an American automobile firm it has been typical for technical specialists only to be evaluated within their specialty. As a result, the program manager responsible for development of a new vehicle has had no real control over specialists, who have tended to see design problems in terms of protecting their specialist interests, as illustrated in Figure 7.5.

The lean design approach starts with a strong product champion, the large-project leader, who has specialist personnel assigned for the duration of the project. There is continuing coherent leadership and an emphasis on bringing out and resolving problems and conflicts early in the design process.

The project leader's evaluation of performance is recognized as important to the future careers of all the technical personnel involved, so there is a real incentive to get on with the project rather than fight for sectional interests. An engineer's professional life is seen as being a participant in a series of project teams, rather than as a technical specialist working in a technical hierarchy who is occasionally brought in to give specialist advice.

There are certainly long-term concerns about lean production. As with many areas of manufacturing employment, ongoing automation

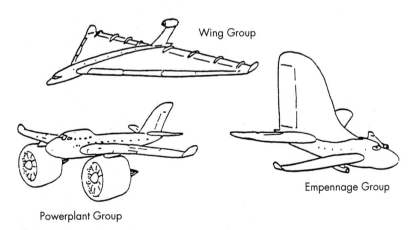

Figure 7.5 Distorted Departmental Perceptions of Aircraft Requirements.
Original artwork by Alex Revel.

The Toyota Production System

The Toyoda family were successful innovators in the textile machinery business in Japan in the late nineteenth century. In the 1930s, with government encouragement, they set up a vehicle manufacturing plant near their home town of Nagoya. Their initial specialty was trucks for the military. The plant manager, Kiichiro Toyoda, visited the Ford plant in Detroit in 1929 to learn about mass production. His nephew, Eiji Toyoda, who succeeded him as manager, made his own pilgrimage to Detroit in 1950.

In Japan in the postwar period, the government prohibited direct foreign investment in the motor vehicle industry, and high tariff barriers encouraged the emergence of a local industry. There were no "guest workers" or recent immigrants prepared to undertake the fragmented tasks offered by the traditional model of mass production. The strongly unionized Japanese work force, encouraged by supportive labor laws, was able to insist that workers would not be laid off when demand dropped, as was the American practice. Employment was effectively for life. This gave manufacturers a strong incentive to train employees and use each person for a variety of tasks.

The market was small, and firms such as Toyota had to be able to make a variety of cars using a minimum of costly capital equipment, particularly the huge stamping presses that made body parts from sheet steel. These presses used large, heavy dies to shape the parts and the dies had to be precisely aligned. Changing dies typically took a full day, so there was a strong incentive to punch out a large number of each part at a time. If there were problems with a part, there was either a great deal of waste, or extra time and effort were needed to make the parts fit. Also, while the dies were being changed, the press workers were idle.

Part of the solution that Toyoda and his chief production engineer, Taiichi Ohno, adopted, was to work towards economic manufacture of small numbers of parts. They developed simple techniques to speed up die changes, and simple adjustment mechanisms to ensure proper alignment. The production workers were then trained to make the die changes. By the late 1950s, the time for a die change was down from one day to three minutes! This made small-quantity production economic, saving on inventory costs and ensuring that problems with a part would be found before too many had been made. To make this system work, they needed a skilled, flexible, and highly motivated work force. If workers failed to anticipate problems before they occurred and take the initiative to work out solutions, the factory could easily come to a halt.

Chapter 7 • Management in Engineering

Part of the basis for lifetime employment was an agreement that workers would be flexible in their work assignments and would initiate improvements, rather than just respond to problems. The workers were grouped into teams, with a team leader who did routine production tasks as well as coordinate the team. The team eventually took over housekeeping, minor tool repairs, and quality checking. Ohno also set aside a regular time for the team to suggest ways of improving the process. (In the West this feature became known as "quality circles.") In collaboration with the (relatively fewer) industrial engineers, quality circles became an integral part of a continuous improvement process, *kaizen* in Japanese. Ironically, the quality message expounded by W. Edwards Deming was being ignored in his native America but taken seriously in Japan (Deming 1986).

When the team started taking responsibility for quality, it signaled a profound change from traditional mass production. The emphasis had been on "moving the metal," keeping the production line moving at all costs, and fixing problems in a "rework" area at the end of the line. This approach did not resolve problems but deferred them, and problems were compounded as faulty vehicles moved down the line. Even in the 1990s, rework in a traditional plant typically took 20 percent of the plant area and 25 percent of total production hours.

Ohno's response was to put a cord over every work station and instruct the workers to stop the line if a problem emerged that they couldn't fix. Then the team would all work on the problem. Ohno went further. Instead of treating errors as random events, simply to be repaired and ignored, he instituted a system called "the five why's." The line workers were taught to analyze production problems, continuing to ask "why" at each level of the problem until it was tracked back to its ultimate cause. The effect was to build in quality as the automobile is assembled, rather than attempt to pick up errors by inspection and fix them in rework before the vehicle leaves the plant. Quality increased sharply.

There was also a need to develop the same level of quality awareness in suppliers. Toyota built long-term relationships with their suppliers, based on tough but close cooperation. The last part of the process was to coordinate the flow of parts to the assembly line. The signal for a supplier to make the next batch of parts became the empty parts container returned from the production line. This extremely challenging step took away all the safety nets. With no float in the system, problems have to be resolved, rather than put aside. This is stressful, and it has certainly been difficult to change the culture in a traditional plant to accept this approach.

of plants will eliminate many of the present jobs. It is not clear where people who now go into such jobs will find work in the future. Is there really any alternative?

■ High-Performance Workplace Systems

Perhaps there is an alternative to eliminating jobs. One organization that has been working in this area for some years, with significant success, is the Work and Technology Institute (WTI) in Washington, D.C. In a major project report (Kaminski et al 1996), they outline an approach described as *high-performance workplace systems.*

The essence of the WTI approach is that, in the present manufacturing environment, a focus on downsizing and eliminating jobs is not advantageous for either individual enterprises or the country as a whole. Direct labor costs are only 8 percent to 14 percent of total costs, so the potential savings are very limited. Improvements that can result from participative work redesign, including improved quality, faster throughput and response times, and reduced inventories, are likely to be much more significant. WTI argues for moves towards production methods that "mobilize the intelligence, commitment, knowledge and creativity of the workforce in deeply participative operations." Flattened organizational hierarchies require a committed work force to benefit fully from advanced production and information technologies.

How can workers be persuaded that enthusiastic participation in redesign of work is in their interests? For Japanese workers in Toyota, it took a commitment from the company to lifetime employment, ongoing training, and a very hands-off style of management. Strong union presence was critical, both in convincing the firm that it needed to be serious about job stability, and in giving employees confidence that it was worth their while to take a serious role in ongoing redesign and improvement of the production process.

One of the studies in the WTI report, carried out from 1987 to 1995, was of a Ford Motor Co. project in Wayne, Michigan. Ford planned to add a stamping plant to an existing assembly plant. This approach was in line with the logic outlined earlier—minimizing the stamping plant's separation from the production line. In the process, Ford was looking to sign a modern operating agreement (MOA) with the United Auto Workers (UAW), representing the workers in the new facility. The clear and strong commitment of Ford management to making the project work, even if they had to give up a lot of detailed, day-to-day control, was essential.

Without it the effective worker participation essential for success could not have been achieved. The workers had to be empowered to take control of much of the process. They had to see that their ideas and their serious suggestions would be listened to and acted on.

Instead of a lengthy, legalistic document with dozens or hundreds of narrow job classifications (the classic confrontationist union response to a Taylorist fragmentation of work), Ford wanted an "enabling" agreement. In 1988 a conceptual agreement was signed that included: teams; payment for knowledge; one classification for production workers; three umbrella classifications for skilled trades; and an attendance standard.

No one was suggesting that the process would be easy. Part of the difficulty was capturing the skills and knowledge of the workers in the existing plant while changing the work culture. A critical step was involving the workers seriously in the design and layout of the new plant. Early and effective communication was essential.

The union leader in the existing plant was eager to have the changes work, and appears to have been very effective at bringing worker concerns about the changes out into the open and getting the workers themselves to suggest how the problems could be resolved. One example was the role and selection of team leaders. Concern that the leader would simply be a "company stooge" was allayed by an agreement that teams elect their own leaders. A six-month period between elections was set to give the process stability. Instead of being brought into a plant about which they had had no say, and having to deal with making it operate effectively, workers were consulted right from the beginning of the design phase, through detailed implementation and operation of the project. When the plant was upgraded in 1995, production workers played an important role in the practical assessment and selection of new production equipment.

By the end of the study, the new approach seemed to be working well. One key decision was that in each team, job rotation would be compulsory, so that workers shared good and bad jobs equally. Many teams apparently ran well without supervision; indeed one difficulty for older-style supervisors was letting the teams sort themselves out. Substantial training was provided, and considerable care was taken to ensure that any suspicion of favoritism was avoided. For example, for the zero defects session, the first twenty trainees selected were those expected to be most resistant to the ideas presented. This group was then asked to select the next group for training, an arrangement that continued until everyone had been through the process. One of the key union leaders said that the best advice he could give other union leaders becoming involved with team concepts was: "Don't lie to employees. Tell them what you know, and if you don't know, tell them that." His conclusion on the change to team working was: "It's the most rewarding thing I've done in my life." (Kaminski et al. 1996)

What implications does this discussion have for professional engineers? The team skills discussed earlier are essential for designing high-performance work situations. Engineering designers who understand the possibilities may find it a great relief to be permitted to recognize the knowledge that resides on the shop floor, and to operate in design situations where they can draw on it in a collaborative rather than a confrontational way. The approach outlined here seems consistent with the principles of human system design outlined in Systems Theory and Systems Engineering in Chapter 2. Designers who do not learn to recognize the possibilities and benefit from them may not have much of a future; nor will the workplaces they design.

Traditional Versus High Performance Design

Cherkasky (1997) summarized the essential differences between what he called *technocentric design*, broadly defined as the traditional mass production approach, and *high-performance design*, the direction Ford was moving in the Wayne, Michigan plant.

In *technocentric* design, the emphasis is on efficiency and cutting costs; the work focus is on narrow tasks; there is hierarchical top down control; and information is concentrated centrally with management and engineering staff, often remote from the production facility, who exclusively make the design decisions. Technologies are expected to compensate for human error, with human activity expected to adjust to the technologies. This is an all-too-familiar picture of much modern engineering design practice. It optimizes some individual elements, rather than the total system.

In *high-performance design*, the emphasis is on optimizing overall performance and quality; work focus is on broad work responsibilities; there is reliance on workers' discretion and judgment, with engineers and managers providing support and resources; and information is dispersed freely throughout the organization. Workers ensure smooth operation of complex technological systems, and participate in all phases of design. Technology is used that maximizes the effectiveness of the knowledge and skills of workers.

CONCLUSION

Although all the skills and knowledge that we have described in this chapter will continue to be needed for successful management of engineering, they do not guarantee success. The new work models discussed by Gee et al. (1996) suggest that, in a world of global competition, success will go to organizations that have clear goals and are made up of individuals and teams who keep learning. The organizations in which they work will continue to need to be structured, and restructured, to support their learning and to take maximum advantage of organization-wide knowledge.

Engineers, however traditional their own firms may be at present, need to be aware of where organizations in general are likely to be heading into the twenty first century. Meyer (1997) insists that the lessons from high-tech companies in Silicon Valley are increasingly relevant, even essential, for all businesses. Their effects will be felt first in the knowledge industry, where engineers, whatever their specialization, should see themselves as belonging. Some of those lessons are:

♦ Everyone in the organization must be creative. Innovative thinking should not be left up to the R&D people—it must permeate the whole culture of the organization. Managing such a company then requires willingness to take risks and being prepared to learn from failure.

♦ Organizations must be "aligned"—decisions and actions must reinforce each other, resources and capabilities must match demand, and management's time horizon must always be a few steps ahead of current operations.

♦ Strategic plans are vital, but must not be set in concrete. Instead of using historical data to make projections and then making a virtue of sticking to those plans, organizations must be prepared to modify their plans and adjust their corporate behavior as they go along. Planning thus becomes more an exercise in exploring a range of futures scenarios.

♦ The most important resource for any business is its own people. Knowledge workers will increasingly refuse to put up with poor management. People will be given a stake in the company, most likely in shares. They then have an added incentive and something to lose if the company doesn't continue to develop.

Innovative organizational behavior will be necessary for survival. Managers will find that change is not an option: it is "change or die." It will be exciting to watch the varying forms that organizations will take on over time, and the roles that engineering managers will continue to play in them.

Our main theme in this chapter has been that engineering management and engineering education need to develop and change to improve the effectiveness and productivity of engineering-based enterprises. We believe that engineers will take a leading role in developing sustainable approaches to meeting broad social needs. At the same time, engineering-based enterprises must focus increasingly on being internationally competitive, with an orientation towards export as well as domestic markets.

Discussion Questions

1. What sort of management responsibilities do you expect to have during your professional career? How will you prepare yourself for them?

2. Arrange an interview with an engineering manager in a firm for which you have worked (or intend to work). Review with her/him what proportion of working time was spent in the preceding week on each of the basic managerial tasks. How does s/he make time for important but not urgent managerial work (such as long-term planning)? What sort of managerial style does s/he have? Does the style change to suit the situation?

3. Draw up a formal organization chart for a firm you know well. As well as lines of authority, show functional connections between groups. How realistic is the model? Are there obvious cases where the formal structure is ignored? Why? How many levels of management are there? Given that extra levels hamper communication, how could the number of levels be reduced? Who would oppose such a reduction and how could their concerns be resolved?

4. You have been asked to introduce Total Quality Management into a medium-sized engineering firm. How would you go about it?

5. How would you reform the administration of either (a) your scholarship scheme, or (b) your university? Start by stating the aims of your reforms.

6. What are EEO principles? How would you implement them in either (a) a new engineering design office, or (b) an existing manufacturing firm? What major difficulties would you have to overcome?

7. Government agencies with a substantial engineering capacity have moved towards a marketing rather than an engineering orientation. What implications does this have for engineers?

8. Discuss the influence of cultural characteristics, including ethnicity and social class, on shopping for clothes or other consumer items.

9. You are considering setting up an engineering consultancy, Sustainable Assessments Inc. It is to focus on socially and environmentally responsible practices, processes, and products. Discuss your marketing strategy. What sorts of clients might you want, and what sorts of skills will your company need?

10. What do you think were the major contributing factors to Apple Computer's loss of market leadership, and to its resurgence in the late 1990s?

11. Do you think Microsoft produces innovative products, or merely successful products? Support your answer with evidence. To what do you attribute the success of Microsoft products?

12. Do you feel Microsoft should be weakened by the U.S. Justice Department, as was done to the Bell system and AT&T? Why or why not?

13. If you were Bill Gates, what markets would you move Microsoft into today?

14. Hooker (1989) suggests that the future is not just something that happens to us, but something we need to design. Describe a social aim and then list some of the social and technical steps we might take to implement it.

FURTHER READING

Adams, J. L. 1993, *Flying Buttresses, Entropy and O-rings*, Harvard University Press. This is an overview of engineering with useful material on broad engineering management issues.

Two important writers on modern management approaches, any of whose works is worth reading, albeit critically, are Peter Drucker and Tom Peters. Good starting points are:

Drucker, P. 1979, *Management: Tasks, Responsibilities and Practices*, Pan, London.

Peters, T. J. 1994, *The Tom Peters Seminar*, Vintage/Random House, NY.

A thoughtful critique of Drucker and Peters and the "fast capitalism" they advocate is given in:

Gee, J. P., Hull, G. Lankshear, C. 1996, *The New Work Order: Behind the Language of the New Capitalism*, Allen and Unwin, Sydney, Australia.

REFERENCES

Adair, J. 1989, *Great Leaders*, Talbot Adair Press, Guildford.

APACE 1992, Micro-hydro power in Ghatere village, *Appropriate Technology for Community and Environment (APACE) Newsletter*, August, UTS, Broadway, Sydney pp. 5-8.

Babcock, D. L. 1991, *Managing Engineering and Technology: An Introduction to Management for Engineers*, Prentice Hall, Englewood Cliffs, NJ.

Beder, S. 1998, *The New Engineer: Management and Professional Responsibility in a Changing World*, Macmillan, South Yarra, Victoria, Australia.

Beer, D., et al. 1997, *Guide to Engineering Writing*, John Wiley & Sons, NY.

Cherkasky, T. D. 1997, Linking Labor and Engineering to Enable High Performance Design, *Technology and Society at a Time of Sweeping Change: Proceedings of the 1997 International Symposium on Technology and Society*, Glasgow, June 20-21, IEEE, Washington, D.C., pp. 24-31.

Clegg, S. 1998, Globalizing the Intelligent Organization: Learning Organizations, Smart Workers, (not so) Clever Countries and the Sociological Imagination, *Professorial Address*, University of Technology, Sydney, July 8.

Deming, W. E. 1986, *Out of the Crisis*, Massachusetts Institute of Technology, Center for Advanced Engineering Study, Cambridge, MA.

DePree, M. *1992, Leadership Jazz*, Dell Publishing, NY.

Drucker, P. F., Nakauchi, I. 1997, *Drucker on Asia: A Dialogue Between Peter Drucker and Isao Nakauchi*, Butterworth-Heinemann, Newton, MA.

Fayol, H. 1949, *General and Industrial Management*, Pitman, London.

Gee, J. P., Hull, G. Lankshear, C. 1996, *The New Work Order: Behind the Language of the New Capitalism*, Allen and Unwin, Sydney, Australia.

Harris, C. E., Pritchard, M. S., Rabins, M. J. 1995, *Engineering Ethics: Concepts and Cases*, Wadsworth, Belmont, CA.

Harris, C. E., Pritchard, M. S., Rabins, M. J., Harris, C. E., Jr. 1997, *Practicing Ethical Engineering*, IEEE, Washington, D.C.

Hearne, J. J. 1982, *Marketing for Managers*. Edward Arnold, Melbourne.

Hooker, C. 1989, Future Studies: Its Role in Education, In: Slaughter R. A., Ed., *Studying The Future*, Commission for the Future, Canberra.

Kaminski, M., et al. 1996, *Making Change Happen: Six Cases of Unions and Companies Transforming Their Workplaces*, Work and Technology Institute, Washington D.C.

Kotler, P., Chandler, P., Gibbs, R., McColl, R. 1989, *Marketing in Australia*, Prentice Hall, Sydney.

Kotter, J. 1990, *A Force for Change: How Leadership differs from Management*, Free Press, NY.

Levitt, T. 1960, Marketing myopia, *Harvard Business Review*, July-August.

Mathews, J. 1989, *Tools of Change: New Technology and the Democratisation of Work*, Pluto Press, Sydney.

McGregor, H. T. 1997, Documentation for Global Engineering Practice, *Proceedings, Fourth World Congress on Education and Training, World Federation of Engineering Organisations*, November, Sydney, IEAust, Canberra.

Meyer, C. 1997, *Relentless Growth : How Silicon Valley Innovative Strategies Can Work in Your Business*, Free Press, NY.

Mintzberg, H. 1975, The Manager's Job: Folklore and Fact, *Harvard Business Review*, July-August.

Parkin, J. V. 1994, Judgmental Model of Engineering Management, *Journal of Management in Engineering*, 10 (1), American Society of Civil Engineers, Engineering Management Division, January-February.

Peters, T. J. 1992, *Liberation Management*, Fawcett, NY.

Peters, T. J. 1994, *The Tom Peters Seminar*, Vintage/Random House, NY.

Petre, D. 1998, *Father Time: Making Time for Your Children*, Macmillan, Sydney.

Rosier, G., Whitmore, H. 1988, *Marketing and Selling Engineering Services: A Continuing Education Course for Professional Engineers*, Programme Learning, Sydney (video and workbook prepared for the IEAust).

Senge, P. M. 1990, *The Fifth Discipline: The Art and Practice of the Learning Organization*, Doubleday Currency, NY.

Sprouster, J. 1987, *Total Quality Control: The Australian Experience*, Horwitz Grahame, Cammeray, Australia.

Thompson, J. 1967, *Organizations in Action: Social Science Bases of Administrative Theory*, McGraw-Hill, NY.

Vaughn R. C., Borgman S. R. 1999, *Legal Aspects of Engineering*, 6th Ed., Kendall/Hunt Publishing, Dubuque, Iowa.

Wightley, A. C. 1997, Reverse Engineering—A High Risk Opportunity, *Proceedings, Materials and Manufacturing in Mining and Agriculture Conference*, June 17–18, Sydney.

Wilde, D. J. 1997, Productivity and Satisfaction of Project-Based Learning Teams, *Second International Conference on Teaching Science for Technology at Tertiary Level*, 14–17 June, KTH, Stockholm.

Womack, J. P., Jones, D. T, Roos, D. 1990, *The Machine that Changed the World: The Story of Lean Production*, Rawson Associates, NY.

Wynne, B. 1989, Frameworks of Rationality in Risk Management: Towards the Testing of Naive Sociology, In: Brown, J., Ed., *Environmental Threats: Perception, Analysis and Management*, Belhaven Press, London & NY.

TWO BRIDGES: FAILURES OF ENGINEERING MANAGEMENT?

Engineering projects can be very large and complex. This Australian case study shows the importance for such projects of teamwork, cooperation, and clear lines of authority and responsibility appropriate to the project in hand, all combined with a sufficient level of technical competence. Without all these ingredients, underpinned by good communication and sound documentation, the success of major engineering projects must always be in doubt.

Kings Bridge, Melbourne

One July morning in 1962, a semi-trailer drove onto the western span of the Kings Bridge crossing the Yarra River, which flows through the city of Melbourne. When it reached span 14 there was a loud rending noise, and the span suddenly collapsed, dropping 30 centimeters. The immediate cause of the failure was found to be cracking that had started in welds at the bottoms of all four of the steel girders of the span. The cold winter morning had reduced the fracture toughness of the steel enough for the cracks to propagate vertically upwards in an almost explosive manner. Subsequent inspection of these and other beams in the bridge showed paint primer in a number of the welding cracks, indicating they had been faulty before they had left the fabricators.

The fifteen-month-old bridge was closed for months. As is usual in such serious situations in Australia, a Royal Commission was set up by the state government; it was headed by Mr. Justice Barber of the Victorian Supreme Court. It traced the history of the disaster, and documented contributing factors.

In 1957, the Victorian County Roads Board (CRB) had accepted a tender from Utah Australia Ltd. for construction of a proposed new bridge across the Yarra. The design was to use an exciting new construction material: low-alloy, high-tensile-strength steel. This material does not cost much more than conventional mild steel, but is much stronger, reducing both bridge dead weight and total material costs. The extra strength comes from small amounts of alloying elements, and careful control of production and heat treatment. The material, however, was then so novel that none of the major parties involved (not the CRB, nor Broken Hill Pty. Ltd. [BHP], who were to make the steel, nor the major local steel fabricator, Johns and Waygood Ltd. [J&W], that was awarded the steel contract) were familiar with the requirements for making and working it.

The people who did know about the material were the welding electrode manufacturers. They drew attention to a new problem for Australian steel fabrication, hydrogen embrittlement. Unless welding electrodes were kept in sealed containers, or in ovens at 150 °C, the flux coating on them absorbed moisture. At welding temperatures this moisture dissociated and the free hydrogen dissolved in the weld metal, making it brittle. As the weld cooled and shrunk, this brittleness made it liable to crack. Yet the warnings fell on deaf ears.

Management of the Kings Bridge Project was assigned by the CRB to a corporation set up for this purpose. This agency had neither the technical competence within its own personnel for effective supervision, nor the authority to procure and put in place someone with that knowledge. It was assumed that everyone knew their business and that all that was required of the corporation was coordination. Unhappily, this assumption was not justified.

The basic structural elements of the bridge were a large number of steel I beams. Each I beam was fabricated from a vertical web plate, to which horizontal flanges were welded. A doubler plate was required to increase the effective cross-section of the flanges in the middle part of the span, where the bending moment was at a maximum. Cracking occurred at welds across the ends of the doubler plates. In fact there was no structural reason for these welds: they were only to seal the ends against water penetration, and could have been replaced by a mastic sealer. The detailed welding procedure could not have been worse, from the point of view of maximizing the likelihood of the welds cracking. The Royal Commission Report (1963) concluded that:

> The cracks were caused by the unfamiliarity of the fabricator (J&W), with the problems of welding low alloy steel and the quality of the steel supplied by BHP, much of which was so high in carbon, and so unexpectedly variable, that even an experienced fabricator would have had difficulty in welding it.

The report was a litany of incompetence. The steel manufacturer's quality control procedures were demonstrably inadequate. Tests conducted to establish that the steel met the low temperature toughness requirements were simply repeated until a specimen was found that passed! Identification procedures were intended to show which plates came from which "heats" (batches) of steel, but were ineffective. The report noted in one instance that test results "are so surprising ... that it is suspected a plate from another heat has been wrongly marked." The welding electrode manufacturer's instructions for electrode handling were ignored. Even inspections of the finished beams were ineffective. The Royal Commission found that:

> The form of contract which was entered into ... was unsuitable in that it failed to provide the necessary over-all supervision of the various aspects of the work and was to some degree responsible for the absence of a proper coordination of the project.

To sum up, all but one of the parties involved were inadequate technically. Even worse, they thought they knew what they were doing, but did not. What then made disaster inevitable was that the overall project management was ineffective. It was unable to prevent problems occurring and unable to recognize them when they did.

West Gate Bridge, Melbourne

Three years after the Kings Bridge debacle in 1965, steps were taken to ensure effective technical management when the city's next major bridge was proposed. This project was to improve access to central Melbourne from the southwest. It was entrusted to the Lower Yarra Crossing Authority and one of the world's foremost bridge designers, the British firm of Freeman Fox and Partners (FF&P), was chosen to design the steelwork and control the technical aspects of the work. A Melbourne consulting engineering firm, Maunsell and Partners, was appointed as joint engineering consultants, controlling all administrative matters and acting as local contacts.

(continued on next page)

Construction began in April 1968. There were soon problems with low morale and poor cooperation. Industrial unrest led to a change in the main steelwork contractor. Construction costs, originally estimated at $42 million, finally approached $200 million.

On 15 October 1970, the first steel span on the western bank, then nearing completion, sagged and then fell on the construction offices, pushing over one pier in the process. Thirty-five men were killed and several more were seriously injured.

A Royal Commission was set up, again headed by Mr. Justice Barber; it completed its report in 1971. The technical issues here were more complex, but essentially the central problem was that the method of construction required more accurate working than the fabricators, John Holland (Constructions) Pty. Ltd., (JHC) achieved. The ultimate collapse was caused by ill-conceived efforts to correct misalignment between the two supposedly parallel halves of one of the bridge spans. JHC had not been involved with the steel bridge works from the beginning. The original contract was signed with World Services and Constructions Co. Ltd., (WSC) a subsidiary in Australia of Werkspoor Utrecht NV, based in the Netherlands. WSC had serious industrial relations problems with the steel erection. The Commission (1971) concluded:

It is plain enough that the company [WSC] failed to place in control of the operation anybody with the ability to manage the labour situation. It is true that the union members behaved irresponsibly and much of the industrial strife was unavoidable by WSC, but we are firmly of the opinion that far more could have been done by its management to improve the situation... the organisation, particularly the lower-echelon supervision, was inadequate. There were insufficient foremen, and the organisation of the work was inefficient.

Some of the steelwork delays actually resulted from uncooperative behavior and errors and omissions by FF&P, who were responsible for the supervision as well as the design of the steelwork. However, Maunsell and Partners, who were responsible for administration of the steelwork contract, were apparently unaware of these FF&P shortcomings, and WSC were eventually replaced by JHC for this work. This created more problems. The Royal Commission concluded that, as the work progressed "the JHC management became over-confident. They ceased to seek or follow advice... leading to errors that could have been avoided."

The JHC site staff also proved inadequate. Their chief was frequently absent, and the section engineers were comparatively junior and inexperienced. There was constant bickering between the JHC men and the FF&P engineers. These problems were compounded by the Authority failing to define clearly the respective functions and areas of responsibility of the two firms.

The Royal Commission found that FF&P, the designers of the steel spans and the consultants directly responsible for supervising their erection, "bear a heavy burden of responsibility for the failure of the bridge. The shortcomings were widespread, and as much organizational as technical."

It was unclear to the Royal Commission that any adequate design calculations existed. It *was* clear that the strength of the bridge components was not adequate for the actual construction loadings. FF&P also failed to specify in detail the method of construction, leaving it up to the builder. Maunsell and Partners were consistently unable to get satisfactory answers from FF&P, the bridge designers. Attitudes all round were extremely uncooperative.

FF&P's organizational shortcomings included appointing relatively inexperienced engineers and then failing either to brief them adequately or to support them by arranging effective communication with the London office—which was itself poorly organized: "There was a failure to insist upon proper information flowing from the site to London."

After substantial redesign, work on the bridge was resumed in February 1972 by a new construction consortium. The 2.6 kilometer long West Gate bridge finally opened in November 1978 (Toakley 1986).

The Kings Bridge Royal Commission report summed up the overall management shortcomings of the organizations involved in both these engineering disasters:

We have also asked ourselves whether we have been judging these incidents by standards that are too high for ordinary mortals to reach and whether we are demanding higher standards of compe-tence from engineers than we would from doctors, lawyers or other professional men ... engineers generally—and in this case certainly—do not work as individuals in a consultant-client relationship, they work as a team. This certainly brings with it problems of communication and organisation but it also means that individual engineers are supported by others who can help and check their work. From this standpoint ... it is the various organizations within which they were working that we regard as being collectively responsible for the unhappy state of affairs that we have been investigating.

These two bridge failures were ultimately failures of engineering management.

Acknowledgement

The material in this case study quoted from the Royal Commissions into the Failure of the Kings and West Gate Bridges is reproduced with the permission of the Speaker of the Legislative Assembly of the Parliament of Victoria.

References

Report of the Royal Commission into the Failure of Kings Bridge 1963, Government Printer, Melbourne.

Report of the Royal Commission into the Failure of West Gate Bridge 1971, Government Printer, Melbourne.

Toakley, A. R., Ed. 1986, *Redesign of West Gate Bridge*, Road Construction Authority, Melbourne.

Development and Technology Transfer

O ur main focus in this chapter is on technological relationships between mature industrial countries, particularly the United States, and the rest of the world. Because of its industrial strength and its role as a major originator of technology, the United States is central to the transfer of technology between industries and between countries. Our emphasis here is on transfer of technologies between countries. We continue the theme of wealth creation from the perspective of development and technology transfer.

We begin by considering the needs and concerns of countries that are looking to develop economically. A brief overview of a selected group of countries allows us to canvas some alternative definitions of development. The assumption that advancing technology and a high rate of technological innovation automatically ensure prosperity and social and political liberation is open to question. We discuss the limitations of some current Western technologies for developing countries and explore aspects of underdevelopment and the linkages between rich and poor countries. Finally, we discuss some of the global issues that are becoming increasingly important in engineering, with an emphasis on the perspective of the firms and individuals involved in transferring technology, particularly the role of engineers in the process.

Technology transfer is important for the creation of national and individual wealth. The choice of technologies and their social impacts are of critical importance in both developed and developing countries. Rich and poor nations alike need effective national technology policies that recognize clearly where national interests lie. However, technology policies alone are not enough, because the real aim is to promote economic benefits, not technology itself. The more profound shaping of technology transfers comes from other economic and social policies, for example protection of infant industries. Governments need to recognize and encourage the commitment to the country that is implicit in local ownership and control. All government policies need to be considered from the perspective of their impacts on technology and associated national well-being.

The crucial issue here is the extent to which, in both the short and long term, local organizations enhance both the economy and the environment. At the same time, the already global character of trade requires national policies to be framed in ways that encourage the rapid development of internationally competitive enterprises. The questions discussed in this chapter highlight our ethical responsibilities to help others in need and shed light on issues that are relevant to the maintenance of a high quality of life in any nation. The ways in which engineers can help to improve quality of life, particularly for the less affluent, are equally significant. They are reflected in some of the directions in which engineering careers are already moving.

An Economic Definition of Development

Like a number of other key terms in this book, development describes both a process and its results. Development is originally an economic term (see Key Concepts and Issues box). GDP/p.c. is the standard measure used to define economic development; however the United Nations Development Programme (UNDP) recognizes that official exchange rates do not attempt to assess the internal purchasing powers of currencies. In order to rank countries in terms of GDP, it uses measures developed by the United Nations International Comparison Project (ICP), "purchasing power parities (PPP$)," to compare the relative domestic purchasing powers of currencies. "Purchasing power parities" are based on "... the number of units of ... [a country's] currency required to purchase the same representative

Development

A typical economic definition is:

> growth in material well-being as measured by the average annual income of the people of a country (its gross national product (GNP) divided by the population).

(Francis and Mansell 1988)

The relationship between GNP and GDP was discussed in Chapter 5. Within the accuracy of the data available, particularly for less-developed countries (LDCs), the terms are effectively interchangeable.

This definition of development suggests that it is a matter of degree, an index by which we can rank countries and compare their relative positions. The conventional approach is then to put some arbitrary value on this index as the division between (economically) developed countries and LDCs. Although LDCs are commonly referred to as developing countries, for many of them this is more a polite euphemism than a description of their immediate prospects.

Some commentators go further, suggesting that an "appropriate" level of resource use should be defined, based on all countries having reasonably equitable access to world resources on a sustainable basis. On this model, countries are overdeveloped when their citizens on average consume many times the world mean level of resources per head. This type of argument is particularly relevant to issues such as limiting world emissions of greenhouse gases (discussed in Chapter 9).

GDP/p.c. is a rather limited measure of quality of life. The inclusion of an activity in GDP does not depend on whether the activity is beneficial, but only on whether or not it goes through the market. For example, commercial provision of personal care adds to the GDP, but the same care within the family is not counted. Damage to communal resources (such as the environment) is ignored, so that the GDP figure tends to overstate the value of productive activity. Official figures tend to understate effective income for countries where much of the population grows its own food, rather than buying it through the cash economy. In these countries official data may also be of rather limited reliability.

---

basket of goods and services ... that a U.S. dollar (the reference currency) would buy in the United States." (UNPD 1998, p. 220) An overview of the development situation worldwide is given in Table 8.1.

TABLE 8.1

					Military Spending as % of Education and Health Spending 1990–91 (Military
Country	Estimated Population (Millions) 1995*	Real GDP/ Capita (PPP)$ 1995†	Overall HDI Ranking 1995‡	Gender-Related Development Index/Gender Empowerment§,‖	Spending as % GDP 1996)¶

	UNDP Development Data for Selected Countries				
Canada	29.4	21,916	1	1 7	15 (1.6)
France	58.1	21,176	2	7 31	29 (3.1)
Norway	4.3	22,427	3	2 2	22 (2.4)
United States	267.1	26,977	4	6 11	46 (3.6)
Iceland	0.3	21,064	5	4 6	...
Finland	5.1	18,547	6	5 5	15 (2.0)
Netherlands	15.5	19,876	7	12 9	22 (2.1)
Japan	125.1	21,930	8	13 38	12 (1.0)
New Zealand	3.6	17,267	9	8 4	16 (1.3)
Sweden	8.8	19,297	10	3 1	16 (2.9)
Spain	39.6	14,789	11	19 16	18 (1.5)
Belgium	10.1	21,548	12	14 19	20 (1.6)
Austria	8.0	21,322	13	15 10	9 (0.9)
United Kingdom	58.1	19,302	14	11 20	40 (3.0)
Australia	17.9	19,632	15	9 12	24 (2.2)
Switzerland	7.2	24,881	16	18 13	14 (1.6)
Ireland	3.5	17,590	17	27 21	12 (1.1)
Denmark	5.2	21,983	18	10 3	18 (1.7)
Germany	81.6	20,370	19	17 8	29 (1.7)
Greece	10.5	11,636	20	20 51	71 (4.8)
Italy	57.2	20,174	21	23 26	21 (2.2)
Israel	5.5	16,669	22	22 32	106 (12.1)
Singapore	3.3	22,604	28	29 42	129 (5.5)
Korea, Rep of	44.9	11,594	30	37 83	60 (3.3)
Chile	14.2	9,930	31	46 61	68 (3.5)
Argentina	34.8	8,498	36	48 ...	51 (1.5)
Mexico	91.1	6,769	49	49 37	5 (0.8)

TABLE 8.1 *(continued)*

					Military Spending as % of Education and
Country	Estimated Population (Millions) 1995[*]	Real GDP/ Capita (PPP)$ 1995[†]	Overall HDI Ranking 1995[‡]	Gender-Related Development Index/Gender Empowerment[§,‖]	Health Spending 1990–91 (Military Spending as % GDP 1996)[¶]
Thailand	58.2	7,742	59	40	60	71 (2.5)
Malaysia	20.1	9,572	60	45	45	38 (4.2)
Brazil	159.0	5,928	62	56	68	23 (2.1)
Russian Federation	148.5	4,531	72	53	...	132 (6.5)
Iran	68.4	5,480	78	92	87	38 (5.0)
Indonesia	197.5	3,971	96	88	70	49 (2.1)
Philippines	67.8	2,762	98	82	46	41 (2.0)
China	1,220	2,935	106	93	33	114 (5.7)
Egypt	62.1	3,829	112	111	88	52 (4.5)
Viet Nam	73.8	1,236	122	108	...	... (4.0)
Solomon Islands	0.4	2,230	123	109	...	... (...)
Iraq	20.1	3,170	127	127	...	271 (8.3)
Papua New Guinea	4.3	2,500	129	119	91	41 (1.5)
Pakistan	136.3	2,209	138	131	100	125 (5.7)
India	929.0	1,422	139	128	95	65 (2.8)
Nigeria	111.7	1,270	142	133	...	33 (3.5)
Bangladesh	118.2	1,382	147	140	80	41 (1.7)
Ethiopia	56.4	455	169	158	...	190 (2.0)
Sierra Leone	4.2	625	174	163	...	23 (5.9)

[*] (pp. 176-177, 200).
[†] Purchasing power parities (PPP$) are based on the actual relative domestic purchasing powers of currencies in 1995 (pp. 128-130).
[‡] Human Development Index (HDI) ranking takes into account life expectancy at birth, adult literacy rate, and mean years of schooling as well as average income; ranking is out of 174 (pp. 128-130).
[§] Based on a study for 163 countries of the difference between female and male HDIs (pp. 131-133).
[‖] Based on a study for 102 countries of inequality of political representation, managerial, professional and technical employment, and income distribution (pp. 134-136).
[¶] Figures for 1990/91 and 1996 respectively (pp. 170-171, 197).
Source: UNDP Human Development Report 1998.

The UNDP 1998 *Human Development Report* covered 174 countries (including provisional GDP figures for 18 of them). Table 8.1 shows data for forty-six selected countries.

Despite some technical limitations, the data in Table 8.1 figures indicates the scale of the problems of underdevelopment.

Until the end of the Cold War, it was usual practice to refer to the Western Alliance of economically advanced capitalist countries (the West) as the "first world," to the Eastern bloc or socialist countries as the "second world," and to the economically less developed African, Asian, and Latin American countries as the "third world" (Todaro 1977, p. xix). More recently there has been a move towards identification of just two groupings—the rich, industrialized north and the poor south. Using realistic "purchasing power parity" values, the 1995 average per capita income in the "north" was $16,337, whereas the "south" had an average of $3,068, and the least-developed countries averaged $1,008. The estimated world population in 1995 was 5.627 billion, of whom people in the industrial north made up 1.233 billion, whereas in the south there were 4.394 billion; the population of the least economically developed (poorest) countries totaled 543 million. Despite our reservations about these labels, they are in virtually universal use and we employ both usages in this book (UNDP 1998).

A substantial industrial base is recognized as a central feature of development. The information revolution notwithstanding, a service sector is still widely seen as lacking credibility without a manufacturing industry base to service. Although this industrial basis for the division between them seems to be generally agreed, the north/south divide also has significant historical, political, and economic dimensions. A few of the developing countries, notably those with substantial oil reserves such as the United Arab Emirates, have come to be financially well off. However, as the Gulf War and more recent concerns with Iraq indicate, their economic and political stability are still relatively uncertain and enjoyment of the benefits of oil wealth is uneven. Although the Soviet Union was recognized as part of the north, some of the states into which it has divided seem likely to be recognized as LDCs and thus part of the south. China and India are accepted as part of the south.

Social Implications of Underdevelopment

During the twentieth century there was remarkable success in reducing poverty, great enough to demonstrate the possibility of eradicating extreme poverty from most of the world in the first few decades of the twenty-first century. Between 1960 and 1995, child death rates in developing countries were halved, and malnutrition rates declined by almost a third. The proportion of children in primary school rose from less than half to three-quarters. Because the social and personal returns on basic literacy and numeracy are very high, this is a particularly encouraging statistic for future development. The proportion of rural families with access to safe water increased during the same period from one in ten to three in four. The advances were worldwide. Since the late 1970s, China and other states, with a total population of 1.6 billion, have halved the proportion of their people living below the national income poverty line.

On the other hand, poverty is still very widespread, and there have been setbacks. Population growth has barely slowed, and three-quarters of the world's population now live in LDCs. The disparity between rich and poor countries is sharp, with 78 percent of the world's people earning only 18 percent of its income. At least 1.3 billion people in rural south and east Asia and sub-Saharan Africa (and not simply in the poorest LDCs) still live in absolute poverty (UNDP 1997).

Diversion of funds, personnel, and natural resources to military uses presents major problems for the rich countries, but it has been a disaster for the poor. A telling indicator of the priority placed by governments on human development has been the ratio of spending on the military to spending on health and education combined (see Table 8.1). In 1990–1991, for all the nations of the north together, this ratio was 33 percent, compared with 63 percent for the south. For the United States it was 46 percent (down from 173 percent 30 years earlier), with military spending around 3.6 percent of GNP in 1996. The same ratio for Japan was only 12 percent, and Japan's postwar constitution formally limits its military spending to 1 percent of GNP. (For domestic political reasons, however, it seems likely that the figures for military spending are underestimates for Japan and overestimates for the United States.)

At the other extreme, in 1991 Iraq spent 2.7 times as much on military expenses as on health and education, whereas its neighbor Syria spent 3.7 times as much. Israel's military spending fell encouragingly over the decade to 1995, from 21.1 percent to 9.2 percent of GDP, although it rose again in 1996. The highest military spending figure in 1996 was 27.2 percent of GDP for North Korea, a level of spending that would have contributed to the difficulties it has had in feeding its people.

Worldwide research and development (R&D) on technologies aimed at solving the problems of LDCs is still minimal, whereas military R&D has for many years run at around half total world R&D. At the height of work on the Strategic Defense Initiative (Star Wars), American military research spending was said to have reached 70 percent of total American research spending. The budget figure for 1998 for U.S. military R&D was still over $41 billion, more than 53 percent of all federal R&D.

Universities in developing countries could help with research into the development of more appropriate technologies for their fellow citizens. However, their staff commonly see themselves as part of the north's academic system. They tend, in their own research focus and even in the curricula they follow, to reflect international interests and trends. Many of the academics are expatriates who need to be able to move back to their own countries when their three-year contracts are up. Thus there has been a tendency for them to follow an agenda that emphasizes high-tech R&D, to the exclusion of work on locally important but small-scale projects, such as the improvement of handling, processing, and storage facilities for local food production and marketing.

Balanced development requires careful selection of the most suitable technologies. The role of NGOs in assisting governments in developing countries to make wise choices of technologies is increasingly recognized. One example, the work of the APACE (Appropriate Technology for Community and the Environment) group, based in Sydney, is discussed in Case Study 8.1. The depth of its approach to supporting development at the village level is notable. To a solid technical and training capability, the APACE leadership also brings unusual political and social insight, and an ability to work effectively at village, national, and international levels. The contrast between the impact of electrification on the village of Iriri, described in Case Study 8.1, and Mae Chaem, described in Case Study 8.2, is remarkable. In Iriri the villagers initiated and controlled the process, and it increased village solidarity. In Mae Chaem, the process was planned and implemented by others, and has encouraged the villagers to abandon traditional beliefs and relationships and move increasingly towards consumerism.

▮ NGOs and International Action

To continue with the example of APACE, such NGOs have an important and continuing role in international discussions on global environmental issues. An example of one such input was through a Special Session of the United Nations General Assembly (UNGASS) in June 1997. A major series of negotiating meetings accompanied and preceded it and came to be known as the "Rio +5" or "Earth Summit +5" process. It was the culmination of action beginning at the UN conference on Environment and Development (UNCED) at Rio de Janeiro in 1992, and proceeding through the Human Rights Conference in Vienna in 1993, the Population

Conference in Cairo in 1994, the Women's Conference in Beijing in 1995, and the Social Development Conference in Stockholm in 1996.

At UNGASS, NGO advisers were generally part of official delegations. There was not a parallel NGO event as there had been, for example, at Rio, and there were few NGOs present overall, whereas governmental political representation was said to be unprecedented for a UN session. A personal perspective on the process was provided by the President of APACE, Paul Bryce, an Australian academic with considerable practical experience of supporting development work. He was a nongovernment representative at UNGASS, accredited through the Commission for Sustainable Development. Bryce's report back to APACE shows the extent to which the intervention of NGOs has broadened the consultation process (Paul Bryce 1997, personal communication). It also gives us a first-hand impression of just how difficult the process is. His summary included the following points:

1. There is a serious divide between north and south. The meeting exacerbated this divide, with very few northern NGOs demonstrating practical first-hand experience of the issues concerning southern peoples. The need for leadership among NGOs, and for a broader appreciation of the diversity of needs and challenges that appear "on-the-ground" in any technology transfer, was apparent.

2. A corresponding divide between north and south existed among governmental delegations that was more apparent than at earlier events. The increasingly frugal approach to assistance for the south (through overseas development assistance) by most north nations was seen by the "Group of 77" (the south nations) as a lack of commitment to the process. As a percentage of GNP, U.S. official development assistance fell particularly sharply from 0.23 percent in 1985/86 to 0.12 percent in 1996. Canada's aid fell from 0.49 percent to 0.32 percent over the same period. Only the nordic countries held the percentage essentially constant (UNDP 1998, p. 196).

3. Only a quite small group of nations had a sympathetic stance on the role of the nongovernment sector in international negotiations, such as those with the World Trade Organization.

4. The human and environmental effects of multilateral, economically focused interventions, such as economic globalization and structural adjustments programs, will continue to be bleak without significant inputs from civil society. Public encouragement is vital.

IRIRI: MICROHYDROELECTRIC POWER IN THE SOLOMON ISLANDS

A successful example of technology transfer has been the microhydroelectric installation at the village of Iriri, on the island of Kolombangera, in the Solomon Islands group in the southwestern Pacific.

In 1974, the island's botany was identified as being of international importance because of the distinct banding of rainforest types on the slope of the extinct volcano. It was recommended that a complete sector of the island, from coast to summit, be preserved as international heritage. Instead, a foreign logging company soon after clear-felled all the accessible rainforests on the island, except for the land owned by the Voko people, who mostly lived in Iriri village. They refused to allow logging. What was unique about this village of 120 people? Why were they able to take such a stand in the face of logging company and church pressures to accept a one-time payment for their forests to be bulldozed?

An important characteristic of the village social structure was a respected traditional leadership, in Joseph Ghemu and Solomon Kana. A notable younger resident at the time was Joini Totua, who had for some time been active in resisting deforestation. The leaders of Iriri were determined to find ways in which the community could become an interesting and satisfying place for their young people to live, with attractions to balance the pull of the "bright lights" of the Solomons capital, Honiara. They had worked out a community development strategy that, among other things, would maintain control of their forest resource.

Electrification of the village was identified as a key requirement of this strategy, but rather than become dependent on imported diesel fuel for electricity, Iriri sought to acquire and install a small hydroelectric power system.

Technical assistance was requested from APACE, an Australian nongovernment organization working in the area of small-scale, appropriate technology. A number of Australians, including staff, students, and

5. The event generated additional political momentum in some nations, notably the United States and the United Kingdom. It was clearly a momentous event in the view of many developed nation governments. It encouraged serious expectations for progress on climate change targets at Kyoto later in 1997. (In the event, only direct intervention by the U.S. Vice-President kept up the momentum in the Kyoto talks, with ratification by the U.S. Senate of binding limits on Greenhouse Gases still far from certain.)

Another NGO, Transparency International (TI), which works specifically on focusing attention on the issue of corruption, was discussed previously (p. 373). As well as assisting governments and others to discourage

graduates of the University of Technology, Sydney (UTS) worked with the villagers to see that the necessary equipment was developed and installed. All the villagers were involved in the decision making, working closely with the Australians. A central part of the process was training the villagers in the technical skills needed to install and maintain the system. Training was done both at UTS in Sydney and in Iriri. The detailed technical steps involved were shown in Table 7.1 and Figure 7.3 (pp. 360–361).

Village control of the project was essential for its success. The limit on the power capacity of the system is that, in the driest month of the year, the power capacity of the nearby stream is only three kilowatts. This effectively limits electricity use to lighting and a cool room and freezer. There is enough power either to light the village hall, or to provide one light in each of the approximately thirty homes. During the day power is available to operate some machinery, including wood working and saw milling to meet village needs. Community participation in fund raising before installation, in deciding how power will be used, and in the

training that is an integral part of the actual installation, are all essential to the success of this type of microhydroelectric installation.

The village approach to the provision of food has also focused on self-reliance and sustainability. This reflects their success in resisting further deforestation and maintaining the integrity of their land. The position they took has required both technical and political skills. Joini Totua's insight and leadership skills were more widely recognized: he subsequently became Minister for Education and Minister for Lands and Agriculture in the Solomons Government.

The Iriri project highlights the potential for effective cooperation in technology transfer between NGOs and well-led local communities in developing countries.

References

Bryce, P., Bryce, D., Irons, C. 1994, Rural electrification and technology transfer, *Conference on Technology Transfer in Remote Communities*, 6-7 April, Murdoch University, Perth.

Waddell, R. 1993, *Replanting the Banana Tree: A Study in Ecologically Sustainable Development*, Sydney, APACE.

and prevent corruption, TI publishes an annual Corruption Perception Index, based on surveys of international business people's perceptions of the countries in which they operate.

Work by NGOs appears to have been increasingly effective in discouraging direct sales by multinational corporations to governments at Cabinet level (with their serious potential for corruption). NGOs have also had particular success at empowerment of the poor, particularly women, although not always with happy endings. There have, for example, been violent clashes when poor people confronted local power blocs. Advocacy is clearly the NGOs' greatest strength, and they have an important role in providing models of participatory development activity (UNDP 1997, pp. 94–105).

MAE CHAEM: TECHNOLOGY TRANSFER IN RURAL THAILAND

How do newly transferred technologies affect traditional village life? This case study illustrates some of their effects on a rural village in Thailand.

Mae Chaem is about sixty miles southwest of Chiang Mai in northern Thailand. It is typical of rural villages in the kingdom. The economy is based on rice growing, the villagers live in wood houses with only two or three rooms, cooking is still done over an open fire. At the same time, power lines cut through the trees, cars travel over dirt roads, and from inside the wooden houses the sounds of radios and televisions can be heard. In rice farming, modern machinery has replaced the water buffalo and artificial fertilizers are the norm.

Technology transfer always gives rise to questions of what technologies are appropriate and beneficial in a developing nation, and what will be the outcome of the transfer. In Thailand, especially in rural areas, technical assistance has commonly been for communication and transport. The modernization of Thailand began in 1962, with the first of five 5-year plans for economic and technical development. The Second National Economic and Social Development Plan, implemented in the late 1960s, called primarily for the building of rural roads and the electrification of village areas. These two developments resulted in numerous other technologies coming to the villages, including cars, television, and electric lights. The focus here is on electrification, which caused the greatest change in the social life of the village.

Even before electrification became widespread, electric products were already having an effect. In the first half of the twentieth century, movie going became a popular pastime, introducing the villagers to a world moving at a much faster pace and producing goods they had never known they wanted. On their own, the movies were not invasive enough to cause significant change but they probably encouraged the adoption of other technologies.

Electrification of rural areas began in the 1960s and 1970s and continues today. One of the first uses of the new power was the fluorescent light. Being able to see and do work after the sun has set has advantages, but are they outweighed by the disruptions caused to the daily schedule? In Mae Chaem, the roosters start crowing at 4:30 in the morning, signaling the beginning of the day, and a lot of work is done before breakfast. Thai villagers commonly divide the day into "before breakfast" and "after breakfast." With fluorescent lighting, people are extending their day past dark, shifting the traditional concept of the day. Some of their folklore has been affected. Traditionally, people believed in night spirits and deities. Electric lighting has demystified the darkness, and belief in such entities has faded.

NGOs active in overseas aid are devoting much larger proportions of their efforts to village-level development projects. Even some of the more traditional NGOs, including church-based organizations, are shifting their emphasis towards longer term relief measures. The scale and importance

Fluorescent lights are not the only electrically driven technology to disrupt village life. Perhaps the greatest changes have been made by television. In 1979 only 1.3 percent of Thai villagers reported having a TV. By 1980 nearly a quarter had a black and white TV and 16.6 percent had a color TV. By way of comparison, even in 1988 only 4.0 percent owned an electric stove, 15.4 percent had a refrigerator, and 0.5 percent a washing machine. Television watching has become a popular night-time activity, replacing the sharing and interaction of the villagers. As a result, village life is less communal. Families in Mae Chaem turn on the television instead of walking to a neighbor's house to visit. Although TV watching itself has disrupted the intimacy of the village, the images the villagers see on the screen have been far more disruptive. Television has introduced them to material consumerism through the commercials and the lifestyles of the characters portrayed in shows.

Coupled with changes in the farming industry, this has shifted the economy from a barter to a monetary system. Even in their traditional market, cash is now the usual form of payment. Farmers who used to save their surplus now spend it on luxury items and consumer goods. Increased focus on material wealth has encouraged many young people to leave the villages to work in Bangkok or smaller cities such as Chiang Mai. They only return when they are older, making the village increasingly a shell of young and old.

These are only two of the technologies brought to the village by electricity, but their cultural impact on the village has been immense. Village life in Mae Chaem, like many other villages, has become consumerized and has moved away from the traditional lifestyle and belief system. The further influx of technology will most likely cause an even greater abandonment of tradition and the traditional way of life. Sixty years ago the major inland city of Chiang Mai is reported to have looked like Mae Chaem. With technology now coming at an even faster pace, it may be only a few decades before Mae Chaem develops into a city the size of Chiang Mai.

Source

Based on work by Katherine D. Steel, Stanford University, 1997.

References

Dahm, B., Gotz, L., Eds. 1988, *Cultural and Technological Development in Southeast Asia*, Nomos Verlagsgesellschaft, Baden.

Ketudat, S. 1990, *The Middle Path For the Future of Thailand: Technology in Harmony with Culture and Environment*, East-West Center, Honolulu, Hawaii.

Stewart, C. T., Yasumitsu N. 1987, *Technology Transfer and Human Factors*, Lexington Books, Lexington, Massachusetts.

Tomosugi, T. 1995, *Changing Features of a Rice-Growing Village in Central Thailand*, Center for East Asian Cultural Studies for UNESCO, Tokyo.

of NGO activity is discussed by Rollason (1991). Internationally, NGOs send billions of dollars annually in aid to developing countries, independent of government to government assistance. This aid is particularly important because of the recognized effectiveness of NGOs in reaching the poor and mobilizing them to take action on their own behalf. Goulet (in Todaro 1977, p. 96) shows why this is so important:

> The prevalent emotion of underdevelopment is the sense of personal and societal impotence in the face of disease and death, of confusion and ignorance as one gropes to understand change, of servility towards men whose decisions govern the course of events, of hopelessness before hunger and natural catastrophe...Chronic poverty is a cruel kind of hell; and one cannot understand how cruel that hell is merely by gazing upon poverty as an object.

We clearly need a more comprehensive understanding of development than simply the economic dimension. Before we go on to broaden our definition, we will look briefly at the historical origins of underdevelopment.

THE DEVELOPMENT OF UNDERDEVELOPMENT

The most coherent model of the processes that generated overdevelopment and underdevelopment is probably the Baran-Frank thesis, named for its originators Paul Baran (1973) and Andre Gunder Frank (1978). In Chapter 1 we looked briefly at the colonization of the rest of the world by European nations. This process of colonization started in the sixteenth century, and was essentially complete by the end of the nineteenth. The twentieth century saw colonial independence, and by the 1960s colonialism had been largely replaced by a postcolonial world order, but with worrying signs in many parts of the world of neocolonialism. Baran argued that one effect of colonization on the colonial powers themselves had been to strengthen their merchant capitalists and weaken the forces pressing for transformation from merchant to industrial capitalism (Wheelwright, 1976).

Four main periods can usefully be distinguished: mercantilism (1500–1770); industrial capitalism (1770–1870); imperialism (1870–1940); and, currently, what might be called monopoly capitalism (World War II to the present). Chapter 1 provides more historical detail, but essentially the thesis of "the development of underdevelopment" is as follows:

◆ By the end of the fifteenth century, there were civilizations in America (Aztec, Inca), Africa (Benin, Zimbabwe), and Asia (India, China) that, in terms of cultural and technological development generally, were at a level comparable to western Europe. However, the West did have a lead in military technology and organizational skills, particularly for large-scale warfare, developed over centuries of wars in Europe.

◆ During the mercantile period, the West used its military lead to conquer and enslave much of the rest of the world. In the process it destroyed potential competitors and potential resistance to colonization. It also accumulated huge amounts of capital, some of which was used to finance the development of industry in the Western powers.

◆ The technical gap between the colonial powers and the colonies increased rapidly during the period of industrial capitalism. This in turn made it easier to extend colonial domination, because the Western technical lead increased. The canoe and the flintlock were no match for the steamship and the machine gun.

◆ A whole chain of exploitative relationships were established, running all the way from the capital city of the colonial power through to the countryside in the colony. A major function of this set of relationships was to transfer part of the colony's economic surplus to the imperial metropolis.

◆ Despite most former colonies gaining formal political independence after World War II, this exploitation, involving a net transfer from the poor to the rich countries, continues to the present day. It contributes to the development of the already developed countries and inhibits self-sustaining and self-perpetuating growth in the former colonies.

◆ The power relationship between the "metropolitan centers" and the "periphery" has been sustained in cultural forms well into the postcolonial era.

◆ An important part of the process has been the export of ideology from the centers of imperialism. As recently as the 1960s and 1970s, the key element of this export was technological pragmatism, which proposed that North American technology and know-how could solve all the world's problems, if only people would allow the Americans to apply it without interference. We will look at some of the shortcomings of the technological approach to solving the problems of development later in this chapter.

Western invasion and conquest often destroyed the social structure on which local agriculture and trade had been based. This was frequently accompanied by population collapse, resulting from introduced diseases, especially influenza, measles, smallpox, and syphilis. The Native American population of North America is estimated to have fallen from 80 million to 10 million, and that of Mexico from 10 million to 1 million. Indigenous populations tended to recover as stable colonial rule was established and they developed immunity to introduced diseases. The richest precolonial areas suffered most.

Baran (1973) spelled out the impact on India. Until the eighteenth century, India had been relatively advanced; it was internationally competitive in both production methods and industrial and commercial organization. In the period from Clive's victory at the Battle of Plassey in 1657 to the Battle of Waterloo in 1815, a period that Baran notes was critical to the development of British industrial capitalism, an estimated £500 million to £1,000 million worth of treasure was taken by Britain from India. This compared with the total value of all the joint stock companies in India in 1800 of £36 million. In the early decades of the twentieth century, Britain was still estimated to be appropriating annually over 10 percent of India's gross national income. Even this figure understated the situation, because it did not include the effects of the unfavorable terms of trade imposed by Britain. During the colonial period, the sources of Indian wealth narrowed. India was reduced from being a great manufacturing and agricultural producer to being a supplier of raw materials for the looms and factories of Great Britain, and a market for British manufactures. The entire framework of Indian society was broken down. The sectarian and social divisions that were systematically encouraged led eventually to the horrors of partition and continue to this day. Although Baran (1973, pp. 277–285) did not idealize India's pre-British past, he pointed out that there would now be:

> an entirely different India … had she been allowed—as some more fortunate countries were—to realize her destiny in her own way, to employ her resources for her own benefit, and to harness her energies and abilities for the advancement of her own people.

In a colony, the social structure did not always collapse. In some countries where the existing social structure was strongly hierarchical, shrewd colonizers such as the Dutch in what is now Indonesia (and the British in northern Nigeria) simply placed themselves at the top of the pyramid. They used existing structures to maintain order and implement their demands.

The returns to the colonial powers were immense. During the nineteenth century, as much as one-third of the Dutch budget was financed by the culture system: the Dutch told the Indonesians what to plant and how much, creating an export agricultural sector that was essentially part of the Dutch economy, operating side-by-side with domestic subsistence agriculture. In the process they destroyed the social integrity of Indonesian village society. Before the Dutch came, there had been rich and poor in rural communities, but the gap was contained, because wealth stayed largely within a community, and the rich customarily used some of the surplus to assist the needy poor. Although the same formal structures remained, with the removal overseas of most of the surplus there was much less opportunity for reinforcement of village solidarity and community feeling. Java was left with the worst of both worlds. Its population increased at the sort of rate that was experienced in industrializing countries, but there was no local capital accumulation to finance technological advancement. Java's population grew from 7 million in 1830 to 41.7 million in 1930, while its domestic agriculture was increasingly locked into low-productivity, labor-intensive methods (Geertz 1963, pp. 50–53, 69).

The contrast with Japan is striking. Japan escaped colonization, largely because it was seen by the colonial powers as not worth the effort of conquering. As a result, it did not suffer "the development of underdevelopment" and was able to undertake capitalist development of its own. Inkster (1991, Ch. 7) notes that gold and silver accumulated during the feudal period were used to finance the critical drive towards industrialization from 1868 to 1881. Systematic political action was taken by the Meiji regime to eliminate surviving feudal privileges and monopolies. Measures to provide modern infrastructure followed rapidly. Japan's high rates of industrial and economic growth reflected increasing productivity, in both agriculture and industry, as a result of advanced technology being adopted from abroad. Adaptation of technology to local conditions and resources was incorporated in the selection and transfer process. Inkster (1991, p. 204) and Morris-Suzuki (1994) emphasize that this adaptation activity was important in the rapid adoption and dissemination of the new technologies and in the success of Japanese industrialization.

Where do the North American Free Trade Alliance (NAFTA) countries fit into this picture? It is worth remembering that, before they became independent nations, the United States and Canada were colonies of Britain. Mexico was a colony of Spain. The United States became independent in the late eighteenth century, Mexico in the early nineteenth. Canada was a well-treated Dominion rather than a harshly exploited colony, and became a confederation in 1867, although it maintained ties with Britain. Former colonial status should only be a temporary hurdle to overcome. However, unequal power relations between officially independent nations are a fact of life. The U.S. relationship to Latin America has been characterized by some observers as neocolonial. Overt and covert U.S. military and civil intervention in the governance of other countries in the western hemisphere, particularly in what are seen to be the interests of U.S. business or U.S. citizens, was claimed in practice as a right under the Monroe Doctrine in 1823, and continues to the present day. The U.S. relationship with Mexico under NAFTA has certainly encouraged industrial growth in Mexico, although the growth in Mexican wages still lags behind that in productivity.

When new technologies are adopted, they can be important agents for social change. In Papua New Guinea, colonial policy required that villagers be persuaded to work as indentured laborers when plantation agriculture was introduced. When not enough "volunteers" were forthcoming, a poll tax was levied to force villagers into the cash economy. Another source of money was the diversion of land and energy from growing food to growing cash crops. In the New Guinea highlands, with the move from sweet potatoes to coffee, this exacerbated existing pressures on land and increased inter-group tensions, sometimes to breaking point (Amarshi et al. 1979).

Case Study 8.3 shows how the specific characteristics of a culture, in this case Native American, can help a society to adapt to technological change rather than be destroyed by it.

Foreign Aid and the Debt Crisis

Which way do resources flow today? Is there still a net flow from the poor countries to the rich? What about the aid from the rich countries to the poor? In practice, aid does not balance the equation, and the net flow is to the rich countries. Part of the explanation is that much foreign aid is in the form of loans, rather than "untied" grants. Unless the use of a loan generates enough income to cover repayments, it eventually leaves the borrower worse off. In LDCs the problems have been increased by the specific character of much of the imported equipment and by associated shortcomings in its technical support, which led to underutilization and reduced equipment life, as discussed in Case Study 8.4. In addition, the terms of trade have declined for major exports from the south. Compared with prices for elaborately transformed manufactures, prices have fallen for primary products such as metals, minerals, wheat, coffee, and sugar.

The pattern of net transfer of financial resources changed between 1980 and 1986, from a peak flow into the poor countries of $35 billion in 1981, to a flow outwards of $29 billion in 1986. By the late 1980s, the Philippines and Mexico were spending more of their central budgets on servicing their debts than on social services. In terms of the proportion of their exports of goods and services spent on servicing debt, their position improved significantly between 1980 and 1994, with the percentage falling from 51 percent to 35 percent for Mexico, and from 29 percent to 22 percent for the Philippines. Even so, in 1994 the total debt of all the developing countries was estimated at $1,444 billion, about 38 percent of their GNP. In order not to destabilize an economy, overseas borrowings need to be used for sustainable wealth creation, not to fund prestige projects or to swell personal overseas bank accounts. Capital flight can be a significant part of such problems. Between 1962 and 1986 in the Philippines, it equaled 80 percent of the outstanding debt. From the mid-1970s to 1990, capital flight equaled at least half of capital borrowings in Mexico and Argentina (UNDP 1991, p. 5).

The major economic problems that began to emerge in east Asia (Japan, Indonesia, Korea, Thailand, and Hong Kong in particular) towards the end of 1997 were related more to inappropriate internal banking and financial dealing than to external issues. If the substantial international rescue packages encouraged more realistic and arms-length financial practices, these economies could be expected to emerge from the crisis with a stronger foundation for long-term growth. The bailout was widely seen as rescuing reckless investors at public expense and high social cost. There was concern that austerity measures imposed by the International Monetary Fund as a condition of the bailout might even be counterproductive, leading to deflation and bankruptcies of inherently sound enterprises. The crash showed that it no longer makes sense simply to prop up ailing banks and other enterprises: when they eventually collapse they may bring down the whole economy.

A Broader Definition of Development

Development probably should be seen as a multidimensional process (see Key Concepts and Issues box). This is clearly relevant to all countries, not only those that are less economically developed.

As can be seen in Table 8.1, the UNDP in its *Human Development Report* has attempted to quantify further the first two of the points in the box, preparing a human development index (HDI). This recognizes the importance of economic growth in carrying out the agenda for change, but specifies that such growth needs to be participatory, well-distributed, and sustainable. It must center on people: development of the people, by the people, and for the people. Essential prerequisites are an adequate level of social, political, and economic stability, and an institutional structure that empowers individuals and groups.

UNDP reports recognize that the readily measurable elements of development are much narrower than the overall concept. The HDI takes in three components: longevity; knowledge (literacy and years of schooling);

NAVAJO SOCIETY AND TECHNOLOGICAL CHANGE

The entire history of the Navajos has been one of adaptation. Several hundred years ago they adapted Pueblo building materials, farming habits, and desert survival skills. When the Spanish arrived in the 1500s, the Navajo began using horses and metal tools. Under the Mexican occupation, they began using guns and other contemporary technologies. This century, despite a slow start, the Navajos have been exposed to and adapted to nearly every aspect of American technological culture.

The Navajo Indian tribal lands cover nearly 25,000 square miles in Arizona, Utah, and New Mexico. In 1920 there were few wagons on the Reservation. Most people still traveled much as they had for generations, carrying their goods and possessions on horseback. Yet by 1950 the wagon was the main form of transportation. Even at that stage, automobiles were so rare that there were only four paved roads through the reservation, totaling around 205 miles, mostly on the periphery. But by 1973, there were 1,370 miles of paved roads. Over 60 percent of all Navajo households on the reservation owned a car or truck, and it is estimated that the vast majority of people had access to a motor vehicle. Where such rapid change has destroyed other technologically primitive societies, the social fabric and cultural cohesiveness of the Navajos do not seem to have faltered.

Any society that can embrace nearly one thousand years of technological change and adapt along with the new technologies without being destroyed by them must have a flexible structure and some luck. For the Navajo, that luck was living on land that no European settlers wanted, and thereby avoiding the disastrous wars, diseases, displacements, and depopulation that befell most of their native brethren. The Natchez were destroyed by wars with the French. The Mandan were wiped out by epidemics. The Choctaw and Seminoles were forcibly removed from their tribal lands and resettled over one thousand miles away. The Sioux were decimated by the destruction of their main food source—the buffalo. The fact that the Navajos' home was in the arid plains and mountains of the American southwest protected them in the late nineteenth and early twentieth centuries, when the U.S. government attempted to distribute Indian reservation lands to white settlers. Few people took the chance of starting a new life in a place where so little would grow and live, and the Navajo remained undisturbed for several more decades.

When the Great Depression of the 1930s and military service in World War II finally forced the Navajos to confront mainstream American culture, they adapted more readily than the more structured eastern Indian tribes. Most eastern tribes had tightly inte-

and decent living standards (giving progressively lower weighting to income above the poverty line). The index ranking for many of the countries studied in this book is presented in Table 8.1. The overall HDI ranking for 1995 shows Canada with the highest ranking, the United

grated, well-defined political, social, and religious units. When catastrophe struck an eastern tribe and important leaders died, or when new technologies replaced old and valued ones, religious and cultural ceremonies broke down. When the tribe was no longer able to carry out its traditional ceremonies, the knowledge and practices died out, and the tribe was in danger of losing its identity. For example, when some of the twenty-four clans of the Osage tribe died out, religious ceremonies structured in twenty-four parts to honor each of the clans broke down. When trucks and guns were introduced, old hunting ceremonies passed out of existence. When the buffalo of the Great Plains were hunted to near extinction, the Sioux, Hidatsu, and other plains tribes lost their traditional and ceremonial source of food and clothing. What makes the Navajo remarkable is that their cultural identity was not dependent on particular artifacts or people. Being "Navajo" is not threatened by tennis shoes and doughnuts. The main religious/ceremonial precept of the Navajos is the concept of "walking with beauty," a concept that directs them to work with nature and their environment, instead of against it. But it can be carried out from a truck, from horseback, from an office, and from a cellular phone just as well as it can from fields of blue corn.

The Navajo remain a population separated from the rest of American society by a strong tribal identity. The traditional multigenerational family household has not broken down, nor has the traditional economic backbone of the Navajo—sheep herding. And although it is true that nontraditional family households do exist, and that wage-labor is replacing sheep herding as the primary job for the younger Navajos, their institutions are adaptable and their political, religious, and social identity does not appear to be in danger.

The flexibility of the Navajo Indians allowed them to adapt well to twentieth century American technology, despite their relative isolation up until the 1930s. Because a wagon replacing traditional horseback transportation did not threaten their ceremonies, their families, or their livelihoods, it was accepted as a tool to make life easier. Thirty years later, the same held true for the automobile. Trucks and cars have been integrated into Navajo daily life, and Navajo culture remains strong.

Source

Based on work by Daniel Kramer, Stanford University, 1997.

References

Bailey, G., Bailey, R. G. 1986, *A History of the Navajos*, School of American Research Press, Santa Fe.

Reno, P. 1981, *Mother Earth, Father Sky and Economic Development*, University of New Mexico Press, Albuquerque.

Weiss, L. D. 1984, *The Development of Capitalism in the Navajo Nation*, MEP Publications, Minneapolis.

States is fourth, Japan eighth, the United Kingdom fourteenth, and Australia fifteenth. The highest-ranking developing countries were the island nations of Cyprus and Barbados at twenty-three and twenty-four. Mexico was at forty-nine, well above the Russian Federation at seventy-two.

MOTOR VEHICLE IMPORTS TO PAPUA NEW GUINEA

Papua New Guinea (PNG) is a small island nation just north of Australia. It includes the eastern half of the island of New Guinea and the islands to the north and east and its terrain is extremely rugged.

One of the major types of technology transfer to PNG has been machinery associated with transport, including outboard motors, motor vehicles, and heavy equipment. Transfer has taken place in a rather chaotic fashion over a long time. Before independence in 1975, it was common for a European expatriate who wanted to use motor vehicles or heavy equipment to take up an agency for a particular "make" and import a relatively small number of units, together with the minimum of spares to carry out routine servicing. Sales to the public were a bonus. When the expatriate went home, the agency was sold.

In the six years to 1974, nearly 16,000 cars and station wagons and 24,000 utilities, trucks, and buses were sold in PNG. They included at least 117 different makes and 250 make and size combinations. Internal transport in PNG is difficult, and for practical purposes most of the nineteen districts (now provinces) needed to be self-sufficient.

In over 600 cases, however, there were fewer than ten of a particular make and size of vehicle registered in a district. The small numbers of each type led to major problems in:

- spare-parts supply and cost;
- training of mechanics, drivers, and spare-parts staff; and
- maintenance and servicing.

Problems were increased by the lack of mechanical expertise of the local owners. The results included high running costs, short vehicle life, and a serious waste of foreign currency. In 1974, the PNG government, responding to these concerns, set up a Royal Commission to review and resolve the problems. Its focus was on "Standardisation of Selected Imports."

The inquiry generated political flak. The prospect of restraint on the imports of Mercedes Benz and Volkswagen motor vehicles apparently upset the West German government, and the brief of the Commission was changed from "Standardisation" to "Rationalisation," to avoid trade retaliation against PNG coffee. It was therefore ironic that the inquiry revealed that:

One of the problems with this ranking scheme is that national averages conceal huge internal differences. In sub-Saharan Africa, for instance, three in four urban dwellers but only one in four rural dwellers had access to safe drinking water. This is a widespread problem in LDCs. Life expectancy for the poorest group in Mexico was fifty-three years, compared with seventy-three years for the top income group. The female literacy rate in Pakistan was less than half the male rate.

- The Honorary Consul for West Germany was the Port Moresby Mercedes and Volkswagen dealer. VWs were, however, being imported not from West Germany, but from Brazil.

- On the basis of VW data, the average retail price of slower moving spare parts needed to be at least six or seven times landed cost.

- VW advised that the minimum number of one make and model of VW in one place to provide a reasonable service and spare parts base was: "50,000 to 100,000"—more than the total number of motor vehicles in the country!

The scale of motor vehicle sales was too small for vehicle assembly to be economic, let alone for vehicle manufacturing. There was not even a base for a wrecking and recycling business. There was, however, a thriving business in building local bodies for PMVs (Public Motor Vehicles). These were a one- to two-ton truck chassis with a covered rear compartment that had seats around the edge, and space in the middle for goods being brought to or from market. These vehicles met the need for group transport from villages into town, and were much more in line with PNG cultural attitudes and economic needs than automobiles.

The Commission significantly improved the situation by applying pressure on importers to improve staff training and spares availability, and to reduce the variety of units imported. The best of current PMV practice was also written into registration requirements, improving rollover protection. The preferred technical solution to the excessive variety of makes and models was to restrict imports to an absolute minimum range. However, this was politically unacceptable. Proposals by the Commission for more comprehensive action were seen variously as empire building or as a threat to trade links, and were not approved by the PNG cabinet.

Many of the typical technology transfer problems faced by LDCs are highlighted by this case study. Where purchasers of equipment are unfamiliar with the technology, specific provision needs to be made for training. The focus on rapid model changes, driven by marketing concerns in rich countries, is especially inappropriate for small third world countries.

Reference

Johnston, S. F. 1980, Rationalisation of motor transport equipment in Papua New Guinea, *Proceedings, Institution of Engineers Australia [IEAust] First International Conference on Technology for Development*, Canberra, pp. 32–38.

Efforts to disaggregate the HDI index are clearly important, and are proceeding. Two steps are indicated in Table 8.1. The first step was to separate male and female data for the countries where such data was available. Gender disparities were found to be widest in Saudi Arabia and Oman, where male literacy was 71 percent, compared with a female literacy rate around 50 percent, and with females getting only 10 percent of earned income. A similar but less extreme pattern seemed common to

Core Values of Development

◆ *Life-sustenance:* the ability to provide basic necessities, including food, shelter, health, and protection. Rising per capita incomes, the elimination of absolute poverty, and greater employment opportunities constitute the necessary, but not the only, conditions for development.

◆ *Self-esteem:* having a sense of worth and self-respect, of not being used as a tool by others for their own ends. Many societies that may have possessed a profound sense of their own worth now suffer from serious cultural confusion from contact with economically and technologically advanced societies. Once the prevailing image of "the better life" includes an abundance of material possessions and a Western lifestyle as essential ingredients, it becomes difficult for the materially underdeveloped to feel respected or esteemed (cf. Case Study 8.2 on p. 408).

◆ *Freedom:* from alienating material conditions of life; and from social servitude to nature, ignorance, other people, misery, institutions, and dogmatic belief. Freedom implies an expanded range of choices for societies and their members. Wealth gives greater control over nature and the physical environment. Wealth confers the freedom to choose greater leisure, to have more goods and services, or to deny the importance of these material wants and live a life of spiritual contemplation.

Source: (Todaro 1977, pp. 96–98.)

most Arab states. Elsewhere, literacy rates seemed broadly comparable, and the big difference was in income distribution. In Latin America, Chile, and Argentina had very low earned income figures for females, at only 22 percent of the total.

Average female life expectancy at birth was greater than that for males everywhere but the Maldives. In industrial countries it averaged 7.5 years longer, compared with 2.9 years in developing countries. In eastern European countries, with high levels of education and income equality, the gender development ranking was significantly higher than the overall HDI ranking.

Although recognizing its importance, the UNDP reports have not really attempted to offer an adequate treatment of Todaro's third point, freedom. They emphasize that human development is incomplete if it does not incorporate freedom. However, although promising efforts are being made to analyze the available data systematically (e.g., by Humana 1986), there are still major problems with both the comprehensiveness of the data and its assessment.

The UNDP indices offer indicative, rather than absolute, assessments of the overall situation and of the relative performances of the countries studied. Nevertheless, they provide a valuable overview of current problems of human development.

THE RANGE OF TECHNOLOGIES

Before we look at the details of industrial technology transfer, Forbes (1995) suggests that it is useful to discard a few myths about technology (see Key Concepts and Issues box).

Key Concepts and Issues

Some Myths About Technology

◆ Technology is not applied science, although much current industrial technology does require a significant level of scientific understanding.

◆ Technology is not an end in itself, but a means to the end of meeting needs and adding value, as we noted in our discussion of technology policy: the test of a good technology is economic, not technological.

◆ Technological self-reliance is a virtue, but it must not be raised to the status of a categorical imperative. Domestic production that involves making a second-rate item uneconomically, rather than importing a first-rate one, only makes sense if there is a clear and relatively short path to economic production of a first-rate item.

◆ Technology is not well-understood and is not easy to transfer. Technology is more than products, processes, and systems; it is embedded in people. It includes public knowledge (patent specifications, or recipes for various activities) and private knowledge (blueprints and other proprietary documentation). Although the processes of ISO 9000 quality documentation are aimed at going beyond these levels, and capturing and codifying even the tacit knowledge people have (such as just how much glue to apply to two surfaces that have to be joined), this is a very difficult task. There is often no real alternative to expert demonstration and commentary.

■ High Technology

Typically, the characteristics of modern high technology reflect the United States, German, or Japanese context in which it is developed, and the aims of those who fund and support its development. Intended to meet the needs of large and sophisticated markets, it assumes highly skilled support staff, effective management, and ready availability of capital. Despite high capital investment, it produces relatively few jobs. For instance, the chip fabrication plants mentioned in Case Study 8.5 cost over a $1 billion to provide direct employment for 800 people, an average of $1.2 million per job.

Labor-exclusivity can be a serious problem even in developed countries. In poor countries, where capital is scarce and the unemployed and underemployed are numerous, it may be disastrous. The problem is often compounded by a requirement for high-quality material inputs, as Case Study 8.7 (at the end of this chapter) illustrates—the variability of locally available raw materials led to plastic feedstock being imported into China from the United States.

The scale of production of much of this technology has historically been too large for countries without well-developed mass markets. However, there may not be any realistic alternative for some requirements, such as urban electricity supplies or air transport. Computer-based production technologies tend to be more flexible than older mass production technologies, but there are still problems with availability of capital and suitable labor. Sophisticated technical and organizational support is required, including high-quality raw material inputs and stable and well-controlled energy supplies. This is not the typical industrial context of poor countries.

■ Intermediate Technology and Appropriate Technology

The capital cost of providing a workplace is important. It varies widely. In open-cut mining and some high-tech manufacturing operations it is of the order of $1 million per job. At the simplest level, where it only involves the cost of a basic hand tool for digging or cultivating, it is of the order of $10.

The founder of the British Intermediate Technology Development Group, the late E. F. Schumacher, argued for increased attention to an intermediate level of technology, significantly more productive than basic hand tools but involving capital outlays that poorer communities could afford. This would now mean capital outlays of between $1,000 and $10,000 per job. The concept was aimed at LDCs, but it seems to have relevance to addressing unemployment and underemployment in some western countries today (Schumacher 1973; Johnston 1976).

Other important technologies on this scale include a variety of hand-operated water pumps and low-cost water storage techniques that make small-scale settlements feasible in remote areas in Africa and Australia, and the solar-recharged battery operated radio/telephone described in the example below (see box, An Example of Appropriate Technology).

The key element in the suitability or otherwise of a technology is probably its compatibility with potential users. What is the machine-user interface like? What inputs are needed to keep it working? Will the long-term users appreciate the need for these inputs and be able to supply them? In considering equipment for LDCs, it may be misleading to assume that users have the sort of preparation for looking after sophisticated machinery that many Americans get in their early lives by servicing bicycles and cars, or by playing with video games and other electronic toys.

Kirpich (1987) emphasized the need for choice of the most suitable technologies. Compatibility with sociocultural factors is important. Some high technologies, for example in communications, are preferred, but unnecessary complexity or demands on servicing or operating skills are inappropriate. There are significant differences between projects and between countries. Cost is very important, and the shortage of skilled personnel can be critical (Morgan and Morgan 1986). Even for less affluent communities, however, "appropriate" is not synonymous with low technology. Solid-state devices can be robust and reliable. Willoughby (1990, p. 15) defines an appropriate technology as: "a technology tailored to fit the psychosocial and biophysical context prevailing in a particular location and period."

One of the difficulties in analyzing the development and transfer of technology is the enormous range of *sociotechnical systems of production and use* involved. At one end of the range are small-scale activities, often in remote areas, and problems such as communications, drinking water purification, or handicraft and agricultural production for small groups. At the other end are large-scale industrial activities, with which engineers are generally more familiar. The whole spectrum of technology transfer is important, but the approaches required at each end of the spectrum are rather different, and it is essential to be clear where the focus is at any time.

BRINGING THE SEMICONDUCTOR INDUSTRY TO SINGAPORE

This case describes the planned and deliberate approach to the process of technology transfer taken by one small newly industrializing country. It suggests that careful planning can pay off.

Rapidly developing east Asian countries have identified active technology transfer as an effective means to accelerate industrialization and create wealth. The government of Singapore, in particular, has realized that keeping abreast of world trends in technology is an important way to promote economic competitiveness; the rapid growth of the semiconductor industry in Singapore shows this strategy at work.

Singapore is a small island nation off the southern tip of the Malay peninsula. Since independence from Malaysia in 1965, a major priority has been industrialization aimed at improving the national standard of living. The birth of the local electronics industry followed. U.S. multinationals were encouraged to establish testing and assembly facilities in Singapore. The first investor was Texas Instruments in the late 1960s. The electronics industry grew rapidly, setting the stage for the technology-intensive semiconductor industry to become a key part of the country's industrial culture. This was part of a concerted plan to tap into the flourishing world electronics market and to usher the nation into a new era of high value-added manufacturing activities.

The initial focus was on importing advanced technology. At the same time there was a drive to create a unique institutional network involving both public and private sectors. This network was designed not only to receive technology, but to absorb it, adapt it, diffuse it, and later improve on it, in the context of indigenous technologies and engineers. The approach differed from traditional, more passive forms of technology transfer, in that the impetus and industry linkages were not created and directed by advanced companies in developed countries but were initiated by the developing country itself.

In practice the network was achieved through patient, but intensive application of "technology leverage." Multinationals such as Hewlett-Packard, Siemens, Linear Technology, and Motorola are still in Singapore, but over thirty years local companies have also gradually emerged. Through statutory boards such as the Economic Development Board of Singapore (EDB) and the National Science and Technology Board (NSTB), the government continues to play a major role in the development of the industry. Assisted by the United States and Taiwan, the EDB has established an industrial culture that is distinctly "semiconductor-friendly." In 1990 the EDB set up a Cluster Development Fund (CDF) to support cluster development, with the semiconductor industry as a key cluster. The EDB has encouraged both the importation of expertise and the establishment of joint ventures with multinationals. The Institute for Microelectronics, also locally established in 1990, is actively involved in research and development.

Another development in this industry has been the heavy investment in wafer fabrication facilities ("fabs"). The EDB envisioned twenty such plants being in operation by the year 2000. By 1997 four companies had nine wafer fabrication plants either operating or under construction in Singapore. The government had a significant stake in three of the four companies.

Other improvement plans for this industrial network include the $200 million "Semiconductor Process Capability Development" scheme launched in late 1996, aimed at improving processes in wafer fabrication, and the "Semiconductor Manpower Development Initiative" aimed at encouraging local university graduates to pursue R&D careers in the field of semiconductors.

This very deliberate, locally initiated method of technology transfer was necessary for a number of reasons. Singapore's comparatively tiny physical size and population made success "on the first try" critical, especially in dealing with highly capital-intensive industries such as the semiconductor industry. Also, the fact that Singapore was a latecomer to this industry (like most of the emerging southeast Asian economies) made it imperative that success came quickly. These conditions were seen as justifying heavy government intervention.

The same model for technology transfer cannot readily be duplicated in other developing countries. A unique industrial and social culture facilitated the entry of the semiconductor industry into Singapore. The high-end technology involved was seen as attractive by a mainly English-speaking population already in the throes of "Westernization." They saw anything high-tech and from the West as helpful in setting the country on the road to achieving developed-nation status. Singapore already had a proven advanced manufacturing capability, with well-established assembly and testing operations. Coupled with political stability, this encouraged multinationals to invest heavily in operations, and to enter confidently into partnerships with the government. Singapore's good telecommunications networks, roads, ports, and airport provided ideal conditions for technology transfer.

This apparently effortless progress was not without costs and risks. Some have speculated that the exponential growth in this industry was only due to the sheer volume of resources channeled in. Future availability of these resources might be in question, especially in a country as small as Singapore. Moreover, as a still-small player in the global market, local industry performance is severely affected by price fluctuations. For example, a dip in the sales of DRAM chips in the past few years has adversely affected the local industry's profitability. Even more worrying is the general susceptibility of regional economies to speculation and attack. When, in 1997, southeast Asia experienced arguably its worst economic crisis in recent times, regional currencies fell as much as 65 percent to 70 percent. Although the least affected in the region, Singapore still faced potential problems with increasing import costs and lack of investor confidence. These are circumstances largely beyond the government's control. They are typical of the risks of engaging in a high-cost, highly volatile industry. It is not clear that Singapore has credible alternatives.

Sources

Based on work by Shannon Quek, Stanford University, 1997.

References

Mathews, J. A. 1995, *Technology—Industrialization in East Asia: The case of the semiconductor industry in Taiwan and Korea*, Tzong-shian Yu, Taiwan

Economic Development Board of Singapore 1997, "Investment News, October 1996." *Singapore Inc.*, Online. Internet. Nov. 25, http:// www.singapore-inc.com.

Ng, K. 1997, "Semiconductors Asia/Pacific." Dataquest, June 30, Online, Internet, Nov. 22, http://www.dataquest.com.

An Example of Appropriate Technology

The range of technologies that the authors would describe as appropriate can be illustrated by reference to one particularly elegant use of technology: a transportable communications system for Aboriginal outstations in Central Australia.

It uses a lightweight mast about ten meters tall, which can easily be erected or taken down by a group of people working together. At the top of the mast is a tuned radio aerial. At adult head height (to minimize playful interference by children) the original version had a closed box containing a two-way radio transceiver and a car battery to power it. Operation was by a timed push button, which gave ten minutes operation before needing to be pushed again. The battery was charged by a photovoltaic solar panel mounted two-thirds of the way up the mast. If children (or adults) flattened the battery, it worked again the next day. When the group moved, the communications system was loaded onto a truck and moved with them.

The most recent version of this communications system includes a coin-operated mobile telephone instead of the radio transceiver. An early problem was that the coin box filled up because the nearest person authorized to empty it was likely to be several hundred kilometers away. Authorizing one member of the community to empty the box resolved that problem. There was initially some confusion caused by the variation of telephone rates over the course of a week, (driven by urban commercial demands that seemed rather unreal, hundreds of kilometers from the nearest town). However, people quickly came to terms with such arbitrary constraints.

TECHNOLOGY TRANSFER

We will start at the large scale end of the spectrum, which is particularly relevant to countries that are commonly described as newly industrializing countries (NICs), for example Brazil, India, Korea, Mexico, and Singapore. In general these countries are *technology followers*, working somewhere behind the technology frontier. The technology frontier is defined by the *technology leaders*, who face the expensive problems of sorting out fruitful directions from dead ends. Technology leadership changes from time to time, but arguably at the end of the twentieth century the United States led in aerospace and electronics, Japan led in automobile manufacture and Switzerland led in food processing.

In essence, wealth creation for economic development is the process of adding value to goods or services, using labor and other resources in an "effective" way. The openness of the market in which the item is to be sold is relevant to defining what is "effective." With falling tariffs and increasingly open international competition, manufacturing effectiveness increasingly has to be defined in terms of world best practice—a target that keeps moving. This is why the encouragement of creativity is so important. A country becomes relatively less developed if its technology does not change while the technology of the rest of the world does. This happened to a significant extent in many countries around the world in the 1970s and early 1980s.

Countries and individual firms keep up with changing technology by developing technology themselves and/or by transferring it from somewhere else. To be able to afford to do this, they need to reinvest some of the profits from existing activity, or to borrow. The basic limitation on the effectiveness of borrowing as a strategy is the need to service the loan. Failure to ensure adequate levels of return leads to increasing indebtedness. In practice, the more successful technology transfers also need to be able to support the less successful ones.

▪ Indigenous Technology Capability

Failure of attempts to transfer technology in the 1950s and 1960s led to attention being paid to the technological capacity of intended recipient countries, their "indigenous technological capability" (ITC). This applies at the level of both the nation and the individual firm. Forbes (1995) describes ITC as involving a dynamic interplay between three factors: *access*, *endowment*, and *effort*, all in the context of the surrounding policy environment. The policy environment and local cultural factors are important in helping or hindering the process, but the critical element in successful technology transfer appears to be the individual firm.

To be effective in technology transfer, the firm needs international access to technology, partly because it is generally much cheaper to buy technology than develop it oneself, even if that were possible. The firm's endowment is in terms of entrepreneurship and the education of its workers (particularly basic literacy and numeracy). These must be adequate, because the critical aspect of technology transfer is getting the information into the heads of the local workers. Some higher technical and managerial education is important, provided the people getting it are prepared to put their education to work at home rather than abroad. Entrepreneurship is about getting the firm's employees, its human capital, to make development happen. It gives the firm much more control and a much stronger bargaining position if the ITC is sufficient to allow for unpackaging a technology (as discussed below). Case Study 8.7 at the end of this chapter illustrates some of these issues for rural China.

Finally, what the firm needs is indigenous technology learning capability. This is the ability to learn how to use the technology effectively. Incremental innovation, particularly process innovation, is the key initial task. The R&D department has an important role in solving the more difficult production problems: it is the unit that can learn about new ideas and new technology on behalf of the firm, and feed the information into the organization where it is most useful. This means that the R&D unit of a technology follower must be physically close to the production unit and in continuing contact with production staff.

Effective transfer of technology is based on people learning the skills to operate and develop the technology. It must therefore take into account the skills available and the general context into which the technology is being transferred. The idea of leap-frogging older technologies in order to "kick-start" development is appealing; however, to be successful it requires a very detailed understanding of all the requirements for infrastructure and training support, as well as access to resources and the commitment to provide them. As discussed later in the chapter, it is clearly a medium- to long-term proposition.

■ Unpackaging Technologies

For a country to be able to control an imported technology and benefit from local improvements to it, the technology must be "unpackaged," that is, the technological content must be separated out from the associated capital equipment and supplies of overseas materials and people.

Grant (1980) pointed out that where technology is brought in as a package deal together with foreign investment and foreign control, "the technology remains under the control of the multinational corporation offering it, making it inaccessible for use in generating innovative or high-value exports." The difficulties this has created for Papua New Guinea are discussed in Case Study 8.4 (see p. 418). The issue is not only one for developing countries. It is instructive to consider the tough approach the Japanese took in developing their industries after World War II. Inkster (1991, pp. 278–284) described how, assisted by coherent national technology policy coordination, Japanese companies moved from imitation in the early 1950s to leadership by the 1980s. Key factors in this approach included the identification and national support of strategic industries, the large scale of operations (underpinned by a well-protected domestic market), and strong linkages between production, trade, and financing.

Vernon (1989) summarized the lessons from research on technology transfer that he saw as most helpful to guide future policy making. He directed this guidance particularly to developing countries, but most of it seems equally relevant to developed countries and individual firms. His advice included the following points:

- Countries that develop a strong capacity to search out and evaluate foreign technologies are usually able to acquire the technologies they need on satisfactory terms. Failure to search out and evaluate is likely to lead to costly errors.
- Developing countries do not need to start with a strong scientific community to be able to identify and master the technologies they need. Close ties between industry and the scientific community only become important for moving towards more advanced levels of technological capability [this is consistent with Kline's model of innovation; see Figure 6.2 on p. 287].
- The most important single criterion for the choice of a technology is probably scale, which is essentially a matter of likely sales. These in turn depend on the size and structure of the local market and prospective exports. Both these factors are strongly affected by government policies.
- Learning processes within the firms involved are critically important.
- Increasing technical capability in an economy demands increasing organizational flexibility, including the development of technological networks among firms.
- The effectiveness of organizations in meeting the conditions listed so far depend particularly on the commitment by managers to remain competitive and on their confidence that the environment, both nationally and within the organization, is unlikely either to block their projects or to allow others to seize the rewards for them.
- Once firms have adapted products or processes effectively to the local situation, they are likely to find that there is a substantial demand for them in other countries.

In many countries, failure to deal effectively with multinationals in the importation of technology has been essentially a political problem. It has often reflected an absence of effective national policies, in turn demonstrating a lack of cooperation between government and local firms in dealing with the issues. Without strong government support, only the largest companies are likely to have the strength to force the unpackaging of technology. An effective national technology policy can be a valuable first step towards managing technology transfer in the national interest.

Case Study 8.5 (see p. 424) shows how the government of Singapore took the initiative in the development of computer-based industry there, and indeed how it continues to encourage higher levels of activity in this area. Case Studies 4.4 (see p. 192) and 8.7 (see end of this chapter) describe how the Chinese government has organized the introduction of the automobile manufacturing industry. The plant at Hebi, described in Case Study 8.7, was deliberately located in a remote area to promote development there. However, even in Shanghai there were tensions between the Chinese and Volkswagen (VW) over the type of vehicle to be produced. The Chinese wanted a more modern vehicle that would be competitive with Japanese imports. VW insisted on starting with the rather basic Santana until the local parts supply situation improved.

As Case Study 8.4 also shows (see p. 418), difficulties inevitably arise when technology transfers involve skills or other resources, such as spare parts, which are not readily available in the recipient country. Unless training and other support issues are addressed effectively, once the foreign technicians who installed and commissioned the equipment have gone home its demise is simply a matter of time.

Part of the difficulty of implementing these insights is that technology is not homogeneous, but is subject to significant variations in scale, capital intensity, maturity, and cost sensitivity. It may also be more or less science-based. As we can see from the difficulties of firms in industrial nations, dealing with technology development and transfer requires an understanding of the specific characteristics of the industry, and skilled change management in the organizations involved.

∎ Technology for Village Development

Successful technology transfer at the village level must be a long-term process, rather than a single-use "magic pill." In the authors' view, the essence of village development is to empower the community and to assist its members to make good choices on their own behalf. The technology transferred needs to be community-based, and community structures need to be in place to supply and maintain essential knowledge and skills. Aid project managers need to review the project context, ensuring that traditional technologies are not unnecessarily destroyed and that the community has a real opportunity to consider the likely benefits and costs of all possible options.

As Case Studies 8.1 and 8.2 indicated (see pp. 406–409), at the village level small-scale, people-centered technologies are more likely to be beneficial and to have minimal negative consequences than grand projects imposed from above. Unfortunately, there seems to be a natural bureaucratic bias towards large-scale projects. The Japanese approach to process innovation is relevant to technology transfer generally: there are very many small "treasures" to be found. Such treasures are small advances that together add up to large overall improvements, and are only available to operators who understand the operation of the technology and are in a position both to make and to benefit from changes.

The aim of development aid should be to provide sustainable technology transfer that enhances its recipients' ability to control their own destinies. The most likely ways are by:

- strengthening community participatory structures;
- transferring knowledge, skills, and processes, rather than simply products; and by
- designing for an equitable distribution of benefits and costs in the community that adopts the technology.

Increasingly important technical and social questions for LDCs are raised by the issue of communications policies. In a world increasingly driven by and dependent on information technologies, LDCs need to make the most of information opportunities that emerge. They certainly cannot afford to ignore global digital technologies. They will join the ranks of the information poor if they simply brand these technologies as a developed-nation conspiracy and the work of the devil.

■ Cultural Factors

Technology transfer has cultural as well as technical and resource aspects. Even in transfers to relatively developed countries, these cultural issues can present serious difficulties. In considering the cultural aspects of technology transfer, McGinn (1991) distinguishes between micro contexts, for example within an individual firm, and macro contexts, which may include an entire regional or national society, or even the whole world. His primary focus is on macro contexts, and he suggests that five key dimensions need to be taken into account to develop an adequate understanding of the causes and consequences of specific technological developments. These are:

1. the practitioner dimension (attitudes, intentions, and resources);

2. the technical dimension (need to meet broad or specific technical demands);

3. the political-economic dimension (public health, national prestige, potential profitability);

4. the cultural dimension (divided into immediate factors such as short-term budgetary factors, and background factors, specifically the cultural system of the society, including typical levels of education, and religious attitudes and beliefs); and

5. the environmental dimension (the natural environment, including emerging or perceived problems such as the greenhouse effect).

McGinn calls the combination of the last two items on this list the "cultural-environmental system." He takes the analysis further by using an anthropological approach to explore subsystems of the cultural-environmental system. This more detailed level of analysis allows the exploration of the interaction of particular technological developments with the various facets of the society (McGinn 1991, pp. 52–70). He goes on to develop a heuristic model of the relationship between technical change and ensuing social change. He calls this the IDUAR model of the process of adaptation, adoption, or rejection of technologies. The model highlights five variables in the process, each of them linked by feedback loops to the society's Social-Cultural-Environmental System. These variables are:

1. the Innovation itself;

2. the Diffusion process;

3. the pattern of Use;

4. the Adaptation the society and the users make in adopting the innovation; and

5. the Resistance to some or all of the uses of the technology.

In this model, "Use" refers to both user groups and uses, and "Resistance" is a function of users, uses, diffusion and adaptation) (McGinn 1991, pp. 98–101 and personal communication, 1999).

McGinn's models are helpful for thinking about both developing and developed countries. Even in industrialized countries such as the United States, social and cultural factors can make transfer of production technology difficult. Effective transfer may require quite different management attitudes and practices, forgetting old ways and learning new ones. It may also require the presence on the production floor of skilled technicians or even professional engineers to provide continuing training and skills development for the operators. This level of technical support seems likely to be important for the continuing process innovation that characterizes the best Japanese practice. In third world countries, the characteristics of the transferred technology may be very much less compatible with existing skills and support resources.

ADVERSE OUTCOMES FROM TECHNOLOGY TRANSFER

The authors of this book are in no way antitechnology. Clearly, without our present technological systems, the world could not support anything like its present population. These systems are essential to our very survival and to our hopes for improvement. However, the point we need to make here is that technologies have significant social, economic, and political impacts that we cannot safely or responsibly ignore. These impacts need to be addressed in the social, political, and economic spheres by some appropriate combination of individual, corporate, community, and government action. Bryce et al. (1994) highlight some of the negative consequences of inappropriate technology transfers to LDCs on the Pacific Rim, particularly at the smaller scale end of the transfer process. In both rich and poor countries, negative consequences of technologies can include:

- generating unemployment by reducing labor demands;
- increasing economic inequalities, making the rich richer and taking work away from the poor;
- increasing power disparities and creating social disruptions, through both of the above;
- creating dependencies, as a result of all three of the above and as a result of increased expectations, creating requirements for special raw materials and support services, and destroying traditional skills and processes;
- creating new pollution problems;
- reducing the stock of nonrenewable resources; and

◆ creating and increasing disparities between nations, manifested in heavy debt burdens.

One of the more widely publicized examples of technology transfer took place in the late 1960s and early 1970s in Asia, and to a lesser extent in Africa and Latin America. High-yield hybrid varieties of wheat and rice were heralded as a "green revolution." These crops were bred to respond to heavy fertilizer applications and to give much higher annual yields than traditional varieties. The rice also needed precise control of water levels.

Because these crops ripened over a shorter period than traditional varieties, they required much more rapid harvesting, typically over a period of one week rather than six. This had the effect of denying traditional, economically important employment to poor laborers. The financial support provided for mechanization of harvesting eventually eliminated the need for laborers. The better-off peasants, who could justify and support the cash outlays involved, expanded their holdings at the expense of the poor, who became poorer. Although the "green revolution" crop varieties increased food production, they also increased the gap between rich and poor. The point here is that the capital requirements and operational characteristics of the new technology led to an increased level of inequality (Glaeser 1987).

There appear to have been unexpected (although, in hindsight, not entirely surprising) additional problems with these high-yield crop varieties. Studies in the mid-1990s indicated that their low uptake of minerals and production of vitamins was giving rise to malnutrition problems in people largely dependent on them.

More generally, during the "development decades" of the 1960s and 1970s, many attempts were made to increase GNP in the poorest countries by transferring high technologies to them from more economically developed countries. Where this was done without adequate consideration of the appropriateness of the technologies or the availability of local support systems, it failed. As Todaro (1977, p. 95–96) pointed out, a number of the developing countries that experienced high per capita income growth rates during the 1960s actually experienced a decline in employment, equality, and in the real incomes of the bottom 40 percent of their populations. The reasons for the failure are now clear. As our comments on the "green revolution" illustrated, the criteria used for the development of new technologies can lead to an inherent tendency to increase existing economic inequalities. This is not, of course, just an issue for LDCs. Developed countries have typical annual productivity increases around 2 percent to 3 percent. Unless their economies grow at higher rates than this, there is no net creation of jobs. Particularly during economic down-turns, employers focus on process innovation to maintain profitability and competitiveness. The resulting changes in technology can significantly boost productivity (output per worker), with the potential to further increase unemployment and the associated inequalities between those with and those without jobs.

Table 8.2 highlights factors involved in government-to-government aid. Bilateral aid commonly carries a more or less explicit requirement that much of the money is spent on products or services from the donor country. This "tied aid" can impose the wrong type and/or scale of technology on the recipient country. The intending recipient country makes the requests, but the potential donor country decides which proposals it will support, often on the basis of its own domestic priorities.

Although there have been instances of multinationals promoting grossly inappropriate technologies in the developing world, negative effects of technology transfer are not the result of a conspiracy by Western technologists. They commonly reflect some combination of inappropriate criteria for developing technology, unsatisfactory processes for selecting technology and inappropriate choice of technology. The donor country, the receiving country, and the technical advisers all have responsibility for ensuring good selection of technologies and systems to transfer and adapt them.

A "cultural cringe" can encourage adoption of less suitable but more technically exciting or glamorous foreign solutions over more suitable

TABLE 8.2

Structural Problems in Government-to-Government Technology Transfer		
Issue	Context	Outcome
Tied aid	Growing national debt, reduced local power	Inappropriate technology, imposed solutions
Imbalance in technology	Importation of "experts" through foreign agencies	Noncontextual solutions, imposed solutions
Technological cringe factor	Local encouragement of foreign solutions	Inappropriate imposed solutions
Weak linkages between government and people	Inadequate contextual knowledge	Inappropriate imposed solutions
The "aid" process itself	Government-to-government agenda, donor country priorities	Products rather than processes, imposed solutions

Source: Paul Bryce 1997, personal communication.

(lower technology and/or local) ones. (Cultural cringe has also been a factor in some notoriously inappropriate marketing of products such as infant formula to LDCs.) Where government officials do not have a detailed understanding of either the technology involved or the technical and social context for which it is intended, they may get it dreadfully wrong. An example of this was a combined solar/wind powered electrification project for a village in southern Sri Lanka. Neither the government officials not the donor agency involved appears to have known that for two months of the year, the weather in the region is both too overcast (not enough sun) and too still (not enough wind) for solar and wind to be a reliable year-round energy source. Clearly, the people in the village were not consulted effectively.

Another issue that came to be associated with aid and other projects in LDCs was the problem of outright corruption, as opposed to poor or unsatisfactory decision making. The projects most susceptible to corrupt practices were those involving large-scale and complex technology, particularly military technology. (As a number of defense/bribery scandals have demonstrated in recent years, the problem is not limited to poor countries.) One NGO, Transparency International (TI), suggests that corruption led to serious distortions in the priorities of aid requests from some LDCs. TI has been helping "honest governments" combat corruption by setting up "islands of integrity," that provide examples of better practice and opportunities for honest officials and suppliers to work in more acceptable ways (see box Dealing with Corruption, p. 372). The problems of corruption were highlighted by well-publicized cases in Europe in the mid-1990s. Contract managers from overseas projects, where bribery had been condoned, were caught bringing the same approaches to domestic projects, there was a sense of shock and outrage (Transparency International 1997).

So far in this chapter, we have been looking at technology transfer from the perspective of developing countries and their people. We now want to change the focus and look at these problems and issues from the point of view of firms involved in technology transfer, and particularly from the perspective of engineers involved in making the transfer a success.

GLOBAL ENGINEERING PRACTICE

Globalization has been a dominant theme of the 1990s, but what does it mean? Firstly, it describes the widening and deepening of international flows of trade, finance, and information into a single, integrated global market. It is also a prescription for opening up national and global markets in the belief that this will produce the best outcome for human welfare.

What has happened so far? World markets have certainly opened up. Average tariffs on manufactured imports fell from 47 percent in 1947 to 6 percent in 1980, and are planned eventually to fall to 3 percent. With the lowering of trade barriers, global trade exceeded $4 trillion a year by 1994 and continued to grow by 6 percent a year.

Financial flows grew even faster. Flows of foreign direct investment reached $315 billion in 1995, a six-fold rise over the figures for 1981–1985. In the same period world trade increased by just over one half. Flows of money grew very much faster than required by trade. In twenty years the daily financial flows in the world's foreign exchange markets increased by 1,200 times, from around $1 billion in the mid-1970s to $1.2 trillion in 1996! Most such flows are now essentially speculative and parasitic rather than value-adding, with serious destabilizing effects on currencies.

Technological change is also affecting the nature of investment. As Case Study 8.6 shows (see p. 442), the manufacture of high technology products can move across borders with increasing ease. This threatens to break traditional links between high productivity, high technology, and high wages. Worker productivity in Mexico rose from one-fifth to one-third of the U.S. level between 1989 and 1993, but the wage gap narrowed much more slowly (Mexican wages are still only one-sixth of U.S. wages).

Is a global culture emerging? By 1997 there were more than 1.2 billion TV sets around the world. The United States exported more than 120,000 hours of television programming a year to Europe alone, and the global trade in programming grew by more than 15 percent a year. Popular culture, not only in sport, clothing, and food but also in types of crime, increasingly transcends national and cultural barriers (UNDP 1997, pp. 82–83).

During the 1990s, one use of the term "global" was to express the idea that modern technologies were making the typical middle class experience of people in the West universally accessible. This vision probably started with Marshall McLuhan's idea of the Global Village, where powerful communications technologies put everyone in touch with everyone else, but we saw earlier in this chapter how far from reality that idea was (McLuhan and Powers 1989). A rather more realistic restatement might be: rich and poor people alike, everywhere, can now see and hear whatever is thought worth carrying by those who control the media!

▪ Global Image and Reality

What people see and hear from a distance is unlikely to give a complete picture: it may not be balanced, but it is clearly powerful. This power was illustrated by the role of West German television in the reunification of

Germany and the fall of the Berlin Wall. East Germans were continuously exposed through their television sets to the images of their fellow Germans in the West, often their relations, living a life of comparative affluence to which they could not aspire. The fact that many of the features of East German life that East Germans took for granted–free education, health care, social security, a high level of gender equality–were not necessarily available in West Germany did not become apparent until the wall fell and the two Germanys, separated since World War II, were reunited. The vision projected had been so attractive that it would have been difficult to think critically about it. (In retrospect, many West Germans who warmly supported reunification, have found the cost of modernizing the East German economy rather chilling.)

Some of the same concerns are illustrated by the example of automobiles in China (see box, Personal Transport in China).

Personal Transport in China

Given the long-standing western love affair with the automobile, it is not hard to imagine how the Chinese might feel about cars. Until the early 1990s, most cars in China belonged to companies or government agencies, and all drivers were professionals. By the end of 1997, there were more than 2.5 million private car owners, a number that was rising by 26 percent annually. Nationwide there were 3,000 driving schools with one million drivers enrolled. Even with only one automobile or truck for every one hundred people, traffic was heavy. (By comparison, the U.S. rate was then seventy-five vehicles (including fifty-six cars) per one hundred people.)

In 1994, the Chinese government officially designated the automotive industry as one of the country's "pillar" industries. The government's official goal was to establish three major domestic car manufacturers by the year 2010, with each of the assembly lines turning out 400,000 cars annually. Government planners see development of the motor vehicle industry as providing badly needed jobs in manufacturing. It is also creating jobs in road building, and helping to preserve jobs in the massive but struggling steel sector. Multinational car manufacturers seem only too anxious to cooperate with the Chinese Government's plans if it means they can include Chinese ventures in their global operations and strategies (see Case Studies 4.4 and 8.7.).

Of course, there are costs. The road toll for 1997 was estimated at around 66,000 deaths and 174,000 injured. Five Chinese cities, including Beijing and Shanghai, are on the World Bank's list of the smoggiest cities in the world. Some

Global, National, and Local Issues

The increasing effectiveness of engineering practice in the twentieth century has made it both possible and necessary for engineers to think and work on a global scale. The environmental impacts and implications of some technologies can only be addressed at this level. The holes in the Earth's ozone layer at the North and South Poles and the greenhouse effect are probably the most obvious and best known examples of problems. Action on containing ozone depleting gases (Case Study 9.1) and agreements on allocation of broadcast frequencies in the electromagnetic spectrum show that strong global collaboration to control the effects of technology is possible.

action has been taken—leaded petrol will be banned nationwide by the year 2000. Petrol adds to the problems from burning coal, which provides three-quarters of China's energy. There are more than one million deaths a year from urban pollution: sulfur dioxide from fossil fuels burned in China causes billions of dollars worth of damage, and acid rain as far away as Japan and Korea. Cleaner coal-burning technology that removed the sulfur would help in the short term, but would not reduce the greenhouse effects from the carbon dioxide produced. Part of the problem is that China's industries, particularly heavy industries, use energy only half as efficiently as modern Western plants.

Moreover, this consumption for the sake of supporting employment in production seems the very opposite of moving towards sustainability. An increasing concern is that bicycle lanes, once inviolable, and often separated from urban motor vehicle exhausts by rows of trees, are starting to be used for motor vehicle traffic and parking. This is making the bicycle, one of the most sustainable of all urban transit modes, steadily less practical for Chinese commuters.

The Chinese are certainly serious about investment in highway infrastructure, although Western experience is that road spending is never high enough to solve congestion problems for long. In the last few years of the twentieth century the Chinese government planned to spend $60 billion to build over 100,000 kilometers of new highways, including almost 6,500 kilometer of multilane expressways. There seemed every prospect that the Chinese would simply repeat the mistakes made in the West, with their major cities moving rapidly to gridlock.

What does all this mean for individual engineers? How can they expect a global economy to change the way they live and work? How will individuals cope with the increasing variety of demands on them to work in a global context?

We suggested some answers to these questions in Chapter 7, when we discussed the capabilities that modern engineers need. The pace of engineering work has increased. Even within English-speaking countries, modern electronic communications have changed the ways engineers work. In international consulting firms, it is now routine for a project to be undertaken by two or three teams, distributed around the globe, with project documentation passed along with the daylight, like the baton in a relay race.

Engineers need to understand the uses, and the limitations, of modern communications technologies. Before it is possible for design teams to work effectively at a distance, team members need to spend time face-to-face building mutual understanding and rapport, so that they can interpret the nuances of communications. E-mail is a particularly satisfactory means of communication, partly because the measured pace it imposes seems appropriate to dealing with design problems, and partly because the decision-making process is then automatically documented.

Global working using electronic communications is becoming routine. Successive drafts of chapters for this book circled the globe by e-mail, from Sydney, Australia to Leicester, United Kingdom, and Palo Alto, Stockton and later Pasadena, California. The different backgrounds of the authors helped us to appreciate how material would impact on students and professionals from a variety of cultures, minimizing confusion resulting from local idiom and ensuring that our message would be less likely to be misunderstood. This process may be a useful model for the preparation of project documentation that needs to be clear to people in a number of countries and for whom English is not necessarily their first language.

Engineers will need to learn to work effectively in other countries and cultures. International studies and study abroad programs already exist that can help with this. It is important to learn about culture as well as language, because the way different peoples approach problems and their resolution vary greatly. So do the rules of polite behavior. The Erasmus Program in the European Union has been a major force in encouraging exchange within the EU. Case Study 8.6 illustrates some of the reasons why corporate activity has become global. It is a first-hand account of the challenges involved in the transfer of manufacturing technology to Mexico within a wholly-owned IBM manufacturing plant, and describes the experience of an engineer whose experience has been wider than he ever dreamed of while he was a student!

One result of the increasing effectiveness of engineering practice in the twentieth century is that in some engineering areas there are now strong technical forces pushing us towards a single, global market. Products and systems such as large passenger aircraft and modern electricity turbogenerators are so costly to develop and produce that only a global market can provide the economic returns necessary to justify the next generation of designs, particularly if the scale of this equipment needs to be significantly larger than at present to increase unit efficiency.

Another obvious example of the need for global-scale projects is the development and manufacture of the core technology for modern computers. The high-density integrated circuits (ICs) used in today's PCs are truly awesome. Given the microscopic size of the elements being built into the ICs, dust particles can play havoc with the process. Therefore, these components must be produced in clean rooms, in which the air is incredibly pure. Stringent Class 1 clean room standards mean that airborne contamination is kept below one particle per cubic foot (thirty-five particles per cubic meter). In comparison, an average office contains 1 million particles per cubic foot (35 million particles per cubic meter). The best of these clean rooms can filter particles as small as 0.001 micron (millionths of a meter), less than the size of bacteria, viruses, and the particles found in cigarette smoke. Top of the line fabrication plants (called "fabs") that use these clean rooms can now cost over $1 billion and employ more than 800 people.

In the clean rooms, automated production systems build up ever-finer patterns of transistors on fingernail-sized pieces of single crystal silicon. During the early 1990s, 0.5 micro-meter lithography was practiced in 16-million-bit DRAM fabrication. Today, half-billion-dollar super fabs churn out CMOS integrated circuits containing over 4 million useable gates, 256-million-bit DRAM, and sixty-four-bit microprocessors many times again more powerful than the world's supercomputers of a decade ago. The thickness of the conductive paths etched on the Intel Pentium II microprocessor, which contains 7.5 million transistors, is of the order of 0.25 micron. Case Study 8.5 (see p. 424) describes the development of the semiconductor industry in Singapore, a country that has specialized in this area.

What are the limits to integration? Time after time, as the so-called limits approach, they are exceeded. Developers now use light to create the tiny circuit patterns, and as the circuits have become smaller, so have the wavelengths of the light that are used. Today, ultraviolet light is used to make the PowerPC and Pentium chips that run our home computers. However, there are serious difficulties in going beyond 0.25 micron technology. Developers are now looking at x-ray lithography. The higher

establish and maintain these connections, our engineers literally cross the globe every day. Today's engineers need languages, people and communication skills, and other talents that have not traditionally been seen as closely associated with engineering.

As offshore manufacturing, international projects, and joint ventures become more common, many more "global engineers" will find themselves in unexpected situations in interesting parts of the world, such as lost and unable to communicate in a Mexican taxi cab!

Source

Case contributed by Richard Bjorn, Stanford University, 1997.

resolution of this technology comes from the extremely short wavelength of x rays, of the order of 0.01 to 1.0 nanometers, and their high penetration ability, arising from the transparency of most materials to this region of the spectrum. ICs exposed with x-ray lithography have been demonstrated with resolutions approaching 0.13 micron. Chip production facilities are not needed in every country, indeed only a few plants are needed to supply world demand.

At the same time, there are constraints on the possibility of a single product line meeting needs around the globe. We have already discussed some of them. We noted earlier McGinn's systematic exploration of the interaction between technology and culture. Products, processes, and systems do not come into being in a social vacuum, nor are they used in one. As Kline pointed out (1995, p. 174) products are not only part of sociotechnical systems of production, but also of sociotechnical systems of use. What is the relevance of these two grand-sounding phrases here? The first reminds us that technical systems can only be created, sustained, and developed in a commensurate social context. Further, it highlights the fact that the continued viability of modern technical systems depends on the existence of supporting capabilities and attitudes in the community in which they are used. To demonstrate this point, let us consider for a moment the sorts of infrastructural and other support required by a technology that is increasingly taken for granted by engineers and other professionals in economically and technically developed countries: the internet. Even a relatively small software operation, such as MMArt in Bulgaria, can have an effective global presence because of the promotion and distribution facilities offered by the internet.

Although access to computers and the internet is now available to some people in most countries in the world, it can also be seen as creating a new type of inequality between the information rich and the information poor. Look for a moment at some of the constraints:

- Including those in developed countries, no more than 20 percent of the world's expected 6 billion people in the year 2000 are likely to have ready and affordable access to a stable supply of electricity and a clean telephone line.

- The capital involved is significant—probably of the order of $2,000 per installation. This compares to average total annual incomes in the mid-1990s of $2,900 across the developing world, and around $965 for the more than half a billion in the least-developed countries.

- Broad political concerns. Given their potential drain on scarce foreign reserves, it is no surprise that the governments of some developing countries see computers, literally or metaphorically, as the work of the devil.

- The availability of the necessary language and technical skills to write and read manuals, and sort out technical difficulties, is a problem even within technically oriented universities in developed countries. Who can help those remote from such support? (Once people are "on line," support may well be available over the internet, but starting may be difficult.)

- Software needs to be available in the appropriate language. Even in English, significant details of usages and spellings vary, for instance the preferred date format, mm/dd/yy in the United States, but typically dd/mm/yy in the United Kingdom and in British Commonwealth countries. For languages that do not use a roman alphabet, either the entry cost is higher or the effectiveness of the computer is reduced, or both.

Broadening access to modern information resources such as the internet has become a high priority. Governments have a place in the process, as do community-based facilities. Engineers have an important social role to play in supporting such essential and technologically intensive developmental activities.

CONCLUSION

We have seen in this chapter that levels of technology and technological change are key factors affecting economic growth. An important part of the professional development of engineers is their becoming aware of positive roles they can play in assisting with more equitable allocation of resources and in enhancing the quality of life. Such recognition is essential to the effective discharge of engineering responsibilities and required under professional codes of ethics. Engineers have an important and valuable professional role to play in development. Sensitivity to the attitudes, opinions, and needs of the community, and to political realities and possibilities, is essential to maximize the benefits of new technology and minimize social and other costs. The opportunities for professionally and personally satisfying achievement seem limitless.

Engineering is now commonly a global activity, with design, manufacture, and construction distributed around the world. High-level communication skills have become as essential as high-level technical skills in engineering activity. The engineering profession of the twenty-first century will be an increasingly interesting and diverse group, with a virtually limitless range of challenges.

DISCUSSION QUESTIONS

1. What problems do developing countries face in reconciling their needs for large-scale and small-scale technologies—for example, urban and rural electricity supplies? How will the consequences of failure vary?

2. Compare the difference in the effects of the electrification of Iriri village in the Solomon Islands discussed in Case Study 8.1, where electrification was a community initiative and Mae Chaem in Case Study 8.2, where electrification came from outside. What conclusions can you draw about technological change and village cohesion? What measures might be used to increase cohesion?

3. Is the maintenance of village culture a desirable goal? Or are technological advances more important? Are these the most relevant questions to ask about Case Studies 8.1 and 8.2?

4. Discuss the proposition that unless a society has an adequate technological base of skill, understanding, and resources, engineering cannot assist it.

5. Choose an example of technology transfer and discuss the social, political, economic, and ecological costs and benefits it has had for the people and the society involved.

6. Discuss the differences between high standard of living and high quality of life. Illustrate by way of examples of cultures that emphasize one or the other.

7. Discuss some of the assumptions associated with the term "development" as presently used.

8. What is normally meant by "appropriate technology"? How appropriate actually is this term?

9. Ivan Illich (1974) suggests that anyone traveling faster than about eighteen kilometers per hour is using more than their equitable share of available energy. Review the basis for his argument and then discuss how the average speed might reasonably be increased, and by how much.

10. What do you see as the most important impacts that recent technology-based innovations (e.g., the internet, tele-working, mobile phones) are having or are likely to have in the places where you live or work? In what ways are they positive, and in what ways are they negative? To what extent do they change the way people work and play?

11. How might the increasingly global nature of engineering affect your professional career in engineering? How will you prepare yourself for the challenges involved?

Further Reading

Marx, L. 1987, Does improved technology mean progress? *Technology Review*, January, pp. 32-41, 71. Marx explores the relationship between development and technology.

Schumacher, E. F. 1973, *Small is Beautiful: A Study of Economics as if People Mattered*, Blond and Briggs, London. This book raises questions on technology transfer and the appropriateness of technology that were taken up by the Intermediate Technology Development Group (ITDG), started in London in 1965 by E. F. Schumacher and George McRobie.

McRobie, G. 1981, *Small is Possible*, Jonathan Cape, London. This work is a continuation of Schumacher's efforts.

Vernon, R. 1989, Technological Development: The Historical Experience, *Economic Development Institute of The World Bank, Seminar Paper No. 39*, The World Bank, Washington, D.C. This is a succinct review of the background to many of the issues discussed in this chapter.

References

Amarshi, A., Good, K., Mortimer, R. 1979, *Development and Dependency*, Oxford University Press, Melbourne.

Baran, P. 1973, *The Political Economy of Growth*, Penguin, Harmondsworth.

Bryce, P., Bryce, D., Irons, C. 1994, Rural electrification and technology transfer, *Conference on Technology Transfer in Remote Communities*, 6-7 April, Murdoch University, Perth.

Forbes, N. 1995, Technology in Newly-Industrializing Countries: Managing Innovation in Nations and in Firms, *Internal Report*, Industrial Engineering and Engineering Management Program in Science, Technology and Society, Stanford University.

Francis, A. J., Mansell, D. S. 1988, *Appropriate Engineering Technology for Developing Countries*, Research Publications Pty. Ltd., Blackburn, Victoria.

Frank, A. G. 1978, *Dependent Accumulation and Underdevelopment*, Macmillan, London.

Geertz, C. 1963, *Agricultural Involution: The Process of Ecological Change in Indonesia*, University of California Press, Berkeley.

Glaeser, B. 1987, *The Green Revolution Revisited: Critique and Alternatives*, Allen & Unwin, Sydney.

Grant, P. 1980, The international transfer of technology for secondary processing projects in Australia, *Australian Mining and Petroleum Law Journal*, no. 2, pp. 234-263, summarized in *Engineers Australia*, August 22, 1980, pp. 17-18.

Humana, C. 1986, *The World Guide to Human Rights*, Facts on File, New York, quoted in UNDP 1991, p. 19.

Illich, I. 1974, *Energy and Equity*, Calder and Boyars, London.

Inkster, I. 1991, *Science and Technology in History: An Approach to Industrial Development*, Macmillan, London.

Johnston, S. F. 1976, Intermediate Technology: Appropriate Design for Developing Countries, *Search*, 7, pp. 27-33.

Kirpich, P. Z. 1987, Developing Countries: High Tech or Innovative Management? *Journal of Professional Issues in Engineering*, 113, (2), American Society of Civil Engineers, pp. 150-166.

Kline, S. J. 1995, *Conceptual Foundations for Multidisciplinary Thinking*, Stanford University Press, Stanford, CA.

McGinn, R. E. 1991, *Science, Technology, and Society*, Prentice-Hall, New Jersey.

McLuhan, M. Powers, B. R. 1989, *The Global Village: Transformations in World Life and Media in the 21st Century*, Oxford University Press, New York.

Morgan, P. R., Morgan, A. J. 1986, The Engineer in a Very Small Developing Country, *ASCE, Journal of Professional Issues in Engineering*, 112, (1), pp. 21-33.

Morris-Suzuki, T. 1994, *The Technological Transformation of Japan*, Cambridge University Press, NY.

Rollason, R. 1991, Preface, In: Laurie Z., Ed., *Doing Good: The Australian NGO Community*, Allen & Unwin, North Sydney.

Schumacher, E. F. 1973, *Small is Beautiful: A Study of Economics as if People Mattered*, Blond and Briggs, London.

Todaro, M. P. 1977, *Economics for a Developing World*, Longman, London (includes quote from Goulet, D. 1971, *The Cruel Choice: A New Concept in the Theory of Development*, Athenaeum, NY).

Transparency International (TI) 1997, WorldWide Web site at *www.transparency.de.*

United Nations Development Programme (UNDP) 1991, 1993, 1997, 1998, *Human, Development Report 1991, 1993, 1997, 1998,* Oxford University Press, Oxford.

Vernon, R. 1989, Technological Development: The Historical Experience, *Economic Development Institute of The World Bank, Seminar Paper No. 39,* The World Bank, Washington, D.C.

Wheelwright, E. L., Stillwell, F. J. B., Eds., 1976, *Readings in Political Economy,* Australia and New Zealand Book Co., Sydney, especially Wheelwright, E. L., Under-development or revolution? The Baran-Frank thesis.

Willoughby, K. W. 1990, *Technology Choice: A Critique of the Appropriate Technology Movement,* Intermediate Technology Publications, London.

GENERAL MOTORS JOINT VENTURE IN RURAL CHINA

This case study describes some of the challenges in operating a manufacturing plant in China. Differences in culture, infrastructure, and language have to be handled carefully if an overseas manufacturing operation is to succeed. It also describes some of the problems encountered by a joint venture, including technology transfer from the parent plant, difficulties with engineering and manufacturing, and the lack of suitably educated and trained employees.

Over the past few years General Motors (GM) has been developing automotive components plants in China.

Delphi Packard Electric Systems (Delphi-P) is one of six divisions of Delphi Automotive Systems, a separate business sector of the GM Corporation. Delphi-P employs more than 90,000 people in thirty-one countries, including 4,000 people in China. With headquarters in Ohio, it supplies full system automobile wire harnesses and components to GM as well as other car manufacturers, and currently has 24 percent of all automotive wiring and power signal distribution in the world. Delphi-P was the first GM division to launch assembly operations in China, with three joint ventures and one wholly-owned plant. Metal and plastic wire harness components are manufactured in the Delphi-P Baicheng and Hebi plants; whereas final wire harness assembly is done at Shanghai and Guangzhou plants. Of the four plants, Hebi, in the heart of China, has had the most obstacles to overcome.

Delphi Packard Electric Hebi

In November 1994, a joint venture was set up with a local firm, Hebi Automobile Electric Equipment (HAEE) Factory, a small state-owned manufacturing plant, to localize automotive component production in China. Delphi-P owns 60 percent and HAEE owns 40 percent. The Chinese government played a large role in helping to establish the joint venture. Delphi-P was strongly urged to place the joint venture in Hebi, in hopes that the presence of a multinational company would be able to help develop the local area. Delphi-P brought management expertise, employee training, and millions of U.S. dollars to refurbish and retool the outmoded Hebi facility, which by late 1997 employed over 800 people.

The Hebi plant manager has been with the joint-venture plant since it opened. He is a German who relocated to Hebi after several years of working at Delphi-P's primary manufacturing and engineering design facility in Wuppertal, Germany. In just a few years as plant manager of the Hebi facility, he has been faced with countless challenges unique to Chinese culture and infrastructure. An added challenge was communicating in English, the second language for both himself and the Chinese translator! Running the Hebi plant is a constant struggle and progress is slow.

Hebi is a rural area located in the Henan Province in central China. It seems almost untouched by modern influences and conveniences. When Delphi-P staff first visited Hebi in the late 1980s, few locals had ever seen a foreigner. Even now, only a handful of foreigners live there. The quality of living is dramatically different from the West. There are public wash-houses, a single telephone for each block, and open-air markets to buy food for the day because most people do not have refrigerators. Coal is the primary source of fuel for heating and cooking; power and water stoppages are common, sometimes occurring several time per week. Power can be out for up to an hour. Clearly, the underdeveloped infrastructure presents problems in basic utilities for the plant. The Hebi facility has generally been able to get around the power problem by tying into two different power grids and switching between them when there is a blackout. However, switching takes a few minutes, which is a problem for Hebi's injection molding operation, which depends on constant operating temperatures for manufacturing consistent parts.

Transport is also a problem. The nearest airport is four hours away, and road conditions are poor. Incoming material shipments are frequently delayed, even lost for days. Sending anything within a developing country like China is somewhat of a gamble.

Central China, and Hebi in particular, is an area with strong old-style nationalist and communist attitudes. Delphi-P encountered some initial resistance from the Chinese employees. Many of the locals disliked the idea of a large foreign firm coming into their town and workplace and telling them what to do. The plant manager believes that resistance will diminish as Delphi-P provides more job opportunities, more overall wealth, and general improvements to the community.

Technology Gap

Delphi-P brought new technologies, sophisticated equipment, and unusual new practices to the joint venture. In a town where only the elite own refrigerators, high-technology computers and fax machines were a marvel. Even compared with the Delphi-P plants in the larger cities of China, such as Shanghai and Guangzhou, there was an enormous technology gap and lack of experience in the local personnel.

An intensive training program was necessary to get the Hebi personnel up to speed on the new technology. As a technology fol-lower, the plant was able to buy rather than develop state-of-the-art technology. However, there was then a need for the basic understanding and fundamentals of the technologies to be learned and internalized. The Hebi plant is equipped with the latest high-powered PCs, operating Windows 95, and several sophisticated analytical tools, but next to each computer in Hebi's finance department there is still an abacus.

Technology Transfer

Successful adoption of new technologies from the North American and German engineering groups was a major challenge at Hebi. Differences in time zones made it difficult to contact personnel in other countries, so most communication was by fax, voice message, and electronic mail, none of them particularly reliable means of communication from China. The language barrier caused significant confusion and delay: translation of technical documents was difficult and time consuming; a standard Hebi staff meeting took three times as long as its counterpart in the United States, because everything had to be translated and clarified several times; most staff meetings ended in utter chaos. A seemingly small task often escalated into a major ordeal. Amazingly, things are eventually accomplished, and progress is made ... slowly.

Most of the component designs come from Delphi-P's North American and German design centers. The primary functions of the Hebi engineers are to insure efficient operation of the manufacturing equipment, update component and tool designs, and help sort out which products would suit local needs. Manufacturing engineers oversee the process problems in the plant and design engineers maintain the component designs and transfer part drawings into AutoCAD r13 formats. Manufacturing

(continued on next page)

departments are divided by process groups: injection molding, metal stamping, plating, slitting, and forging.

Most of the local engineers are graduates of local colleges and their level of engineering skill is insufficient for the new high-technology manufacturing processes. Ongoing training is provided and many Chinese engineers are sent to the German and U.S. facilities for several months of training. Engineering experts from other plants are also brought in to help with specific problems and training requirements. Hebi engineers typically take on more tasks than their counterparts in U.S. plants. The lack of specialized engineers (i.e., metal and plastics material engineers) make it necessary for engineers to take on responsibilities outside their particular area.

Manufacturing problems are encountered in Hebi that would not be found in a Western facility. Problems with communication, availability of raw materials, and manufacturing quality are only a few of the everyday challenges. Plastic feedstock has to be imported from the United States because local material is too variable—another example of the inadequacy of the local infrastructure. Equipment is of very varied quality and age, and mostly poorly laid out. Raw material properties vary more than in the United States or Germany, affecting press operation and increasing the amount of rework. Not surprisingly, a major challenge is changing old habits and ways of doing things, and developing a consciousness of the importance of quality. For the most part, the plant manager recognizes areas that need improvement but can only attack one major problem at a time.

Conclusion

It is clearly possible to run a successful manufacturing operation in China. Delphi-P's Shanghai and Guangzhou plants have not had as many problems as the Hebi plant, and are currently thriving. As final assembly plants, they have more resources provided by the parent company. The Guangzhou facility, established in 1996, is wholly-owned and was built from the ground up with the latest technology and carefully selected personnel. The Shanghai plant has been ISO-9002 certified and has won several excellence awards. It has a larger engineering staff than Hebi, as well as more management and engineering support personnel.

The Hebi plant has had further to go, and less monetary and technical support than some of its sister plants, so the plant manager and his staff have to be more resourceful and imaginative to make Hebi a success. The strategy in Hebi has been to:

◆ improve plant facilities;
◆ invest in equipment; and
◆ invest heavily in employee training.

The goal is to make the plant competitive in the global market and eventually self-sufficient, run by locals capable of maintaining and improving the plant.

Source

Based on work by Ruth Kim, Stanford University, 1997.

References

Wood, A. Delphi Packard leads GM's drive to China. *Business Journal of the Five-County Region*, November, 1997.
Delphi Automotive Group Website: http://www. delphiauto.com/electric.

9

Energy and Sustainability

The principles and ethics of human law and conventions must not run counter to those of thermodynamics.

— Professor Sir Frederick Soddy (1877–1956), British radiochemist, discoverer of isotopes

Energy is the capacity to do work. An important aspect of engineering has always been the harnessing of the energy sources in nature to the service of humanity. Energy availability has been essential to continuing economic growth and the process of wealth creation. In this chapter we review energy sources, and explore patterns of energy use. We argue that quantifying all the costs, including environmental costs that are at present largely ignored, will encourage more efficient energy use. We take electricity generation as a specific technology that requires greater attention by policy makers. Low prices for off-peak electricity have encouraged use of this high-quality energy for low-quality applications such as space and water heating. We suggest the need for a balance between regulation and reliance on market mechanisms and outline roles for government and the market in energy planning and in energy policy development.

In responding to mounting concern about our natural resources and the environment, engineers are looking for fresh insights from ecology. An ecological perspective takes a holistic approach and challenges the basic assumptions of some other disciplines presented in this book, particularly economics. It leads to what is becoming an increasingly

453

central concern, the issue of long-term sustainability. Around the world, engineers are being challenged to take a lead in the move towards more sustainable practices.

Sustainability issues will be taken as seriously in the boardrooms of major corporations as in environmental organizations. Much of the present focus is on reducing the environmental impacts and increasing the efficiency of current technologies to meet legislation. There is a need to move beyond these changes to develop and apply technologies that will be sustainable in the long term.

Natural Resources

The comfortable medieval view of the Earth as the center of the cosmos, with boundless natural resources put there for humans to exploit for their own benefit, was very appealing. What courage it must have taken to challenge this view and accept that the Earth is just a relatively small rocky lump, orbiting a minor star in a by-way of the universe! Unfortunately, we still appear to need frequent reminders that our planet is a fragile and distinctly finite environment.

For our material and most of our energy needs, for our very survival, we rely totally on a layer of air a few tens of kilometers thick above the surface, on oceans a few kilometers deep, and on a layer of soil whose thickness is commonly measurable in centimeters. These thin layers, commonly referred to as the biosphere, are being subjected to a range of increasingly powerful human stresses. This is what ecologically sustainable development is ultimately about: ensuring the survival of our species and others while meeting human needs and maintaining the systems that enable life on this planet to continue.

Trade and industry policies everywhere have assumed that economic growth will continue indefinitely. Such growth would be based to a significant extent on increasing consumption of material resources, including energy. We have been reluctant to question whether these increases can continue.

It is important to look at how far our present approach to technology is part of the problem, and to explore how technology can contribute to generating solutions. Broader issues about longer-term sustainability are outlined in the Key Concepts and Issues box. The picture is not all "doom and gloom," and some positive options are explored in this chapter. Some of the immediate, practical questions we attempt to answer include:

Sustainability

Sustainability is increasingly being accepted as a central issue. Sustainability involves much more than greening our practices and supporting the environment. Communities need to be empowered to become sustainable: this may involve the reduction of nonrenewable resource consumption or the limitation of population growth; it should involve education and result in an improvement to the quality of life. Sustainability implies practical recognition of the need for global and intergenerational equity; equity must be the underpinning principle for global development. Sustainability is essentially an ethical concept that focuses on the long-term survival of communities.

In practice, although engineers need to be aware that sustainability involves social, political, economic, and environmental aspects, their professional contribution will be made in technical areas, including the conservation and more effective use of resources, especially energy.

There are many definitions of "ecologically sustainable development" (ESD). One widely accepted definition is that of the World Commission on Environment and Development: "development that meets the needs of the present without compromising the ability of future generations to meet their own needs." We explore various approaches in this chapter and consider the changes in resource use they might require.

Questions we pose include:

- ♦ To what extent does consumption of nonrenewable resources, particularly fossil fuels, cause environmental damage?
- ♦ How extensive might such damage be? Would it be irreversible?
- ♦ Can scientific studies give us early warning of these problems? If they do, will policy makers respond promptly to the warnings? What can be done?
- ♦ What levels of resource use will allow us to pass on at least as rich a legacy to our children and grandchildren as the one we ourselves inherited?

- ♦ How are the uses of natural resources, energy generation, and waste products connected now? How might we wish to change their connections in the future?
- ♦ What sorts of limits are there to resources?
- ♦ What constraints should and can be placed on resource exploitation?

- What sorts of relationships exist between resource owners and users?
- What happens to the wastes generated in the processes of exploitation of resources?
- Do those who benefit from resource exploitation suffer the effects of the wastes produced?
- What sorts of threats are posed to humans and to other life forms on Earth?
- How can the conflicts associated with these issues be addressed at local, national, and global levels?
- How do we secure equity between countries and between generations?

Some powerful trends and basic assumptions are challenged in this process. Technological development in Western economies has historically emphasized increasing labor productivity. A more recent emphasis has been on miniaturization, to which many portable consumer items owe their existence. Along with this trend has been a move away from user-serviceable items, towards throw-away products with a relatively short life that is often determined by fashion, or by technological rather than physical obsolescence.

Long-term solutions to the problems of waste have been slow in coming. There is clearly a role for engineers in reducing the resource costs of commodity production. (One promising direction is the provision in some new personal computers for upgrading performance by unplugging the existing microprocessor and plugging in a more powerful model.)

There is an increasing tension in our society between emphasis on quality of life and standard of living. The latter, like economic growth, is a convenient, quantifiable concept; however, as discussed in Chapter 8, it only takes account of items passing through the market. It ignores aesthetic and other less readily quantifiable costs and benefits, such as being able safely to breathe the air and drink the water. As we noted in Chapter 8, neither of these apparently obvious "rights" can be taken for granted. Even in rich countries such as the United States, contamination of groundwater can be a serious problem. The concept of quality of life seeks to take account of these less tangible, less readily quantifiable benefits. It values satisfaction of human desires rather than simple economic appetites.

Our increasing awareness of the limits to economic growth has focused attention on ways of improving the quality of life, without increasing resource consumption. Barbara Ward's 1966 model of "spaceship Earth" has been reinforced by images of the Earth from space. The Club of Rome report *The Limits to Growth* (Meadows et al. 1972) raised very sharply the unsustainability of existing approaches to the Earth's resources and environment. The methodology and the detailed conclusions of this pioneering study were very effectively criticized (e.g., Freeman and Jahoda 1978). Even so, the central message, that we must preserve and conserve our planet, is increasingly accepted, and we are starting to clarify the scale and types of problems to be addressed. Engineers need to be involved in the associated debates, partly to ensure that they are well-founded technically, and partly because their skills and competencies are essential to generating genuinely sustainable solutions.

Energy Resources and Use

Energy is the most basic and essential of all resources. All the energy we use on Earth comes from fission or fusion of atomic nuclei, or from energy stored in the Earth. The latter is predominantly fossil fuel, but includes geothermal and tidal energy. The problem with both fission and fusion is that they have dangerous radioactivity as a side effect. In any case, to date we have not been able to make fusion reactions on Earth self-sustaining. Fortunately we have a working fusion reactor, the Sun, at a distance of around 146 million kilometers, which minimizes radioactivity problems. Solar radiation is our paramount energy source.

■ Energy Production

A curious usage in the energy industry is the term "production," used for winning fossil fuels from the Earth and converting them into a suitable final form of energy supply. If one takes the availability of fuels as given and concentrates on the enormous task of "producing" the energy from them as efficiently as possible, it may not be unreasonable. Given the knowledge, experience, courage, and risk often associated with the processes involved, the term is certainly understandable. The usage dates from a time when resources were considered to be essentially unlimited, and ignores the fact that the miners and drillers are not really producing these fuels, but extracting them.

The unlimited availability of cheap fuels can no longer be taken as given. Use of the term "production" tends to obscure the profound qualitative difference between nonrenewable extraction of fossil fuels and the use of renewable energy sources. Pricing of resources that took into account the cost of synthesizing them when economically viable resources have been depleted would see quite different approaches to energy policy. Even where official data includes figures for renewables, it usually understates the contribution of solar energy because it does not include nonmarket uses of solar energy such as passive building lighting, heating, crop drying, and outdoor laundry drying.

In 1950 about two-thirds of the world's electricity came from thermal (steam-generating) sources and one-third from hydroelectric sources. By 1995 thermal sources still produced 61 percent of the power, but hydropower had declined to just under 20 percent and nuclear power accounted for 17 percent. Geothermal and renewable energy were about 2 percent of the total. The growth in nuclear power slowed in some countries, notably the United States, in response to cost overruns and concerns about safety. Nuclear plants generated 19 percent of U.S. electricity in 1995; in France, the world leader, the figure was 77 percent.

According to the Department of Energy, in 1996, total U.S. energy production was 76.6 exajoules (EJ) (72.6 quadrillion Btu). Total consumption was 100 EJ (93.81 quadrillion Btu), 29 percent more than the U.S.-extracted production. The United States accounted for 24 percent of the world's energy production.

Fossil fuels provide 85 percent of the energy consumed in the United States. Of the fossil fuels, coal provided 31 percent of the U.S. energy production, natural gas 27 percent, and crude oil 20 percent. Table 9.1 summarizes U.S. fossil fuel usage and reserves. Nuclear energy provided 10 percent of the United States total, and renewable energy sources provided 8 percent.

Of the total energy consumed in the United States in 1996, 33.9 EJ (32.13 quadrillion Btu), or 34 percent of the total, was used to produce electricity. In order of significance, the sources used for the production of electricity are coal, natural gas, crude oil, nuclear, hydroelectric, and geothermal. Fossil fuels produced 67 percent (58 percent from coal, 9 percent from natural gas, and 2 percent from oil). Nuclear energy provided 22 percent of electrical energy and hydroelectric sources 11 percent. However, much of this electricity is used for heating and cooling at temperatures close to ambient. These are not effective uses of a premium form of energy. Good matching of energy sources with end-use needs is important for efficient use of limited resources. It minimizes the output of carbon dioxide and waste heat, both potential threats to the stability of the biosphere.

Coal, natural gas, oil, and nuclear electricity are all associated with significant pollutants or hazardous byproducts. These resources are also nonrenewable, meaning that the world will, sooner or later, completely

TABLE 9.1

U.S. Fossil Fuel Energy Usage and Reserves, 1996*			
	Coal	Crude Oil and Petroleum products	Natural Gas
Domestic use	19.9	33.9	21.4
Exports (imports)	2.2	(19.1)	(2.7)
Annual extraction	21.4	13.0	18.5
Economic reserves	7,034	142.0	182.0
Years of reserves	353	4.0	8.0

* Given in exajoules (EJ).
The basic units used here are joules (J), liters (L), tons (t), and watt-hours (Wh), together with their multiples. The standard SI (system internationale) multiplier prefixes (not all used here) are:
kilo (k) = 10^3 (thousand); mega (M) = 10^6 (million); giga (G) = 10^9 (U.S. billion); tera (T) = 10^{12} (U.S. trillion); peta (P) = 10^{15} (U.S. quadrillion); exa (E) = 10^{18}.
A handy guide for energy equivalencies is that a gigajoule (GJ) is 278 kilowatt-hours, or 0.948 million Btu, or the energy in thirty liters of gasoline.
The EJ is a convenient unit in which to express national energy use. One EJ is a thousand million GJ or 1.0551 quadrillion Btu. One kilowatt-hour = 3,600 kJ = 3.6 MJ—enough to run a computer for three hours or a 100-watt light bulb for ten hours. A barrel (of oil) is 158.987 L, with a chemical energy content between 5.88 and 6.15 GJ (5.57 to 5.83 million Btu). (The traditional U.S. engineering energy unit is the Btu, equal to 1055.1 joules.)
Source: Energy Information Administration.

consume the energy source, provided that is allowed to happen. Nuclear power stations can be used as "breeder" reactors to produce further fissionable material, but technical and cost problems (and concern about nuclear weapons proliferation) make this an unattractive option at present. Increasing use of approaches such as life cycle analysis (LCA), which seeks to identify and take account of all the costs involved in conversion, including environmental impacts, may accelerate the change to more renewable energy sources. It can also be politically sensitive, because increasing the price of staples hurts the poor much more sharply than middle or upper classes.

To make sense of the discussion of energy issues, it is convenient to distinguish between primary energy and end-use energy (see Key Concepts and Issues box).

■ Laws of Thermodynamics

Energy is neither created nor destroyed, but can be transformed from one form to another. The First Law of Thermodynamics is concerned with the quantity of that energy and its conservation. Rudolf Clausius (1822–1888) formally stated the First Law in 1850, based on insights by

Primary and End-Use Energy

Primary energy includes:

- ◆ chemical energy, stored in the fossil fuels of the coal we dig and the oil or gas we drill for;
- ◆ nuclear energy, in the radionucleides we mine and process to use in nuclear power plants;
- ◆ geothermal energy, stored in hot materials in the Earth's crust;
- ◆ tidal energy, due to the Earth's rotation;
- ◆ solar energy, in sunlight as it falls on the earth, or is transformed into:
  1. kinetic energy of wind or waves;
  2. potential energy, due to its elevation, of water for generation of hydroelectricity; or
  3. chemical energy of plant biomass, such as fuel wood, or bagasse (from sugarcane).

Primary energy sources may be convenient for handling and stockpiling, but are not necessarily convenient for use in industry, commerce, or the home. It may be necessary or convenient to convert some types of primary energy to other forms, in which they are finally used as end-use energy. End-use energy includes the forms of energy delivered to end users:

- ◆ chemical energy stored in processed fossil fuels that are burnt by the user in transport or other machinery: gasoline, distillate, aviation and power kerosene, liquefied petroleum gas (LPG), compressed or liquefied natural gas;
- ◆ chemical energy stored in processed fossil fuels that are burnt by the user for space or water heating or cooking: natural gas, heating oil, kerosene, coke;
- ◆ chemical energy of plant biomass, such as fuel wood, burnt by the user;
- ◆ solar energy, in sunlight as it falls on the earth and is used in solar water heating; or
- ◆ electricity delivered to an end user.

End-use energy is generally used for heating and cooling ventilating air; heating water for washing and cooking; moving people and goods up and down (hoists, elevators, escalators); moving vehicles, including ships and aircraft; heat treatment processes in industries such as metals, glass, ceramics, and cement; operating machinery, appliances, computers, and so on. Whenever energy undergoes a change in form, losses occur. Primary energy becomes end-use energy plus conversion losses. Figure 9.2 shows the U.S. usage pattern.

J. P. Joule (1818–1889) and Lord Kelvin (1824–1907). It established the equivalence between the various forms of energy.

Clausius also observed that it was impossible to transfer heat energy from a low temperature source to a sink at higher temperature without doing work. Lord Kelvin independently concluded that it was impossible to take heat from a high temperature source and use all that heat to produce an equivalent amount of work. These are alternative statements of the Second Law of Thermodynamics (which is attributed to Clausius). The Second Law of Thermodynamics is about the quality of energy and its degradation during processes. It is concerned with the generation of *entropy* (the lost opportunities to do work) and has become a powerful tool in the analysis and optimization of heat engines, processes, and complex systems.

If thermal energy is used to produce an output in the form of work, then that work output can be determined by subtracting any heat wasted or rejected from the heat input. In any real process heat is wasted, so it is necessary to have a definition of efficiency that quantifies losses in the conversion process. In accordance with the Second Law, the thermal efficiency of energy conversion in a process or cycle is given by dividing the work output by the heat input. An alternative Second Law expression for the thermal efficiency is found, using the First Law, by subtracting the heat wasted from the heat input and dividing the result by the heat input.

A further useful tool for analyzing these lost opportunities to do work is *exergy* or available energy. The exergy of a system is the maximum amount of useful work that could be obtained from the system at a given state in a specified environment. The process is undertaken and work extracted until the output is maximized; the system has then reached a state of thermodynamic equilibrium with its environment. This is known as the dead state. The exergy destroyed is proportional to the entropy generated and is a measure of the irreversibility of a process. This results in a more precise definition of Second-Law efficiency as the exergy recovered divided by the exergy supplied.

The Laws of Thermodynamics are fundamental to an understanding of the environment. The First Law is about conservation and tells us that energy is already conserved. But the useful work potential of that energy, the exergy, is not conserved. The problem of energy sustainability therefore hinges not on energy conservation but on exergy conservation. Real processes are irreversible and therefore waste exergy. The Second Law tells us about the quality of states in which energy is found and the processes involved in its transformation. It leads us to match the energy source as closely as possible to the end use. It tells us that once the exergy is destroyed it can never be recovered and gives us the means to quantify that loss, the irreversibility, taking place in the process. The Laws of Thermodynamics lead us to focus not on energy *per se* but on the useful work potential of that energy, its exergy.

■ Energy Quality

An important consideration in the choice of the optimum form of energy for a particular application is energy quality. There are three main criteria for assessing the energy quality of different types of energy sources:

1. The ease of extracting the energy. The Second Law of Thermodynamics implies that, for transforming thermal energy into useful work, it is most efficient to extract the energy from a small quantity of material at a high temperature. High-temperature heat is therefore high-quality heat.

2. The ease of transporting and storing the energy. Chemical energy stored in hydrocarbons, especially petroleum and its derivatives, is relatively straightforward to transport and store. Problems do, of course, arise, regularly resulting in leaks and spillages (although usually much less dramatic than the Exxon Valdez) and occasionally in serious fires (particularly associated with lighter hydrocarbons such as natural gas and LPG).

3. The ease of transforming the energy from one form to another. Electricity is attractive because it is extremely easy to transform to mechanical or other forms such as light, or to quite high-temperature heat. Electricity has become a preferred end-use form despite the high percentage of energy loss involved in converting most types of primary energy to electricity, and the cost, difficulty, and losses involved in storing it.

Transforming chemical energy into electricity, while keeping the losses as low as practicable, requires sophisticated plant and equipment. The Second Law of Thermodynamics gives the theoretical upper limit on the efficiency of a thermal (heat) energy transformation cycle, the Carnot efficiency. This limiting efficiency can be found by dividing the difference between the hottest and coldest temperatures in the cycle by the hottest absolute temperature. For modern coal-fired electricity generation plant with steam temperatures reaching 540°C and cooling water around 40°C, the theoretical maximum efficiency is just over 60 percent. In practice, losses bring this down to around 40 percent. Distribution losses reduce the overall coal to end-use electricity conversion efficiency by about another 5 percent. Overall U.S. electricity generation efficiency is around

34 percent. Nuclear power plants typically operate at lower maximum steam temperatures than coal-fired plants, so they are even less efficient. Many new installations are of combined cycle plant, described in the box Matching Electricity Supply and Demand. The latest of these have thermal efficiencies approaching 60 percent and lend themselves well to operation with natural gas.

The historical improvement of steam power and related energy conversion technologies, from reciprocating engines through steam turbines to the current exploitation of the combined benefits of steam and gas turbines, is represented by the power-plant efficiencies shown in Figure 9.1. Exponential growth and the sigmoid curve were discussed in Chapter 6

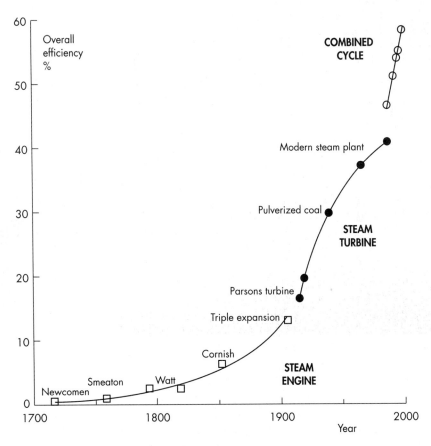

Figure 9.1 Historical improvement of steam power and related energy conversion technologies. *Source:* Partly based on McVeigh 1984.

and this is a good example of those patterns. Electricity generation authorities were locked into a pure steam mindset; this resulted in a leveling out of station efficiency at around 40 percent for thirty years, giving a sigmoidal efficiency-time curve. We argued in Chapter 6 that returning to exponential growth required a qualitative breakthrough. In this case, the combined cycle was the breakthrough that took power-station efficiencies back onto an exponential improvement curve. In addition to big cost savings this change is giving a major reduction in greenhouse gas emissions per megawatt-hour of electricity.

In processing plants where heat as well as electricity is required, the overall energy efficiency of the system can be increased by burning fuel (increasingly, natural gas) to generate steam at higher temperature and pressure than is required by the process. The steam first expands through a turbine to generate electricity. The cooler and lower pressure steam is then used to supply heat for use in industrial processes or space heating. This approach, described fully by Horlock (1997), is called cogeneration (of heat and electricity).

Cogeneration uses otherwise wasted heat productively. This makes cogeneration facilities much more efficient than conventional power plants, which simply dump their waste heat into the atmosphere or a nearby body of water. Although a typical utility power plant converts about 34 percent of its fuel's energy into useful electricity, a cogeneration plant can utilize 80 percent or more of its heat input productively. About 10 percent of electricity generated in Europe is from cogeneration plants.

Some forms of high-quality energy can be converted directly into electricity, without the use of a heat transformation cycle. One example of this approach is the fuel cell technology discussed in Case Study 9.3 at the end of this chapter.

Consideration of energy quality and matching of energy sources with end uses highlights the inappropriateness of using electricity for low-temperature space and water heating. An exception to this is the use of electricity to operate a heat pump. Later in the chapter we will show that in most locations low-temperature heating, up to 80°C, is already a straightforward and economically attractive application for solar energy.

U.S. Energy Use—Energy Information Administration

To appreciate the patterns and potential problems of U.S. energy use, we need to know the available resources, the energy use pattern, and the problems associated with these patterns. Table 9.1 shows known and

Matching Electricity Supply and Demand

Good matching of energy supply with demand is not easy to achieve, because demand for end-use energy varies considerably with time. Much energy usage is seasonal, a function of heating or cooling loads in homes, offices, factories, and public buildings. The pattern and degree of variability reflect geographical location and contain a random element relating to the weather. There are also variations throughout the day, corresponding to the hourly usage patterns of people and industries. These temporal variations are especially important for electricity usage. As a first approximation, the required electricity supply can be divided into a relatively constant base load and a highly variable peak load.

In general, electricity does not lend itself to storage. Chemical storage in the form of batteries is too expensive for large-scale use. One means of storing electrical energy is pumped storage. Surplus base-load electricity from thermal power stations can be used to pump water from lower reservoirs to higher ones. When electricity demand again exceeds supply this water is allowed to run down through turbines, converting most of its potential energy back to electricity. There are losses at each step, resulting in an overall efficiency of only around 70 percent, and such schemes are net overall consumers of electrical energy. However, they do make effective use of cheap surplus power from base-load plants.

For the most part, it is necessary to match electricity supply to demand by appropriate selection and running of generation plant. Thermal power has the major difficulty that the slow thermal response of the components makes it difficult to turn on and off quickly. Because hydropower facilities do not have significant thermal lags, they can cope well with load fluctuations and are well suited to peak load operation.

Gas turbines are particularly flexible in operation. They can run on a variety of fuels, including oil and natural gas, and the fuel can actually be changed during running. They can be started and shut down within minutes, and can be installed and commissioned much more rapidly than steam power stations. Open cycle gas turbines (which most are) do not need a supply of cold water for cooling or condensation purposes. Unfortunately gas turbines have not been as efficient as steam turbines and this has tended to limit their application to peak load operation and other short duration functions.

(continued on next page)

Matching Electricity Supply and Demand *(continued)*

The gas turbine can be used in conjunction with a steam turbine to produce a combined cycle that results in a very efficient unit. The gas turbine can sustain high combustion temperatures, and the "waste" (exhaust) heat from the gas turbine is then used to raise steam for a steam turbine cycle. Combined cycle thermal efficiency in new units approaches 60 percent. The widespread availability of natural gas enhances the attractiveness of gas turbines and the combined cycle approach is naturally popular for new installations.

Large coal-fired or nuclear thermal power stations are reasonably efficient, but relatively inflexible in operation. They cannot be started or shut down quickly or frequently. They are thus best suited to base-load operation. However, because the new combined cycle plant is significantly more efficient, in countries such as Britain it is being used for base-load operation with coal plant being brought in to meet the peak demand. To some extent this is a function of how the market is organized, but there are clearly trade-offs between generation efficiency, mechanical reliability, and life-cycle costs.

It is important to recognize the traditional organizational character of the electricity generation industry. In most countries it has had a bias towards large-scale, centralized responses. Power generation by burning coal and from nuclear reactions requires large complex plant but it is also necessary to locate coal-fired stations in coalfields and nuclear stations remotely from centers of population. The economic importance of improved thermal efficiency has driven the tendency for power stations to be designed with large turbogenerator units. This follows from the beneficial effect on plant efficiency of increased Reynolds number (the product of flow velocity and unit size, normalized by the kinematic viscosity). In turbulent flows, the losses in turbine and associated boiler plant decrease monotonically as unit size is increased. A doubling in blade chord gives approximately 13 percent reduction in specific fuel consumption if the fluid dynamic design is performed correctly (Gostelow 1984). In the last two decades, unit sizes have leveled out at around 660 megawatts (MW) and there has been renewed interest in combined cycle plants that are smaller but make maximum use of the best features of gas turbines and steam turbines.

Electricity suppliers are obviously aware of the need for a range of strategies (and hardware) to meet variable demand patterns. They can influence demand patterns by offering special tariffs for using electricity at off-peak times, but this only achieves a modest leveling of the variable daily load.

expected U.S. fossil fuel energy reserves in terms of years of use at present levels. In 1997 the Energy Information Administration announced that, based on an annual economic growth rate of 2.1 percent (rather than the 1.9 percent previously assumed), the emission of carbon dioxide and other greenhouse gases from energy use would grow faster than was previously expected. This will make it more difficult for the United States to implement President Clinton's 1997 proposal to cap emissions of these gases over the next ten to fifteen years. Carbon emissions in 1990 were 1,336 million metric tons, and the new estimate is that this would correspond to a 2010 level of 1,803 million tons. To trim this back to 1990 figures would require a reduction in emissions of close to 35 percent.

U.S. primary and end-use energies are shown in Figure 9.2. The distinction between primary and end-use energy highlights a problem with the handling of energy data that systematically undervalues the contribution from some renewable energy sources. Hydroelectricity comes into both primary and end-use tables as essentially the same quantity of electrical energy. To obtain an equivalent value for primary energy as electricity generated from fossil fuel one would need to work back from the end-use table, using a multiplier of about three. Working with exergy data would give a clearer picture.

On the basis of known reserves, Table 9.1 shows the unsustainability of present oil usage. Because of the cost of exploration the oil industry tends to project oil reserves only about a decade ahead of its proven

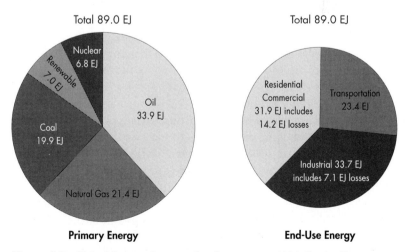

Figure 9.2 United States primary and end-use energy, 1996. *Source:* Data from Energy Information Administration.

capacity, the situation is less critical than the data indicate. Even so, the desirability of moving towards greater use of natural gas because of its greater availability is clear. There will be changes in the fine detail as exploration proceeds and prices change, but the figures cited here give a reasonable indication of the medium-term picture.

On a per capita basis, the consumption of energy in the United States is about twice that of western Europe and Australia and about forty times that of Asia, excluding Japan. U.S. primary energy consumption has almost doubled in the last thirty years. Over one-third of U.S. primary energy comes from oil and gas, and we can expect to start to experience shortages of these within a few decades.

▮ Government Planning

The model shown in Figure 9.3 was developed by the Australian government as a basis for its planning for sustainable development. An important and often neglected point it raises is the issue of externalities. In marketing and other commercial analyses, environmental damage is often treated as an externality, that is, a cost that is not carried by the business, but by society generally. This reflects a major failure of the market approach, in that the environment is then treated as a free good. As a result, it tends to be overused, because most economic decisions are made by the operators of private companies that do not pay for environmental benefits or directly absorb the essentially social costs of environmental degradation. They thus have no direct economic incentive to conserve the natural capital stock (WCED 1990, p. 32). Good energy planning includes developing mechanisms that ensure, firstly, that environmental damage is minimized, and secondly, that any damage that does occur is charged against the organization doing the damage and thus implicitly benefiting from it.

The steps set out in Figure 9.3 seem to be an excellent way to start. They move us towards recognizing real supply and environmental costs and encouraging energy users to increase the efficiency of their energy use. They can be summed up as "no regrets" policy options—changes that will not cost very much, indeed may produce substantial savings, and that we ought in any case to implement.

Diesendorf and Kinrade (1992) note that an insight from marketing is useful here. They suggest expressing wants in terms of energy services. People do not want a kilowatt-hour of electricity, or a megajoule of gas, but energy services such as hot showers or cold drinks. These services can

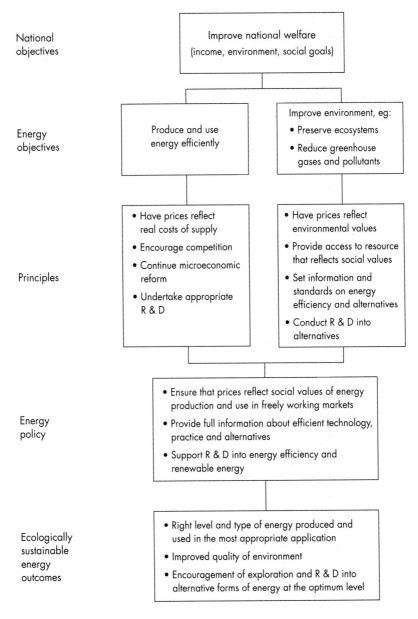

Figure 9.3 Model Implementation Plan for a Sustainable Energy Policy. *Source:* DPIE 1991, Issues in Energy Policy: An Agenda for the 1990s. Commonwealth of Australia copyright reproduced by permission.

be provided in a variety of ways, using a variety of energy sources. Some approaches will be much more energy-efficient than others. Even the simple step of introducing labels showing the relative energy efficiency of appliances encourages and enables purchasers to take energy efficiency into account in their purchasing decision making. It also encourages manufacturers to improve equipment performance.

▮ Strategies for Change

Local decentralized energy systems could reduce energy losses because of their shorter distribution paths. Smaller systems have shorter lead times and so can respond more effectively to unexpected changes in demand patterns. It is clear from recent experience that, when demand ceases to grow at historic rates, large systems are committed, years in advance of need, to building unnecessary plant and equipment.

Thermal power units above 660 megawatts have encountered significant problems in fluid dynamics and materials technology, with resulting maintenance difficulties. These limitations are only likely to be resolved by substantially increased R&D, funded on a global basis. A new generation of larger turbogenerators would be significantly more efficient in its consumption of fossil fuels, resulting in a reduction of greenhouse gas emissions for the same fuel type and electricity output. A strong case can be made on environmental grounds for increased size in generating plant. This is analogous to the case for increased unit size in transport, with public vehicles being favored over private cars, and a new generation of large aircraft over smaller aircraft. These measures are beneficial on grounds of fuel efficiency and reduced pollution, provided they meet a genuine need and do not simply increase consumption. Secure, reliable energy supply is important. Balanced national and international policies must be based on thorough evaluation of all the alternatives, including co-generation and combined cycles.

Moves over recent years towards increased flexibility and market orientation of energy suppliers have not necessarily challenged their basic assumptions and attitudes. Historically, the brief given to most electricity generation authorities was to supply electricity at lowest cost. They commonly focused on supply and ignored the users and the demand side of the market. This approach missed important opportunities to make the energy industry safer and more sustainable. We need to contain electricity demand in a planned and systematic way while still providing equal or better levels of energy services.

One strategy that could be adopted much more widely is least-cost planning. It suggests that the cheapest way to meet energy needs is to open the market up, not just to energy producers, but also to energy conservers. Investment in energy conservation is often more economic than investment in new generation capacity, providing money is available on the same terms. Energy audits of commercial, industrial, and domestic users can pay major dividends in this way. This approach is particularly attractive in periods of high unemployment—retrofitting insulation, high-efficiency lighting, and solar heating to existing premises can be cost-effective, and is labor- rather than capital-intensive. By the end of the century, Pacific Gas and Electric in California is looking to conserve the equivalent of 2500 megawatts from energy efficiency increases.

Apart from lifestyle changes that could have dramatic effects on energy consumption, theoretical studies have indicated the potential for energy efficiency to be improved by a factor of two. Automobile engine efficiencies can be improved, and major savings result from reduction of vehicle mass. We have seen in an earlier chapter the importance of tough clients in improving designs. Experience shows that where governments set clear and reasonable long-term targets, manufacturers can and do meet them. Less overlighting, overheating, and overcooling of buildings and the recuperation of waste heat in industry (possibly for domestic use) could further reduce electricity and oil consumption.

Even in the mid-1970s, cogeneration (of electricity with process steam) provided 12 percent of electricity to industry in West Germany and 4 percent in the United States. This process could easily provide U.S. industry with half its own electricity needs. In buildings the savings on heating and cooling more than pay for any extra costs incurred by the use of passive climate tempering systems and effective insulation. These modifications could save the United States up to one-third of its total national energy bill. This huge potential energy saving is largely being ignored at present.

Probably the most threatening energy problem is that of liquid fuel for transport. U.S. dependence on oil fell from about 45 percent of energy consumption in 1960 to about 38 percent in 1996. This can be attributed to engineering improvements in engines and other key automotive technologies and to the relative costs of coal and natural gas having fallen substantially. There are huge reserves of natural gas and its combustion can be relatively low in greenhouse gas emissions, so we should be pricing to encourage its use in transport as an alternative to oil. In the long term the probable shortages of liquid fuels for transportation remain a problem.

Converting solid or gaseous fuels into liquid fuels involves long and costly fuel-conversion chains; with existing technologies these have high losses. Producing liquid fuel from our largest energy resource, coal, is inefficient, very capital-intensive, requires large supplies of water, and creates large amounts of greenhouse gases. The high capital and environmental costs of these conversion technologies, with their long lead and pay back times, encourage us to look hard for alternatives.

There has long been enthusiasm for solar-energy-generated hydrogen as an ideal, sustainable energy source. It has particular promise as the energy source for fuel cells for direct production of electricity. An interesting recent development in the transport field is the apparent promise of high-efficiency fuel cells. Case Study 9.3 at the end of the chapter explores the prospects for their use for personal transport. One attractive aspect of fuel-cell-driven automobiles is that fuel cells could be used at their destinations to generate electricity for domestic or commercial use.

The use of hydrogen as a fuel for transportation and power generation presumes an extensive supply infrastructure. If hydrogen succeeds its impact will be dramatic. Hydrogen production is likely to be large scale and centralized, probably relying on nuclear power.

Possible supplements and eventual alternatives to oil for transport include power alcohol from sugar or cassava. The typical present cost is two or three times that of petroleum products. Relying on the price mechanism is likely to see us developing alternatives after we have found we desperately need them. A much more economic and secure approach is to plan for a measured and unhurried transition. The three main transitional strategies towards more sustainable solutions to our energy problems, are:

1. lifestyle changes, such as using public transport, walking, cycling, and car pooling;
2. conserving present energy sources; and
3. developing alternative (and preferably sustainable) energy sources.

The competitive market-led approach to energy supply adopted in the United States and more recently in Britain has had significant advantages. In Britain it coincided with the "dash for gas" and the introduction of relatively efficient and environmentally friendly combined cycle units. However the generators are answerable to their shareholders to return a profit and short-term effects dominate. Questions of security of supply and of social and environmental costs tend not to be considered. More particularly all incentives are to promote energy use rather than to conserve it. For example total energy systems, effectively designed and implemented, can make much more efficient use of energy sources, but have been opposed by utilities.

Future energy policy will be influenced as much by environmental factors and by security of supply considerations as by technical developments. In a rapidly changing world it is difficult to argue against a balance of diverse supply sources: to ensure this takes a degree of government planning and political will. An energy policy that is market-led will not accommodate medium- or long-term programs. It would give severe problems for the builders of large dams, such as the 18,000 megawatt Three Gorges project in China, where long lead times require strategic rather than market-led decisions.

Energy supply authorities need a mandate for conservation rather than promotion of energy use. Fiscal incentives to use energy more efficiently will be necessary. Well-targeted and properly costed subsidies for installation of energy-conserving devices would be a positive step.

Fells and Horlock (1995) have advocated energy planning that prescribes a mix of energy sources for electricity supply and a policy that contains a strong energy efficiency element; this should also extend to transportation.

Pricing policies with flatter rate structures, the pricing of energy supplies on the basis of the long-term cost, costing of systems over their whole operating life, and recognizing all the costs involved, including social and environmental ones, would go a long way towards achieving rational policies.

Sources of Renewable Energy

Although renewable energy technologies are ultimately likely to be the most sustainable, the words sustainable and renewable are not synonymous. "Spaceship Earth" is a complex system that works by feedback and thrives on adaptivity, as do its inhabitants including the humans. The planet has always been in a state of dynamic equilibrium, not least due to climatic changes, and this is why the system approach, described earlier, is necessary.

The various energy resources described in this chapter represent some available options. We do have choices and sustainability comes in a rich variety of forms. There is room for all viable options, including fossil fuels, at least for a while. It has been noted that sustainability has many definitions. These naturally tend to reflect the interests and values of the proposers. A strong case can be made, for example, on grounds of global equity, that developing countries have as much right to development as the West and that this involves rapid expansion of electricity supply from coal-burning power stations. This development is happening and will continue and is being encouraged on sustainability grounds (Preston 1995).

Since the industrial revolution we have mostly been living off solar energy stored over geological time in fossilized animal or vegetable deposits (hence fossil fuels). Unfortunately we are using them very much faster than they are being replenished. To use a financial metaphor, we are living off capital rather than income.

As might have been inferred from Table 8.1 (see p. 400), per capita energy consumption around the world is very variable. With only 4.7 percent of the world's population, the United States annually consumes approximately 25 percent of fossil fuels used in the world. The United States now imports about one half of its oil (25 percent of total fossil fuel) at an annual cost of approximately $65 billion (USBC 1992a). Proving oil reserves is costly, and United States oil companies have limited themselves to a ten to fifteen years horizon. The cost of finding and exploiting new domestic reserves continues to rise, and our best guess is that there seem to be grounds for real concern about availability and cost of domestic supplies by about the year 2020. Natural gas reserves are expected to last somewhat longer. In contrast, coal reserves have been projected to last approximately one hundred years, based on current use and available extraction processes. Sooner or later, U.S. residents will need to turn to renewable energy for some of their energy needs.

In 1996, renewable energy sources accounted for 10 percent of the United States total domestic energy production, or 8 percent of the total energy consumed. The main renewable energy sources used in the United States are biomass, hydroelectric, geothermal, solar, and wind. Of the total U.S. energy production, biomass provides 4.0 percent, hydroelectric power 3.8 percent, geothermal power 0.4 percent, solar power 0.1 percent, and wind power 0.06 percent. Only hydroelectric sources produce significant renewable electricity in the United States, supplying 11 percent of the total. Together, all other renewable energy sources (led by geothermal), provide 0.4 percent of the U.S. electricity.

For sustainability, we need to move away from living on our energy capital (fossil fuels) and towards living on our energy income, solar radiation. Several orders of magnitude more solar radiation reaches the Earth than we are currently using as energy. As well as direct solar radiation, this includes wind, wave, water, and biomass energy. Biomass includes firewood. Most of the people of the world cannot afford electricity and have no prospect of doing so. For them, the basic energy requirement is firewood for cooking. Fuel wood and charcoal provide two-thirds of all energy needs in Africa and one-third in Asia, the equivalent of 5.5 million barrels of oil a day, 80 percent of it used for cooking. For these energy users, the real energy crisis is the increasing effort needed to collect wood and its limited supply (Chandler, in Brown 1985, p. 165). Even in relatively prosperous countries such as Thailand, with its rapidly developing manufacturing sector, there are tens of thousands of villages with no likelihood of electricity in the foreseeable future. What prospects do they have? In Case Study 8.1 (see p. 406) we looked at one option, the use of microhydroelectric systems.

Engineers can use their engineering science background to bring particularly helpful and important perspectives to problems in this area. Traditionally, engineers have tended to focus on the hardware aspects of energy issues. This approach is necessary but not sufficient. To be most effective engineers also need to take time to review the assumptions underlying the technology and to clarify precisely the problems it is intended to solve. What practical steps can we take to move from living on our energy capital to living on income? What energy income is available?

▮ Hydroelectric Power

Although biomass and hydropower together account for 93 percent of the total energy contribution from renewable energy sources in the United States, only hydropower provides significant electrical energy. Hydropower uses the energy of flowing water to turn a turbine, which rotates a generator to produce electricity. Most high-output hydropower facilities use large impoundment dams. Hydropower plants played a major role in spurring industrial development in the late nineteenth and early twentieth centuries. By the 1930s, hydropower provided 30 percent of U.S. electrical generating capacity. However, the growth of nonrenewable generation sources slowly eroded the hydropower capacity share to its current 12 percent of total U.S. electrical energy.

Considerable potential still exists for obtaining additional capacity from hydropower resources. The Federal Energy Regulatory Commission (FERC) estimates that existing U.S. hydropower capacity of more than 72,000 megawatts could be nearly doubled through a combination of new site development and equipment upgrade at existing plants. There is also a significant potential for development of small hydropower facilities.

Nevertheless, hydropower development has slowed in recent years because of environmental concerns and more stringent regulatory and operating requirements. The 1986 Electric Consumers Protection Act (ECPA) significantly increased the time and cost of licensing hydroelectric projects.

■ Biomass

The use of energy embodied in organic matter (mainly plants) is the most important fuel for most developing countries, where it provides about 35 percent of total energy. In the United States it provides about half of the renewable energy, or 4 percent of total U.S. energy needs, generating almost 10,000 megawatts of electric energy. Wood fuels provide the bulk of this generation (66 percent), followed by municipal waste (24 percent), agricultural waste (5 percent), and landfill gas (5 percent).

Wood is the leading biomass energy resource for power generation, primarily because of its use as a boiler fuel in the lumber and pulp and paper industries. The U.S. lumber industry meets close to 75 percent of its energy needs through direct wood combustion, whereas the pulp and paper industry has a 55 percent aggregate fuel contribution from wood. In general over 70 percent of biomass power is produced in cogeneration schemes that also provide process heat; these are mostly small-scale independent power plants.

Municipal waste is the second largest source of biomass power in the United States, generating more than 2,000 megawatts of electricity and providing steam for industrial uses. More than 526,000 metric tons (580,000 tons) of municipal waste are generated in the United States each day, with at least three-quarters going to landfills. With landfills nearing capacity, charging higher costs, and adopting stricter regulations, many localities have turned to waste-to-energy (WTE) systems as a disposal alternative—an estimated 15 percent to 20 percent of municipal waste is burned for energy.

Agricultural waste plants are the third largest biomass generators in the United States, producing another 575 megawatts. These plants use such diverse feedstocks as bagasse (sugarcane residue), rice hulls, rice straw, nut shells, crop residues, and prunings from orchards and vineyards. Finally, more than one hundred power plants in thirty-one states burn landfill-generated methane. More than 10 percent of the 6,000 existing landfills in the United States are expected to require methane collection systems to comply with federal regulations on hazardous emissions from landfills. Methane is also a potent greenhouse gas, and this may provide greater impetus for landfill methane projects in the future.

■ Solar Energy

Passive solar energy for lighting and heating is probably the most effective and benign energy source. Passive solar energy systems use little mechanical hardware and require little or no electrical energy to operate;

indeed they are so technically straightforward that they tend to be neglected. For temperate-zone home design, the basic requirement is to orient buildings with a long axis running east-west, using large, equatorial-facing windows and wide eaves. These eaves let sunshine into the house in the winter, when the sun is lower in the sky, but exclude direct sunlight in the summer, when the sun is high. Suitable planting of shade trees can protect the building from late afternoon sun in summer. Also important are good insulation of the roof, and preferably also the walls; where frost conditions do not preclude a concrete ground slab this provides some day-to-night heat storage, also known as thermal mass.

Use of solar energy instead of fuel combustion, particularly for simple applications such as low and medium temperature water heating, reduces costs and environmental impact, provided the solar collectors are well-designed and made. Applications include water and space heating for residential and commercial buildings. Flat plate solar collectors are particularly suited to swimming-pool heating. The optimum arrangement for reliable hot water supply is probably gas-boosted solar heating. Solar energy is already distributed over most of the Earth during daylight hours, with an intensity of more than one kilowatt per square meter on a surface inclined at the best angle for the time of year. Tax credits available during the late 1970s and early 1980s led to thousands of solar heating system installations across the United States. Installations waned after the tax credits expired in the mid-1980s.

In countries as climatically and industrially different as Australia and Britain, about 40 percent of net energy usage is for heat at temperatures below 80°C. This figure includes individual industries such as food processing. Temperatures below 80°C are easily achieved using flat-plate solar heaters with selective collector surfaces. A smaller proportion of heat is required at temperatures between 80°C and 150°C, temperatures readily achieved by more sophisticated collectors (such as units with evacuated tubes to suppress conduction and convection losses, located in reflective troughs to concentrate the sunlight on the tubes).

Electricity can be generated directly in solar thermal systems that collect the thermal energy in solar radiation and use it to directly or indirectly heat water to produce steam. These are high-temperature applications for which conventional flat-plate collectors are inadequate and focusing collectors are needed. The leading solar thermal electric technology is the parabolic trough, which focuses the sunlight on a tube that carries a heat-absorbing fluid, usually oil. The fluid is circulated through a boiler, where its heat is used to boil water to steam to drive a turbine and generate electricity. More than 350 megawatts of parabolic-trough electric generating capacity is operating in California's Mojave Desert, connected to the

Southern California Edison Company (SCE) utility grid. These projects represent more than 95 percent of the world's solar electric capacity. Focusing collectors using selective surfaces (which absorb visible light well, but do not reradiate infrared easily), are expected to produce power at around 7 cents per kilowatt-hour. This is close to competitive, especially when it coincides with summer, air-conditioning-based, peak demand.

More localized and direct conversion of solar energy to electricity is achieved using solid-state photovoltaics. Photovoltaic (PV) cells, also called solar cells, are one of the most benign forms of electricity generation available. They can be used for stand-alone systems with no fuel or cooling requirements and no operating emissions or noise. However, because much of the current cell technology uses crystalline semiconductor materials (similar to integrated circuit chips), production costs have been high. Technology improvements have reduced PV power generation costs from a prohibitive $1.50 per kilowatt-hour in 1980 to a range of 30 to 40 cents per kilowatt-hour today.

Several collaborative programs were initiated in the late 1990s between the federal government (through the Department of Energy) and the PV manufacturing industry to develop lower cost PV manufacturing processes. The electric utility industry also joined with these entities to identify current, cost-effective, utility markets for PV systems, providing a near-term market pathway for further PV cost reductions.

▌ Wind Power

Wind power has been used for thousands of years to pump water, grind grain, and power sailing vessels. By 1900, 100,000 windmills were operating in Denmark, and by 1916, over 1,300 of these were generating electricity. More than 6 million windmills were built in the United States over the last century. Most were abandoned when the rural electrification program supplied cheap electricity.

The United States is well endowed with favorable wind energy sites with average annual speeds over twenty kilometers per hour (12.4 miles per hour). There are extensive geographically suitable and economically viable U.S. wind resources within sixteen kilometers of transmission lines. The best sites are in mountainous areas, in the Great Plains, along the coastlines, and around the Great Lakes. Major existing clusters of wind turbines are in especially windy spots, such as the Altamont Pass in California. Smaller commercial wind farms now operate in Massachusetts, Hawaii, Vermont, and Montana. Wind power has been widely used in remote areas for many years. There was a major boom in wind generation of electricity in the United States in the 1980s, with a slump towards the end of the decade. Ashley (1992) suggests this was because:

> The wind generator is exposed to more extreme conditions than was at first thought. Only about one third of the wind turbines installed in California—the hub of U.S. wind farm activity—during the decade proved successful. Poor load prediction, inadequate detailed design and low quality control had led to failed turbine blades and overloaded gear boxes, brakes and yaw drives.

Even so, there are today about 15,500 wind turbines in California, producing about 1650 megawatts. The problems encountered have been essentially engineering failures, much the same sort of problems as experienced by nuclear power producers (but with less dramatic consequences). The failure rate of early flat-plate solar collectors for water heaters was similarly high.

Wind turbines capture the wind's energy, usually through two or three blades mounted on a shaft and turning a generator to produce electricity. Because the wind is generally slower close to the ground, turbines are mounted on towers.

Energy available from the wind is proportional to the cube of the wind speed, so even small improvements in the wind regime can be important. Good detailed design and proper product development are essential. Over its expected life of thirty years, rotating at sixty rotations per minute for 5,000 to 6,000 hours a year, the unit goes through around one thousand million load cycles. This is like a motor vehicle traveling hundreds of thousands of kilometers per year, with minimal maintenance, for decades.

Improved turbine design and operation have contributed to a reduction in wind energy generation costs from 25 cents/kilowatt-hour in 1980 to around 7 cents/kilowatt-hour for today's commercial machines in the best locations. The current wind generator design preference is for horizontal axis units, with two rather than three blades. This configuration simplifies designing the blades to flex rather than to impose dynamic loads on the drive train.

A new class of wind turbines is being developed with advanced airfoils designed to stall progressively from the hub outwards as the wind speed increases, automatically limiting the tendency to overspeed. There are other improvements in electronics and controls; some turbines will incorporate variable speed operation. The new generation is expected to further reduce the cost of wind energy to 4 to 5 cents/kilowatt-hour.

Utilities are looking to use wind power to save fuel costs and reduce emissions, and for peak load leveling. The long-term contribution to the American grid is expected to be at least 10 percent—about the same as hydroelectricity.

Reported 1996 government support for direct R&D work in wind power ranged from less than $1 million in Australia, Canada, Finland, New Zealand, Norway, and Sweden, through $1 million to $7 million in Denmark, Germany, Greece, Italy, Japan, the Netherlands, Spain, and the United Kingdom, to $31 million in the United States. These countries support their wind power projects through investment subsidies, tax incentives, and premium price payments for the energy produced.

The countries listed above generated 8,500 gigawatt hours of electricity during 1996 and installed 1,840 new wind turbines. The trend has been toward machines of higher rated capacity. Installed wind turbines are generally performing well, with few operational difficulties. Lightning strikes and icing are the main operational problems. One concern is the significant number of bird deaths, resulting from the birds flying into the turbine blades. No major problems were reported on the integration of output into the electrical distribution systems.

In this industry, careful choice of site is as important as suitable machines. Similarly, for successful integration of wind energy into national supply systems, government incentives for both installation and operation of machines need to be well designed. The unfortunate experience with many wind farms in California demonstrates this point all too well. Investors were paid large subsidies for installing machines, rather than for operating them effectively. At the same time, although utilities were required to purchase the electricity generated, insufficient attention appears to have been paid to encouraging utilities to adjust their supply and demand regimes so as to benefit significantly from wind generation. In 1996 California announced it would no longer subsidize its wind power industry.

Until the environmental costs of nonrenewable energy sources are recognized, wind energy (even at 5 cents/kilowatt-hour) will be economically unattractive. In many countries low-cost conventional generation (around 4 cents/kilowatt-hour) is based on cheap fossil fuel and surplus capacity. Even Denmark's modest carbon and environmental taxes tip the economic balance in favor of wind, which is planned to supply 40 percent of the country's energy by the year 2030. In countries where premium buy-back prices make the generation of electricity by wind power economically viable, the main constraint on the rate of development is the difficulty of obtaining planning consent for projects. Environmental concerns about wind turbines are a planning issue in Denmark, Germany, Italy, the Netherlands, Sweden, the United Kingdom, and the United States. Only Germany reports integration into the electricity distribution system as a potential constraint.

Ecological Studies

It is essential that unintended, and possibly unforeseeable, side effects of our interventions in natural systems are minimized. This requires us to look to another discipline that impacts on engineering, ecology, to understand how these natural systems work.

Ecological studies consider the impacts of resource consumption and of the technological systems associated with it. They challenge the assumption that economic growth based on increasing resource use can continue indefinitely. Ecological perspectives confirm the need to move towards sustainable approaches.

The way we use resources and technologies is affecting the ecosphere. We can see how the ecosphere is affected by looking at four major areas of ecological concern, two in the atmosphere and two elsewhere.

Key Concepts and Issues

Ecology

Ecology is defined by Webster's Dictionary as: "the branch of biology dealing with the relations and interactions between organisms and their environment, including other organisms."

The thin skin of air, water, soil, and plant and animal life around the Earth is often called the ecosphere (we referred to it earlier as the biosphere). Commoner (1990) differentiates it from what he calls the technosphere, the human-made system of production. Some of the informal laws that govern the operation of the ecosphere are given in Table 9.2. As we might expect, these laws have a systems focus. They emphasize the interconnectedness, consistency, and mutual dependence of living organisms. They also highlight the problems caused when natural systems are subjected to overloads or rapid changes.

Commoner points out that, in contrast with the ecosphere, the technosphere changes very rapidly. Synthetic organic chemistry is only 150 years old, but it produces a huge array of compounds not found in nature. There are no natural enzymes to break these compounds down, so they persist in the environment. Most plastics, for example, are not biodegradable. As examples like the chlorofluorocarbons (CFCs) illustrate, even apparently benign new compounds are potentially dangerous intruders in the ecosphere (see Case Study 9.1, p. 484).

TABLE 9.2

How the Ecosphere Works	
Informal Law	**Application in the Ecosphere**
1. Everything is connected to everything else	1. The ecosphere operates as a network, in which each organism may play a number of roles: producer, consumer, habitat, and prey.
2. Everything has to go somewhere	2. The ecosystem is made up of closed cycles: the soil cycle involves plants growing, using nitrates from the soil and carbon dioxide from the air. Plants are eaten by animals, which return carbon dioxide to the air and organic compounds to the soil, where microorganisms convert them back to available nitrates.
3. Nature knows best	3. The ecosphere is self-consistent; the components have evolved over millennia and are mutually compatible. They do not normally poison one another or get very far out of balance with each other.
4. There is no such thing as a free lunch	4. Any distortion of a natural cycle leads unavoidably to harmful effects. In the biological as well as the economic sphere, the "free lunch" is actually a disguised debt. In the ecosphere (as in the money economy) it often happens that debts are not paid by those who incur them.

Source: Based on Commoner, Barry 1990, *Making Peace with the Planet*, Pantheon Books, New York. Copyright, reproduced with permission.

■ Holes in the Ozone Layer

Ozone is a form of oxygen in which the molecules include three atoms of oxygen (O_3) instead of the usual two (O_2). It is formed in the atmosphere by the action of ultraviolet radiation on oxygen (O_2). Within the stratosphere there is a region about three-kilometers thick with a particularly high concentration of ozone. This region is referred to as the "ozone layer." It acts as a barrier to protect the Earth from the harmful effects of hard (high-energy) solar ultraviolet radiation, which could otherwise damage plants and animals. In people this radiation can cause sunburn, skin cancer, and cataracts in the eyes.

Chemicals that find their way into the upper atmosphere can decompose ozone. The most worrying of these chemicals are the chlorofluorocarbons (CFCs). Since their development in the late 1920s (or, as Commoner might put it, since their addition to the technosphere), they have been widely used as refrigerants, aerosol propellants, plastic packaging foaming agents, cleaners, and solvents. CFCs are nontoxic and almost inert, and were thought to be benign. In the mid-1980s it was realized that their chlorine content was responsible for the formation of a "hole" in the ozone layer over Antarctica. There was concern that hard ultraviolet radiation coming through this hole to the Earth's surface might damage plant and animal life on or near the surface of the Antarctic oceans. There was also concern about how rapidly and how far north the hole might come to extend. We have seen an astonishingly rapid international response, although there are still questions about who should pay the associated costs (Flavin 1992; French 1997). Problems with CFCs and the international response to them are discussed in Case Study 9.1.

The dangers associated with damage to the ozone layer have been recognized as real and serious. However, there is still some debate about the urgency of the threat posed by the second issue we will treat here, the greenhouse effect. Although, strictly speaking, the term should be "enhanced greenhouse effect" this is cumbersome and "greenhouse effect" is universally used. There is a marked reluctance to take decisive action on this issue, partly because of the far-reaching character of the problem and the controversy over how real and urgent it is.

■ The Greenhouse Effect

If the Earth were a bare rocky sphere without an atmosphere, like the Moon, it would have a temperature range from 100°C during the day to −150°C at night. The average surface temperature would be about −18°C. The relatively thin blanket of air around the Earth raises the average temperature of its surface (and the layer of air just above the surface) by 33°C, to about 15°C.

The predominant wavelength of the energy radiated by a hot object depends on its surface temperature. For the Sun, this is about 6,000°C. As a result, 50 percent of solar radiant energy is in the infrared, with a wavelength greater than 0.7 micrometers (μm or microns). Another 40 percent is concentrated in the narrow band from 0.4 to 0.7 micrometers, the "visible" part of the spectrum, with the peak at 0.47 micrometers. The remaining 10 percent of the sun's radiation is in the ultraviolet, below 0.4 micrometers. Most of it is absorbed in maintaining the Ozone Layer.

The Earth's atmosphere absorbs part of the infrared, but does not absorb the visible radiation or the shorter part (0.7 micrometers to 4 micrometers) of the infrared. Clouds reflect some of this radiation, but the balance transmits light and warmth to the land or sea on which it falls. We probably developed organs that "see" this part of the spectrum precisely because it is the main one striking the Earth.

CFCs AND THE MONTREAL PROTOCOL

When the hole in the ozone layer was discovered over Antarctica, there was worldwide concern. In response, the Montreal Protocol was signed in 1987 by around forty countries; the goal was to limit the use of one of the main causes, chlorofluorocarbons (CFCs). With the discovery of another hole over the Arctic and growing appreciation that reducing consumption would take some years to have the desired effect, the phase-out dates for CFC replacement have been moved forward a number of times. In 1986, United States use of CFCs was 364,000 tons, and world production peaked at almost 1.3 million tons. By 1994 world production had fallen by 80 percent and U.S. production had ceased (French 1997).

The peak in ozone depletion was expected to occur around 1998 to 2000. The seasonal hole over Antarctica is unlikely to disappear before the year 2045. Full restoration of the ozone layer will take even longer (EPA 1997).

As one of the countries that seemed most directly affected, Australian progress in phase-out has been well ahead of the world average, apart from CFC use as cleaning fluids and solvents. Australian usage in 1992 was still 9,324 tons per year, 63 percent of it for refrigeration or air conditioning. This compared with 14,601 tons if Australia had slowed its reduction to the same rate as the rest of the world, and 20,466 tons if it had

still been using CFCs at 1986 rates. Significant technical difficulties and costs were involved, particularly if change was to be accelerated. The scale of the problem in Australia alone was indicated by the estimate that it involved:

◆ 4,000 water chiller air conditioning units in high-rise buildings;

◆ 4 million motor vehicles containing CFC-12 (also known as the refrigerant R-12);

◆ 5.8 million domestic refrigerators using R-12; and

◆ 350,000 food shops or supermarkets with either cold storage or refrigeration equipment (Mills 1992).

Thus a nation that, right from the beginning of the alarm about the hole in the ozone layer, was highly motivated and well equipped technically to address this problem, still faced a daunting task. And it was something it could not accomplish alone; yet, for the rest of the world, the picture was much bleaker.

The refrigerant typically used in large centrifugal compressors was R-11, which has three chlorine atoms per molecule, and a correspondingly severe effect on the ozone layer. It is no longer used. R-12, with two chlorine atoms per molecule, was the most common refrigerant. It was phased out by the end of 1995. The open types of com-

---

Because the surface of the Earth is at only a few degrees Celsius, it reradiates energy in the infrared. It is the infrared, mostly in the wavelengths between 4 micrometers and 100 micrometers, with which we are mostly concerned here. Water vapor in the inner layer of the atmosphere, the troposphere, absorbs strongly in the band from 4 micrometers to 7 mi-

pressors using R-12, for example in automobiles, are more tolerant of changes in refrigerant, so changing them over was more straightforward technically. R-22, the other common CFC-type refrigerant, only has one chlorine atom per molecule, and seems likely to continue in use into the next century. It is a potential replacement for R-11 in the short-term, but the condensing pressure for R-22 may be as much as 15 percent to 20 percent higher, so that condensers may need redesign and replacement.

Another possible refrigerant is ammonia, NH_3, but it is toxic in air and therefore cannot be used in commercial or domestic buildings. It is also incompatible with copper or brass pipes and fittings, so these must be replaced with steel pipes before it can be used in places where it is suitable, such as food storage.

Ideally, a new family of refrigerants that did not contain chlorine would have been developed, which could just have been dropped into refrigeration units as a replacement. However, although R-11 can be replaced by R-123, and R-12 by K-132 or R-152 (which has flammability problems), there are substantial practical engineering problems associated with these changes. All aspects of operation must be checked. Most large chillers in buildings are driven by hermetically sealed electric motors, immersed in the refrigerant. The lacquer that insulates the motor windings may react with the replacement refrigerant, in which case the motors must be rewound with new insulating lacquers before the changeover. Lubricants compatible with CFCs are not usually suitable for the new families of refrigerants, so they too need replacement.

Even when the technical issues involved in changing refrigerants have been resolved, there can be substantial cost and inconvenience with the changeover, particularly in facilities such as hospitals that operate continuously. The fact that the process is proceeding shows that the seriousness of the problems associated with the hole in the ozone layer and the need for prompt action have been accepted.

The ozone issue illustrates the global character of ecological problems. It has obvious implications for engineering practice in the future. Refrigeration is a technology of which engineers are rightly proud. It makes an important contribution to improving the quality of life by retaining food quality and reducing wastage, but an unintended side effect of the development and use of synthetic refrigerants containing chlorine atoms is that their release has caused significant damage to the ecosphere.

References

EPA (NSW Environmental Protection Agency) 1997, *NSW State of the Environment Report 1997*, NSW Government, Sydney.

Mills, S. 1992, "Mixed response to CFC phaseout," Engineers Australia, 13 November, pp. 22–27.

crometers, and the carbon dioxide in the atmosphere absorbs strongly between 13 micrometers and 19 micrometers. Energy radiated out by the Earth in these two bands is absorbed in the atmosphere and this heat is then reradiated, still in the infrared. The part that goes back towards the Earth keeps it warmer than it would otherwise be. There is an analogy with the glass in a greenhouse, which is transparent in the visible spectrum, but absorbs and reradiates in the infrared. Gases that produce this warming effect are known as greenhouse gases.

Between these bands in the infrared, from 7 micrometers to 13 micrometers, is a "window" that allows more than 70 percent of the energy radiated from the Earth's surface to escape into space.

Awareness that human activities are increasing the amount of carbon dioxide in the atmosphere has led to concern about a greenhouse effect. Over the 10,000-year period until 1850, the base level of carbon dioxide in the Earth's atmosphere has been shown to have been around 270 parts per million (ppm). By 1957, mostly as a result of humans burning fossil fuels, especially coal, but also from destruction of tropical forests, the level had risen to 315 parts per million. By the early 1990s, it was around 350 parts per million. The rate of increase represents a little less than half the additional fuel carbon burned during the period. Plants grow more vigorously in air richer in carbon dioxide, and some of the rest is taken up through photosynthesis by plants, including marine plankton. Some may be dissolving in the oceans. Even so, given present fuel consumption patterns, the atmospheric carbon dioxide level will double by the 2080s. Because of the resulting increased absorption in the 13-micrometers and 19-micrometers band and reradiation in the atmosphere, this doubling might be expected to raise global mean temperatures by 2°C.

Other gases being released into the atmosphere by human activities absorb heat in the 7-micrometers to 13-micrometers band. These include methane, CFCs, ozone, and nitrous oxide. CFCs have been used as refrigerants and aerosol propellants, and are better known as the chemicals responsible for the thinning of the ozone layers over the North and South Poles. They are also extraordinarily potent greenhouse gases. A single molecule of either of the two most common CFCs has the same greenhouse effect as 10,000 molecules of carbon dioxide!

Methane is generated by bacterial action in paddy fields and in the digestive tracts of sheep and cattle. It is also released from oil and gas fields. In the late 1980s, methane had reached an atmospheric concentration of 1.7 parts per million, and was rising at a rate of 1.2 percent per year. Nitrous oxide, principally from nitrogen-based fertilizers, was building up at 0.3 percent per year from a base level of 0.3 parts per million. The presence of these other gases could move the date of the temperature rise that would be expected due to carbon dioxide doubling alone, backward from the 2080s to the year 2030 (Gribbin 1988).

Not everyone agrees that the greenhouse effect is real. Some see it almost as a conspiracy. One skeptic (Singer 1992, p. 19) has argued that:

Greenhouse warming is an imaginary, "hyped" situation, that is being used by a variety of people for a variety of purposes, simply to promote whatever they happen to be interested in, whether it is enforced energy conservation, redistribution of income in the world or as a vehicle for social change.

The issue has certainly been used as a political weapon, for example by the British government of Margaret Thatcher in its attacks on the coal miners and its arguments in favor of nuclear power. Greenhouse computer simulations still clearly need a great deal of development, including the inclusion of the detailed effects of natural perturbations, such as volcanic eruptions. Modeling of the greenhouse effect is still relatively crude and there cannot be said to be a scientific consensus on how much temperature rise might be expected and on what the local effects might be.

In Australia, Brian Tucker, the chief of the Commonwealth Scientific and Industrial Research Organisation's Atmospheric Research Division, does see the threat as real and serious. He has argued that it is important neither to ignore nor to exaggerate the problem (Tucker 1992). The concerns he raises illustrate the need for a clear understanding of both the evidence and the assumptions underlying the arguments; they also show the importance of being able to tell which is which. The basic facts seem to be as follows:

- fossil fuels are a major contributor to the greenhouse effect, but other sources include the gases from burning forests and other gases such as methane and CFCs, which are generated by human activities;

- although there would be some melting of nonpolar ice, sea levels would rise because of ocean water warming and therefore expanding. The best estimates of likely sea level rises have changed very little in recent years;

- although overall average temperatures could be expected to rise as a result of a greenhouse effect, in a complex system such as the Earth's atmosphere the effects would not be uniform, so that extreme temperatures in some regions could well fall.

Should we really be worried about the greenhouse effect? The likely results are not clear, and are not likely to be uniform across the globe. The effects of increases in levels of greenhouse gases have to be balanced against the likely increases they will cause in cloud cover. They also have to be balanced against influences that can cause cooling. Dust at high levels in the atmosphere, such as resulted from the eruptions of Mount Pinatubo in the Philippines, can cause cooling. One concern about the effect of the use of nuclear weapons is that dust carried high into the atmosphere by nuclear explosions might cause a "nuclear winter."

Global climate, represented by average surface temperatures, is by no means static and never has been. It is likely that human evolution has to some extent been driven by the need to adapt to climatic changes. There is interesting evidence that, some 65 million years ago, one or more large meteors or similar bodies collided with our Earth; the ensuing cooling may well have resulted in the extinction of large dinosaurs and provided the conditions that were eventually to result in the emergence of the human species. Other important long-term climatic changes have been associated with geological effects upon the climate. Over geological time, continents were formed and drifted over the face of the earth. The upward thrusting of the Himalayas, for example, has had a significant effect on the climate of the entire planet.

Shorter, more periodic variations, with a wavelength of the order of 100,000 years, appear to be caused by perturbations in the earth's orbit around the sun. These would seem to cause long cold spells, of around 80,000 years duration, which at their peak could be characterized as ice ages. There is evidence that they have alternated with warm spells that last about 10,000 years, and that we are currently 10,000 years into the most recent warm spell!

Life on earth, and particularly human life, has survived enormous climatic changes. People are killed every day by adverse climatic affects in some parts of the globe. These can encompass extremes of low or high temperatures, of dry or wet conditions, of high wind velocities, and other effects. Most people are either healthy enough or lucky enough to adapt and survive. Are we prepared to abandon the rest? As professional engineers, and as members of the human race, we have a responsibility to minimize the adverse consequences of nature and of the technologies we develop. The "precautionary principle" discussed later in this chapter suggests that we need to recognize the possibilities of long-term or irreversible effects and take timely action to minimize the likelihood of their occurring.

▌ Population Pressures

Stabilizing human population pressure and limiting human consumption of nonrenewable resources are essential steps in moving towards global sustainability. During the twentieth century, world population will have grown from 1.6 billion to 6 billion. One urgent priority is to slow population growth, which drives many environmental and social problems, and in 1997 population growth was running at 88 million a year. World population growth did slow sharply between the 1960s and the 1990s. One of the leaders in the change, Indonesia, saw average fertility fall by 41 percent during this period. In the same period infant mortality also fell sharply, from 133 to 57 infant deaths per 1,000 births.

The International Conference on Population and Development in Cairo in 1994 was the third of a series of UN population conferences that began in 1974. Flavin (1997, p.16) argues that at the Cairo Conference, largely as a result of nongovernment organization (NGO) input, the links between population growth, social inequity, material consumption, and environmental degradation finally began to be addressed effectively. The Conference recognized that unless urgent social needs were met, fertility levels were likely to remain high. As a result, the plan of action it produced called for efforts to empower women, reduce poverty, and make education, health care, and economic opportunity more accessible. Many family planning programs now focus on raising the status of women, an approach that has been shown to be very effective in generating voluntary limitation of population increase. More rapid reductions in population growth seem to require much more coercive approaches. When the Chinese leaders became desperately concerned with continuing population growth, they adopted stringent family planning targets and effectively limited women to one, or at the most two, children. Despite the seriousness of the world situation, such an approach is not generally seen as an acceptable model.

The effects of population size on global sustainability cannot be considered adequately without recognition of the levels of consumption of resources in the countries involved. Flavin (1997, p. 19) notes that: "... the annual increase in the United States population of 2.6 million people puts more pressure on the world's resources than do the 17 million people added in India each year." Regardless of the ultimate level at which the world's population stabilizes, a sustainable world economy will require industrial countries to develop less resource-intensive lifestyles and less polluting technologies. Flavin cites detailed German studies indicating that industrial countries could reduce their energy and material consumption by a factor of four while still increasing their standard of living (Weizsäcker et al., 1997). Flavin also suggests that the example set by the rich countries will set the pattern that will be decisive for the world as a whole.

■ Resource Depletion

Exergy decreases as energy is dissipated to a higher entropy state by economic activity. High-exergy resources, such as fossil fuels, are depleted and become unavailable to future generations. The values that different generations assign to resources cannot be reconciled by traditional market economics. It is quite unsound to imagine that the price mechanism can cope with these disparities (Georgescu-Roegan 1971). The most easily accessible resources tend to be used first, whereas sustainability considerations would see us minimizing future regrets.

The supply and use of fossil fuel energy is the most urgent aspect of the broader problem of resource depletion. Some other issues are discussed briefly below.

Soil degradation is a major and increasing problem. Worldwide during the last fifty years, an area of land about half the size of the United States has been so degraded by human activity that it is effectively dead and useless. The main causes of this were overgrazing, deforestation, and unsustainable agricultural practices. In the early 1990s, the World Resources Institute and the United Nations Environment Program funded a land-degradation survey. It found that some 10.5 percent of the most productive soils had been damaged since the last major study in 1945. Many millions of hectares were beyond repair, and more than a billion hectares could only be restored at great cost.

The destruction of rainforests has become a major concern, particularly as we have started to appreciate their astonishing biological diversity. Deforestation rates have accelerated, with about 9 percent of tropical forests being cleared in the 1980s alone. Although much attention has been focused on the Amazon region, which includes more than 6 million square kilometers of tropical forest, tropical and temperate rainforest in other countries is also under threat. The scale of destruction in Brazil is dramatic, with more than 250,000 square kilometers cleared in recent times. Exploitation of the hydroelectric potential of the region is also controversial, with just two dams flooding almost 48,000 square kilometers of forest. The Brazilian authorities suggest, with justification, that there is a degree of hypocrisy in demands by the industrialized countries to stop logging and other use of the resources of the Amazon region. Having cut down their own forests long ago, they now appear to be seeking to stop poor people from making a living by doing the same, as well as blaming the poor for greenhouse problems. The Brazilians point out that their contribution to greenhouse problems is insignificant compared, for example, with the effect of the more than 500 million motor vehicles in the world, which emit some 56 percent of the carbon dioxide that goes into the atmosphere.

At first glance, ecological concerns such as soil degradation and rainforest destruction may appear of little relevance to engineers. However, as creators of many of the technologies involved, engineers have been part of the problem. Engineers will need to be involved in the solutions.

Reduce, Reuse, Recycle

Reduce, reuse, recycle has been a major slogan of the environmental movement for many years.

- ◆ "Reduce" has given rise to what is now a major planning approach, "source reduction," which is one focus of what has become known as "cleaner production," discussed separately below. The aim of source reduction is to eliminate waste production—to design products, processes, and systems that do not have waste associated with them. An obvious target is the billions of essentially single-use plastic bags handed out every year. Source reduction is particularly important for processes that otherwise would produce toxic or other intractable wastes. Their disposal is difficult, but biological processes seem promising. Reduce also implies more economical, less wasteful approaches to resource use.

- ◆ "Reuse" focuses attention on making things to last, to be used more than once. Some of the values implicit in the design of throwaway products are starting to be challenged. For example, the revival of the Harley Davidson motorcycle seems to have been partly attributable to the desire for user-control. Most operators of Harleys were then able to do much of their own servicing (although this seems no longer to be the case). This was a positive example of technology being adapted to meet a user requirement for reusability and straightforward maintenance (Pirsig 1974).

- ◆ "Recycle" is intended to conserve both energy and material resources. It includes separating the potential waste stream into categories that are valuable and ordered enough to be economically used for making new products. Glass, aluminum, paper, and some plastics are obvious recycling targets. One promising Australian technology already being exported to the United States is for the recycling of road surface materials.

▪ Waste Management

Waste management is not one problem, but many. Under this heading we should consider disposal of radioactive and toxic material, as well as domestic, commercial, and industrial waste. The scale of each of these problems is different, but the preferred approaches have much in common. The world market for waste-management technologies has been estimated at around $300 billion.

The technical challenges involved in separating plastic from the waste stream, sorting it, and successfully recycling it are formidable. Under pressure from consumers and because of government interventions, the plastics industry has made a serious start on the problem, but much remains to be done, and there are major engineering challenges in redesigning manufacturing plants and their products to minimize waste.

Waste reduction and recycling are being taken seriously worldwide. This is indicated by a new emphasis in manufacturing engineering on design for disassembly. Design criteria for BMW and Mercedes Benz automobiles now specifically include attention to ensuring that the materials in them can readily be separated for recycling.

SUSTAINABILITY

The central approach by which governments around the world are starting to address ecological issues in a positive and proactive way is summarized in the principles of sustainability, also referred to as ecologically sustainable development (ESD).

One implication of ESD is that the process of environmental assessment should be approached much more broadly as part of resource management, rather than each project being considered on an individual basis in isolation from all others. A resource management plan for an entire region would come first. Based on the sustainability of the resources identified in the region, this plan would set the parameters for individual projects in an area, providing guidelines for environmental, infrastructural, and other issues. Individual environmental impact statements would have to fit into the framework provided by regional plans, which would require much more interdisciplinary knowledge than is at present available. Detailed base line studies of local flora and fauna should be routine. Much better understanding is also needed of the dynamics of local ecosystems, which are often highly specific.

In late 1990, the Australian government set up a number of working groups to formulate policies to integrate environmental policies into governmental decision-making processes. The basic principles it assumed in its approach to ESD are summarized in Table 9.3. They raise some interesting issues. Are economic and environmental goals essentially compatible or incompatible? How can the value of wilderness be assessed? Is it

TABLE 9.3

Sustainable Development: A Governmental Approach	
Principle	**Notes**
1. Integration of environmental and economic goals for all policies and activities	1. Assumes that economic growth and environmental protection are essentially compatible. Where conflicts do arise, costs and benefits of the alternative approaches should be assessed and compared.
2. Ensuring that environmental assets are appropriately valued	2. Assumes that resources will not be wasted if people have to pay the proper price for them. Pricing structures should be adjusted accordingly.
3. Providing for equity within and between generations	3. How fairly are the nation's assets, including both natural resources and other wealth, distributed? How do we compensate future generations if we use up nonrenewable resources?
4. Dealing cautiously with risk and irreversibility	4. Recognizes problems of possible long-term or irreversible damage. Argues that although some environmental damage might be worth the risk, provided that the benefits warrant it, it might also be worth some economic cost to prevent future damage.
5. Recognizing the global dimensions and irreversibility	5. Some problems go well beyond national boundaries, such as ozone holes, and require international responses. Also a basis for considering the merits of accepting some environmental degradation in Australia rather than elsewhere in the world.

Source: Commonwealth Government 1990, *Ecologically Sustainable Development: A Commonwealth Discussion Paper,* AGPS, Canberra. Commonwealth of Australia copyright, reproduced by permission.

simply a matter of the preparedness and/or of the ability of citizens to pay? An approach that assigns some value to environmental qualities is clearly preferable to one that declares the problem to be too hard and therefore simply values the qualities at zero.

As Beder (1996, pp. 4–6) noted, this approach focused on the needs of human beings. It did not guarantee that interests of other forms of life on Earth would be taken into account. The Rio declaration took a similar position (UNCED 1992, p. 1).

Principles for sustainable development endorsed by mainstream environmental groups are summarized in Table 9.4. They are clearly more demanding than the principles set out in the government position. They indicate concern for all forms of life, rather than simply being human-centered.

TABLE 9.4

Ecologically Sustainable Development: Key Environmental Principles

Principle	Notes
1. Intergenerational equity	1. The present generation should pass on to the next an environment that is at least as healthy, diverse, and productive as the one we enjoy.
2. Conservation of biodiversity and ecological integrity	2. This should be a fundamental constraint on all economic activity.
3. Maintaining or enhancing natural capital	3. Only taking income that can be sustained indefinitely, taking account of the need to maintain biological diversity, healthy environments, fresh water supplies, and productive soils.
4. Anticipatory and precautionary policy approach	4. Erring on the side of caution. Placing the burden of proof on intending developers to show that technical and industrial developments are ecologically sustainable.
5. Social equity	5. A key consideration in developing social and economic policies.
6. Limiting natural resource use	6. This should be done both by the capacity to supply renewable resources and to assimilate wastes.
7. Qualitative development	7. Economic and social policy focused on quality of life rather than simply increased use of resources.
8. Pricing environmental values and natural resources	8. This should be at levels that recover the full social and environmental costs of their use and extraction. Many of these values cannot really be priced in monetary terms, so these policies will be part of the broader framework of decision making.
9. Global perspective	9. This ensures that a country does not simply move its environmental problems elsewhere.
10. Efficiency of resource use	10. This must be a major policy objective.
11. Resilience	11. Economic policy needs to be resilient to external economic or ecological shocks. This implies a need to move beyond a natural resource-based economy.
12. External economic balance	12. This is important to minimize pressures to deplete natural capital, which could undermine moves towards an ecologically sustainable economy.
13. Community participation	13. This will be a vital prerequisite to a smooth transition to an ecologically sustainable society.

Source: After: ACF 1992, Vol. 20, No. 1, p. 2.

■ Government Action

The Framework Convention on Climate Change was agreed upon at Rio in 1992 and negotiated and ratified by over 160 countries over the following two years. The Treaty took effect on 21 March 1994 and is legally binding on the countries that ratified it. The ultimate objective of the Convention is to stabilize greenhouse gas concentration in the atmosphere at a level that would prevent dangerous interference with the climate system. It does not specify what these levels might be. The main responsibility, and the largest share of the bill, fall on the rich countries, which made a commitment that, as a minimum, they would seek by the year 2000 to return their emissions to the levels of 1990. The Kyoto conference in 1997 was aimed at strengthening commitment to positive action.

The question of equity raises the important question of compensation. How serious are we about equity? As we discussed in Chapters 3 and 4, the present distribution of wealth and power in the United States is unequal. History has shown that gross levels of inequality encourage social instability and breakdown. Reducing inequality can also help to minimize environmental damage.

Many countries adopted national Greenhouse Response Strategies in 1992, and updated them in 1997. Even voluntary programs for cooperation between government and industry and commerce have had some significant successes, with reductions in energy use of 15 to 75 percent having been achieved in manufacturing programs and 50 to 75 percent in lighting of commercial buildings. For the Kyoto targets to be met, reducing greenhouse gas emissions to sustainable levels will need to be integral to the planning, design, and implementation of every type of engineering activity. The buying and selling of energy will incorporate trade in entitlement to produce carbon dioxide emissions.

How can we most usefully debate environmental issues? As Commoner (1990, p. 16) points out, the notion of rights is not very helpful. After one side in the debate has asserted the right of a tree to exist, and the other side has asserted the right of a logger to cut it, how can discussion proceed? Respect for species other than our own is a more productive approach. It is also consistent with the attitudinal changes needed in the engineering profession, including a much greater awareness of, and responsiveness to the broader contexts in which engineering activity is carried out. Case Study 9.2 at the end of the chapter, on the building of the Trans-Alaska oil pipeline, illustrates just how far we have already come and how much attitudes towards the environment, in both the community and the engineering profession, have changed in the last few decades.

Many situations arise in which the constraints and requirements of available energy resources, ecological considerations, and the growth of the economy can appear to be mutually incompatible. Conflicting demands and perspectives are increasingly reflected in national and international tensions. In this situation the engineer has a key role to play. The most important future role for engineers will be to manage these situations; no other professional group is as well placed to develop a balanced understanding of the conflicting technological interests. By deftly satisfying or transcending these conflicting objectives, the project engineer of the future will ensure that society continues to address equity concerns. Johnston (1997) summarizes the extent to which current engineering education in Australia addresses sustainability issues. He also discusses some promising educational initiatives intended to strengthen attention to sustainability.

One distinctive contribution the engineer can bring to any discussion of the environmental impact of large-scale energy use is an understanding of the biosphere as a system with inputs and outputs that can, in principle, be quantified. It is inconceivable that ten billion human beings could enjoy comfortable urban lifestyles on this planet without causing very significant changes in the biosphere. The challenge is to first understand the dynamics of the biosphere as a system and then create global political structures that will allow us to manage those changes in an acceptable way.

Considerations of ecology, economics, and energy need to be reconciled. There is a danger of these disciplines operating separately from each other, as if each was on a flat plane, without vision. To make progress, a higher perspective is needed and this is precisely what an engineering education should provide. It must also be admitted that such a lofty perspective is precisely what engineering education has mostly failed to provide. To take full advantage of these opportunities engineers will need to take every opportunity to broaden their education and experience and resist temptations to specialize excessively. Humility and a sense of proportion and humor will also be important.

▮ Transitional Technologies

Sustainability may be most useful not as an absolute concept but a relative one for comparing options and scenarios. A simplistic model for an absolutely sustainable state would involve the extinction of the human race—an unacceptable and essentially trivial solution, for which it may, in any case, be too late. Thoughtful human intervention may now be an essential condition for sustainability. The human condition, and the design solutions humans have adopted, have always been part of a dynamic system. We learn by trial and error, by feedback, we adapt, we make deals, and we compromise. Over the next few decades different countries will have different growth rates, different expectations, different rights. The world will need to become more sophisticated in negotiating and trading these rights.

It is axiomatic that only ecologically sustainable products, processes, and systems have a long-term future. The real question is how soon they can start to have a major impact. Groups that move quickly and deliberately to research, to design, and to develop sustainable activities may gain an important, possibly critical, competitive advantage. For countries that wait for others to take sustainability seriously and to develop the relevant technologies, including wind, wave, and solar energy conversion, there will be continuing costs, rather than exportable benefits.

One fundamental difficulty with renewable energy sources that engineers need to address is that the exergy of their energy is very low. One has therefore to process colossal quantities of the energy-containing material, usually air or water, in order to replace each 500-megawatts generating unit in a conventional power station. This requires large quantities of land, capital investment, and hardware (fabricated from energy-intensive manufactured materials such as steel and cement). This initial investment needs to be taken fully into account in assessing the implications of a future without fossil fuels or nuclear power. It could be argued that these considerations have already hampered the implementation of renewable energy schemes and they certainly present a challenge to engineers in the future.

Although they will only be a part of the solution, devices that make more efficient use of resources will be increasingly important and increasingly appreciated. Designing and developing them should present a major challenge for the engineering profession. National technology policies should include real government support for applied research and for full commercial development of technologies that use fossil fuels sparingly, and can ease the transition to soft technologies. The sophisticated use of coal offers important possibilities. Techniques are becoming available to cream off valuable liquids and gases before burning the rest. In conjunction with electricity generation, this approach could offer effective conversion technologies at rather more modest costs.

Fluidized bed combustion offers exciting possibilities for using alternatives to fuel oil for heating, and can be adapted to very large or very small systems. Natural gas will be an important transitional fuel to reduce greenhouse gas emissions while sustainable approaches are developed.

Much of the discussion on integrating the use of solar energy with existing energy supplies is related to questions of peak loading and energy storage. There is only really a problem with electricity, because the large storage capacity within gas and liquid energy supply systems enables them to accommodate uneven loadings. Policies need to be formulated to move electricity use towards continuous loads and away from uneven loads such as domestic cooking, which can be handled by gas or liquid fuel.

Conclusion

Sustainability may have been ignored because it was seen as a "soft" subject without hard numbers. Engineers would seek to influence the field by measuring and controlling—the question is what? Clearly much needs to be done to quantify the various sustainability options. This needs a control parameter. The most suitable would appear to be exergy. By performing a rigorous exergy analysis on all available sustainability options engineers will be in a position to make relatively uncontroversial day-to-day decisions.

On both national and international levels, the work of professional engineers is central to the process of innovation and the development of new technology. Engineers have a major responsibility for the conservation and enhancement of the natural and built environment. Associated with this role is an ethical responsibility for the development and promotion of ecological sustainability. Many national professional engineering associations have been actively raising these issues since the 1980s. They are now moving to help put these ideas into practice with the publication of practical guides for implementing sustainable practices. Ecologists, from the relatively conservative Paul and Anne Ehrlich (1981) to the more radical Barry Commoner (1990), insist on the need to limit economic growth. They see the rapidly increasing demands on natural resources and loads on the natural environment as unsustainable. The World Commission on Environment and Development reached the same conclusion (WCED 1990). It detailed the extent of the environmental crisis and pointed out the need for action to reverse some of our present resource usage and waste disposal patterns. Given their major roles in these areas, engineers will need to be actively engaged in developing solutions.

Industrial societies need to move towards the use of technologies that rely on renewable energy flows (Lovins 1977). Systems need to be diverse, with networks of small local energy sources matched and scaled to local end-use needs. This will make for systems that are both flexible and accessible to user control. All these characteristics can help reduce adverse environmental impacts. The problems associated with sustainable development are socially, politically, and technically challenging. Engineers working in this area need to take a lead in the interdisciplinary teams for developing technologies and products that will allow us to move towards more stable and equitable resource distribution and use.

DISCUSSION QUESTIONS

1. Why is it ecologically inappropriate to use electricity for low temperature heating?
2. Why is the location of the boundary of the system important in consideration of energy efficiency?
3. What are the key environmental issues for the next decade?
4. What problems are involved with: (a) the enhanced greenhouse effect; (b) the ozone layer? What is being done about them? Is it enough?
5. What is Environmental Impact Assessment (EIA)? What is the process involved? What is the EIA supposed to do? How effective is it? Are all issues negotiable?
6. What is the difference between a marginal and a fundamental conflict? Give examples of each type.
7. A major concern of developers is that they should have clear guidelines for acceptable development. Why are such guidelines not always (often) available? Should "best current technology" always be acceptable?
8. Different forms of energy conversion are appropriate for the disparate geographical regions. Describe, with reference to factors such as resource availability, environmental constraints, quality of energy, and demand fluctuations, the most appropriate energy form for several different locations.

FURTHER READING

Beder, S. 1996, *The Nature of Sustainable Development, 2nd. Ed.* Scribe, Newham, Victoria. A very accessible and well-balanced presentation of both the global perspective and Australian policy formation on sustainability.

Çengel, Y. A., Boles, M. A. 1998, *Thermodynamics: An Engineering Approach*, McGraw-Hill. A good thermodynamics text is essential for anyone wishing to make further progress in studying energy and sustainability. This one is also readable!

Diesendorf, M., Hamilton, C. 1997, *Human Ecology, Human Economy*, Allen and Unwin, St. Leonards, NSW, Australia. Deals with many of the issues raised in this chapter.

IEAust (Institution of Engineers, Australia) 1997, *Towards Sustainable Practice: Engineering Frameworks for Sustainability*, IEAust, Canberra. Offers practical guidance for engineers.

State of the World, the annual report produced by the Worldwatch Institute in Washington D.C., gives an excellent picture of the progress being made worldwide towards sustainability.

Weizsäcker, E.U. von, Lovins, A.B., Lovins, L.H. 1997, *Factor Four: Doubling Wealth, Halving Resource Use: The New Report to The Club of Rome*, Earthscan Publications, London. The factor approach provides a clear test for the adequacy of technologies proposed as sustainable. The authors argue that resource productivity needs to grow at least fourfold over the next fifty years. They provide practical examples that this can be done in some areas simply by systematic application of current technologies. (A higher improvement factor may be needed, and this would require much more radical changes. "Factor Ten" approaches might see biotechnologies playing a much more central role.) Further material on this work is available at: *www.wuppertal-forum.de/factor four/*.

REFERENCES

Anon "UNCED: The Earth Summit" 1992, *Connect*, XVII, (2), June.

Ashley, S. 1992, Turbines Catch Their Second Wind, *Mechanical Engineering*, ASME, NY, pp. 56-59.

Australian Conservation Foundation (ACF) 1992, *Habitat* (special issue on ecologically sustainable development) February.

Beder, S. 1996, *The Nature of Sustainable Development, 2nd. Ed.* Scribe, Newham, Victoria.

Brown, L. R., Ed., 1985, *State of the World 1985*, Worldwatch Institute, Norton, NY.

Commoner, B. 1990, *Making Peace with the Planet*, Pantheon, NY.

Commonwealth Government 1990, *Ecologically Sustainable Development: A Commonwealth Discussion Paper*, AGPS, Canberra.

Department of Primary Industries and Energy (DPIE) 1991, *Issues in Energy Policy: An Agenda for the 1990s*, AGPS, Canberra.

Diesendorf, M., Kinrade, P. 1992, Integrated Greenhouse Policies for Energy and Transport, *IEAust Seminar on Australia's Greenhouse Policy*, 26 June, Canberra.

Ehrlich, P., Ehrlich A. 1981, *Extinction: The Causes and Consequences of the Disappearance of Species*, Ballantine Books, NY.

Flavin, C. 1992, Building a Bridge to Sustainable Energy, In: *State of the World 1992*, Earthscan, London (Worldwatch Institute), pp. 27-45.

————, 1997, The Legacy of Rio, In: *State of the World 1997*, Earthscan, London (Worldwatch Institute), pp. 3-22.

Freeman, C. Jahoda, M. Eds., 1978, *World Futures: The Great Debate*, Universe, NY.

French, H. F. 1997, Learning from the Ozone Experience, In: *State of the World 1997*, Earthscan, London (Worldwatch Institute), pp. 151-171.

Fells, I., Horlock, J. 1995, UK Energy Strategy, In: Rooke, D., Fells, I., Horlock, J., Eds., *Energy for the Future*, Spon, London.

Georgescu-Roegan, N. 1971, *The Entropy Law and the Economic Process*, Harvard University Press.

Gostelow, J. P. 1984, *Cascade Aerodynamics*, Pergamon, Oxford.

Gribbin, J. 1988, The Greenhouse Effect, *New Scientist*, 22 October, (Inside Science: special supplement).

Horlock, J. H. 1997, *Cogeneration: Combined Heat and Power*, Krieger, Malabar, FL.

Johnston, S. F. 1997, Sustainability, Engineering and Australian Academe, *Journal of the Society for Philosophy and Technology*, 2 (3/4), pp. 80-101, <http://scholar.lib.vt.edu/ejournals/SPT/v2_n3n4html/johnston.html>

Lovins, A. B. 1977, *Soft Energy Paths: Towards a Durable Peace*, Penguin, Ringwood, Victoria.

McVeigh, J. C. 1984, *Energy Around the World*, Pergamon, Oxford.

Meadows, D. L., et al. 1972, *The Limits to Growth*, Universe Books, NY.

Mills, S. 1992, Mixed Response to CFC Phaseout, *Engineers Australia*, 13 November, pp. 22-27.

Pirsig, R. M. 1974, *Zen and the Art of Motorcycle Maintenance: An Inquiry into Values*, Bantam Books, NY.

Preston, G. T. 1995, Fossil Power Generation and Sustainable Development, In: Rooke, D., Fells, I., Horlock, J., Eds., *Energy for the Future*, Spon, London.

Singer, F. 1992, Bad science and the art of deception, *Engineering World*, (3), June, pp. 16-19.

Tucker, B. 1992, CSIRO Combats Greenhouse Myths, *Engineers Australia*, June 12, pp. 16-19.

Ward, B. 1966, *Spaceship Earth*, Columbia University Press, NY.

WCED (World Commission on Environment and Development) 1990, *Our Common Future*, Oxford University Press, Melbourne.

Weizsäcker, E.U. von, Lovins, A.B., Lovins, L.H. 1997, *Factor Four: Doubling Wealth, Halving Resource Use: The New Report to The Club of Rome*, Earthscan Publications, London.

TRANS-ALASKA PIPELINE

Among the largest projects ever undertaken by private industry, the Trans-Alaska Pipeline (TAP) illustrates how development interacts with environmental policy; and how conflicting interests and values lead to differing conceptions of sustainability, and hence the possibility of different policy outcomes. The decisions concerning TAP were influenced by sociocultural factors; the political and social structure, ideologies, and cultural values affected every step of the decision process from approval, through implementation, construction, and operation.

The 1968 discovery of a large oil resource on the North Slope of Alaska was good news, not only for the oil companies, but also for the state of Alaska. The proposed pipeline extended across federal land from Prudhoe Bay on the North Slope to Port Valdez in the south. The project required the construction of almost 1,300 kilometers of 1.2-meter (48-inch) tubing, twelve pumping stations, a control terminal at Valdez, a 580-kilometer haul road, bridges, and a multitude of other components that would require extensive materials and labor. To Alaska, this translated into more job opportunities, a boom in businesses and industry, and economic independence. Furthermore the revenue from the 380 million liters (2 million barrels) a day the pipeline was expected to carry would make Alaska a very wealthy state!

Eight oil companies banded together to form one joint company: Alleys Pipeline Service Company: (Aleyska). Before they could start construction on federal land, they had to file for a right of way in 1969. To Aleyska's surprise, instead of a speedy approval, it took five years before right of way

was granted. They had not taken account of the changing social scene in the (post-Vietnam War) United States. A growing environmental consciousness was reflected in legislation; Congress passed the National Environment Protection Act (NEPA) in that same year (1969). This act required the government to "prepare a detailed environmental impact statement" for any major project undertaken. Three environmental groups quickly took advantage of this provision and sued the oil company and government for failure to prepare an adequate impact statement.

The government was especially wary because the pipeline posed not only an environmental but also a cultural threat. The Native Alaskan way of life and culture were closely intertwined with their environment. In 1966 and 1967, the Alaska Federation of Natives had filed land claims on federal land. This froze all federal Alaskan lands until the issue was resolved, adding to the delays to Aleyska's plans. Finally, in 1971, the government gave the Native Alaskans what was seen as a very generous settlement.

The Alaskan landscape posed many natural challenges to the design and construction of TAP. Earthquakes and extreme changes in temperatures, from 32°C (90°F) in summer to −62°C (−80°F) degrees in winter, were only two of the concerns. Both Aleyska and the government assured the public that all possible action would be taken to address environmental concerns and minimize the risk to the environment. In fact, the thirty-four-page list of stipulations released by the Interior Department was described "as the most rigid controls ever imposed upon a private construction project by the government." (Berry 1975, p. 111) These stipulations were designed to protect the interests of the public and environment by ensuring safe construction and operation procedures (GAO 1975, p. 37). The Secretary of Interior also set up a separate organization, headed by the "Authorized Officer" to review and approve all plans and monitor implementation. The General Accounting Office (GAO) was to conduct regular audits (GAO 1975, p. 30).

However, both the oil companies and the government had placed heavy reliance on technology to provide acceptable environmental protection. In an era of exponential technological growth, the power of technology seemed unlimited and seemed to hold solutions to all environmental problems. An oil company advertisement promised "that environmental disturbances will be avoided where possible, held to a minimum where unavoidable, and restored to the fullest practicable extent . . . the pipeline will be the most carefully engineered and constructed crude oil pipeline in the world." (Berry 1975, p. 146) The environment was seen as just another machine; once they disturbed it, they could simply fix it with their omnipotent technology: the problem and its consequences would go away. Their confidence in the power of technology convinced the pipeline builders and the public that the environmental risks were minimal.

The finished pipeline is indeed a masterpiece of engineering technology. Many engineering techniques were employed to ensure environmental protection during construction. For example, prior to construction, a work pad was created along the right of way to protect the vegetation mat from construction traffic. (If the vegetation mat covering the permafrost had been dam-

aged, it would have resulted in disastrous water flows and resulting erosion when the permafrost melted.) Furthermore, complex engineering systems were designed in order to ensure safe operation. Among the measures to protect the environment were:

◆ A system of 142 block and check valves was installed along the pipeline in case of oil spills and leaks (automatically preventing reverse flow of oil) (GAO 1977, p.12).

◆ Earthquake monitoring systems were installed.

◆ Pressure relief system at each pump station to detect excessive static and surge pressures and divert them out of the pipeline before problems occurred.

◆ A vapor recovery system to prevent oil fumes in storage tanks escaping into the atmosphere.

◆ A communication system designed to interlock and coordinate all these systems (GAO 1977, pp. 12-13).

However, despite the technology and strict government regulations, environmental problems arose and persisted. In its 1975 and 1977 audits, the GAO emphasized the inadequacy of Aleyska's Quality Assurance control program. It reported that not only were problems identified in 1975 still unsolved in 1977, but that "actions taken did not prevent additional nonconformances from occurring" (GAO 1977, p. 14). Furthermore, after TAP was completed in 1978, it suffered two oil leaks in 1979. With such advanced technology, why did the pipeline fail to work effectively and safely?

The GAO and Aleyska admitted a lack of foresight and planning. In a 1978 GAO summary report, Aleyska offices reported

(continued on next page)

that they had not counted on the numerous additional difficulties they encountered. Their estimated base budget had been 6.4 billion dollars; by December 1977, they had already exceeded that by 1.5 billion dollars (GAO 1978, p. ii). Aleyska also admitted that not all engineering problems had been solved prior to construction. The 1978 GAO report observed that if Aleyska had obtained "more site-specific geotechnical data, Aleyska could have … improved the engineering and system design and reduced the number of subsequent design changes" (GAO 1978, p. 18). Aleyska also underestimated the amount of labor needed. State law required preferential employment of Alaskans, which meant hiring unskilled labor that had to be trained. This caused further delays. Due to Alaska's dramatic seasonal changes, certain parts of the project could only be constructed at specific times of the year.

The GAO recognized Aleyska's key shortcoming as a failure to resolve problems, or "nonconformities," in a timely manner. But this is a problem embedded in the very culture and philosophy of oil companies: profit from oil is the main goal and the environment is merely a means to attain this end. Take, for example, erosion control. The Federal monitors in the 1975 GAO audit cited Aleyska's failure to implement erosion control plans, resulting in the structural failure of the work pad. The Authorization Officer had spelled out the problem and a list of actions necessary to reduce the risks. Yet in 1976, Federal monitors found the same erosion problems. According to Federal monitors, "these conditions arose because Aleyska assigns a higher priority to pipeline construction than to erosion control work." (GAO 1977, p. 15) The oil company's main priority was to finish the pipeline and prevent delays that would mean further costs. They believed they had a responsibility to the American public to provide oil. As James Gallow, vice president of Humble oil, declared in his speech at the twentieth Alaska Science Conference, … "the industry must *of necessity* broaden its search for new oil and gas deposits to all corners of the continent;" the implication was that, regardless of cost, oil must be provided.

The government shared the belief that the American public had an unquestionable right to oil. Senator Morton, head of Interior, asserted: "We have a duty to the American people to make sure that construction and operation of this pipeline cause minimum damage and operation to the environment, while we constantly bear in mind that the Nation needs the oil" (GAO 1975, p. 77). In the end, government decisions on TAP were dictated by fear of an impending energy crisis: the need for energy was used to justify environmental risks. The government overruled the NEPA on July 17, 1974, granting right of way for construction. They decided the energy crisis was more important to the national interest than upholding the law. Oil companies fed this fear with a $4 million marketing campaign on the theme "A country that runs on oil can't afford to run short."

Thus the question was no longer a matter of whether a pipeline should be allowed; the question became when and where? This attitude was evident in the evaluation of alternatives. Several independent studies concluded that a trans-Canadian pipeline was more economically and environmentally advantageous. They were ignored because the Canadian pipeline would take several years longer to build, and officials felt that the energy crisis simply couldn't wait that long (Cicchetti 1972, pp. 177-119).

If the systems had been implemented and maintained properly, the technology may actually have minimized environmental damage. But there was unquestioning faith in technology. The Government thought that if it simply made a regulation, the technology would take over and fix all their problems. But just as technology by itself does not create our problems, it cannot be our only solution; technology is not some irresistible force that comes into play, leaving us to deal with its consequences as best as we can. Our values and beliefs play a large role in directing and shaping the role technology plays in society, and likewise technology has long-term effects on society. In the case of the Alaska pipeline, there was a whole range of other social and economic impacts (e.g., influx of workers, increase of population) on Alaskan lives that have not been discussed here.

As the population of the world has grown, so has our consumption of energy resources. Sustainable management is becoming a vital issue for maintaining human life on earth. Although stricter regulations and more advanced technology are needed, they are not enough: a whole shift in social ideology is needed. The oil companies' "take whatever you can use" philosophy must be discarded. A joint effort between government, private industry, engineers, and members of the community will be needed to develop the solutions that will move us towards sustainable management.

Source

Based on work by Lan Tran, Stanford University, 1997.

References

Berry, M. C. 1975, *The Alaska Pipeline: The Politics of Oil and Native Land Claims*, Fitzhenry and Whiteside, Canada.

Cicchetti, C. J. 1972, *Alaskan Oil: Alternative Routes and Markets*, Resources for the Future, Washington, D.C.

Galloway, J. H. 1970, Petroleum Industry Policies for Protecting the Environment, In: Rogers, G., Ed., *Change in Alaska: People, Petroleum and Politics*, University of Alaska Press, pp. 112-120.

GAO (United State General Accounting Office) 1975, *Trans-Alaska Oil Pipeline—Progress of Motivation through November 1975*, GPO, Washington, D.C.

GAO (United State General Accounting Office) 1977, *Trans-Alaska Oil Pipeline—Information on Construction, Technical and Environmental Matters through Spring 1977*, GPO, Washington, D.C.

GAO (United State General Accounting Office) 1978, *Lessons Learned from Constructing the Trans-Alaska Oil Pipeline*, GPO, Washington, D.C.

THE HYDROGEN AGE? THE RISE OF FUEL-CELL TECHNOLOGY

After a century of neglect and decades of unrealized promise, fuel-cell technology appears close to practical reality. Recently, the fuel cell—a device that converts hydrogen directly into electricity in a clean, highly efficient electrochemical reaction—has taken on new interest because of its potential application as an alternative energy source for automobiles and in power generation. The technology is revolutionary, providing greater fuel efficiency, the possibility of a renewable energy source, and zero emissions. Riding the wave of environmental awareness, proponents of the fuel cell are excitedly proclaiming the next century as "The Hydrogen Age." However, several social, economic, and political stumbling blocks may impede the progress of fuel-cell engines. Indeed, many experts agree that fuel-cell adoption is no longer a question of competent technology, but is rather an issue of industry commitment and consumer acceptance.

History and Technical Background of the Fuel Cell

The fuel cell dates back to 1839, when William Grove discovered that he could produce electricity from hydrogen and oxygen through reverse electrolysis. However, the technology was not applied until the early 1960s, when NASA began using fuel cells to power space expeditions. In the 1980s and 1990s, the Canada-based firm Ballard dramatically reduced fuel-cell costs and increased flexibility, making it technologically possible for fuel cells to compete with the internal combustion engine (ICE). In 1997, Germany's Daimler-Benz—parent firm of Mercedes-Benz—set a timeline for the industry with a statement that it would produce 100,000 fuel-cell engines per year by 2005. Other automakers such as Toyota, General Motors, Ford, and Chrysler have also entered the fuel-cell race.

A fuel cell consists of a fuel supply (hydrogen or hydrogen-rich fuels), an oxidant (oxygen), and two electrodes (an anode and a cathode) separated by a polymer membrane electrolyte. In a catalytic reaction, hydrogen atoms release electrons at the anode and become hydrogen ions in the electrolyte. Electrons released at the anode flow through an external circuit to the cathode and can be used to drive electrical devices. Electrons and hydrogen ions combine with oxygen at the cathode to form water vapor, releasing some heat. The process is two to three times more efficient than an internal combustion engine.

Five different fuel-cell technologies are currently competing for the market of the future: Proton-Exchange Membrane Fuel Cells (PEM), Alkaline Fuel Cells, Phosphoric Acid Fuel Cells, Molten Carbon Fuel Cells (MCFC), and Solid Oxide Fuel Cells. Because of its high performance-to-weight ratio, the PEM fuel cell is favored to power automobiles and become a commercial success. The MCFC fuel cell runs on natural or coal gas at a temperature of 650°C with a thermal efficiency of 54 percent to 60 percent and is favored for power generation.

PEM fuel-cell vehicles powered by pure hydrogen create a useable electric current with no emissions except pure water. However, there are serious concerns over the cost and availability of a hydrogen fuel infrastructure. As a result, fuel cells have been developed that use "impure" liquid fuels, including methanol, ethanol, and gasoline, all of which "carry" the hydrogen needed for fuel cells. Compressed natural gas is also being investigated. On-board reformers are needed for all of these alternatives except methanol. Because the reactions involved take place on activated surfaces, clean fuel is particularly important. Fuel choice is one of the most difficult issues in fuel-cell implementation.

Fuels for the Future

A fuel cell based on sustainable energy sources could have exciting implications: dramatic reductions in air pollution, acid rain, ozone depletion, and greenhouse gas emissions; a propitious geopolitical shift of power as global dependence on Middle East oil withered away and a realigned trade balance; the rise of quiet, clean, decentralized electric generators whose form fitted their function—small enough to power a car, large enough to power a town.

The hydrogen age may still arrive, but since Chrysler Corp. announced in January 1997 that it had developed a fuel cell that could extract hydrogen from gasoline, critics are concerned that fossil fuels such as gasoline and methanol, once regarded as transitional fuels on the way to pure, renewable hydrogen, will become permanent features of fuel-cell use.

Gasoline fuel cells are still expected to be better than today's internal combustion engine (ICE) vehicles in terms of fuel economy and environmental friendliness, with about twice the fuel economy and so emitting half the greenhouse gases. Even better is methanol, with 2.5 times the fuel economy of an ICE using gasoline, and 47 or 11 percent of the greenhouse pollutants, depend-

ing on whether it is made from natural gas or biomass. Hydrogen is (at least in theory) most impressive of all, with a notional 3.0 liter/100 kilometer (80 miles per U.S. gallon), a factor of improvement of 2.8 over gasoline ICE's. If formed by solar processes, hydrogen can achieve genuine zero emissions over the end-use and upstream life cycle. Unfortunately, there are major practical difficulties in handling and storing hydrogen (which only becomes liquid at temperatures close to absolute zero). Table 9.3.1 gives a detailed comparison of potential fuel cell inputs.

Enabling Factors for the Fuel Cell

Why, after lying dormant since 1839, has the fuel cell finally emerged to challenge conventional fossil fuel power sources? Better fuel efficiency, lower emissions, and a noiseless engine make the fuel cell an attractive subject for laudatory media coverage and an almost noble champion of cleaner living. It is no wonder that big business has gotten into the fold, with nine out of the top sixteen automakers in the world aggressively investing in fuel-cell technology.

In today's politically correct atmosphere, a firm's image as a technological leader and friend to the environment may be as strong a motivation as potential sales. Another driving force for fuel cell research and development is state and federal legislation, which is progressively imposing stricter regulations on vehicle emissions. In California, clean air rules require that by the year 2003, 10 percent of all cars sold in the state must be zero-emission vehicles. There has also been generous government financial support for alternative fuel vehicles.

Economically speaking, America's use of fossil fuels is costly, amounting to over $500

(continued on next page)

billion per year. Dependence on oil and foreign suppliers is also a source of concern, having already provoked two major U.S. recessions. With oil imports increasing and energy demand rising sharply in less developed countries, the potential for military conflict over oil is high. Hydrogen-powered fuel cells represent a way to cut economic and environment costs, reduce foreign dependence, and increase long term sustainability.

Inhibiting Factors for the Fuel Cell

Fuel cells face a number of challenges. Replacing the ICE will not be easy. Conventional cars have a century of history and familiarity behind them, and a well-developed infrastructure that includes the $200 billion investment by the oil companies in gasoline distribution systems.

In itself, fuel-cell technology is adequate: range is comparable to conventional vehicles and is much better than today's battery-powered electrics. Less maintenance should be required over the life of the vehicle because the fuel cell has no moving parts. Energy efficiency is better than gasoline ICE's or battery-powered electrics. However, public support has so

TABLE 9.3.1

	Fuel Choices at a Glance		
Fuel Choice	Fuel Economy L/100 km (mpg)	Relative Greenhouse Gases (%)	Relative Fossil Fuels (%)
Today's ICE	8.4 (28)	100	100
Gasoline fuel cell (FC) low efficiency high efficiency	5.6–3.7 (42–64)	73 48	73 48
Ethanol FC corn biomass	not available	60 0	37 3
Methanol FC natural gas biomass	3.4 (69)	47 11	58 6
Hydrogen FC natural gas biomass Solar	2.95 (80)	37 17 0	40 6 0

far been weak. Despite environmental concern, in 1998 the public was not yet ready to pay the $4,000 to $7,000 premium car makers are asking for alternative fuel cars. Consumers also appear unwilling to sacrifice performance or convenience for environmental compatibility. Gasoline fuel cells may indeed hasten the market development of the technology by taking advantage of the existing infrastructure. Methanol would fit in relatively easily. Hydrogen may have to wait a little longer.

Source

Based on work by Glenn Davis, Stanford University, 1997.

References

Ballard Power Systems, Daimler-Benz AG, and *Wired* websites.

Energy Innovations (Executive Summary), a report by Alliance to Save Energy, American Council for an Energy Efficient Economy, Natural Resources Defense Council, Tellus Institute, and the Union of Concerned Scientists, June 1997.

Advantages	Disadvantages
Infrastructure is in place, history and familiarity with ICE, continually improving fuel economy and lower emissions.	Thermodynamic limits: ICE can't convert all energy into propulsion, fossil fuels are finite resources, causes 60% of urban air pollution.
Could use existing infrastructure and pumps, familiar fuel.	Relies on nonrenewable energy, still has harmful pollutants.
Lower CO_2 and hydrocarbon emissions, formed from renewable crops.	Increased fuel vaporization.
Lower CO_2 and hydrocarbon emissions, cuts ozone production in half	Highly toxic, invisible flame, lower energy content.
Renewable, high-energy efficiency, low/zero emissions.	Public perception is linked to Hindenburg disaster, difficult to distribute and store.

▌Case Study 9.3

10

Towards a Philosophy of Engineering

[**P**hilosophy is] . . . looking at things which one takes for granted and suddenly seeing that they are very odd indeed.

— *Iris Murdoch (1919–), British novelist and philosopher*

Wrestling with philosophy can help us to understand the scope and boundaries of engineering practice and the assumptions underlying our profession. Not to do so can leave us at the mercy of those whose views of engineering may be quite different from our own. As engineers we need to think through and express the aims and aspirations of the engineering profession ourselves, in our own terms. Philosophical issues may seem obscure and even irrelevant to practical professionals who are inclined to take things very literally, but they are important because they underpin our attitudes and ideas. We encourage the reader to work thoughtfully through the material presented here.

In this chapter the development of Western philosophy is briefly summarized and some branches of philosophy are outlined. There are deep divisions within Western academic approaches to philosophy. A long-standing schism still exists between empiricism and rationalism. A more recent split is that between the logical positivism of the Anglo-American world and the existentialism of Continental Europe. These divisions are deeply ingrained in national attitudes. They can be traced back to the Renaissance and can still have undesirable effects today in education and industry. For the past few decades, engineering education has been dominated by engineering science, with its roots in the philosophical tradition of logical positivism. Within that tradition, but more as interpreters than defenders of it, are philosophers of science such as Karl Popper and Thomas Kuhn.

We consider the relationships between the bodies of knowledge involved in the practice of science, technology, and engineering in our society. How adequately do the assumptions and attitudes expressed in the philosophy of science meet the needs of engineering? We emphasize the need for a philosophy of engineering to help clarify the values and priorities on which engineering judgments are based. Such a philosophy could help us to recognize the ethical and social issues inherent in professional practice, and to build them into the discourse of engineering. We discuss the likely shape and content of an acceptable philosophy of engineering and how it might differ from a philosophy of science or technology.

The widely held view that science and technology are value-free is no longer sustainable. As engineers we are constantly called upon to make value judgments and we must recognize this as an essential aspect of our professional practice.

WHAT IS PHILOSOPHY?

Is it useful to try to condense a discussion about "the meaning of life," or even "what is meant by engineering?" into a chapter of a book for engineering students and practicing engineers? Philosophers have offered a variety of responses to this sort of question. We mention only two challenging, but very different replies. The first is from Lord Ashby (1966, p. 85):

> A student who can weave his technology into the fabric of society can claim to have a liberal education; a student who cannot weave his technology into the fabric of society cannot claim even to be a good technologist.

Our sympathy with that point of view is obvious. A second response might take the form of "a shot across the bow" from the logical positivist school discussed in detail below, and ask for definition of the terms "meaning," "life," and "engineering." Like it or not, we would immediately be into a philosophical discussion. This kind of linguistic analysis has been a dominant trend in philosophy this century, and has had a surprisingly wide impact. In particular, it has affected the way engineers see and conduct their profession.

During the two-and-a-half thousand years the term has been in use, philosophy has itself had a wide variety of meanings. It is now commonly used in two quite distinct senses. The first describes people's commitments and values, their outlook on the world. The second is the disciplined study of beliefs about being and knowledge, and about the general nature of things (Flew 1984). Although we recognize the importance of the first sense, the second is the one we will emphasize here.

Our search for a philosophy of engineering addresses both long-term issues, such as the choice of desirable directions for the profession, and urgent practical questions, such as developing satisfactory curricula and educating and training engineers effectively. These challenges require an appreciation of what engineering science and design are, and how they fit together into the practice of engineering.

SOME BRANCHES OF PHILOSOPHY

In philosophy, as in other disciplines, some words take on special meanings that may change with time. Some terms have a variety of meanings— for example, "rationalism" can range from applying thought to explain natural phenomena, to ignoring nature entirely, and proposing to proceed from thought alone. The Key Concepts and Issues box includes some definitions of philosophical terms and mentions some branches of philosophy that are particularly relevant to engineering.

THE ORIGINS OF WESTERN PHILOSOPHY

Philosophy is historically important to engineers because it attempts to explain the world on the basis of rationality rather than myth. It can help us to appreciate both the extent and the limitations of our knowledge. At the same time, by stimulating and supporting our moral imagination, it can help us to think through the likely implications of our actions.

The first philosophers of whose work we have useful records were from classical Greece. The roots of the term philosophy are *philos* (friend or lover) and *sophos* (expertise or wisdom, or even science). Philosophy seems to have started as criticism of then-current religious beliefs. The father of philosophy is taken to be Thales of Miletus, who attempted to explain the world on the basis of rationality rather than myth. We mentioned other early philosophers in Chapter 1.

In the late twelfth and early thirteenth centuries, the works of Aristotle and many others of the classical period were translated into Latin and

Ontology

Ontology is concerned with the fundamental nature of the world and of being, in particular with which principles or categories are primary, and what derives from them. The basic division in approaches to ontology is between those who start with ideas, or mind, or the consciousness of some supreme being (an ontological approach called "idealism") and those who assert the primacy of matter, of physical reality (an approach called "materialism"). Putting this in a slightly different way, in philosophical terms *materialism* means, not a greedy lust for material possessions, but rather an approach to the world that takes matter as primary, as opposed to *idealism*, which takes ideas or mind or consciousness as primary, and physical reality as secondary and derivative. Philosophically, idealism relates to ideas rather than ideals.

The materialist approach to knowledge is based on practice (from the Greek *praktikos*, active), defined as purposeful human material activity, particularly in production. Materialism is characterized by its recognition of the existence of the object of practice, most broadly the material world, as independent of human (or even divine) consciousness. It sees theory as intimately connected with practice, on which it is based. The relationship is summarized in the propositions that "practice without theory is blind; theory without practice is barren."

The rulers of societies and their supporters have tended to be remote from the processes of production of goods and services (as were most of the Greek and medieval philosophers), and seem generally to have taken an idealist position. Deeply held religious ideas and beliefs can be important in holding societies together. They also play a role in legitimizing the status quo. It is useful to remember that it is the winners (not just in wars) who write the history in any society. In thinking about ontology it may help to consider whether we as individuals are principally stimulated by things, ideas, or people. Such basic attitudes may tend to color our philosophical outlook.

Logic

Logic deals with the rules governing reasoning and inference. It is, broadly:

> the study of the structure and principles of reasoning, of sound argument... in its narrower sense, it is the study of the principles of deductive inference, or of methods of proof or demonstration... [its] aim is to make explicit the rules which are implicitly recognized as rules according

to which arguments ought to be constructed, at the same time pointing out any anomalies that may appear in the process (Flew 1984).

Engineering scientists have traditionally placed great emphasis on logic. It is important, however, to recognize that logic is neither the only reasonable basis for accepting a proposition, nor the only way of thinking. Given a repeated experience, even without apparent underlying reasons (logic) for the repetition, it is reasonable to anticipate further repetition. The process of induction, of inferring a general law from observed particular instances, is important in bringing out what may be considered implicit in such instances (i.e., in creating what may, on testing, turn out to be new knowledge). However, induction does not follow the rules of formal logic. This is important for engineers because overemphasis on logic tends to stifle creativity.

Ethics

Ethics, the basis of our transactions with society, is an essential foundation for professionalism. Philosophically speaking, it is:

> an investigation into the fundamental principles and concepts that are or ought to be found in a given field of human thought or activity. Being a branch of philosophy it is a theoretical study. As such it differs from "ethics" in the lay sense ... in that any actual body of ethical belief ... will be intended to be a practical guide to living and not merely an exposition and analysis of certain theoretical doctrines (Flew 1984).

We can distinguish between normative ethics, the investigation of how individuals and organizations ought to behave where moral considerations are involved, and conceptual inquiries about the logical form of morality, such as the question of whether moral judgments are objective or subjective. These conceptual approaches, also known as meta-ethics, are relevant to deciding the sorts of arguments that can appropriately be used to support or refute any given belief.

Dialectics

This term (from the Greek "discourse" or "debate") has had rather a checkered history. It seems to have started as a method of seeking knowledge by question and answer. This approach suits some problems, for example exposing inconsistency or humbug, or uncovering matters where we already have the

(continued on next page)

necessary information but have not thought it through properly. It is not helpful in matters such as empirical scientific inquiry. The current meaning of a dialectical approach is that it seeks to analyze change in terms of contradictions, i.e., tensions and conflicts between different elements or forces in an argument, a situation, or a society.

Epistemology

This is the branch of philosophy concerned with the theory of knowledge. Central issues include the nature and derivation of knowledge and its scope and reliability, all of which are of practical importance for engineering theory and practice.

Three main types of knowledge are commonly recognized: (1) factual knowledge (knowledge that); (2) practical knowledge (knowledge how); and (3) knowledge of people, places, and things (knowledge by acquaintance). For engineering the second type of knowledge is particularly important, as is the related issue of how we acquire it.

Heuristics

Heuristics are procedures to discover or design something, "rule-of-thumb" approaches that may solve a particular kind of problem, but offer no certainty of success. They may depend on induction or on previous experience and are discussed in more detail in Chapter 6.

---

became freely available in western Europe. They fostered a revival in education, and the universities founded in that period became centers of philosophical discourse. Many great philosophers appeared in the thirteenth century. Thomas Aquinas (ca. 1225–1274), for example, established a synthesis between Christianity and Aristotelian logic. William of Ockham (1285–1349) formulated the principle of economy of thought (entities are not to be multiplied beyond necessity), commonly known as Ockham's Razor. A modern equivalent is the KISS principle (keep it simple, stupid!).

With the Renaissance, the availability of printing led to popular languages taking over increasingly from Latin as vehicles for thought. Along with the decline in the power of the Inquisition to censor heretical ideas, this opened the way for more nationalistic philosophers such as Niccolo Machiavelli (1469–1527), Sir Francis Bacon (1561–1626), Thomas Hobbes (1588–1679), and René Descartes (1596–1650). To break the domination of the authorities of the time, these daring minds called ideas themselves into question (Roszac 1986). Political theory became more important, and theology less so. Bacon and Descartes were mainly interested in physics and mechanics, but found that they needed to draw on philosophy to provide a sound basis for their ideas.

In Bacon and Hobbes can be seen the roots of British empiricism, which was to be a cornerstone of the industrial revolution, and that persists to the present day. Bacon in particular displayed skepticism about invented ideas and *a priori* knowledge (which could be "known" to be true without reference to experience); he considered that the only worthwhile knowledge was that discovered by experiment or observation of the natural world. Bacon formulated the principles of scientific induction and introduced the objective style that we tend to take for granted in modern research.

Descartes is generally considered to be the founder of modern philosophy. In particular, the school known as rationalism is associated with him. It holds that truth resides in our minds, to be discovered by reason. Combining important elements of previous great philosophers with mathematical genius, Descartes almost single-handedly overturned the excessive deference to authority that had shackled medieval thought. He was, for example, the inventor of analytic geometry and Cartesian coordinates. Doubting everything, and resolving to accept nothing on the basis of authority or convention, Descartes finally arrived at the celebrated "proof" of his own existence: *Cogito, ergo sum* (I think, therefore I am).

In the late seventeenth century, Sir Isaac Newton ushered in the "age of enlightenment." Three philosophers who interpreted the significance of Newton's achievements and laid new foundations of thought were John Locke (1632–1704), Bishop George Berkeley (1685–1753), and David Hume (1711–1776). Between them they established empiricism as scientific orthodoxy. Their stated aim was to apply the methods of natural science to human nature. It didn't quite work like that. Hume's denial of a necessary order of nature and of a unified self precipitated a philosophical crisis that was not resolved until the work of Immanuel Kant (1724–1804). The empiricists' bold skepticism about all inherited ideas played an important role historically in liberating scientific inquiry. Their reductionist conception of knowledge, however, undervalued both the role of the imagination in the creation of ideas and the importance of ideas in the creation of knowledge.

Immanuel Kant, a German professor, discussed both the sensory and the *a priori* elements of knowledge, suggesting that "objects must conform with knowledge." His argument was that the order and regularity in objects, which we call nature, is introduced by the observer. This idea is still helpful in cognitive psychology in analyzing how people construct theories. Kant's approach to knowledge, *critical idealism*, addressed both the extreme rationalism of Gottfried Leibniz (1646–1716) and the excessively agnostic ("nothing is certain") empiricism of Hume. Kant offered a new definition of philosophy as "the science of the relation of all knowledge to the essential ends of human reason."

Georg Hegel (1770–1831), a Berlin professor, represented a return from Kant to a broader and more rationalistic idealist viewpoint. For Hegel "the rational is the real" and "the truth is the whole." Hegel's approach included the notion of a continuing dialectical process, in which truth is not eternal, but evolving and historical. This was important because it brought into consideration the changing conditions of human society, providing an important foundation for the thinking of Marx. Hegel used the idea of estrangement in a metaphysical sense. Marx extended this notion with his ideas of alienation based on people's relationships to productive activity.

Marx also applied the dialectical approach to the study of history, arguing that the starting point needed to be material reality, rather than ideal types and models. Marx is often described as "the last of the classical philosophers." He argued was that it was only possible to understand a situation by seeing how it had come about, and recognizing the internal tensions that might generate change. This was why his philosophy came to be described as "dialectical and historical materialism." His basic concerns were perhaps best summed up by his proposition (inscribed on his tombstone in Highgate Cemetry, London) that: "The philosophers only interpreted the world in various ways; the point, however, is to change it."

Marx emphasized the essential place of practice (or "praxis") in developing understanding and knowledge. He argued that practical involvement in a range of working class movements was an important basis for his exploration and understanding of the dynamics of social class and social change. Needless to say, this approach did not make him or his philosophy popular with the powerful, including those who funded universities and research. However, the power of his approach was such that (apart from his emphasis on praxis) it has been widely adopted, although not always acknowledged, in the humanities and social sciences.

The dialectic received renewed attention in a specialist work by Bhaskar (1993), who argues that the dialectical process is "the pulse of freedom."

PHILOSOPHY IN THE TWENTIETH CENTURY

∎ Existentialism

The roots of existentialism can be seen in the nineteenth century with Arthur Schopenauer (1788–1860), Søren Kierkegaard (1813–1855), and Friedrich Nietzsche (1844–1900), who wrote with color, depth, and irony in reaction to the rationalist Hegelian philosophy of the day. Their theme was the primacy of the concrete (physical reality) over the abstract, the subjective (personal reality) over the objective, and free will "free from the perplexities of intellect," in Nietzsche's words. Making the point that we can never have enough information to act with certainty (a situation familiar to every practicing engineer), Kierkegaard wrote: "It is perfectly true as philosophers say, that life can only be understood backwards. But they forget that other proposition, that life can only be lived forwards." (Goldman 1990)

An introduction to the richness of existentialist writing can be gained from Wilson's *Outsider (1963)*, or from works by Fyodor Dostoevsky (1821–1881), Hermann Hesse (1877–1962), or Albert Camus (1913–1960). Better still to read Nietzsche's *Thus spake Zarathustra* (1961), or discover the haunting beauty of T. S. Eliot's poetry.

Existentialism became influential in continental Europe during the second quarter of the twentieth century, partly as a reaction to World War II. There have been various strands to existentialist thought, reflecting the backgrounds and skills of the philosophers involved. The common starting point was that *being* must take precedence over *knowledge*: "existence precedes essence" or "subjectivity is truth." Christian existentialists such as Karl Jaspers (1883–1969) and Paul Tillich (1886–1965) were concerned with realizing the human being and with God as the ground of all being.

Jean-Paul Sartre (1905–1980) put the viewpoint of atheistic existentialism quite bluntly. For Sartre (1957), because there is no God, human beings have not been defined for any particular purpose. Thus, man exists before he can be defined by any concept. He appears on the scene and then, afterwards, tries to define himself and determine what he will be: "he is also what he wills himself to be after this thrust towards existence. He is nothing else but what he makes of himself." This is Sartre's way of defining subjectivity: "Existentialism's first move is to make every man aware of what he is and to make the full responsibility of his existence rest upon him." Sartre argued that we are free to choose how we will live, so that to claim that we are compelled by our situation, our nature, or our role in life to act in a particular way would be to exhibit bad faith, to betray our own possibilities. This appears to be as close as Sartre came to asserting moral standards by which our conduct could be judged.

If existence precedes essence, and if we exist and fashion our image at the same time, that image is valid for everyone and for our own age. Thus our responsibility involves all mankind. Solipsism is the philosophy that "I alone exist." Existentialism is different, emphasizing that every existential possibility leads a person to a relationship beyond his or her own self. This necessary communication is something that most existentialists find difficult and frustrating. Sartre summarizes this situation as: "Man is nothing else than a series of undertakings that he is the sum, the organization, the ensemble of the relationship that makes up these undertakings."

Existentialism, then, like engineering, proceeds by projects. The relevance to engineering practice of the existentialist approach to reality was asserted by Samuel Florman when he titled his enthusiastic discussion of the engineering experience *The Existential Pleasures of Engineering*. Florman saw the essence of existentialism as having two aspects: "rejection of dogma, particularly scientific dogma, and ... reliance on the passions, impulses, urges and intuitions that are the basic ground of our human existence." (Florman 1976, p. xi)

Florman saw the existential pleasures of engineering as lying in the ecstasy of creation, in successfully designing solutions that allow humans greater control over their destiny, in contributing to the welfare of society, and in sharing in great and positive enterprises. He suggested that the golden age of engineering ended with the development in the early 1950s of the hydrogen bomb, which was widely seen as a betrayal of professional ethics and as irredeemably destructive and pernicious. Florman takes a clear ethical position when he insists that the reputation and credibility of the profession depends on its recognizing and consciously taking action to reverse this fall from grace. He believes that the profession must consciously and deliberately re-assert its participation in, and responsibility to humanity.

Case Study 4.5 on the U.S. Interstate Highway System (see p. 202) offers a somewhat different perspective from Florman's. It starts by acknowledging the tremendous social impact of postwar engineering projects. It suggests that problems emerged as conflict between the narrowly defined technical objectives of the engineering profession and the much broader aspirations of the communities on which the impacts of these projects fell. One of the central purposes of this book has been to explore this area of conflict and the lessons it has for the engineering profession.

Lewis Mumford (1964) criticized the impacts of large-scale technologies very strongly, arguing that small-scale association and human-centered technologies were in danger of being eliminated by large-scale organization and institutional regulation. From an engineering perspective, the principal contribution of existentialism may be to open up questions of scale and control. Left to the establishment, with its focus on power, all projects would be large, and control would be centralized and authoritarian. Existentialism challenges each engineer to assume project responsibility and in doing so to engage with individual concerns and develop technologies that maximize personal autonomy.

Pirsig (1974) recognizes the potential conflict between human values and technology, but argues very positively that the way to resolve this conflict is not to reject technology, but "to break down the barriers of dualistic thought that prevent a real understanding of what technology is—not an exploitation of nature but a fusion of nature and the human spirit into a new creation that transcends both."

▌ Logical Positivism

In total opposition to existentialism is the school of thought known as logical positivism. It affirms, firstly, that all knowledge regarding matters of fact is based upon the positive data of experience, and secondly that beyond the realm of fact is that of pure logic and pure mathematics.

Logic and mathematics were held by Hume and other empiricists to be purely formal sciences concerned with relationships between ideas. The positivists, starting with Liebniz, a contemporary of Newton, and including the Viennese physicist and philosopher Ernst Mach (1838–1916), were involved in the development of physics, paving the way for Albert Einstein (1879–1955) by challenging some of Newton's more simplistic notions. Liebniz and Mach were both concerned about effects in mechanics that could not be explained in terms of an absolute Cartesian frame of reference. They also provided an ideological basis for what Sinclair (1977) calls the practicing professions, including engineering, as opposed to the liberal professions, which were pre-industrial in origin. They took the view that theories and laws were merely computational devices for describing and predicting phenomena, rather than explanations of reality: all the mind could really know was its own sensations. Burke

(1985, pp. 294–299) suggests that they played an important role at the time in freeing physics from "mysterious imperceptible substances," opening the way for Einstein and relativity. Donovan (1986) notes their encouragement of the view "that many forms of knowledge can be organized into distinct and autonomous sciences, [so that] there appeared to be no limit to the number of functionally discrete professions that might be created."

Important contributions were made by a group of logical positivists centered on Vienna University in the 1920s and 30s, and known as the "Vienna Circle." They attempted to use logical and mathematical theory to develop and systematize empiricism (partly in order to create a viable alternative to Marxism). Along with Bertrand Russell (1872–1970) and Ludwig Wittgenstein (1889–1951) in Britain, they have had a lasting influence, particularly on English-language philosophy.

Wittgenstein's view was typical. He saw the role of philosophy as being about clear explanations rather than content:

> The object of philosophy is the logical clarification of thoughts. Philosophy is not a theory but an activity. A philosophical work consists essentially of elucidations. The result of philosophy is not a number of philosophical propositions but to make propositions clear (Wittgenstein 1922).

The empiricism of logical positivism is especially evident in its restatement of the fundamental thesis of Hume's philosophy in a form known as the "verification principle." This recognizes as meaningful only those propositions that are verifiable by reference to sensory experience. On this basis, any assertion that claims to be factual has meaning only if it is possible to say how it might be verified. Unfortunately for this approach, the verification principle itself seems to be unverifiable.

Positivism in its most radical form, that of A. J. Ayer (1910–1989), stated that all assertions about moral, aesthetic, and religious values were scientifically unverifiable, meaningless, and therefore nonsensical. In his manifesto, *Language, Truth and Logic*, Ayer (1946) went on to argue that moral judgments were not statements at all, but merely expressions of approval or disapproval.

The basic tension, which has remained unresolved for three centuries, is that between empiricism and rationalism. The resultant divisions are deeply ingrained in national attitudes. They can be traced back to the Renaissance, and can still have undesirable effects on education and industry. From an engineering perspective, empiricism might be defined as a refusal to use theory, even when it would save much wasted effort. Tesla's criticism of Edison in Case Study 6.1 (see p. 280) makes precisely this point. Rationalism, which discounts practice entirely, would be a disastrous engineering approach. Our point here is that theory and practice are both important, and an overemphasis on either is regrettable and damaging for professional practice.

The influence of logical positivism has declined, probably because its very narrow approach to reality now seems rather dated and irrelevant, whereas less dogmatic and more flexible approaches have been seen as more fruitful. Logical positivism would exclude insights offered by the sociology of science on the way that scientific activity is carried on in our society, including how problems are chosen and resources allocated. Even after the contributions of Popper and Kuhn (discussed below), the prevailing orthodoxy among philosophers of science still tends to be empiricism. Some researchers in engineering and science do still concentrate solely on either experimental observation or theoretical prediction, quite oblivious to the potential value of the other. This is often a question of temperament and training. If anything, the power of computers has exacerbated this tendency, which may well present a problem in the management of engineering work. It is nevertheless conceivable that the computer could have a role in bridging the gulf between observation and theoretical prediction. For instance, sophisticated techniques for graphical representation permit the results of computations and laboratory measurements to be presented in identical and readily comparable forms. In this context, newly available technology may perhaps help to settle a dispute that philosophers have failed to resolve.

▮ Postmodernism

As the twentieth century advanced, there could be heard, with increasing frequency and intensity, declarations concerning the decline and fall, the deconstruction and collapse, of virtually every one of the West's great intellectual and cultural engagements: theology and religion, art, literature, and even science itself. From a boundless confidence in our own powers and capacity for certain knowledge and mastery over nature, the postmodernists assert that in the late twentieth century mankind has suffered a profound crisis of confidence, loss of spiritual faith, uncertainty, and a mutually destructive relationship with nature. The contemporary mind is seen as increasingly bereft of established certainties (there is no "true" meaning), in profound flux, yet at the same time open in ways it has never been before, with greater self-responsibility and capacity for creative innovation (Tarnas 1991, pp. 393–399).

The postmodernists argue that all human understanding is interpretation and that no interpretation is final (or even necessarily particularly meaningful). Although perhaps familiar, this is not an approach that seems likely to appeal to an engineer attempting to decide whether or not a particular design is ready to go into production. The terms postmodern, poststructural, and postindustrial tend to be used interchangeably by critics, who focus on what they see as the common themes and implicit aims of the postmodern approach: antirationalism or an insistence on the basically disordered nature of reality.

The authors do not find either the general attitudes or the detailed arguments of postmodernists such as Jacques Derrida (1930–) compelling or fruitful. However, we do see two postmodernist concepts, *discourse* and *deconstruction*, as extremely useful tools for exploring the way the professional culture of engineering is constructed. A discourse (Foucault 1970) is "a set of possible statements that produce the meaning and values of a cultural formation," in our case, the profession of engineering. Each area of specialization develops and values its own discourse, its own body of shared assumptions and special terminology, at the same time as it limits or disallows others.

We have shown throughout this book that both technology and engineering practice are at least partly socially constructed. We have suggested some shortcomings in the way in which, as a consequence, engineering education has commonly been constructed. In Chapter 3 we looked at some of the limitations of the present culture of engineering. We believe that at least some parts of this discourse need to be deconstructed (that is, dissected and critically analyzed), and then reconstructed along more inclusive lines. Later in this chapter we discuss the extent to which the discourses of science and business have been allowed to dominate the discourse of engineering.

Engineers have often been quite scathing about the failure of academic philosophers to provide intellectual guidance. Davison (1977, pp. 130-131) writes:

> Academic interests, quite unwittingly, have appropriated Western society's birthright, and, by specializing in purely technical questions, have emasculated and trivialized philosophy. Past cultures have decayed ... because they have lacked the intellectual equipment needed to understand their condition.

A very positive development that provides a basis for insight and intellectual leadership and allows the discussion to move forward is the recognition of a number of separate disciplinary areas within philosophy, including the philosophy of history and the philosophy of education. One of the most important for our purposes is the philosophy of science.

THE PHILOSOPHY OF SCIENCE

Before the seventeenth century, "science" included anything to be learned, ranging, for example, from sewing to horse riding. The meaning became much narrower and more specific during the periods of the scientific and industrial revolutions. During the seventeenth and eighteenth centuries, "science" came to be used for systematic knowledge, including that associated with craft activity, which we would now describe as technology. Rapp (1989, p. x) argues that: "Before the time of the Industrial Revolution ... technology was just taken for granted ... [and] not regarded as deserving special theoretical investigation." In the nineteenth century the meaning changed again as "science" replaced the older term "natural philosophy" for the body of knowledge generated by the people we now call scientists (particularly physicists) (Layton 1976). This continuing tendency to narrow the meaning of two of the fundamental terms used here, science and philosophy, and the absence of an adequate definition for technology have tended to generate confusion in the literature.

Physical and social scientists base their claim to be scientific on an organized, disciplined approach to research in which proposed theories and concepts are open to testing: that is, they should be able to be disproved. On this basis a science would be defined as an accumulation of knowledge that has been tested and will continue to be.

The logical positivists took a very static position, restricting themselves to the formal analysis of scientific arguments and disregarding the processes of conceptual change and the development of science. Two philosophers of science who helped to bring the dynamic issues of change and development to the forefront of the debate were Popper and Kuhn.

▌ Karl Popper

Karl Popper's interest was in the philosophy of science, and in how scientific knowledge grows. Although broadly sympathetic to the empiricist position, he insisted on the central role in scientific work of falsifiability by experience.

Popper (1902–1994) rejected the view that science advances merely by accretion of knowledge. He emphasized the way an existing theory was rejected and replaced by a new one. Many scientists (and engineers) would see philosophers' attempts to choose between empiricism and rationalism as a barren, irrelevant, and inappropriate attempt at oversimplification. Experiment and theory are intimately related, and both are clearly essential for a comprehensive understanding of any real phenomenon. Popper (1963) argued that scientific knowledge progressed by unjustified anticipations, guesses or tentative solutions to problems. These he called conjectures (they are more usually called hypotheses). Obviously if there were no control mechanism, these conjectures would proliferate. There needed to be a critical testing process. The tests were attempted refutations. The conjectures might survive these tests, but they could never be positively justified; they could never be established as certainly true, nor even as probably true.

Criticism of conjectures was, Popper asserted, the way that we became more thoroughly acquainted with a problem. By highlighting mistakes, this approach also emphasized difficulties and made us propose better solutions. The refutation of a theory was seen to be a step forward, clearing the way for a better theory. As the learning process proceeded, knowledge grew, although paradoxically we could not prove this. The truth was not at all obvious, but very elusive. Propositions could be shown to be false, but could not be proven correct. Knowledge proceeded by the elimination of error. Humility and willingness to be corrected were the most important hallmarks of a scientist. The image of a scientist tended to be that of a disinterested searcher after truth. To be scientific was, among other things, taken by Popper to mean being objective and open minded. Albury (1983) has shown how difficult this ideal is to achieve in practice.

▮ Thomas Kuhn

Kuhn (1922–1996) offers useful insights for the practice of both science and engineering. In his landmark book, *The Structure of Scientific Revolutions*, Kuhn (1970) produced a model of how basic researchers really work, although he omitted consideration of some of the social pressures on them. He described the central role of a "paradigm," involving the growth both of a discrete scientific model or problem network, for example Newtonian physics, and of its community of practitioners. The vitality of the paradigm was in its open-endedness, that is, the extent to which further work was left to new entrants into the field. Areas outside the paradigm were implicitly defined as not being suitable areas for serious scientific research.

The basic phenomena of the particular scientific field were satisfactorily explained by the paradigm, and scientists in that field could then "push on to more concrete and recondite problems." Kuhn suggested that, during its early years, a science proceeded without any general agreement on explanations for the basic phenomena observed to occur in its area. For example, until the middle of the eighteenth century, there were almost as many views about the nature of electricity as there were important electrical experimenters. The achievement of a paradigm was thus a very large and positive achievement. It opened the way for further solid advances, on the basis that "truth emerges more readily from error than from confusion."

In receiving a paradigm, the scientific community committed itself, consciously or not, to the view that the fundamental problems had been solved once and for all. The paradigm thus implied a deep commitment to a particular way of viewing the world and of practicing science within it. It defined both the problems available for pursuit and the nature of acceptable solutions to them. Paradigms were not usually permanent; on the contrary, the development pattern of a mature science was usually from paradigm to paradigm. An example was the move from Newton's model of the physical world to Einstein's. Paradigms were mutually exclusive. Thus, if any group developed a paradigm, it could only have that one.

Once a paradigm was established, what was there for scientists in that area to do? Basically, to make it a better and better fit to the natural world of which it was a model. The development of a paradigm was thus basically the justification of existing hypotheses. It provided a constrained dynamic for very detailed and specific research, and for the development of very specialized and sophisticated research equipment.

Gradually, however, the paradigm became ossified; nonconformists were excluded and the life force was drained. An era of crisis dawned; the old regime tottered, but a new one had not yet emerged. Much as with an exhausted political system, a pioneer would propose a new paradigm and a scientific revolution would be under way. The scientists who were too entrenched in past doctrines were bypassed. The revolution would have included the identification of anomalies challenging the old paradigm. The new paradigm would be formulated in such a way as to include both the phenomena explained by the old paradigm and the material accumulated as anomalies to it.

Kuhn's contribution can be summarized as the claim that science has advanced not only in a linear fashion but also by discrete revolutionary jumps. The model-building role of the paradigm is central to this, with science described in terms of communities of practitioners. In mature sciences, where there are received paradigms, teaching is characteristically by textbooks, rather than from current research papers, and the textbooks typically differ mainly in level and in pedagogic detail, rather than in substance or conceptual structure.

Successful basic research demands both a deep commitment to the status quo, and a preparedness to recognize where this does not provide an adequate explanation—that is, a preparedness to innovate. Scientists are trained to operate as puzzle-solvers from established rules, but they are also taught to regard themselves as explorers and inventors who know no rules except those dictated by nature itself. The result is an acquired tension, partly within the individual and partly within the community, between professional skills on the one hand and professional ideology on the other. That tension, and the ability to sustain it, are important to the success of science.

ENGINEERING AND RELATED AREAS OF KNOWLEDGE

The study of practical activities like technology and engineering has unfortunately been somewhat neglected by philosophers. Rapp (1989) suggests four main reasons for this: the domination of the Western history of ideas by theoretical rather than practical or craft issues; the false notion that technology is merely applied science; the fact that technology was taken for granted until after the industrial revolution; and finally, the difficulty of the task. Because of its complex, multidimensional nature, the only way structural models or handy theories of technology can be formulated is by significant over-simplification.

Figure 10.1 shows graphically how we conceptualize the relationship between technology, science, engineering, and engineering science. We see engineering (including its management functions) as entirely included within technology. Significant parts of empirical science are also included within technology. We would describe the overlap between science and engineering as engineering science, a part of both, but not a major or determining part of either (the relative sizes of the circles for engineering and science are not intended to be significant).

It should be emphasized that Figure 10.1 represents a static model. In practice there will be flows of information and resources between categories, as well as time lags and feedback loops. A more dynamic model is needed.

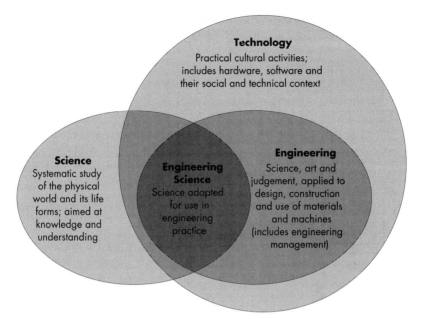

Figure 10.1 Relationships between technology, science, engineering, and engineering science.

The relationships described above are of course subject to challenge, but we believe we have now satisfied an essential philosophical requirement, to state our premises clearly. Whether we are right or wrong, our meaning is clear and Kuhn's proposition applies: "Truth emerges more readily from error than confusion."

An important aspect of philosophical analysis is the attempt clearly to define concepts and relationships. This is particularly important (and difficult) where there is a degree of overlap between the areas to be analyzed, and where extensive mutual borrowing of ideas and practices takes place. A clear understanding of the character of engineering practice is, however, essential for the professional formation of student engineers.

Some working definitions of technology, science, and engineering were given in Chapter 1. We now go on to a more detailed examination of these areas (and of engineering science), and explore how they differ. We will then consider what the philosophy of science and the philosophy of technology have to offer us in the search for a philosophy of engineering.

▌ Technology

When engineering students are asked to define technology, they tend to emphasize the high-tech aspects and ignore the bread-and-butter ones. So, one suspects, do governments when formulating science and technology policy. The emphasis on high technology reinforces the view that technology is scientific and therefore "value-free"; any consideration of human emotions, needs and aspirations is then extraneous, if not irrational. Popper and Kuhn are within this tradition, which has been used to justify structuring engineering courses so as to avoid consideration of the social content of engineering practice, including social values and ethics.

It may be helpful to look briefly at how technology has been seen as interacting with its social context. For instance, Pacey (1983) suggested the concept of technology practice. He drew an analogy with medical practice, which included not only medical instruments, drugs, and the like, but also the social structures and systems in which they were used. Pacey discussed how technology was integrated into society and how particular sorts of social organization were required to use technology effectively. As the Key Concepts and Issues box on Sociotechnical Systems in Chapter 1 shows, Kline (1995) offered a similar approach with his models of sociotechnical systems of practice and sociotechnical systems of use.

Ford (1986) argued that the differences in social organization in Japanese and Australian manufacturing plants were so great that they severely limited the possibilities for successful transfer of technology between the two. This problem paralleled the difficulties in transferring lean production technology from Japan to the United States, described in Chapter 7.

Martin Heidegger (1889–1976) saw the principal danger in the way technology was used in modern industrial societies as associated with the belief that technology was value-free. He argued that we were so caught up in technique, and besotted with the technology itself, that we missed the point, the essence, of the technology. He anticipated contemporary feminist philosophers by seeing our current efforts as aimed at converting everything—all of nature, including its sources of energy—into resources, stored and available for our own use. He feared that this focus on the instrumental and quantifiable clouded our awareness and blinded us to questions of social values. As a result we failed to recognize that the essence of technology was to enhance the quality of life, to support beauty, poetry, creativity, and delight (Heidegger 1993). Although the authors find Heidegger's detailed argument unconvincing, we sympathize with his concerns.

The word "technology" is an amalgam of the Greek words *techne*, meaning art or craft, and *logos*, meaning word or speech. Technology first appeared in English in the seventeenth century, and was used to signify a discussion of the applied arts. By 1909, *Webster's Second International Dictionary* narrowed it to: "industrial science, the science or systematic knowledge of the industrial arts, especially of the more important manufactures." During the twentieth century, the meaning broadened again.

By 1961, *Webster's Third New International Dictionary* was defining it as "the totality of means employed by a people to provide itself with the objects of material culture." Many engineers would now regard that definition as hopelessly dated and inadequate because the focus in recent decades has been more on systems than objects.

In 1954, Jacques Ellul offered an even broader definition. He suggested that the French equivalent term, *la technique*, meant "the totality of methods rationally arrived at and having absolute efficiency in every field of human activity." This rather extreme approach takes the use of the word technology back to its original sense.

The 1996 edition of Webster's Dictionary expands on the earlier approach, defining technology as:

1. the branch of knowledge that deals with the creation and use of technical means and their interrelation with life, society and the environment, drawing upon such subjects as industrial arts, engineering, applied science and pure science or its practice, as applied to industry; applied science.
2. the terminology of an art, science, etc.; technical nomenclature.
3. a technological process, invention, method or the like.
4. the sum of the ways in which social groups provide themselves with the material objects of their civilization.

Definitions of technology by Kline (1995) and MacKenzie and Wajcman (1985), suggest that technology has at least three levels of meaning:

1. Artifacts or hardware: the most basic level of meaning is physical objects made by humans—"artifacts" to anthropologists, "hardware" to engineers.
2. Human activities: a second level includes human activities, the processes involved in making and using hardware. Kline includes in his definition the legal, economic, political, and physical environment in which these activities take place, altogether making up what he calls a "socio-technical system of manufacture."
3. Know-how: a third level of meaning of technology would include the information, skills, methodologies, techniques, and procedures needed to devise, use, maintain, and develop equipment and processes (Kline's "socio-technical systems of use"). Science, the systematic study of people and their natural and built environment, is now an important part, but only a part, of this know-how.

Technology has both technical and social aspects. As stated earlier, we favor a very broad definition that emphasizes its value-laden nature. In this book (although we rephrased them slightly in Chapter 1), we have taken technology essentially to include all five of the characteristics that Monsma (1986) proposed:

- technology is a form of human cultural activity;
- it involves exercising human freedom and responsibility, particularly in choosing problems and in design approaches, that is, it involves making choices in response to normative values, such as those deriving from a belief in God;
- it involves forming and transforming the material world ("the natural creation") and not primarily the sphere of ideas, thoughts or symbols;
- it is done with the aid of tools and procedures, not randomly or whimsically; and
- it is for practical ends or purposes, distinguishing it from art.

▮ Science

Science is defined by the 1996 Webster's Dictionary as:

1. a branch of knowledge or study dealing with a body of facts or truths systematically arranged and showing the operation of general laws: the mathematical sciences.
2. systematic knowledge of the physical or material world gained through observation or physical science ...
4. any of the branches of natural or physical science ...

Kline (1995, p. 9) describes science as "...a group of methods for the purpose of discussing, creating, confirming, disconfirming, recognizing and disseminating truth assertions about nature." He draws a clear distinction between scientific conclusions and religious beliefs, arguing that they fall into different areas of inquiry, with different methodological approaches. He sees religious beliefs as being about values and ethics, rather than about essentially testable and disprovable matters of fact.

Flew (1984, pp. 319–321) asserts that: "Organized empirical science provides the most impressive result of human rationality and is one of the best accredited candidates for knowledge." In principle, the philosophy of science seeks to show wherein this rationality lies, what is distinctive about its explanations and theoretical justifications, how it differs from guesswork and pseudo-science, what basis there is for confidence in its predictions and technologies, and whether its theories actually reveal the truth about a hidden objective reality.

We see the connection between modern science and technology as being close and important, as we have indicated in Figure 10.1. Rapp (1989, p. x) goes so far as to argue that:

> modern science is technology—as far as experimental procedures are concerned. But scientific research does not by itself easily translate into technical practice. To make the transition, all the skill and knowledge of the engineering arts and sciences are needed. Clearly the structure of thinking in the technological sciences, as well as the methodological principles of design and of efficient and purposeful action, exhibit patterns of their own. So they can be by no means be reduced to the philosophy of science.

▌ Engineering

The Engineers' Council for Professional Development has defined engineering as:

> the profession in which the knowledge of the mathematical and physical sciences gained by study, experience and practice is applied with judgment to develop ways to utilize economics, materials and forces of nature for the progressive well being of human kind.

This definition captures many of the qualities of the modern profession. The emphasis on judgment, on constant testing of decisions against knowledge of the laws of nature, on economics, and on human advancement, all characterize the engineering approach to the world. Taking up concerns outlined in the previous chapter, we would want to add to the definition a commitment to ecological sustainability and a recognition of responsibility towards nonhuman life-forms. This is one of the reasons why in Chapter 1 we endorsed the rather more visionary definition of engineering purpose proposed by the Institution of Professional Engineers New Zealand (IPENZ 1993):

> Engineers will translate into action the dreams of humanity, traditional knowledge and the concepts of science to achieve sustainable management of the planet through the creative application of technology.

Engineering has been central to the great economic growth that has characterized the rise of industrial capitalism. Although engineers have been particularly involved in increasing production, as indicated in Chapter 8 they have generally ignored questions of the distribution of the resulting benefits and often even questions of the social character of production processes.

What constitutes engineering knowledge? How do engineers go about analyzing problems and finding possible solutions to them? What sorts of solutions are acceptable?

Although scientific knowledge is now an invaluable analytical tool for the engineer and is often a basis for new developments, analysis does not grapple with the real problem in all its complexity, but deals with more or less simplified models of it. Design is seen as a central engineering activity. Accomplished despite huge gaps in knowledge, it ranges from the development of computer software or consumer goods, to the planning of construction work or the organization of a factory or a production run. The process of design and the importance in it of heuristics (approaches that may solve a particular kind of problem, but offer no guarantee of success) were discussed in Chapter 6 in the context of the process of innovation, the creation and bringing to commercial application of new products, processes, and systems. Engineers play a major part in this process.

The situation of engineers is quite unlike that of "pure" scientists, who are taken to be seeking to address features of an independently existing reality by reasoning in ways that they conceive to be, in principle, universal and context-free. Most engineers accept the applied nature of their activity, and: "... the constraints they must satisfy come from outside engineering: from managerial interpretations of the marketplace, of institutional needs, of political objectives, or of corporate agendas to which engineering excellence as defined by engineers is typically a secondary consideration" (Goldman 1990, p. 133).

The two professions, science and engineering, are very different. At the risk of some oversimplification, one could argue that, over the last century or so, the thrust of science has been towards understanding the nature of things and the causes of their behavior. Its basic question has been "Why?" Engineering is about innovation, the production of commercially viable products, processes, or systems. Its central focus has been "How?"

The high prestige of science and scientists has occasionally led to deliberate confusion between science and engineering in common usage. Vannevar Bush, a professor of engineering at MIT and leader of the U.S. research mobilization in World War II, asserted that, following British attitudes, the American military rated "the engineer [as] a kind of second class citizen compared to the scientist ... So all OSRD [Office of Scientific

Research and Development] personnel promptly became scientists (Reingold and Molella 1976). The same bias against the title engineer has also been evident in Australia, particularly in research-based organizations, where scientific officer has often been the classification used. On the other hand, in the United States and continental Europe engineering is certainly a highly regarded profession.

∎ Engineering Science

Forming a bridge between science and engineering, engineering science is of major importance to engineering, and we need to discuss its origins and recent development. Despite its historic and continuing importance in providing a scientific base for engineering practice, we need to understand that it is only a part, albeit an essential part, of engineering. Failure to appreciate this point has led to serious shortcomings in the teaching of engineering.

The scientific revolution of the seventeenth century ushered in a model of the world as a wonderfully complex but essentially deterministic machine. Technological determinism reflects this type of model, in its assumption that everything that happens is determined by a necessary chain of causation. In the early nineteenth century, insights from thermodynamics suggested that if the Earth was a machine, it was running down. Early in the twentieth century, Einstein re-asserted the importance of the observer, showing that where one stood affected what one saw. Even today, despite the efforts of the pioneers of quantum mechanics and the postmodernists, the machine paradigm still dominates thinking in the physical sciences and, to some extent, the social sciences.

A central feature of the changes to engineering education in Britain in the latter part of the nineteenth century was the development of engineering science, which was seen by the British engineering profession as necessary to be able to compete with "foreigners" on equal terms. W. J. M. Rankine, one of its pioneers and the author of a range of textbooks, saw engineering science as based in the universities, and as intermediate between "pure" science and "pure" practice; it required the development of engineering teaching laboratories and provided a structured framework for design analysis, emphasizing prediction of behavior rather than causation (Buchanan 1985).

Layton (1976) suggests that engineers have their own rather narrow perception of what science is, limiting it to the elements they use: that is, to engineering science. He goes further, arguing that engineers' concern with relationships rather than "causes" reflects their positivist view of engineering science. Pedagogical approaches to engineering science in twentieth-century universities have certainly had a substantial foundation in logical positivism.

■ Computers and Nonlinearity

The advent of the digital computer has had a profound impact on both science and engineering. In many ways it is too soon to assess this impact. The early effect was to reinforce the illusion of determinism. It appeared likely to become possible to compute and predict almost everything. The ability to solve large numbers of simultaneous equations, for example, tended to give the structural engineer a sense of infallibility. We now know that this impression was spurious. The expression "garbage in, garbage out" is all too true. The apparent predictive power of the computer pandered to the rationalist aspirations of engineers, and we were soon hearing of computers replacing wind tunnels, replacing machine tools, and even replacing people. Some of these developments have already happened; others appear impossible to achieve. The new quest is for virtual reality. Quite clearly, in some of these matters technology is setting the pace for philosophy. One aspect of this is the formulation of entirely new theories and world views based on discoveries made as a result of the use of computers. The most famous and significant of these is probably the discovery of deterministic chaos. Most real situations and systems are inherently nonlinear. Such nonlinearities have tended to be beyond the limits of engineering analysis. The tools of what has, somewhat overdramatically, been called "chaos theory," offer the potential for some understanding of nonlinearity and some enhancement of our appreciation of what is random behavior and what only *appears* to be random. In a sense we are glimpsing, for the first time, the real limits to our predictive power. An excellent introduction to chaos theory is given by Gleick (1988).

An outstanding contemporary scientist and philosopher of science is the Belgian mathematician Ilya Prigogine (1984). He has made contributions that are particularly interesting for mechanical engineers. Central to his work is the question of the nature of time. He has been concerned with the contradictions between mechanics, as exemplified by Newton's laws, and thermodynamics, as exemplified by the Second Law of Thermodynamics. He also points out the apparent discrepancy between Darwinian selection in nature, implying greater order, and entropy, implying increasing disorder.

Prigogine saw great hope in new developments in nonlinear dynamics that address some of these questions. Many processes of dynamics that we thought were reversible are, in fact, very sensitive to initial conditions and can exhibit nonlinearities and chaos. Prigogine considered this to be a natural state that corresponds to the introduction of irreversibility. The chaos of dynamics corresponds to the entropy of thermodynamics.

The Newtonian model is still very important in engineering science but, as Prigogine pointed out, it needs to be recognized as a first approximation and fitted into a much broader view of reality. The machine paradigm emphasized stability, order, uniformity, and equilibrium. It focused on linear relationships and closed systems that were essentially reversible. Until recently, such an analysis suited engineers; it was all we could handle with the analytical tools at our disposal.

The new engineering science paradigm needs to deal with disorder, instability, disequilibrium, and nonlinearity. It also needs to be sensitive to the flows of time, even at the simplest level; for example, the recognition in legal liability cases that our knowledge is not complete or static, but progresses. As Kierkegaard noted, hindsight is not the only perspective that needs to be taken into account. We also need to work out how to deal adequately with the relationship between free will (or uncertainty) and determinism. Existentialism may offer some useful insights for this. We need to be sensitive to the roles played by social structures and organizations in encouraging some areas of inquiry and constraining others. These social constraints may be more obvious in hindsight. As an (admittedly extreme) example, Galileo died in 1642 after suffering many years of house arrest for supporting Copernicus in the view that the Earth revolved around the sun. This view was held to clash with Biblical verses such as: "God fixed the Earth upon its foundation, not to be moved forever". Three-and-a-half centuries later, after a twelve-year review, the Church finally rehabilitated Galileo in 1992. Some modern constraints are deliberate and explicit, such as those imposed by ethics committees; others may be more about preventing people from rocking well-established boats. The most serious constraint may be the self-censorship that people impose on themselves (often unconsciously) because of their awareness of how they are expected to think about things.

Warnecke (1993) drew on the insights offered by chaos theory to arrive at new structures and organizational models for manufacturing organizations. He recognized that the characteristics mentioned above—disorder, instability, disequilibrium, and nonlinearity—can also be identified in the corporate world. He then outlined possible new ways of building and controlling "fractal companies" to cope with them.

From "Philosophy of Science" to "Philosophy of Engineering"

Considerable attention has been paid so far to the philosophy of science. First recognized as a separate disciplinary area during the twentieth century, the philosophy of science has provided some valuable insights for engineers. Could the philosophy of science also give us useful guidance in the search for a philosophy of engineering?

Because the practices, purposes, and personnel of engineering and science are essentially different, it seems unlikely that we could just adopt the philosophy of science as a philosophy of engineering. There might well, however, be some features we could adapt.

Max Planck suggested that unless we believe in the enduring reality of external nature, there is no motive for theoretical improvement in science. This attitude is even more relevant to engineering. Bhaskar (1975, pp. 23–24) seems to provide a useful starting point. A materialist, his basic ontological questions (about being, as such, or, first principles and categories) were: "What must the world be like for science [or in our case, engineering] to be possible?" and "What must science [engineering] be like to give us knowledge of real objects and structures that exist independently of us?"

Bhaskar drew the ontological distinction (i.e., between what is primary and what is derived) between the empirical reality we produce, for example by experiment, which he concluded was derived, and the causal law that it enables us to identify. He argued that the latter did not depend on human activity, but on the tendencies of the real things investigated, and was therefore primary.

Bhaskar's ontological approach led to his epistemological position on the character and significance of knowledge. He described scientific knowledge as both taxonomic, about the kinds of things there are, and explanatory, about how they behave. He proposed three progressively more fundamental levels of knowledge: firstly, the invariance of an experimental result; secondly, a correct explanation, based in the nature of the thing studied, for the behavior; thirdly, a real definition of the thing.

We can similarly argue that engineering is a continuing social activity, producing knowledge that is in turn the means for further production of knowledge. This approach to some extent challenges Kuhn's emphasis on a series of incompatible and mutually exclusive paradigms. Bhaskar argued that in the physical sciences a thing could be analyzed and a model of its action demonstrated at each of a series of levels of reality. Thus a Newtonian mechanical model might be seen as a real and useful description of behavior at one level, then new experimental techniques and sense-extending equipment (probably developed jointly by engineers and scientists) would be needed to discover the next level of behavior, for example, the atomic or the very high-speed level.

The approach to epistemology (the origins, nature, methods, and limits of knowledge) taken by Kolb (1984), based on models of experience-based learning, complements Bhaskar's view of knowledge. Kolb defined learning as "the process whereby knowledge is created through the transformation of experience." Knowledge was not, in his view, an independent item or entity to be acquired or transmitted like a bank deposit, but the result of the learning process. Roszac (1986) argued that the bank deposit approach resulted from confusion between data or information, which did not necessarily have human significance, and ideas and understanding, which did. The acquisition of knowledge was essentially a process of building on previous experience. This model of the learning process would explain, for example, the strength of the cooperative approach in engineering education; it involves alternating periods of structured work experience and theoretical study (including reflection on practical experience), so that each type of experience enriches and reinforces the other. This pattern of education, continued throughout life, is important to every practicing professional.

More generally, these arguments and those of contemporary philosophers of science such as Feyerabend (1988), emphasize the difference between education and training. It is vital for engineers to understand the difference and recognize the role and value of both. Adaptability is at a premium in times of rapid change in society and technology, and it is even more important than it has been for engineers to be able to change direction intelligently and rapidly. In 1605, Francis Bacon summed up beautifully the limitations of an overemphasis on technique:

> Another different error is the ... peremptory reduction of knowledge into arts and methods, from which the sciences [areas of useful knowledge] are seldom improved; for as young men rarely grow in stature after their shape and limbs are fully formed, so knowledge, whilst it lies in aphorisms and observations [broad general principles], remains in a growing state; but when once fashioned into methods [specific applications], though it may be further polished, illustrated and fitted for use, is no longer increased in bulk and substance.

Although scientific knowledge is now necessary for engineering, it is not sufficient. Science has come to underpin much of engineering practice, but the usual purposes of engineering and of engineers are quite different from those of science. Scientific results are not supposed to depend on a particular social context, although writers such as Martin (1982) and

The Paradigm Model and Engineers

Although there are significant differences between scientific and engineering paradigms (Constant 1980), there is an obvious need for engineering developments to be analyzed in sociohistorical studies of the type undertaken by Kuhn (1970).

The paradigm model is relevant to us, as engineers, in a number of ways:

- This model gives us an insight into the ways in which the theory and the professional practice of science are different from those of the humanities.

- It reminds us of the style of learning in the engineering sciences that engineering students are initially exposed to and that they continue to need for advancing specialist knowledge. However, study of the relationships between engineering and society involves very different skills and can help students with the transition to the practice of engineering, which is as much art as science, and is very much involved with people.

- Given that engineering draws upon the pool of accumulated scientific knowledge, there is a reciprocal obligation for engineering to foster and nourish basic science. This is now especially relevant to those areas of science (e.g., fluid mechanics) that have been virtually abandoned by the science community and are principally undertaken within engineering. Rosenberg (1972, pp. 170–171) noted that, although improvements in

Wynne (1988) have pointed out that this context is important in the allocation of funding and the selection and organization of the work. Engineering is inherently context-driven. As Goldman (1990, p. 129) points out, engineering results "never lose their particularity and are *explicitly* inseparable from the intentional, contingent, willful, and value-laden contexts of their formulation."

Whatever the central philosophical issues for engineers and engineering may be, they are clearly not identical with those of science. However, as we have seen, the philosophy of science does have a contribution to make to a philosophy of engineering. A philosophy of engineering will need to go on to address the question of the values and priorities that underpin technical choices.

technology have been critical to the advancement of science, the relationship between the two has been badly neglected.

- What are the sorts of paradigms within which we as engineers operate? Many of our paradigms have started to change in the last decade, with the emergence of the personal computer as an increasingly powerful and affordable tool. Broadband communications and powerful databases are starting to change the way companies (and universities) handle and transfer information. Further, how will engineering paradigms need to change as we move towards sustainable management of the environment?

- Barker (1989) points out that the people who create new paradigms tend to be those on the fringes of the dominant paradigm who are not well served by it and have little or nothing to lose by its replacement. This is one of the reasons why interdisciplinary activity is so productive of new ideas. The effect of a new paradigm in an area—such as the invention of the transistor, or the quartz movement for watches—is that everyone is back to square one. Success in the old paradigm is no guarantee for the future. And it is vital to remain open to new paradigms in order to avoid being paralyzed by the old. Barker suggests that we need regularly to ask ourselves: "What sorts of things are not now possible, but would change the ways we do business [or in our case, engineering practice] if they were?"

From "Philosophy of Technology" to "Philosophy of Engineering"

Engineers provide bridges between science and technology, and between technology and commerce. They play an essential part in meeting the material requirements of society and in the generation of wealth. In these processes we see technological innovation as the central, dynamic element. It is therefore important to explore the contribution that the philosophy of technology can make to a philosophy of engineering.

The philosophy of technology is an emerging and increasingly important discipline. Ihde (1993) is a good introductory text for this field. Some of the issues it attempts to address are:

- How does technology evolve?
- How are the choices made as to which potential technologies will be developed and which ignored?
- Who makes these choices?

We have already discussed the process of innovation and the phenomena of technology push and market pull. The former focuses choice in the management of organizations involved in research, design, and development, and the latter limits it to people who are financially able to buy. Cultural factors, significantly shaped by historical tradition, also play an important role in the selection of technologies. As McGinn (1991) demonstrates, in order to benefit from technological processes or systems, we have to adjust to the way they work: we pay a price for our benefits. The price for modern technology includes the specialized division of labor and the standardization of many of the elements that shape our lifestyles. Technology shapes many of our priorities—inevitably, because it produces the material framework within which we live. Technology is thus a phenomenon of both nature and culture. This was indicated in our discussion of the transfer of technology in Chapter 8.

Because technology has been so successful, it has changed the way we live. Along with the intended results (which are in any case not always benign, at least for those at whom we aim the technology) there are often unintended and unexpected secondary effects. Many of these results may not be new phenomena, but aggravations of existing problems, such as salinity, deforestation, the enhanced greenhouse effect.

One of the lessons of history is that, once the first step has been taken, it is difficult if not impossible to stop a development. This is the point of environmental impact studies, and why in genetic engineering, for example, detailed discussion of fundamentals is essential *before* the technology proceeds. Such discussion demands both an understanding of the problem and a sense of responsibility. Rapp (1989) suggests it is encouraging that we have become much more sensitive to these issues. He concludes that we have come to a much deeper awareness of the inherent ambivalence of technology.

Langdon Winner (1977) argues that, regardless of how deterministically we regard technologies, they are clearly not neutral. In *The Whale and the Reactor* (1986) he developed this theme, showing how technologies become cultures or ways of life. Humans are absorbed into technological systems, and the artifacts they create perpetuate political views. In this approach, Winner appears to come close to the views of many contemporary ecologists and feminist philosophers, who stress the patriarchal tendencies of technology.

The discussion of innovation in Chapter 6 concentrated on techno-logical issues, adopting a relatively determinist approach. It did not real-ly challenge Heidegger's argument that technology takes the world as a well of resources. An effective challenge to such technological determin-ism would need to explore the social and economic processes involved in the choice and development of technologies. Goldman (1990) develops ideas within the philosophy of technology that start to explore these processes, and that suggest the likely shape of a philosophy of engineer-ing. Describing technology as "a social process to which engineering con-tributes," he goes on to assert that:

> ... the definition of engineering problems, the determination of the means to be used in solving them, and the identification of what will count as solu-tions, all derive from the institutional content of engineering's practice, not from the knowledge engineers possess, and certainly not from nature. (p. 125)

Engineering, as a part of technology, is involved in changing the world. Because this is so, Goldman suggests that it: "intrinsically and inescapably raises moral, political, and aesthetic questions ... it is implicated as well in the struggles of powerful vested interests acting to control, and to ben-efit from, that change." (p. 137) Marx argued that the aim of philosophy should be to change the world. Goldman sees this proposition as need-ing careful qualification. Goldman argues that it is not all that difficult to change the world, in ways both superficial and profound, but that it is very difficult to ensure that the result of change is an improvement.

Goldman's major focus is therefore on the purposes of engineering, which he illustrates by a discussion of the central role of design: "an ex-plicitly valuational activity, a necessarily non-unique synthesis of a "box" of means, a set of imposed constraints, some natural, most arbitrary, and a fuzzy vision of an end to be achieved." (pp. 131-132) He quotes Edwin Layton, an important writer on the theory and practice of engineering: "From the point of view of modern science, design is nothing, but from the point of view of engineering, design is everything." (p. 149) Given the syn-thetic character of design and the importance to it of visualization, Goldman argues that teaching of the design process has suffered as a re-sult of an overemphasis on scientific analysis in both teaching and research.

Johnston et al. (1996) take this point further, arguing that the discourse of engineering has become "captive" to the discourses of science and busi-ness, which have largely been allowed to define engineering knowledge and forms of practice. They suggest that the domination of engineering education by the discourse of science has limited the types of problems posed and the types of solutions accepted. In the process, instead of open-ended, design-oriented problems, which recognize the social context of engineering and are intended to foster creativity, the emphasis has moved too far towards closed problems with a single correct answer. In industry, they see the domination of engineering organizations by a business rather than an engineering ethic as tending to lead to de-engineering of these organizations and eventually to their inability to carry out central technical tasks safely and effectively.

An important practical issue here is that design teachers are commonly generalists, with substantial industrial experience but few publications, so they find it hard to compete for promotion with engineering scientists, a matter that can be a sensitive point in university engineering schools. Fundamental engineering design information obviously needs to be produced, and many engineering academics take up essential and otherwise ignored problems in the areas of scientific and engineering scientific research. An appropriate balance needs to be struck, with a more conscious partnership between designers and engineering scientists.

PHILOSOPHY, ENGINEERING, AND VALUES

The way in which engineering has been taught for most of this century has been dominated by the engineering science approach. It has stressed the importance of a thorough grounding in the relevant scientific disciplines and emphasized the high-tech research areas, commonly at the expense of serious attention to, or even recognition of, the importance of design process and practice. As the technical content of engineering programs has increased, treatment of broader social issues has tended to be squeezed out. An underlying assumption has been that technology is value-free and that therefore any consideration of human emotions, needs, and aspirations is extraneous, if not irrational.

Many engineering courses have been structured so that they avoid explicit value judgments. Part of the justification for this has been the argument that the organization and management of engineering (and other) activities can be separated from the technical aspects. With increasing responsibility, engineers are then expected to "transfer" from engineering, with a narrow technical focus, to management, where they have to deal with people, including nonengineers. The inadequacy of this approach becomes evident when graduate engineers emerge into the marketplace. The stereotypical engineer has been white, male, conservative, unemotional, and task-oriented, but at work he has had to deal with feminists, militant unionists, environmentalists, politicians, and all sorts of nonengineers, including people who do not act as "rationally" as he would like. Many engineers have failed to recognize that every situation in which they act is value-laden. This failure is a major threat to the standing of the profession. Engineers are constantly called upon to make value judgments, and their preparation must develop in them the necessary skills, sensitivity, and insight.

We see engineering as a creative activity, as much art as science. A philosophy of engineering would seem to be possible and helpful—even essential. It would provide a basis for better education and training and better understanding of the roles of engineers and engineering in our society, all key elements in ensuring that advancing technology contributes to social progress.

A major aim of this book is to clarify the relationships between engineering practice and human values. Because technology embodies the values and aspirations of its creators, it cannot be value-free or ethically neutral. Only by acknowledging its problems and seizing its opportunities will we challenge the assumption of autonomy and develop technologies that respond to social demands. Because we do not have a developed philosophy of engineering, or even of technology, we are not even close to a solution to what could now be the central ethical question for the human race: the control of technology.

The central issue for a philosophy of engineering will probably be clarification of the meaning and significance of engineering activity, both as it might ideally be carried out, and as it actually is. Case Study 7.5 (see p. 392) on bridge (and engineering management) failures illustrates some of the differences between ideal and actual performance of engineering activities.

A philosophy of engineering will need to take account of the fact that engineering is carried out by real, fallible human beings in specific social, cultural, and economic conditions. We see the lack of an adequate philosophy of engineering as a major problem for the engineering profession. Such a philosophy is essential for the assertion of professional identity and the formulation of clear directions for future development.

We have made a start in this chapter. A comprehensive philosophy of engineering will have to deal adequately with the character and significance of engineering practice, the nature of engineering knowledge and the way it is learned, and the roles in engineering of art, aesthetics, creativity, and judgment. The development of such a philosophy is a challenging task, but it is likely to have profound implications for engineering education.

DISCUSSION QUESTIONS

1. Assuming the acceptance of the quotation from Lord Ashby at the beginning of this chapter, elaborate on the concepts of a "liberal education" and a "good technologist."

2. How does your understanding of scientific method fit in with the description of it in this chapter? How relevant is the work of Popper and Kuhn to the practice of engineering?

3. How would you start to develop a philosophy of engineering? What issues would you seek to address? What questions would you ask? Where and to whom would you look for answers?

4. What influences do you think Eastern philosophies, including their values and attitudes, have had on engineering?

5. Is existentialist philosophy really compatible with engineering?

6. Discuss Marx's proposition, that philosophers have explained the world, but the point is to change it. Goldman (1990) went on to argue that the real point was not just to change, but to improve the world. Do engineers usually implement change in the way Marx or Goldman suggested?

7. From your own work experience, of from an example in the literature, discuss an example of how one technology has been preferred to another.

8. Is it possible for us to control technology?

9. Is it desirable to control technology? What sorts of consequences might such control produce?

FURTHER READING

Burke, J. 1985, *The Day the Universe Changed*, BBC, London. This is the book of the television series, and looks at discoveries that changed the way people saw the world.

Florman, S. C. 1987, *The Civilized Engineer*, St. Martin's Press, NY. Florman has written extensively and interestingly on engineering and its context.

Hesse, H. 1951, *Siddhartha*, Bantam, NY. This is the tale of one person's search for meaning in life. Its approach is an interesting contrast to that of the typical engineer.

Ihde, D. 1993, *Philosophy of Technology: An Introduction*, Paragon House, NY. Ihde's book is a readable introduction to this expanding field.

MacKenzie, D., Wajcman, J., Eds., 1985, *The Social Shaping of Technology: How the Refrigerator got its Hum*, Open University Press, Milton Keynes. This book gives exciting examples of how technology is developed and of nontechnical forces that have affected the choice of both civil and military hardware.

McGinn, R. E. 1991, *Science, Technology, and Society*, Prentice-Hall, New Jersey. This is a rich analysis of contemporary issues in these areas that deserves to be widely read.

Noble, D. F. 1977, *America by Design: Science, Technology and the Rise of Corporate Capitalism*, Oxford University Press, Oxford. Noble discusses the institutional structure within which engineers in American industry worked, and their role in its transformation between 1880 and 1930.

Pirsig, R. M. 1974, *Zen and the Art of Motorcycle Maintenance: An Inquiry into Values*, Bantam Books, NY. This draws on Eastern philosophy to review the subjective character, the "quality," of the machines we design, build, and maintain.

Russell, B. 1946, *A History of Western Philosophy*, Allen & Unwin, London. Russell sets the philosophers in their historical and social contexts.

Technology & Culture, a journal published by the University of Chicago, focuses on the history of technology, but also includes useful interdisciplinary material, including some papers cited below.

Winner, L. 1986, *The Whale and the Reactor*, University of Chicago Press, Chicago. Winner explores the practical politics of engineering ethics and decision making in a thorough but very readable way.

Wynne, B. 1988, Unruly Technology: Practical Rules, Impractical Discourses and Public Understanding, *Social Studies of Science*, 18, pp. 147–167. Brian Wynne's work on how technologies are actually applied is wonderfully insightful.

References

Albury, R. 1983, *The Politics of Objectivity*, Deakin University Press, Victoria.

Ashby, E. 1966, *Technology and the Academics*, Macmillan, London, pp. 71-88.

Ayer, A. J. 1946, *Language, Truth and Logic*, Pelican, UK.

Bacon, F. 1605; 1973, *Advancement of Learning*, Dent, London.

Barker, J. A. 1989, *The Business of Paradigms*, videocassette, Charthouse Learning Corp, Burnsville, Minn.

Bhaskar, R. 1975, *A Realist Theory of Science*, Leeds Books Ltd., Leeds.

———, 1993 *Dialectic: The Pulse of Freedom*, Verso, London.

Buchanan, R. A. 1985 The Rise of Scientific Engineering in Britain, *British Journal for the History of Science*, 18 (2, 59), July, pp. 218-233.

Burke, J. 1985, *The Day the Universe Changed*, BBC, London.

Constant, E. W. 1980, *The Origins of the Turbojet Revolution*, Johns Hopkins University Press, Baltimore.

Davison, I. 1977, *Values, Ends and Society*, University of Queensland Press, St. Lucia.

Donovan, A. 1986, Thinking about Engineering, *Technology and Culture*, 27 (4), October, pp. 674-679.

Ellul, J. 1964, *The Technological Society*, translation by John Wilkinson of "La Technique," (1954), Knopf, NY.

Feyerabend, P. K. 1988, *Against Method: Outline of an Anarchistic Theory of Knowledge*, Verso, NY.

Flew, A., Ed., 1984, *A Dictionary of Philosophy*, Pan, London.

Florman, S. C. 1976, *The Existential Pleasures of Engineering*, St. Martin's Press, NY.

Ford, G. W. 1986, The Concept of Skill Formation, *Engineers Australia*, 11 July, p. 40.

Foucault, M. 1970, The Discourse on Language, In: M. Foucault, *The Order of Things*, Tavistock, London.

Gleick, J. 1988, *Chaos*, Cardinal, London.

Goldman, S. L. 1990, Philosophy, Engineering and Western Culture, In: Durbin, P. T., Ed., *Broad and Narrow Interpretations of Philosophy of Technology*, Kluwer Academic, Dordrecht, pp. 125-152. Quotations reproduced with kind permission from Kluwer Academic Publishers.

Heidegger, M. 1993, The Question Concerning Technology, In: Krell, D. F., Ed., *Martin Heidegger: Basic Writings*, Harper San Francisco, San Francisco, pp. 311-341.

Ihde, D. 1993, *Philosophy of Technology: An Introduction*, Paragon House, NY.

Institution of Professional Engineers New Zealand (IPENZ) 1993, *The Pathway to the Future*, Wellington, NZ.

Johnston, S., Lee, A., McGregor, H. 1996, Engineering as Captive Discourse, *Society for Philosophy and Technology Quarterly Electronic Journal*, 1 (3/4), on the World Wide Web at <http://scholar.lib.vt.edu/ejournals/SPT/spt.html>.

Kline, S. J. 1995, *Conceptual Foundations for Multi-Disciplinary Thinking*, Stanford University Press, Stanford.

Kolb, D. A. 1984, *Experiential Learning: Experience as the Source of Learning and Development*, Prentice-Hall, Englewood Cliffs.

Kuhn, T. S. 1970, *The Structure of Scientific Revolutions*. University of Chicago Press, Chicago.

Layton, E. T., Jr. 1976, American Ideologies of Science and Engineering, *Technology and Culture*, 17 (4), October, pp. 688-701.

MacKenzie, D., Wajcman, J., Eds., 1985, *The Social Shaping of Technology: How the Refrigerator got its Hum*, Open University Press, Milton Keynes.

Martin, B. 1982, The Naked Experts, *The Ecologist*, 12 (4), pp. 149-157.

McGinn, R. E. 1991, *Science, Technology, and Society*, Prentice-Hall, New Jersey.

Monsma, S. V., Ed., 1986, *Responsible Technology: A Christian Perspective*, Eerdmans Publishing, Grand Rapids, Michigan.

Mumford, L. 1964, Authoritarian and Democratic Technologies, *Technology and Culture*, 5 (1), pp. 1-8.

Nietzsche, F. 1961, *Thus Spake Zarathustra*, Penguin, UK.

Pacey, A. 1983, *The Culture of Technology*, MIT Press, Cambridge, Mass.

Pirsig, R. M. 1974, *Zen and the Art of Motorcycle Maintenance: An Inquiry into Values*, Bantam Books, NY.

Popper, K. R. 1963, *Conjectures and Refutations*, Routledge and Kegan Paul, London.

Prigogine, I. 1984, *Order out of Chaos: Man's New Dialogue with Nature*, Heinemann, London.

Rapp, F. 1989, Introduction: General Perspectives on the Complexity of Philosophy of Technology, In: Durbin, P. T., Ed., *Philosophy of Technology: Practical, Historical, and other Dimensions*, Kluwer Academic, Dordrecht, pp. ix-xxiv. Quotation reproduced with kind permission from Kluwer Academic Publishers.

Reingold, M., Molella, A. 1976, Introduction to Conference Proceedings on the Interaction of Science and Technology in the Industrial Age, *Technology and Culture*, 17, pp. 624-633.

Rosenberg, N. 1972, *Technology and American Economic Growth*, Harper & Row, NY.

Roszac, T. 1986, *The Cult of Information, the Folklore of Computers and the True Art of Thinking*, Pantheon, NY.

Sartre, J. P. 1957, *Existentialism and Human Emotions*, Philosophical Library, London.

Sinclair, G. 1977, A Call for a Philosophy of Engineering, *Technology and Culture*, 18, October, pp. 685-9.

Tarnas, R. 1991, *The Passion of the Western Mind*, Harmony Books, NY.

Warnecke, H. J. 1993, *The Fractal Company: A Revolution in Corporate Culture*, Springer-Verlag, NY.

Wilson, C. 1963, *The Outsider*, Pan Books, London.

Winner, L. 1977, *Autonomous Technology: Technics-out-of-control as a Theme in Political Thought*, MIT Press, Cambridge, Mass.

Winner, L. 1986, *The Whale and the Reactor*, University of Chicago Press, Chicago.

Wittgenstein, L. 1922, *Tractatus Logico-philosophicus*, Routledge and Kegan Paul, London.

Wynne, B. 1988, Unruly Technology: Practical Rules, Impractical Discourses and Public Understanding, *Social Studies of Science*, 18, pp. 147–167.

Ethics and Professionalism

In the previous chapter we argued the need to develop a philosophy of engineering. This process must include the study of engineering ethics. Typical questions that engineers might ask themselves are: What are the connections between meaning well and doing good? Should we refrain from action because we cannot foresee the consequences? Although reluctance to act has not been typical of engineers in the past, do we need to be more cautious in the future? Would a Hippocratic Oath for engineers make a difference to our behavior? Engineering codes of ethics, such as those developed by major professional engineering bodies, are important guides in defining the social responsibilities of engineering professionals. We explore some of their implications in this chapter.

We go on to consider the responsibilities of professional engineers and the rewards they receive from society for their work. After discussing the term "professional" and considering how its meaning has developed, we look at how professions are differentiated from other occupations. We then describe how the engineering profession is organized nationally and internationally and the educational requirements for membership of the profession.

We conclude this chapter (and our book) with a discussion of the key directions in which we believe the engineering profession needs to move in the future.

Ethics and Social Responsibility

Our understanding of the character and significance of engineering activity is greatly enhanced by a consideration of ethics, the aspect of philosophy that focuses on personal and professional values and the effects those values have on others. The ethical issues we address in this chapter include ways in which the engineering profession can encourage creativity while remaining sensitive to the ways in which our creations affect our fellow humans. Specifically, we challenge engineers to practice their profession in a socially responsible and moral manner.

Engineers change society through the artifacts they create and the processes they develop to create them. What responsibility do engineers have for these changes? Can we lay down hard and fast rules for engineering behavior? Should we try?

Even where the methodology employed in problem-solving exercises is scientific and apparently value-free, neither the selection of the problems to be solved nor the criteria for their solution in fact are. Both are heavily weighted by cultural and economic factors within our society, and of course by the organizations for which we work. In this process, questions of importance to the engineering profession are often ignored, issues such as who chooses the problems or selects the projects. What groups or individuals decide on building a freeway rather than a light rail system? Who sets the criteria for a new dam or a new type of word processor? What input is there from the people who are most affected by such changes? Do the engineers involved have a say? Should they?

The debate about whether technical professionals have responsibility for the outcomes of their work has been going on for centuries. When the Royal Society of London was formed in 1663, its curator of experiments, Robert Hooke (1635–1703), reminded the members that their business was "to improve the knowledge of natural things, not meddling with Divinity, Metaphysics, Moralls, Politicks, Grammar, Rhetorick or Logic."

However, by the eighteenth century there was a widespreadexpectation that the new sciences and technologies would be instruments for carrying out a comprehensive transformation of society. Science had liberated the ideas of those who read and reflected. Since the middle of the nineteenth century, the explicit association of science with broad goals of social and political advancement has weakened. The continuing emphasis on the importance of science and technology, however, has led to a narrow technocratic view of progress. This assumes that a society only needs to ensure the advance of science-based technologies and the rest will take care of itself. The rate of technological innovation is treated as the essential criterion of social progress.

Some engineers today still feel inclined to agree with Robert Hooke's sentiments. It is as though they were living in the seventeenth century with Hooke, before the motor vehicle, television, the atom bomb, the computer, and space travel provided ample evidence that quite dramatic meddling is the inescapable outcome of the work they do. Responsible engineers cannot now sit back and claim they are detached from other human concerns, from personal safety to local and global environmental, economic, and political issues. The ways in which particular technologies tend to incorporate the values inherent in the circumstances of their creation are discussed elsewhere in this book, particularly in terms of production systems and their management. Awareness of these values is important for socially responsible engineering practice. We believe that to be good engineers, men and women need first to be good citizens, to know how their society functions, and to take a proactive role in that society.

Hooke's prescription for natural philosophy in the seventeenth century is inappropriate for engineering in the twenty-first century—the role of engineers is, and always has been, to "meddle." However, our meddling must be beneficial, not harmful. In order to sustain and enhance the quality of life on this planet and ultimately beyond it, we need to understand the context in which we interfere. First we must appreciate what is happening and the implications of our activities. We can then go on to consider how to minimize the negative consequences and maximize the benefits of our work. Encouraging students to develop this sort of critical, reflective approach to engineering practice is an extremely challenging and increasingly central responsibility for engineering educators.

Case Study 11.1 provides one example of genuinely concerned technical professionals. There are a great many local initiatives of this sort among engineers in many countries and their role among their peers in challenging unthinking assumptions and raising awareness is of inestimable value.

SOCIALLY CONCERNED COMPUTER PROFESSIONALS

The computer industry is very competitive; every company is trying to establish itself as the best, taking whatever measures are necessary. Those involved in the industry take their jobs very seriously, and see this pursuit of fortune, fame, and power as legitimate. However, those outside the industry mostly see the antitrust allegations, corrupt business practice, and cutthroat business deals. It is often thought that the industry's relentless pursuit of profit causes the industry to disregard potential societal ramifications. Although this may be accurate for certain enterprises, one group of computer professionals is dedicated to sustaining a socially responsible computer industry. This organization is the Computer Professionals for Social Responsibility (CPSR).

CPSR began as a board consisting of members of the Xerox Palo Alto Research Center and members of Stanford University. This small group began meeting weekly, mostly in a discussion-forum atmosphere. The focus of their discussion back in October of 1981 was the potential dangers of computers, mostly in the realm of military usage. At that time the group had yet to adopt the name, but in over seventeen years their focus has remained the same: to express a concern for the possible (negative) effects of computers on society. The members attempt to predict problems before they happen, and suggest safer, more practical alternatives *before* any major changes can be adopted.

CPSR has expanded to express concerns in many other areas besides military computing. CPSR has an extensive list of concerns that address such issues as the possible misuses of Caller ID, the gender gap in the area of computing, the ethical computer usage issues, and the issues of privacy and civil liberties. CPSR is able to address such concerns because the organization is made up of computer professionals, who know how systems work and are more likely to be able to predict possible misuses.

CPSR and the National Information Infrastructure

One of CPSR's campaigns relates to the National Information Infrastructure (NII)—a topic of discussion since the Clinton administration introduced the NII in February of 1993. As the NII seeks to improve the communication between people, CPSR suggests there are some social, cultural, and economic issues that must be taken into consideration along with the NII. These "Fundamental Principles" are expressed in CPSR's major publication entitled "The NII: Serving The Community." The fundamental principles are as follows:

1. Universal Access
2. Freedom to communicate
3. Vital civic sector
4. Diverse and competitive marketplace
5. Equitable workplace
6. Privacy
7. Democratic policy making
8. Functional integrity

CPSR is suggesting an approach to the NII that will not broaden social gaps and will not create any problems with privacy. It is also recommended that the NII not be under the complete control of any one governing power. It is the goal of CPSR for the NII to be a self-governing body, which fits into the guidelines suggested by the fundamental principles. These principles can be seen as the goals for an idealized NII.

This group of computer professionals seeks to provide a functional model for a NII that is as justifiably beneficial to the average citizen as it is for businesses. The focus of CPSR's efforts in the NII project is to prevent corporate or governmental monopolization of the NII. The goal is to make sure the NII emphasizes communication rather than focusing exclusively on commerce. CPSR also favors not only a NII in which every one has access, but a system in which there is public space available and the public has unrestricted access to government information. CPSR is also aware that precautions must be taken to keep private information secure, and at the same time, on the same information infrastructure, allow unrestricted access to "public" information. Precautions should also be taken to ensure that the NII allows for unrestricted international communication. With all of the declared freedoms of the NII, it is most important that the content of the NII not be censored. For this reason, CPSR supports the Blue Ribbon (Free Speech Online) campaign.

It is refreshing to see an organization addressing the complicated issues of a rapidly changing technological era. Though the organization is still young, it has already made an impact on the "information age." CPSR's activism in computing is an important step towards a society in which the engineer is as concerned with the macroscopic social ramifications of a particular innovation as with the microscopic short-term benefits. If each and every engineer has social concerns as a part of the "engineering thought process," we will not only have a better-educated collection of engineers, but the innovations will be better thought out, with more attention to the potential problems they may raise.

Though it is not realistic to think that a group such as CPSR can foresee all of the potential problems in a weekly discussion forumatmosphere, such a collaborative effort to address these social concerns is a good start. There would be little societal improvement if only a few engineers were to follow the model of CPSR; what is really needed is for every engineer to make a commitment to bringing about technological innovation that improves our society overall. As CPSR points out, it is not good enough to create a National Information Infrastructure without providing equal opportunity for access. Failing to do this will only broaden the already enormous social and economic gaps between the information poor and the rest of society.

If a NII were constructed that ignored globalization, it would isolate the United States from the rest of the world's information. CPSR would then ask, where could people go to take advantage of these innovations if they cannot afford to have a computer in their house? How could we conceive an information infrastructure that has the potential to cause isolation within the United States as well as isolate the United States from the rest of the world?

Simply put, these concepts were overlooked in the original plan for the NII. Fortunately, a collaborative effort such as CPSR does exist, and these issues have been addressed. However, it is not enough for CPSR to think in this manner. Each and every one of the practitioners in engineering needs to take social considerations into account for every new innovation.

For this reason, the heading on CPSR's web-site, which says, "Technology is driving the future; the steering is up to us," is very appropriate. As the computer industry forges ahead, it is important that an influential group of concerned computer professionals is available to guide the innovations in a way that is socially beneficial.

Source

Contributed by Randall Fish, Stanford University, 1997, based on interviews and the Society web site.

WHAT IS ETHICS?

It is useful to differentiate between morals, which center on rules of right conduct for individuals, and ethics, which center on systems of moral principles and rules of conduct for the behavior of groups; in our case the group defined is engineering professionals.

Based on Flew's definition (see Key Concepts and Issues box on ethics) we could suggest a spectrum of codes of behavior with an increasing degree of concern to ensure moral and socially responsible and responsive

Key Concepts and Issues

Ethics

To a lay person, ethics suggests:

"a set of standards by which a particular group decides to regulate its behavior—to distinguish between what is legitimate or acceptable in pursuit of their aims and what is not."

This definition is not specifically concerned with whether the agreed behavior is "right" or "wrong." In philosophical terms, ethics is:

"...an investigation into the fundamental principles and concepts that are or ought to be found in a given field of human thought or activity. Being a branch of philosophy, it is a theoretical study. As such, it differs from "ethics" in the lay sense described above in that any actual body of ethical belief will be intended to be a practical guide to living and not merely an exposition and analysis of certain theoretical doctrines" (Flew, 1984, p. 112).

Ethics is the branch of philosophy concerned with what is morally right and wrong, but there are wide differences in the approaches that different philosophical schools take to ethics. These include:

1. From the perspective of *linguistic analysis*, ethics is concerned with analyzing moral concepts and arguments, which may be scientifically unverifiable expressions of approval or disapproval.

2. *Situation ethics* arose out of existentialism. It is openly subjective, and asserts that there are no absolute rights or wrongs. The focus is on the individual who has a moral problem or choice; the factors considered to be important are authenticity, autonomy, and deciding to act rather than allowing the situation to control the individual.

behavior. The spectrum could run from a code of etiquette through a code of conduct, to a code of ethics. The first of these terms carries an implication that it is social proprieties, rather than a concern for justice or moral excellence, which drives behavior. The second and third terms are often used interchangeably, and carry an implication of moral concern.

Engineers, as educated professionals, should be familiar with ethical theories. They should also be explicitly aware of the values by which they live and how those values conform or conflict with the values inherent in the various ethical theories.

3. *Ethical relativism* draws its principles from the society in which it is practiced. Right and wrong are relative to a particular society; what is perfectly acceptable in one society may be completely repugnant in another, or within the same society at another time.

4. *Deontological ethics* calls its adherents to the observance of absolute moral laws. Hurting others is always wrong, helping others is always right. It may be best summed up with the golden rule, "Do unto others as you would have them do unto you."

5. *Ethical egoism* asserts the claim that every action of every individual is motivated by self-interest. Even philanthropic behavior results in "feeling good." Of course, it may not be in our self-interest to have everyone else acting in their own self-interest.

6. *Utilitarian ethics* judges the rightness or wrongness of an action by its consequences. It moves the emphasis from self-interest to the interests of others. Whatever action brings about the greatest good is the correct action. Some bad is justified by much good.

However, our concern here is with the *application* of ethics rather than its theory. We will concentrate on *normative ethics*, which is the investigation of how individuals and organizations *ought* to behave where moral considerations are involved.

Ethics includes:

- sets of standards by which to distinguish what is legitimate or acceptable in pursuit of our aims and what is not; and
- rules of conduct that are necessary for, or at least an aid to, the purposes of a profession.

Albert Schweitzer once wrote that ethics up to now had been incomplete because it had held that its chief concern was merely with the relationship of man to man. This was, indeed, an enormous mistake; it could only be sustained by a belief that science, technology, and human relationships with the physical world generally are ethically neutral, and also that we as individuals have no moral obligations to the natural world.

Human-Centered Technology

Because modern technology increasingly shapes and influences everything we do, there are huge responsibilities on those who develop it to do so thoughtfully and responsibly. It cannot be assumed that technology itself is ethically neutral, or that it is only the uses to which it is put that raise ethical issues. We explicitly reject this view. Choices have to be made as to what technologies are developed and what are not; for example, whether labor-eliminating or skill- enhancing approaches are to be preferred. We believe that, because technology embodies the values and aspirations of its creators, itcannot itself be ethically neutral. We see technology as essentially ambiguous, with both positive and negative features. Only by recognizing the problems and seizing the opportunities can we reach meaningful answers to ethical questions.

We suggest that engineers are in a unique position to give leadership in establishing the priorities and essential checks and balances our society must achieve in order to prosper. Within the work situation there are important issues related to the way technology is applied in job design that illustrate the ambiguity of technology and that professional engineers need to consider. They may raise ethical questions, particularly because most codes of ethics specify that the professional's responsibility to the community (which includes employees) comes before responsibility to his or her employer.

How might ethics relate to our responsibilities in the design of production jobs? Is designing satisfying jobs an ethical imperative? Rosenbrock (1990) argues that our dominant model of production (and ultimately of people) is still a "machine" one. The machine operates on the basis of cause and effect. It is designed to meet a purpose, but does not itself have purpose, particularly a human purpose. This machine approach is consistent with factory workers being treated as rather inferior robots, to be programmed and replaced as soon as suitable machines have been developed that can perform their particular functions.

Rosenbrock proposes an alternative way of thinking about these issues, based on consideration of specifically human purposes. He characterizes sustained and deliberate aesthetic and creative activities, such as art, craft, architecture, and engineering design, as specifically human; these activities resist formalization and reduction to routine. Rosenbrock advocates the development of a quite different type of technology, in which machines serve people, rather than vice versa. Computer-based technologies would be designed to support and enhance human skills, rather than steal them away.

These challenging concepts may make commercial as well as emotional sense in the new manufacturing environment, with its premium on flexibility and quick responses. Although new technologies may have been intended to reduce the need for operator skill, Rosenbrock argues that the opposite has turned out to be the case. Computer-aided techniques appear to make a human-centered approach economically feasible. Indeed, getting the full benefits from applying programmable technologies to manufacturing may require this approach, as we suggested in Chapter 7 in the section on *high-performance* design.

I NDIVIDUALS AND ETHICAL CHOICES

There is a widely held view that technological systems are autonomous and have a momentum of their own, and that it is simply naive to believe otherwise. Let us look for a moment at the ethical considerations in the development, construction, and deployment of nuclear weapons. One stated aim of nuclear weapons systems was to be able to inflict such terrible damage on an opponent that war would become unthinkable. However, there is a possibility that such systems might become dangerously unstable and wipe out most of the world's population. Might we then consider whether such systems are intrinsically evil? This involves thinking about the character of the choices involved and coming to an ethical view on them. The international ban on weapons for chemical and biological warfare, and the global discussion on antipersonnel land mines, both illustrate the possibility of arriving at such a view. Any engineer, then, who is offered a job that involves designing or supporting such weapons must make an important value judgment.

We have already indicated that individual engineers are constantly called upon to make value judgments. The question for ethics is: "How should these judgments be made?" This is the central question both for individual members of the profession and for organizations within it. To retain their personal integrity, individuals need to be able to make judgments that reflect their own moral values, ethical values they share with other members of the profession, and community values.

As a result of religious or other beliefs, some will wish to recognize a higher authority, higher even than the law of the land. This authority must obviously govern the ethical decision-making process for that person as a professional practitioner, as well as in his or her private life. At the other pole might be the person who merely wishes to know the rules of the game. Such a person may well find that any reasonable code of conduct gives sufficient guidance. The first choice in ethical decision-making is therefore that of the level of authority or engagement.

Everyone must make choices. Martin (1981) suggests that an important element in the education of a professional person should be the development of moral autonomy in making reasoned moral choices. Each individual needs to be encouraged to work towards formulating his or her own views clearly and explicitly, both for the intrinsic value of this activity and because unless individuals are moral beings they are unlikely to take ethical issues seriously and personally.

Many of the authority sources listed above provide straightforward and useful guidelines for making simple, clear-cut, ethical decisions. Harris et al. (1995, 1997) offer useful case studies and practical guidance for analyzing and working through complex ethical problems in engineering practice. However, "no man is an island," and many decisions are not simple. Multiattribute decisions involving a number of parties commonly involve choices between shades of gray. This can move into the sphere of

---

Possible Levels of Ethical Authority

(These may be seen as making up a hierarchy, with differences of opinion about the order.)

- Religious or other beliefs (these are potentially absolute)
- Examples of "great leaders" (Gandhi, Mao)
- Laws of the land
- Codes of ethics of professional bodies
- Recognized customs and ideals
- The golden rule: i.e., do unto others as you would have them do unto you (or even as *they* would have you do unto them)
- One's own conscience (this may or may not be highly developed)
- Situational decisions (for this you need to be ethically "smart")
- None of the above (for this option you may need to move fast and keep moving!)

politics, as Case Study 11.3 demonstrates (see later in this chapter). Many of our most intractable social and environmental dilemmas fall into this category. In such circumstances, the professional should not have to rely solely on his or her personal moral code, but should be able to seek guidance from the profession's code of ethics.

Codes of Ethics

What role can codes of ethics (sometimes referred to as codes of conduct) play in changing the behavior of individuals in organizations? Codes of ethics can provide guidance and help to make people aware of the ethical content of their work, for example, in providing routine but important services to the public. Codes can raise consciousness, but they are most effective if they are developed from the bottom up, and the process of their development shapes the entire culture of the organization.

Harders (in Jackson 1992, p. 8) goes further, arguing that codes of conduct are essentially irrelevant unless ethics is a part of life, a daily way of working. She sees the foundation of ethics as being in personal relations. Senior management has a central responsibility to demonstrate a clear commitment to ethical behavior, for example in its conduct of everyday office politics. If the organizational culture condones subtle (or not so subtle) abuses of people or of proper procedures, then sermons on corruption, misconduct, and conflicts of interest will be received cynically.

Martin and Schinzinger (1996) agree that codes of ethics can help stimulate ethical behavior and give helpful guidance and advice on moral obligations. The codes of ethics of the professions are intended to ensure that the control they exercise is in the interests of the community. Because of their brevity, codes are very general. This breadth and generality has the advantage of offering assistance in a very wide range of situations. However, to be of much practical value, codes need to be supplemented by guidelines and examples of interpretations. Codes can give moral, and to some extent, legal support to professionals striving to act ethically. To date, they have generally stopped short of providing practical guidance or assistance for "whistleblowers"—professionals trying to expose or prevent unethical behavior. The Institute of Electrical and Electronics Engineers (IEEE) does seems to be moving in this direction with its plans to provide support to engineers who are disadvantaged as a result of taking ethical stances. Such assistance is important because of the tendency to attack the messenger rather than the problem.

Codes present a positive image of the profession that, if warranted, can help professionals serve the community more effectively. However, if they are seen as window dressing, they merely increase a persistent level of public cynicism about the professions. Formal codes cannot be the final moral authority for professional conduct or be a substitute for personal moral responsibility. This is partly because codes change and develop as the community changes. It is also because it is easy for conflicting responsibilities to arise from the different tenets of a code.

All major engineering organizations have a code of ethics in place. An international example is the FIET, which is the 11-million strong International Federation of Commercial, Clerical, Professional and Technical Employees. FIET was founded in 1904 in Amsterdam and is now based in Geneva, Switzerland. FIET has around 420 affiliates in over 100 countries. Within FIET there is a special department for professional and managerial staff that represents engineers and scientists. The FIET Code of Ethics includes sections that deal, for example, with sustainability, human rights, and professional integrity.

In spite of the term "American" in their names, the American Society of Civil Engineers (ASCE), American Society of Mechanical Engineers (ASME), and the American Institute of Chemical Engineers (AIChE) are also international organizations. An organization specifically serving the American engineering profession is the National Society of Professional Engineers (NSPE). Although the NSPE code of ethics is the most thorough and general, it is also lengthy; therefore we quote here the more concise code of ethics from the IEEE (see box entitled the Institute of Electrical and Electronics Engineers, Inc.).

The primary emphasis in the IEEE Code, and the codes of most other engineering organizations, is on the engineer's responsibility to society as a whole. The code seeks to address the tensions between responsibility to the community, to the employer, and to fellow professionals. These tensions are real and important. Beder (1993) explored some of them in a review of the operation of the Institution of Engineers, Australia (IEAust) code of ethics. She argued that there was still a tendency for codes to be used to stifle criticism within the engineering profession. She concluded, however, that there was potential for codes to be interpreted broadly enough for them to be used to put pressure on engineers to take account of long-term community interests.

A Hippocratic Oath for Engineers?

In about 400 B.C., Hippocrates codified all the previous contributions to medicine that he considered sound. He is also credited with having formulated the "Hippocratic Oath," an oath that is still taken by medical

The Institute of Electrical and Electronics Engineers, Inc.

Code of Ethics

We, the members of the IEEE, in recognition of the importance of our technologies affecting the quality of life throughout the world, and in accepting a personal obligation to our profession, its members and the communities we serve, do hereby commit ourselves to the highest ethical and professional conduct and agree:

1. to accept responsibility in making engineering decisions consistent with the safety, health, and welfare of the public, and to disclose promptly factors that might endanger the public or the environment;
2. to avoid real or perceived conflicts of interest whenever possible, and to disclose them to affected parties when they do exist;
3. to be honest and realistic in stating claims or estimates based on available data;
4. to reject bribery in all its forms;
5. to improve the understanding of technology, its appropriate application, and potential consequences;
6. to maintain and improve our technical competence and to undertake technological tasks for others only if qualified by training or experience, or after full disclosure of pertinent limitations;
7. to seek, accept, and offer honest criticism of technical work, to acknowledge and correct errors, and to credit properly the contributions of others;
8. to treat fairly all persons regardless of such factors as race, religion, gender, disability, age, or national origin;
9. to avoid injuring others, their property, reputation, or employment by false or malicious action;
10. to assist colleagues and co-workers in their professional development and to support them in following this code of ethics.

Source: (Approved by the IEEE Board of Directors, August 1990, ©1990 IEEE. Reprinted with permission.)

practitioners as they complete their training. The oath summarizes the ethical code for the practice of medicine.

Early in this century, codes of ethics commonly emphasized refraining from criticizing other professionals and upholding the reputation of the profession. This limited discussion of failures or problems because it might reflect badly on other members of the profession. However, complexities and uncertainties are inherent in engineering activity. To avoid repeating failures we need to analyze them openly and reach a detailed and comprehensive understanding of their causes.

The morality of engineering work has been discussed by a number of authors. Thring (1970) posed a "moral scale of machines," from devices for increasing human possibilities of self-fulfillment, through elimination of drudgery and the provision of luxury consumer items, to machines with directly harmful purposes, such as killing and torturing. He suggested a type of hippocratic oath for engineers analogous to that for medical practitioners (see box, An Oath for Engineering Professionals).

It is striking how much this oath asks of engineers. It assumes a great deal of autonomy and dignity for individual practitioners. It affirms that employee engineers are not just cogs in a large and impersonal machine, but individuals who can have an impact on the world.

A variation on this notion of a hippocratic oath has actually been administered on a voluntary basis in Canada (and to a lesser extent in the United States) for decades, as described in Case Study 11.2. Although the concept seems unexceptionable, the style and language of the ceremony have dated rather unfortunately.

What can individuals do in the short term? Thring suggested three ways to apply what he saw as the dictates of conscience. First, choose jobs

An Oath for Engineering Professionals

I vow to strive to apply my professional skills only to projects which, after conscientious examination, I believe to contribute to the goal of co-existence of all human beings in peace, human dignity and self-fulfillment.

I believe that this goal requires the provision of an adequate supply of the necessities of life (good food, air, water, clothing and housing, access to natural and human-made beauty), education and opportunities to enable each person to work out for him (or her) self his (or her) life objectives and to develop creativeness and skill in the use of the hands as well as the head.

I vow to struggle through my work to minimize danger, noise, strain or invasion of privacy of the individual: pollution of earth, air or water, destruction of natural beauty, mineral resources and wildlife.

Source: (Proposed by Meredith Thring, 1970)

higher up the moral scale. Second, struggle to make the technology you work on safer and more sustainable. Third, take every opportunity to rectify the uses to which engineering is being put. Move the profession towards more ethical, sustainable, and equitable practice. He suggested that unless technical professionals change their practices in the directions he proposes, the machines we make may be used to destroy us all.

Perhaps the major weakness of Thring's proposal is that it sidesteps the fact that there are many rather unpalatable and potentially destructive aspects of technology that are politically and economically important. A more complete treatment needs to address these issues squarely. Ensuring national security, for example, needs hard-headed decision making, and the technologies used need to be functional and cost-effective. At the same time, equity issues need to be addressed more effectively; the allocation of resources between competing national priorities raises major ethical issues.

Winner (1990) took up this point. He suggested that the way we have approached the teaching of ethics in engineering courses (when we have thought about it at all) is to treat ethical problems as if they only arose in the context of "relatively rare, narrowly bounded crises portrayed against an otherwise happy background of business as usual." This approach implicitly assumes that technology is usually beneficial. It ignores the fact that the kind of work that engineers do every day itself *inevitably* raises ethical questions. Winner suggested that ethics courses should explore the moral purposes and strength of vocation of the potential engineer; as a person contemplates committing several decades to a profession, some basic issues ought to be addressed. What are the fundamental ends of a life invested in this work? What is the purpose of developing my knowledge and skill in this direction in the first place?

Addressing these issues is essential to laying a solid moral and political foundation for professional life. Winner suggested that two crucial skills needed to be cultivated: political savvy and the capacity for political imagination. With them, students would be prepared for going beyond the question of "how?" They would be ready to address the moral and ethical question of "why?" to proceed beyond sterile discussion about instrumentality and efficiency and take up concerns about human ends.

Winner felt that professionals are needed who are prepared to dream of a better world and invent technologies that enhance the opportunities for equity and democratic control. This deeper consideration of technological purpose, this concern for the welfare of all in the society, is the real challenge for engineering ethics. A focus on "melodramatic rehearsals for whistleblowing" misses the point. Whistleblowing may sometimes

THE IRON RING OF THE CANADIAN PROFESSIONAL ENGINEER

Even though they don't look like Dilbert, most Canadian Professional Engineers have a distinctive look. Why are they so recognizable? Is it the confident stance that says "Yea though I walk through the valley of new technology I will fear no innovation for I am an engineer and I am in the know?" No, the dead giveaway is the hammered iron ring on the little finger of the working hand. This case study introduces the notion of the iron ring and the "Ritual of the Calling of an Engineer." It also considers how successful these practices have been and what engineers from other countries might learn from the Canadian experience.

An Obligation To Society

The iron ring is a publicly visible symbol of the Canadian engineering profession. Most nonengineers regard it as little more than a wearable logo. Few have seen the ceremony in which the iron ring is received. The more public side often consists of unofficial high jinks, funny clothes, informal parades, and sometimes excessive drinking. During the ceremony, participants also take what is called "The Obligation," something many engineers will carry on a small card in their wallets for many years. It is signed by the bearer and reads in full:

> "I ... in the presence of these my betters and equals in my Calling, bind myself upon my Honour and Cold Iron, that, to the best of my knowledge and power, I will not henceforward suffer or pass, or be privy to the passing of, Bad Workmanship or Faulty Material in aught that concerns my works before men as an Engineer, or in my dealings with my own Soul before my Maker.

> "MY TIME I will not refuse; my Thought I will not grudge; my Care I will not deny towards the honour, use, stability and perfection of any works to which I may be called to set my hand.

> "MY FAIR WAGES for that work I will openly take. My Reputation in my Calling, I will honourably guard; but I will in no way go about to compass or wrest judgment or gratification from any one with whom I may deal. And further, I will early and warily strive my uttermost against professional jealousy or the belittling of my working- brothers, in any field of their labor.

> "For my assured failures and derelictions, I ask pardon beforehand of my betters and my equals in my Calling here assembled; praying that in the hour of my temptations, weakness and weariness,the memory of this my Obligation and of the company before whom it was entered into, may return to me to aid, comfort and restrain."

"The Obligation" is an expression of intent, not an oath. It is part of the "Ritual Calling of an Engineer," which culminates in having the iron ring placed on the little finger of the working hand of the engineer-to-be. It is a very moving ceremony whose details the participants are asked not to divulge. Most engineers experience it during the final term of their last undergraduate year at university.

Not everyone who is eligible participates. Some object to the secrecy and exclusion of family, others to the quasi-religious tone or to sexist language such as "working-brothers." But even fine tuning would not satisfy all critics because the iron ring and the "Ritual of the Calling of an Engineer" are expressions of the force of history, symbolism, and ritual, things that are not always appreciated in a largely ahistorical society.

A Historical Legacy

During the 1920s engineers were at the heart of an exciting boom based on new technology and economic growth. Canadian engineers were good, needed, and appreciated. Nonetheless, people such as the University of Toronto's Professor H. E. T. Haultain worried that most engineers lacked both a sense of professional identify and an understanding of societal obligations. Haultain's work around the world as a mining engineer had given him "an appreciation of the need among engineers for a strict sense of ethics, which he emphasized at every opportunity." Haultain wanted engineers' self-image and professional conduct to go far beyond "demolishing 40 beers" and singing "We are, we are, we are the engineers."

In his unpublished memoirs Professor A. E. Allcut, an accomplished artist as well as engineer and noted author, remembers that "for many years the medical profession had admitted young doctors by means of an induction ceremony that includes the Hippocratic Oath and Professor Haultain ... thought that the idea might well be adopted by the engineering profession ... He talked to me about it and finally decided to ask Rudyard Kipling to compose the ritual. Much to our surprise, Kipling agreed to do so ..."

Kipling was an apt choice for the times, a very popular author who admired technology and the manly virtues of those who created and maintained it. He, like many others of his day, was comfortable with ritual and symbolism. The iron ring took its significance from both form and material.

"The ring is one of the oldest human symbolic artifacts; its circular form, with neither beginning nor end, seems to embody perfection. Iron, for its part, is a traditional symbol of utility and strength, and was at the heart of much of the progress for which engineers were gaining prominence." (Ball, p. 46). (The idea that the iron for the rings comes from the Quebec Bridge that collapsed twice in its construction earlier this century is just a myth.)

Professional Significance

Although most presentations take place in educational institutions, the iron ring and its associated ritual are not, in any formal sense, connected with the education of engineers in Canada. Moreover, they are in no way connected with official recognition as a professional engineer. Even though "The Obligation" is important to professional practice, the whole ceremonial is, regrettably, separate from the very institutions (the universities and the professional associations) that should give it living meaning, "fleshing out" what "The Obligation" means in practice.

As well as being very widely used in Canada, the iron ring ceremony has made a modest entry into the United States. Despite its limitations, the Canadian iron ring ceremony is a serious affirmation of the profession's commitment to socially responsible practice.

Source

Contributed by Norman R. Ball, Waterloo University, Ontario,Canada and drawing on:

Ball, Norman R. 1987, *Mind, heart and vision: professional engineering in Canada 1887 to 1987*, National Museum of Science and Technology, Ottawa. This is a history of the engineering profession in Canada.

become necessary, but it needs to be recognized as a last and rather extreme resort in situations where earlier and more positive efforts to improve the situation have failed.

Ethics and the Role of the Expert

It is important for the long-term stability of a democracy that the groups or communities affected by decisions have effective input. In the process, their views have to be balanced against expert opinion. To what extent are experts detached from the problems on which they advise, and thus in a position to make value-free judgments? Should we see each expert as an advocate for a particular position?

The role of the expert witness in a court of law raises important issues. Court cases in the United States are normally run on an adversarial basis, as a contest between two sides. However, despite the fact of being retained (and paid) by one side or the other, the expert is supposed to be impartial. His or her role is commonly that of an officer of the court, with responsibility for ensuring that judgment is based on a full understanding by the court of the facts and their implications.

This is a special case of a general problem. We are accustomed to hearing from experts on the whole range of public policy matters. Traditionally, they are supposed to be neutral, without personal or vested interest in the matters on which they speak, to provide objective advice on the basis of which others can formulate policy. How realistic is this concept of their role? What if experts disagree? Does it mean one is biased (or even that both are biased)? To answer these questions we need to look at how experts actually work. Part of their role is to collect data, "facts," and then to interpret them. Facts do not choose themselves. The selection of relevant evidence is part of an expert's job. They must not destroy or ignore inconvenient data, nor invent data to suit their needs, but they do have to make judgments on what is of central importance and what can safely and legitimately be neglected.

Questions as to what constitutes objectivity and how it is to be achieved are not easy to answer. Part of the problem is that the experts we hear from may be committed to positions in the debate. If much of their working lives have been devoted to mastering and advancing a specialty, some degree of commitment to a particular position on many of the issues involved is almost inevitable. In this case the expert is already a party to the debate, and must be expected to act as an advocate, not as a neutral outsider. This does not mean that the expert will act unprofessionally, but it does mean that experts with alternative approaches are needed to put other positions in the argument.

Objectivity means more than simply working through an argument systematically. The right questions need to be asked, and the debate must be broad enough for all the relevant issues to be properly canvassed. The problem must be well defined before it can be debated effectively. Otherwise the debate inevitably remains unfocussed and inconclusive. Technical controversies are often unstructured, and the adversaries may never confront one another. If we could bring some order to these controversies, their value would be greatly enhanced. Unfairness in a debate is probably inevitable if there is:

- massively unequal funding for the opposing sides;
- domination of research by a few experts;
- career penalty for unwelcome advice or outcome; or
- administrative procedure that imposes secrecy or otherwise severely limits debate.

These are public, identifiable, overt sources of bias that can and should be remedied by public methods. Environmental conflicts need to be handled this way. The proper function of a controversy is actually the identification and evaluation of potential problems. The nature of what is essentially a political debate is that consensus is unlikely; however, a decision may still need to be made. The best decision is then one that can be changed if and when a consensus is reached that it was wrong. It is also necessary to preserve the ability to detect an error in the decision and to correct it if necessary. This may involve a cost, but good decision making is not free. The costs of persevering with a bad decision are liable to be orders of magnitude higher (Collingridge 1980, p. 194).

The essence of good decision making under uncertainty is to keep options open. A classic example is in the development of a new residential subdivision. Although it may be clear that some mass transit routes will be needed, the best form or forms of vehicle to use may not be obvious. As long as a suitable corridor is reserved, the costs of implementing whichever solution is eventually chosen will be manageable, and disruption to the community minimized (c.f. Case Study 2.1, p. 62).

∎ Expert Credibility and Trust

An important aspect of expert advice is the extent to which the community affected is prepared to rely on it. (One aim of most codes of ethics is to ensure that the community should always be able to rely on such advice from professional engineers. However, advice is shaped by the way problems are perceived and posed, so it may not satisfy people who see the issues differently.) In the absence of trust, decisions must be imposed by coercion and the proliferation of rules. In the authors' view, this is an area that will be seen as increasingly important in both more and less economically developed countries. Although technology assessment is very important for developing countries, there is rarely a substantial group of people with both the opportunity and the expertise to carry it out, particularly in advance of commitment to its implementation.

The public credibility of decision making on technological issues was discussed by Bella et al. (1988a, 1988b), in the context of the selection process for a site for the proposed American nuclear waste repository. They argued that most of the institutions of power in the United States are now largely technocratic (i.e., insulated from the political process), and that a new approach to the question of appropriate checks and balances is needed to maintain citizen control over them. They suggested that the professionals charged with responsibility for making decisions tended to have what they described as a technocratic perception of the issues, regarding opposition as the result of "activists" alarming a "misinformed public." This approach might be characterized as the "trust me, I know best" posture.

Nelkin (1975) pointed out that technologies of speed and power—airports, power generation, highways, and dams—tended to become the focus of opposition. This was partly because, in the process of "rational" planning, problems were defined as technical rather than political. An essentially political choice was then buried in a polarized debate between experts. Different evaluation criteria were used by the proponents and opponents. Provided each side could muster credible experts, they neutralized each other, and the argument was back where it really belonged, in the political arena. Nelkin noted that the clients used expertise in a biased way. Its value to them depended more on its utility than its scientific merit.

In modern practice, problems are becoming increasingly complex: a systems approach is now essential to understanding and implementing engineering projects. The world has become a complex dynamic social, political, economic, and environmental system, in which technologies act as both positive and negative amplifiers, changing system behavior. Because of this, new technologies cannot effectively be assessed in isolation, but must be reviewed in terms of their effects on the various system inputs and outputs.

Bella et al. (1988a, 1988b) argued the need for a democratic perception of the issues, emphasizing real consultation with citizens potentially affected by decisions, in a process not dominated by the institution involved in implementation. Their analysis seems widely relevant. Their proposed approach was modeled on an ideal model for scientific practice. This involved open and free review of assumptions and evidence, and free exchange of information with an independent community. It was aimed at addressing the problem of obliging each side to address the issues raised by the other, rather than talking past them or ignoring their concerns in the ways described by Martin (1982).

What Is A Professional?

The concept of the "expert" brings us to the term "professional." Is this model still the most useful and relevant one for engineering? How might we wish to influence the development of the engineering profession over the next decade?

The term "professional" has its origins in the Latin verb *profiteri*, to declare aloud or profess a vocation or special calling; it related to the vow made on entering a religious order. Today this term is commonly used in two different senses. It is contrasted with amateur, to indicate someone who is paid for an activity and typically does it full time; it is also used to indicate thoroughness and competence. Although modern usage stresses the role of learning and expertise it is important to recognize, in the Latin origins of the term, a strong ethical dimension that is still an essential aspect of professionalism.

Professions and professionalism are central features of modern social organization. They provide an important way of managing specialization within society, although they are not the only way, or even necessarily the best (Johnson 1972). "Professional" is a claim to social standing and recognition, based on occupational identity. This élite status is seen as essential to recruiting, retaining, and motivating quality practitioners. Professions seek autonomy to regulate the activities carried out by their members. The corollary is a formal recognition by a profession of its members' responsibilities to society, commonly defined in a code of ethics. Membership of a professional body entails participating in the corporate life of the practitioners. Information and experience are shared and a moral esprit established.

There is great diversity of opinion as to precisely which occupations are professions. The traditional professions have changed in major ways. At the same time, the understanding of what constitutes a profession has changed. In 1933 Carr-Saunders and Wilson summarized the difficulty:

> There is no more agreement about the boundary of professionalism than there is about its value ... There are certain vocations of ancient lineage which by common consent are called professions, law and medicine among the foremost ... There are many other vocations which, though more recently and therefore less firmly established, are nevertheless usually granted professional rank; all such vocations, architecture, engineering, chemistry, accountancy and surveying, for example, require mention.

Nearly seventy years later, this remains a fair assessment of the situation, although the list has expanded considerably. Perhaps a better understanding is gained by looking at how the modern professions arose and how they differ from those in pre-industrial times, and some of the features that characterize modern professions.

∎ Status Versus Occupational Professionalism

In Europe during the Middle Ages, the Church controlled most types of knowledge and education. Men in holy orders made up the original "learned profession," which included specialists on canon (Church) law. During the twelfth century, free associations of scholars came together to form universities. By the beginning of the thirteenth century, the universities had largely taken the place of monasteries as intellectual centers. Although still very much under its influence, the universities were separate from the Church. Universities increasingly became responsible for training the rather small number of professionals who serviced the ruling élite. They taught theology, medicine, and law (the original "pre-industrial" professions), as well as providing practical education for potential statesmen and public servants. A parallel system in the craft guilds trained practitioners to service the needs of the rest of the population.

By the eighteenth century in most of the Western world, the professions provided occupation and a living for the younger sons of the nobility and gentry. Oxford and Cambridge universities made little contribution to the industrial revolution by way of relevant professional education. Their stated object was rather to develop "capable and cultivated human beings" (Elliott 1972, p. 51). Rothblatt (1968) argued that the universities developed an ideology of public service and gentlemanly professionalism to justify their roles in developing, passing on, and promoting the culture and ideology of the upper classes. Universities provided a liberal education that fitted one to be a gentleman, able to make a living without having to resort to manual labor or engage in commerce or trade. Until the beginning of the nineteenth century in Britain, the gentlemanly learned professions continued to be divinity (effectively including university teaching), law, and medicine. Architecture and the military also had high status.

The competitive examination system for university entry was introduced in Britain in the middle of the nineteenth century. It began the process of opening up the professions and the civil service to members of the rising middle class, who could aspire to enhanced status by taking up a profession. By this time, the civic universities in England, Scotland, and Ireland had started to cater seriously for the demands of engineering.

Attitudes to education and the ideology of gentlemanly professionalism are of more than simply historical importance. A bias persists in Britain towards liberal tertiary education, directed towards general intellectual culture, rather than narrowly technical or professional education. In many ways manufacturing and engineering continue to be neglected; the very different attitude to education in America was important to U.S. world technological leadership and to the high status of engineers in the United States.

The nature of the professions changed with the rise of industrial society. The title "professional" was and remains a claim to social status, but the basis for claiming it changed from the upper class, "who one was," to the middle class, "what one did." Abercrombie and Urry (1983) provide an interesting review of the literature of class and professionalism and discuss its ideological content. Despite revolutionary changes in the professions, many of the attitudes and ideals of professionalism today seem to have been carried over from the pre-industrial period. Probably the most important of these are claims for high social status, indifference to commercial considerations, and the right to autonomy and self-regulation.

The satisfaction inherent in addressing demanding technical problems clearly has its own intrinsic rewards, which can engender commitment to the occupation. Early this century, professionalism was promoted as a counterbalancing force against what was seen as the narrow self-interest of business. It is interesting to speculate on whether this anticipated modern attitudes towards social responsibility, or whether it drew on pre-industrial "gentlemanly" attitudes based on status rather than occupational professionalism.

The Characteristics of Modern Professions

The most useful questions may be: To what extent do various occupations exhibit the characteristics of professions? In what ways do occupations try to change as they try to become professionalized? What criteria for professions do they follow? Is there an ideal type on which aspiring professions can model themselves?

Johnson (1972, pp. 22–25) draws attention to the limitations of the "ideal type" approach to defining professions. The characteristics commonly cited are interdependent, and do not have much in the way of theoretical underpinning or even coherence. Johnson argues that they "specify the characteristics of a peculiar form of occupational control." These characteristics relate to the power of professional groups and their claims to status and income, compared with the claims of other occupational groups in the society. Historically, the social status of both the clientele and the recruits have been very relevant to the status of a professional occupation. Today it is no longer clear-cut which groups are and which are not professions. Professionalism is much more a matter of degree. In any case, professions are not static—the way they operate changes with economic, social, technical, and political changes.

Bearing in mind Johnson's qualifications as to their limited theoretical significance, it is still useful to look in detail at some of the characteristics on which claims to professionalism are currently based.

■ Substantial Intellectual and Practical Training

This involves mastery of a body of generalized and systematic knowledge, achieved by a long period of study and certified by a degree or diploma from an accredited institution, usually a university. The knowledge base needs to be broad enough that tasks cannot become a matter of routine, but distinct enough to be clearly defined and defended against adjoining specialist areas. Johnson (1972, p. 19) suggests that the theoretical body of knowledge also "includes a set of assumptions, an implicit or explicit theory, about the way the world is and the way society is organized." This is what Foucault (1970) might have described as the discourse of engineering. This discourse is discussed in detail elsewhere (Johnston et al. 1996). It is important, since it shapes the world-view and social attitudes of practitioners.

Part of the professional formation of practitioners is that they largely come to accept their profession's view of the world, including its ethical attitudes, as their own. The loose control over professionals, especially those in individual practice, implicitly offers opportunities for exploitation. This makes the adoption of ethical attitudes important.

Differences between the norms and values of the professional group and those of the wider society give rise to important issues of social control in three main areas:

1. between the professional and the unskilled client;
2. within the profession; and
3. between the profession and the wider society (Elliott 1972, pp. 11–12).

▌ Professional Autonomy and Responsibility

Professional autonomy implies a degree of acceptance by society of the legitimacy of the claim for control by the professional group. The level of professional autonomy is related to the type of work and work relationships. So is the extent to which professionals are able to impose their own definitions of the relationship between them and their clients. This is in turn related to the power of the professional group in the society and how it compares with the power of the consumers of the professional services.

The increasing scale and complexity of all professional activity is reflected in the rising proportion of salaried employees rather than independent practitioners. The question of compatibility between professional autonomy and employee status is particularly important for engineering, because most professional engineers are employees. There has been concern that employee status is not compatible with the level of autonomy required for professional activity. This is only a special case of the problems of incorporating the professions within the social system generally. It has been argued that in order to be able to establish and implement their own goals in terms of technical competence, professional groups need to be sufficiently insulated from their constituents—that is, from groups or interests in the society that might be in a position to shape them to their own needs (Elliott 1972, p. 104).

The key issues for salaried professionals are the potential for conflict or competition between professional and organizational goals and the means each has to press its claims. Such means can include appeals to reference groups outside the organization. Case Study 2.3 on the space shuttle Challenger (see p. 94) describes a conflict between engineers and their organization. Project engineers were concerned about the booster-rocket joint design, whereas management was concerned about economics, politics, and scheduling. Regrettably, management overruled the engineers, with fatal consequences.

Professionals in large organizations, particularly public ones, may be better insulated from direct commercial pressures than are individual practitioners in marginally economic practices. Pressures towards corporatization and privatization have reduced this isolation. Professionals in larger groups may also be in a better position to concentrate their efforts on specific areas within the profession, rather than trying desperately to keep abreast of developments across a wide disciplinary area.

As the scale of projects and the degree of specialization needed for successful practice increased, engineering consulting firms evolved from partnerships to corporate structures. The management of professionals within these organizations raises important issues, particularly if the managers do not come from the same professional background. This situation has become more common with the increasing use of the multidisciplinary project teams needed to address the ecological and social impacts of projects, as well as the technical aspects.

■ Registration or Licensing

A profession's autonomy is confirmed when the profession itself is responsible, on behalf of society, for testing the competence of practitioners, regulating standards, and maintaining discipline. This responsibility can include control of working conditions, entry to the profession, and even expulsion from it.

The extent of the control a profession is allowed to exercise over its own activities is at least partly a political question, decided through political processes. Control of professional activities by professional organizations such as the American NSPE, and the IEEE, ASCE, or ASME, may extend to an effective and even a legally enforced monopoly on the type of services the profession offers. Such a monopoly is at variance with the ideology of the free market discussed in Chapter 5.

Most developed countries regulate engineering through the licensing of those who desire to practice engineering. There are laws regulating the practice of professions including law, medicine, and engineering in every state in the United States of America, the District of Columbia, and in U.S. territories such as Samoan Islands and Guam. These laws protect the public health, safety, and welfare by ensuring that those receiving licenses to practice have at least met certain requirements of competence, ability, experience, and character. In most U.S. states, laws require a license for engineers who work on private or public projects that involve use by the public. Generally, this requires registration of engineers who consult to or work for government agencies or public utilities.

In the United States, registration is controlled exclusively by the individual state legislatures and the laws vary from state to state. For professional engineers these laws typically require graduation from an accredited engineering curriculum, followed by approximately four years of responsible engineering experience, and finally the successful completion of a written exam. Some states may waive the written exam on the basis of education and experience, but the trend is toward an examination requirement.

The engineering profession is a mobile one, with engineers often practicing in a number of states or countries during their professional careers. Therefore, most states provide for reciprocal licensing, allowing engineers licensed in one state to become licensed without further examination as long as the state that originally granted the license has at least equal minimum standards. Uniformity among state licensing laws is a goal that most engineering societies have been working toward for many years. A model law was developed and has been used as a guide for almost forty years. Currently, the majority of states require that a candidate pass an eight-hour Fundamentals of Engineering written exam (the FE exam), and an eight-hour Principles and Practice of Engineering exam in the applicant's discipline. The NSPE offers interactive, satellite-delivered refresher courses through the National Technological University (NTU) and exam review textbooks through Professional Publications, Inc. Andrews and Kemper (1992) provide support and guidance for engineers seeking registration in Canada.

The North American Free Trade Agreement (NAFTA) has as one of its goals that the United States, Canada, and Mexico implement a mechanism for mutual recognition of engineering practitioners at the professional level.Under a Mutual Recognition of Engineering (MRD) qualifications agreement, individuals will normally require considerable periods of experience as a licensed professional engineer before becoming eligible for mutual recognition. The actual length of time involved will depend on whether or not the practitioner concerned was initially licensed on the basis of an accredited degree.

As a general guide, an engineering practitioner who has completed an accredited degree would need eight years of subsequent experience before becoming eligible for recognition under the provisions of the MRD. Practitioners holding a nondegree qualification recognized in their home country as appropriate for professional engineers would require twelve years of subsequent experience.

Interestingly, the generally held view that higher level management of engineering activity does not require engineering competence has resulted in licensing laws not generally applying to this group. This view is strongly disputed by Lloyd (1991, p. 200), who maintains that:

1. Executive engineers... have to understand engineering systems and processes and make engineering decisions that cross disciplines or functional boundaries.
2. Executive engineers have to advocate engineering policies, decisions, proposals and plans to higher authority, requiring professional understanding of engineering technologies and processes. ...Such representation in the hands of non-engineers... is totally inadequate.

■ A Fiduciary Relationship with the Client

The relationship between professional and client must be based on trust and confidence; a *fiduciary* relationship is essential. This is partly because clients are not generally expected to be able to assess for themselves the quality of the service being provided. Much of the independence and authority of a professional in dealing with a client comes from the fact that his or her primary reference group is made up of professional colleagues.

In engineering, the client should play an important role in the design process. In collaboration with the professionals, the client needs to be able to develop a clear and comprehensive description of the problem. Without such a description, the client's needs are unlikely to be met well, if at all. Gardiner and Rothwell (1985, p. 25) argue that the technologically demanding client can make a major contribution to innovation. In many cases, for example in research work, there may not be a specific client. There may also be conflicts between the interests of the client and those of the community. The code of ethics of the profession can be helpful here, in that it typically calls for primary responsibility towards the community as a whole.

Consulting engineers are often in fiduciary relationships that arise out of their professional associations. One who provides engineering services, or any other services related to the business affairs of another, and maintains a position of superiority and influence over the other person, will be held to be in a fiduciary relationship. Violating that fiduciary relationship through negligence, error, or omission leads to a professional liability, meaning the engineer may be successfully sued.

Engineers, like other professionals, hold themselves out to be skilled and knowledgeable to perform certain duties and services within their area of expertise and, accordingly, they are legally bound to act in an accurate and skillful manner. Engineers who act in a fiduciary capacity are also required to maintain client confidences and remain loyal to the client's interests.

▪ Professional Indemnity

As Case Study 11.3 (at the end of the chapter) describes, in the late 1960s a situation arose at an engineering firm that illustrates the complexity of real-world ethical problems and puts the sorts of issues involved in the whole notion of professional responsibility and indemnity into sharp focus. The failure to resolve those issues satisfactorily as recently as three decades ago did, however, provide some of the impetus for legal moves to protect at least some of the parties involved.

Today, in view of changes to the professional environment, including the introduction of strict liability for products, employee engineers should have an indemnity clause included in their contracts of employment, along the lines outlined in the box, Legal Liability (p. 581).

The Engineering Profession in the United States

The following figures show the composition of the U.S. engineering profession in 1994 by specialty, based on data from the Engineering Workforce Commission and by industry, based on sample data from the U.S. Bureau of the Census.

Figure 11.1 shows the distribution of employment specialties of all U.S. engineers, including both military and civilian employees. Figure 11.2 shows the various industries in which U.S. engineers in the civilian (nonmilitary) workforce are employed. It highlights the different demands for the three largest professional groups of engineers (civil, electrical/electronic, and mechanical).

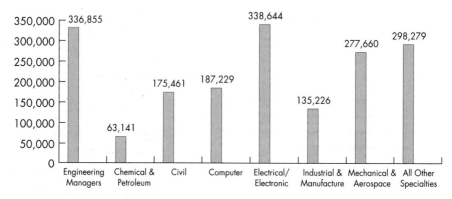

Figure 11.1 Employment of all U.S. Engineers—By Specialties, 1994. (Graphic by John Wayland.

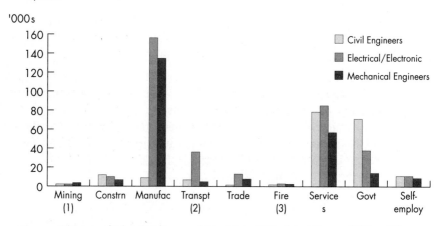

Source: US Bureau of Labor Statistics, unpublished data: US Statistical Abstract 1997, p 610.
Note: (1) Includes oil & gas extraction
      (2) Includes communications & public utilities
      (3) Finance, insurance and real estate

Figure 11.2 Civilian Employment of Engineers by Occupation and Industry, 1994.

PROFESSIONAL ENGINEERING ASSOCIATIONS

As the engineering profession spread around the world and new special-izations emerged during the nineteenth century, practitioners began form-ing professional bodies to support their work and advance the reputation and status of the profession. Other reasons engineers formed associations included: to make business contacts, to learn from one another, to in-crease the influence of their discipline with government and the public,

Legal Liability

Most general advice on legal liability for engineers emphasizes the complexity of the law and the need to obtain independent advice at the earliest opportunity where a problem arises. Engineers should recognize the difference between:

- ◆ criminal liability (for example, under occupational health and safety legislation), where the offense attaches to the individual; and
- ◆ civil liability, which is largely concerned with compensation of injured third parties. With civil liability, the doctrine of "vicarious liability" holds the employer responsible for breaches of duty of employees when committed "in the course of their employment."

In view of changes to the professional environment, including the introduction of strict liability for products, all employee engineers should consider having an indemnity clause included in their contracts of employment, along the following lines:

> The professional engineer acting honestly, diligently and in good faith shall not suffer any loss or damage of any kind by reason of any liability incurred by the company as a result of the conduct of the professional engineer and the company shall hold the professional engineer harmless and indemnify the employee against any loss, claim, cause of action of any kind whatsoever arising out of the employment relationship

Sources: (APEA 1989).

It is important that professional engineering employees ensure that senior management is aware of any potentially dangerous situations within their area of responsibility. This is even more important where insufficient funding or staffing may make it impossible even to identify precisely what risks may exist.

and to experience the camaraderie of like- minded people (always men initially). These reasons still exist today. However, the associations are now much larger, better organized, and more powerful and offer much more to their members.

As professional engineering practice became increasingly global, national engineering societies helped develop international agreements and form international societies that now coordinate international engineering activities and moderate qualification and certification of engineers.

Today there are over 1,200 national and international engineering associations. In this section we describe a few that focus their efforts on international issues, as well as the major U.S. associations. (In spite of the term "American" in their names, the IEEE, ASCE, ASME, and the AIChE are also international organizations.) All of them have easily-found World Wide Web sites, usually in the format <http://www.[initials of the organization].org/>. Many national associations also have useful web sites.

Two important international engineering bodies, WFEO and FEANI, are described briefly in the boxes below.

The World Federation of Engineering Organizations (WFEO)

The World Federation of Engineering Organizations (WFEO) is the worldwide leader of the engineering profession and cooperates with national and other international professional institutions in developing and applying engineering to the benefit of humanity.

The mission of WFEO is:

♦ to provide information and leadership to the engineering profession on issues of concern to the public or the profession;

♦ to serve society and to be recognized, by national and international organizations and the public, as a respected and valuable source of advice and guidance on the policies, interests and concerns that relate engineering and technology to the human and natural environment;

♦ to foster peace and socioeconomic security among all countries of the world through the proper application of technology; and

♦ to facilitate relationships between government, business, and society by contributing an engineering dimension to discussions on policies and investment.

The WFEO Standing Committee on Technology (COT) was established to provide leadership and guidance to the engineering profession worldwide on the promotion and application of sustainable technology, and the promotion of arrangements to encourage sharing and worldwide dissemination of technologies.

Source: (Reproduced with permission, WFEO)

The European Federation of National Engineering Associations (FEANI)

The federation was created in 1951 and adopted its present name, Federation Europeenne d'Associations Nationales d'Ingenieurs (FEANI) in 1956.

Today twenty-seven countries are represented in FEANI. It covers more than 1,500,000 European engineers. FEANI is a founding member of the World Federation of Engineering Organizations (WFEO) and collaborates with many organizations dealing with engineering and technology issues and engineering education. It is officially recognized by the European Commission. FEANI's objectives include affirming the professional identity of the engineers of Europe:

- by ensuring that professional qualifications of engineers of the member countries are acknowledged in Europe and worldwide;
- by asserting the status, role, and responsibility of engineers in society;
- by safeguarding and promoting the professional interests of engineers and by facilitating their free movement within Europe and worldwide;
- by striving for a single voice for the engineering profession of Europe, while acknowledging its diversity; developing a working cooperation with other international organizations concerned with engineering matters and representing the engineers of Europe internationally.

All European engineers are registered by FEANI and in the future, it could be difficult for engineers who are not registered to practice in Europe. The purpose of the Register is to:

- facilitate the movement of practicing engineers inside and outside the FEANI area;
- establish a framework of mutual recognition of qualifications in order to enable engineers who wish to practice outside their country to carry with them a guarantee of competence;
- provide information about the various formation systems of individual engineers for the benefit of prospective employers; and
- encourage the continuous improvement of the quality of engineers by setting, monitoring, and reviewing standards.

Source: (Based on the FEANI World Wide Web site at www.feani.org and reproduced with permission.)

▌Engineering Associations in the United States

Some of the most important of the associations that represent the 1.2 million engineers in the United States are described briefly here.

The American Association of Engineering Societies is a multidisciplinary organization dedicated to advancing the knowledge, understanding and practice of engineering in the public interest. Its members represent the mainstream of U.S. engineering—affecting over 800,000 engineers in industry, government, and education. Through its councils, task forces, commissions, and committees, AAES addresses questions relating to the engineering profession. The Engineers' Public Policy Council (EPPC) is the principal forum responsible for the coordination, development, and implementation of AAES public policy. The Engineering Workforce Commission (EWC) conducts annual surveys of industry and engineering schools and monitors trends in the supply and demand for engineers.

Some of the major technical societies for engineering professionals are introduced briefly in the box below.

Two other important bodies, representing licensed professional engineers and consulting engineering firms respectively, are described below.

The National Society of Professional Engineers

The National Society of Professional Engineers (NSPE) is the U.S. engineering body that represents licensed professional engineers (PEs) across all disciplines. Founded in 1934, the goal was to provide a forum for dealing with the social, ethical, economic, and professional dimensions of engineering. Technical subjects were acknowledged as being better treated by various technical societies. The NSPE strengthens the engineering profession by promoting engineering licensure and ethics, advocating and protecting PEs' legal rights at the national and state levels, publishing news of the profession, and providing continuing education opportunities.

The NSPE serves some 60,000 members and the public through fifty-four U.S. state and territorial societies and more than 500 chapters. Members are generally licensed engineers, referred to as PEs. The stated goals of the NSPE are:

- ◆ To formulate high ethical standards for the practice of engineering and to lead the profession in adhering to these principles.
- ◆ To promote the licensed practice of engineering, the enforcement of uniform licensing laws, and to communicate the importance of licensing.

Major Technical Professional Engineering Bodies in the United States

The Institute of Electrical and Electronic Engineers (IEEE)

Founded in 1884, the Institute is today the world's largest technical professional society. Its stated goal is to promote the development and application of electrotechnology and allied sciences for the benefit of humanity, the advancement of the profession, and the well-being of its members.

The Institute has 230,000 members in the United States and 90,000 members outside the United States.

The technical objectives of the IEEE focus on advancing the theory and practice of electrical, electronics, and computer engineering and computer science. The IEEE sponsors technical conferences, symposia, and local meetings worldwide and publishes nearly 25 percent of the world's technical papers in electrical, electronics, and computer engineering.

The IEEE publishes a monthly journal, IEEE Spectrum, dealing with engineering and society issues, and sponsors an annual International Symposium on Technology and Society (ISTAS).

Source: (© IEEE 1999. Reprinted with permission.)

The American Society of Civil Engineers (ASCE)

The Society, founded in 1852, is America's oldest national engineering society. The ASCE represents more than 120,000 civil engineers worldwide, serving 6,700 members in 142 nations outside the United States and maintaining ASCE Sections and Groups in 22 countries.

The ASCE's Journal of Professional Issues in Engineering Education & Practice leads the world in its broad coverage of engineers' obligations and responsibilities.

Source: (© ASCE 1999, reprinted with permission)

The American Society of Mechanical Engineers (ASME)

The Society was founded in 1880. Today the ASME is a worldwide engineering society with 125,000 members. Its focus is on technical, educational and research issues and it conducts one of the world's largest technical publishing operations.

The American Institute of Chemical Engineers (AIChE)

The Institute was founded in 1908. It helped to establish chemical engineering as a separate discipline. Today AIChE has over 50,000 members whose fields of specialization include petrochemicals, food, pharmaceuticals, textiles, pulp and paper, ceramics, electronic components and chemicals, biotechnology, environmental control and clean-up, and safety engineering.

- To help establish educational standards for engineering and to provide learning opportunities that enable licensed engineers to maintain practice competency.
- To incorporate engineering principles and perspectives in government decisions that protect the public and foster technological development.
- To inform licensed engineers of professional, political, social, economic, and business developments that affect their practice.
- To promote employment conditions that stimulate productivity and encourage culturally diverse populations to practice engineering.
- To stimulate student interest in mathematics and scientific principles that underlie a technological society and to guide those who express an interest in engineering.
- To provide forums that focus these goals on the licensed practice of engineering in construction, education, government, industry, and private practice. (© 1999 NSPE, Reproduced with permission.)

The NSPE maintains that, "Professionalism requires engineers to place public welfare before personal gain. Public trust in the integrity of engineering is essential to the prosperity of our profession. This interrelationship is the foundation of our value system. The core values of this system must be recognized, embraced, enhanced, and sustained through generations." As we have tried to do throughout this book, the NSPE emphasizes the engineer's responsibility to protect the public health, safety, and welfare; to maintain stewardship of the environment and natural resources; and to serve the community.

The Association of Consulting Engineers Council

The Association of Consulting Engineers Council (ACEC) has headquarters in Washington D.C. It is a federation of fifty-one member organizations with over 5,000 member firms that together employ over 180,000 people and design more than $100 billion in capital projects each year. The Association represents the collective views of the engineering profession to U.S. and European engineers and firms, as well as to other engineering institutions and associations in trade and industry.

ACEC members are firms of consulting engineers ranging from sole practitioners, through partnerships to companies and corporations. Member firms offer a comprehensive range of independent consultancy services, from evaluation of the requirement, investigation of engineering and economic feasibility, to design and supervision. On behalf of clients, they can draw up tenders and evaluate bids.

The Association offers guidance and support to potential clients from the private and public sectors, and will suggest a selection of firms meeting the client's specific technical requirements and operating in the appropriate region within the United States or worldwide. It offers its member firms a range of services as one of the benefits of association.

ENGINEERING EDUCATION AND ABET

The importance of a substantial intellectual and practical training for any profession has already been discussed. The Accreditation Board for Engineering and Technology (ABET) is recognized in the United States as the sole agency responsible for accreditation of educational programs leading to degrees in engineering. The first statement of the Engineers' Council for Professional Development (ECPD, now ABET) relating to accreditation of engineering educational programs was proposed by the Committee on Engineering Schools and approved by the Council in 1933. The original statement, with subsequent amendments, was the basis for accreditation until the year 2000. The statement presented here is required of programs beginning in the year 2001.

Engineering education programs may be accredited at the basic or the advanced level; however, a program may be accredited at only one level in a particular curriculum at a particular institution. All accredited engineering programs must include "engineering" in the program title (an exception has been granted for programs accredited prior to 1984 under the title of Naval Architecture.) To be considered for accreditation, engineering programs must prepare graduates for the practice of engineering at a professional level.

The ABET accreditation process is a voluntary system of accreditation that:

- ◆ assures that graduates of an accredited program are prepared adequately to enter and continue the practice of engineering;
- ◆ stimulates the improvement of engineering education;
- ◆ encourages new and innovative approaches to engineering education; and
- ◆ identifies these programs to the public.

ABET Engineering Criteria 2000

Criteria for Accrediting Engineering Programs

Effective for Evaluations during the 1999–2000 Accreditation Cycle

Criterion 1: Students. The quality and performance of the students and graduates are important considerations in the evaluation of an engineering program. The institution must evaluate, advise, and monitor students to determine its success in meeting program objectives.

The institution must have and enforce policies for the acceptance of transfer students and for the validation of credit courses taken elsewhere. The institution must also have and enforce procedures to assure that all students meet all program requirements.

Criterion 2: Program Educational Objectives. Each engineering program for which an institution seeks accreditation or reaccreditation must have in place:

(a) detailed published educational objectives that are consistent with the mission of the institution and these criteria;

(b) a process based on the needs of the program's various constituencies in which the objectives are determined and periodically evaluated;

(c) a curriculum and process that ensure the achievement of these objectives; and

(d) a system of ongoing evaluation that demonstrates achievement of these objectives and uses the results to improve the effectiveness of the program.

Criterion 3: Program Outcomes and Assessment. Engineering programs must demonstrate that their graduates have:

(a) an ability to apply knowledge of mathematics, science, and engineering;

(b) an ability to design and conduct experiments, as well as to analyze and interpret data;

(c) an ability to design a system, component, or process to meet desired needs;

(d) an ability to function on multidisciplinary teams;

(e) an ability to identify, formulate, and solve engineering problems;

(f) an understanding of professional and ethical responsibility;

(g) an ability to communicate effectively;

(h) the broad education necessary to understand the impact of engineering solutions in a global and societal context;

(i) a recognition of the need for, and an ability to engage in life-long learning;

(j) a knowledge of contemporary issues;

(k) an ability to use the techniques, skills, and modern engineering tools necessary for engineering practice.

Each program must have an assessment process with documented results. Evidence must be given that the results are applied to the further development and improvement of the program. The assessment process must demonstrate that the outcomes important to the mission of the institution and the objectives of the program, including those listed above, are being measured. Evidence that may be used includes, but is not limited to the following: student portfolios, including design projects; nationally-normed subject content examinations; alumni surveys that document professional accomplishments and career development activities; employer surveys; and placement data of graduates.

Criterion 4: Professional Component. The Professional Component requirements specify subject areas appropriate to engineering but do not prescribe specific courses. The engineering faculty must assure that the program curriculum devotes adequate attention and time to each component, consistent with the objectives of the program and institution. Students must be prepared for engineering practice through the curriculum culminating in a major design experience based on the knowledge and skills acquired in earlier coursework and incorporating engineering standards and realistic constraints that include most of the following considerations: economic; environmental; sustainability; manufacturability; ethical; health and safety; social; and political. The professional component must include:

(a) one year of a combination of college level mathematics and basic sciences (some with experimental experience) appropriate to the discipline;

(b) one and one-half years of engineering topics, to include engineering sciences and engineering design appropriate to the student's field of study; and

(c) a general education component that complements the technical content of the curriculum and is consistent with the program and institution objectives.

(continued on next page)

ABET Engineering Criteria 2000 *(continued)*

Criterion 5: Faculty. The faculty is the heart of any educational program. The faculty must be of sufficient number, and must have the competencies to cover all of the curricular areas of the program. There must be sufficient faculty to accommodate adequate levels of student-faculty interaction, student advising and counseling, university service activities, professional development, and interactions with industrial and professional practitioners, as well as employers of students.

The faculty must have sufficient qualifications and must ensure the proper guidance of the program and its evaluation and development. The overall competence of the faculty may be judged by such factors as education, diversity of backgrounds, engineering experience, teaching experience, ability to communicate, enthusiasm for developing more effective programs, level of scholarship, participation in professional societies, and registration as Professional Engineers.

Criterion 6: Facilities. Classrooms, laboratories, and associated equipment must be adequate to accomplish the program objectives and provide an atmosphere conducive to learning. Appropriate facilities must be available to foster faculty-student interaction and to create a climate that encourages professional development and professional activities. Programs must provide opportunities for students to learn the use of modern engineering tools. Computing and information infrastructures must be in place to support the

---

An institution seeking accreditation of an engineering program must demonstrate clearly that the program meets the criteria in the box above.

■ Continuing Professional Development

The codes of ethics of all of the engineering associations mentioned here place ethical obligations upon all professional engineers to maintain competence in areas in which they practice. The sort of reflective approach to professionalism advocated by Schon (1983) is also important in continuing professional development. As he points out, "the situations of practice are not problems to be solved but problematic situations characterized by uncertainty, disorder and indeterminacy" (Schon 1983, p. 50).

Although we may try to convince ourselves that we operate professionally on what has been described as the "hard high ground of technical rationality," most of our problems are actually located in the surrounding swamp of everyday reality. The challenge is to learn from

scholarly activities of the students and faculty and the educational objectives of the institution.

Criterion 7: Institutional Support and Financial Resources. Institutional support, financial resources, and constructive leadership must be adequate to assure the quality and continuity of the engineering program. Resources must be sufficient to attract, retain, and provide for the continued professional development of a well-qualified faculty.

Resources also must be sufficient to acquire, maintain, and operate facilities and equipment appropriate for the engineering program. In addition, support personnel and institutional services must be adequate to meet program needs.

Criterion 8: Program Criteria. Each program must satisfy applicable program criteria (if any). Program criteria provide the specificity needed for interpretation of the basic level criteria as applicable to a given discipline. Requirements stipulated in the program criteria are limited to the areas of curricular topics and faculty qualifications. If a program, by virtue of its title, becomes subject to two or more sets of program criteria, then that program must satisfy each set of program criteria; however, overlapping requirements need to be satisfied only once.

Source: (Reprinted with permission from the Accreditation Board for Engineering and Technology, Inc. (© ABET 1998))

our experience how to rescue the most appropriate problems from the swamp, and how to deal with them once we do.

We can draw on systems approaches to come at this issue in another way. Kline (1995) points out that once real systems involve human beings, they are at such a high level of complexity that it is no longer possible to produce anything like complete systems representations for them. The approximations and simplifications involved require us to draw on data from our own and other's experience to make sense of the systems.

Schon argues that deliberate, structured reflection on our experience is the key to the sort of learning required here. Keeping a journal of intentions, events, and emotional responses to both provides the essential raw material. It can help ensure that we learn from our experience and move forward, rather than simply repeating the same experiences over and over again.

The development of managerial skills, especially in communication, is particularly important for professional engineers. Senior professionals have a significant impact on the attitudes of the people they supervise, and the responsibility to assist and encourage their development is explicitly stated in many codes of ethics.

The International Agreement on Undergraduate Course Accreditation, or just the Washington Accord, provides for the mutual recognition by participating organizations of their respective engineering course accreditation processes, and hence of the degrees that are seen as being acceptable academic preparation for entry to engineering practice, and whose quality is assured through these processes. The agreement entitled "Recognition of Equivalency of Engineering Education Courses/Programs Leading to an Engineering Degree" was initiated in 1988. The final agreement was implemented in 1991. It is an agreement between the bodies recognized in each signatory country as the national organization independent of government responsible for the accreditation of undergraduate engineering courses. The six foundation members were:

1. The Accreditation Board for Engineering and Technology, United States.
2. The Canadian Council of Professional Engineers.
3. The Engineering Council, United Kingdom.
4. The Institution of Engineers, Australia.
5. The Institution of Engineers of Ireland.
6. The Institution of Professional Engineers, New Zealand.

The Engineering Council of South Africa and the Hong Kong Institution of Engineers have also been admitted.

CONCLUSION

Drawing the threads together in a final chapter of a book that has ranged so widely is no easy task. Attempting to grasp the portents of a new millennium, as well, positively invites grand generalizations. Our world appears to be approaching a crisis in its global economy, global politics, and global ecology. The accumulation of unresolved problems, international political and military antagonism, and a (perhaps fortunate) lack of central authority are contributing factors to this sense of crisis. Our planet continues to be plundered, leading some to feel that a collapse of the ecosystem is imminent. Decisions are made on the basis of short-term economics and reflect our desire to exploit the environment.

How much credence should we give any "doomsday" scenario? Where do engineers fit into this picture? By some, they are seen as part of the problem. While recognizing and acknowledging their profession's contributions to both problems and advances, engineers can and should see themselves as part of the solution.

Typical engineering codes of ethics assert the need for engineers to be sensitive to ecological concerns (see Tenet 1 of the IEEE Code of Ethics above). The Accreditation Board of Engineering and Technology (ABET), which accredits American engineering degree programs, requires engineering curricula in the United States to include material that strengthens ethical and environmental awareness. At a time when many engineering careers have become international in nature, engineers need to be all the more concerned with these issues. There are breathtaking opportunities for the engineering profession to take a lead, to apply technology creatively to achieve sustainable management of the planet. If the hopes and dreams of humanity are to be realized in the process, engineers surely have an important role to play.

As we have been considering the challenges of professional practice throughout this book, we have seen how engineering has shaped the ways in which our own and other societies have developed. Leadership in engineering practice has gone to countries that have anticipated and recognized critical new fields of technology and have mobilized their resources and developed their capabilities to excel in them. The implications are clear. Excellence requires economic and other policies that recognize the essential role of sustainable wealth generation in the economy. Policies must support rather than hinder innovation and invention. An engineering profession that understands the broad political as well as the technical implications of its activities can play an important part in this process.

Engineers need to understand how the society in which they work is structured and how the processes of choice and control of technology can be handled, so that by their activity they can create meaningful, satisfying, and sustainable jobs for themselves and others. The education of engineers needs to prepare them for involvement in the social, political, and cultural lives of their communities. Consultation with likely users, and with those potentially affected by the implementation of the technology engineers are creating, is important for effective and ethically sound design. A sense of vocation and an insightful and principled approach to their work are essential. As engineers, we have opportunities to develop technologies that maximize control by the users, technologies that reinforce democratic rather than centralist and authoritarian tendencies in our society.

Historically, engineering practice has been driven mainly by functionality and cost-effectiveness, with cost narrowly defined in traditional economic terms. It has been suggested in this book that engineers will also increasingly be expected to meet social needs in the context of sustainable development. The actual definition of social needs will not necessarily be a part of their work, but they will need to appreciate how needs have been determined, and be able to adapt their practice to suit. In this new context it is even more important for engineers to be politically astute and committed to developing options that help people to be more in control of their own lives. This is the real meaning and purpose of wealth creation.

Technology has offered solutions to problems of production. However, it often appears to fail to deal with, or even to exacerbate, problems of distribution, problems that are widely and correctly regarded as political questions. (It would be more accurate to say that it is not "technology" per se that has failed in this regard, but rather that the societies and individual firms that have specified, developed, diffused, regulated, and used the technologies may be said to have failed.) The characteristics of modern technology have been discussed at length. It tends to be capital-intensive, research- and development-intensive, management-intensive, and labor-exclusive. It is also increasingly dependent on inputs of sophisticated materials. There are compelling technical reasons why this should be so. By deploying the higher efficiencies inherent in centralized provision of power, water, and mass transportation, greater economic benefit can flow on to larger numbers of people. Indeed, if we fail to use the most efficient technologies, the cost mechanism tends to restrict their use to an élite.

Industrial societies also need to exploit softer technologies based increasingly on renewable energy flows. Large-scale centralized organizations and facilities cannot possibly reach everyone in our society. There will always be regions and groups of people potentially disadvantaged by their lack of access to the benefits of centralized systems, such as economical rail and air transportation. Systems need to be diverse, and include networks of small local energy sources matched and scaled to local end-use needs. This approach will support systems that are both flexible and accessible to user control.

A truly appropriate technology matches the scale and nature of provision to the need while optimizing users' economic benefit and control. These characteristics can help reduce adverse environmental impacts. Engineers need to take a lead in developing technologies and products that will allow us to move towards more stable and equitable resource distribution and use. Indeed, a principal obligation of the professional engineer working in any area of engineering is to innovate—to lead society from inadequate existing technologies to more satisfactory ones. A prime requirement of the engineer is leadership. Engineering needs to have direction and purpose.

Because we do not have a developed philosophy of engineering, or even of technology, we are not even close to a solution to what could currently be the central ethical question for the human race: the problem of the control of technology. The lack of an adequate philosophy of engineering is a major constraint on the engineering profession's development.

A philosophy of engineering is needed that will deal adequately with the character and significance of engineering activity, the nature of engineering knowledge and why it is learned, and the roles of science, art, aesthetics, creativity, and judgment in engineering. Such a philosophy will provide a better basis for engineering education and better understanding of the potential roles of engineers and engineering in our society. These are key elements in making sure that advancing technology contributes to social progress. The development of such a philosophy is one of the most significant challenges for engineering and engineering education.

We have shown throughout this book that technology is neither neutral nor value-free. The types and levels of technology developed and adopted can increase or decrease our ability to meet human needs sustainably. Technology's potential to be molded towards either democratic or centralized control is an important issue that can have a major influence on its social and environmental impacts. It is no longer acceptable for the engineering profession or the nation to ignore the social and political effects of the technologies we use, or the longer-term economic and resource considerations. We have two or three decades at most to move to an economy in which our consumption of nonrenewable resources is stable and sustainable. The engineering profession has a vital role to play in ensuring that these changes are of maximum benefit to all and that they involve gradual transitions rather than sudden catastrophic disruption. The authors of this book are optimistic that engineers around the globe can and will rise to this challenge.

Discussion Questions

1. What are the characteristics of a modern profession? How have the characteristics of a profession changed since the industrial revolution? To what extent does engineering exhibit the characteristics of a profession?

2. *Jurisprudence* is the study of the science and philosophy of the law. Invent a comparable term for engineering. How might the existence and widespread use of your term encourage consideration of broader issues within the profession?

3. What do you understand by engineering professionalism? Discuss the view that because most engineers are employees, they cannot be fully professional.

4. How might professional engineers in the United States be effectively distinguished in the public eye from paraprofessionals calling themselves engineers? What effects (if any) might this have on the economic and other position of professional engineers in the United States? What effects might this clearer distinction have on paraprofessionals?

5. To what extent are engineers' views and practices determined by their social environment? What implications does this have for the question of engineers' responsibility to society? Does it make it meaningless?

6. How important are codes of ethics for practicing professional engineers? What are some of their limitations?

7. What place do professional ethics have in a technological society? Is it reasonable for a professional body to expect its members to abide by a code of ethics? If so, what sanctions should be available to enforce this?

8. There is a widely held view that our Western technological capability has outstripped: (1) our development of wisdom; (2) our ability to control technology; and (3) our ability to direct technology to meet social needs. Discuss this view, illustrating your argument with examples from the area of engineering in which you are involved. To what extent does this view confirm concerns expressed by Johnston et al. (1996) with the adequacy of the current discourse of engineering?

9. Illich and others argue in *Disabling Professions* (1977, p. 9) that the professions "have gained an ascendancy over our social aspirations and behavior ... and that we have become a virtually passive clientele: dependent, cajoled and harassed, economically deprived and physically and mentally damaged by the very agents whose *raison d'être* it is to help." Their complaint is focused mainly on the "health" and legal professions, but also applies to engineering. Spell out their basic concerns with the engineering profession and then discuss these concerns.

10. What roles should technical experts such as traffic engineers play in public policy decision-making issues, such as airport location or freeway development? What does objectivity mean in this context? What sorts of processes are needed to earn public credibility and trust in the integrity of the decision-making process?

11. What is whistleblowing? Suggest a system for facilitating whistleblowing in an organization where you have worked. How would your system deal with differentiating between real problems and artificial issues invented by disgruntled and/or spiteful staff? How would your system protect the whistleblower in a situation such as the one described in Case Study 11.3?

12. What impact would a hippocratic oath such as Thring proposed for engineering have on engineering practice?

13. Some of the material in this book is considered by the authors to be essential background for competent engineering design practice. Which of the issues raised do you feel are most important for design? Explain and justify your answers.

14. Explain how you will use reflection in your own professional practice. How would you explain to a fellow engineer its relevance in continuing professional development?

Further Reading

Andrews, G. C., Kemper, J. D. 1992, *Canadian Professional Engineering Practice & Ethics*, Saunders College (Holt, Rineheart & Winston), Canada.

Ball, N. R. 1987, *Mind, Heart and Vision: Professional Engineering in Canada 1887 to 1987*, National Museum of Science and Technology, Ottawa. This is a history of the engineering profession in Canada.

Beder, S. 1993, "Engineers, Ethics and Etiquette," *New Scientist*, 25 September, pp. 36-41. A civil engineer teaching and researching in Science and Technology Studies, Beder writes insightfully on ethical issues in engineering. Much of her work is available on her web site at: <http://www.uow.edu.au/arts/sts/sbeder/ghindex.html>

Harris, C. E., Pritchard, M. S., Rabins, M. J. 1995, *Engineering Ethics: Concepts and Cases*, Wadsworth, Belmont, CA. This book includes a multitude of realistic cases.

Harris, C. E., Pritchard, M. S., Rabins, M. J., Harris, C. E., Jr. 1997, *Practicing Ethical Engineering*, IEEE, Washington, D.C. This is a compact and accessible introduction to engineering ethics.

Johnson, D. G. 1990, *Ethical Issues in Engineering*, Prentice Hall, NJ. This is an important reader.

Johnson, T. J. 1972, *Professions and Power*, Macmillan, London. This is a short but effective critique of the theoretical weakness of much previous work on the sociology of the professions.

References

Abercrombie, N., Urry, J. 1983, *Capital, Labor and the Middle Classes*, Allen & Unwin, London.

ABET (Accreditation Board for Engineering and Technology) 1998, World Wide Web site at: <http://www.abet.org/>

Andrews, G. C., Kemper, J. D. 1992, *Canadian Professional Engineering Practice & Ethics*, Saunders College (Holt, Rinehaart & Winston), Canada.

APEA (Association of Professional Engineers Australia) 1989, *Legal Liability Advice for Professional Engineers*, West Melbourne.

Ball, N. R. 1987, *Mind, Heart and Vision: Professional Engineering in Canada 1887 to 1987*, National Museum of Science and Technology, Ottawa.

Beder, S. 1993, "Engineers, Ethics and Etiquette," *New Scientist*, 25 September, pp. 36-41.

Bella, D. A., Mosher, C. D., Calvo, S. N. 1988a, Technocracy and Trust: Nuclear Waste Controversy, *Journal of Professional Issues in Engineering*, 114 (1), pp. 27-39.

————, 1988b, Establishing Trust: Nuclear Waste Disposal, *Journal of Professional Issues in Engineering*, 114 (1), pp. 40-50.

Carr-Saunders, A. M., Wilson, P. A. 1933, *The Professions*, Oxford University Press, Oxford, p. 3.

Collingridge, D. 1980, *The Social Control of Technology*, Pinter, London.

Elliott, P. 1972, *The Sociology of the Professions*, Macmillan, London.

Flew, A., Ed. 1984, *A Dictionary of Philosophy*, Pan, London.

Foucault, M. 1970, The Discourse on Language, In: Foucault, M., *The Order of Things*, Tavistock, London.

Gardiner, P., Rothwell, R. 1985, *Innovation: A Study of the Problems and Benefits of Product Innovation*, The Design Council, London.

Harris, C. E., Pritchard, M. S., Rabins, M. J. 1995, *Engineering Ethics: Concepts and Cases*, Wadsworth, Belmont, CA.

Harris, C. E., Pritchard, M. S., Rabins, M. J., Harris, C. E., Jr. 1997, *Practicing Ethical Engineering*, IEEE, Washington, D.C.

Illich, I. et al. 1977, *Disabling Professions*, Marion Boyars, London.

Jackson, M. W. 1992, *Ethics Almanac III*, June, Public Affairs Centre, University of Sydney.

Johnson, T. J. 1972, *Professions and Power*, Macmillan, London.

Johnston, S. F. 1986, Nuclear Fuel, an Incomplete Cycle—Who Pays for the Power? *Engineers Australia*, 16 May, IEAust, Sydney, pp. 43–48.

————, Lee, A., McGregor, H. 1996. Engineering as Captive Discourse, *Society for Philosophy and Technology Quarterly Electronic Journal, V. 1*. On the World Wide Web at: <http://scholar.lib.vt.edu/ejournals/SPT/spt.html>

Kline, S. J. 1995, *Conceptual Foundations for Multidisciplinary Thinking*, Stanford University Press, Stanford, CA.

Lloyd, B. E. 1991, *Engineering in Australia: A Profession in Transition*, Macmillan, Melbourne.

Martin, B. 1982, The Naked Experts, *The Ecologist*, 12 (4), pp. 149–157.

Martin, M. W., Schinzinger, R. 1996, *Ethics in Engineering, 3rd Ed.*, McGraw-Hill, NY.

Martin, M. W. 1981, Why Should Engineering Ethics Be Taught? *Engineering Education*, January, pp. 275–278.

Nelkin, D. 1975, The Political Impact of Technical Expertise, *Social Studies of Science*, 5, pp. 35-54.

Rosenbrock, H. 1990, *Machines with a Purpose*, Oxford University Press, Oxford (by permission of Oxford University Press).

Rothblatt, S. 1968, *The Revolution Among the Dons: Cambridge and Society in Victorian England*, London, quoted in Elliott 1972.

Schon, D. A. 1983, *The Reflective Practitioner: How Professionals Think in Action*, Basic Books, NY.

Thring, M. W. 1970, Our Responsibility to Mankind, *The Chartered Mechanical Engineer*, September, pp. 348–351, 359. Reprinted by permission of the Council of the Institution of Mechanical Engineers.

Winner, L. 1990, Engineering Ethics and Political Imagination, In: Durbin, P., Ed., *Broad and Narrow Interpretations of Philosophy of Technology*, Kluwer, Dordrecht, pp. 53–64.

THE AIRCRAFT BRAKE SCANDAL

The Cast

Searle Lawson was a young design engineer at the B. F. Goodrich Wheel and Brake Plant in Troy, Ohio. He held a bachelor's degree in aeronautical and astronautical engineering and a certificate in aircraft design technology.

Kermit Vandivier was a Goodrich technical writer who worked with Lawson.

John Warren was the Goodrich engineer who designed the original A7D disc brake. In his early thirties, Warren had an excellent record in aircraft brake design, including brakes for the Boeing 727. After he completed the A7D brake design, Warren handed over the project to Lawson, whose task it was to test brake lining temperatures in Warren's design prior to building the prototype.

Robert L. Sink was the A7D project manager.

Russell Van Horn, manager of the Aircraft Wheel and Brake Design section, was Sink's and Lawson's immediate supervisor, with ultimate responsibility for the A7D proposal and project.

(Bud) Sunderman was chief engineer for the B. F. Goodrich Wheel and Brake Plant. *Senator William Proxmire (Wisconsin-Democrat)* had a reputation for tough stands on government waste.

The Story

In 1967 the U.S. Air Force awarded the prime contract for a new Air Force A7D light attack aircraft to the Ling-Temco Vought (LTV) Co. of Dallas, Texas, which then subcontracted the brake assembly contract to B. F. Goodrich Wheel and Brake Plant. So, in June 1967, Goodrich received a contract to produce 202 four-rotor disc brake assemblies for the new aircraft. Before the Air Force could accept the brake via LTV, B. F. Goodrich had to present a report showing that the brake passed specified qualifying tests. The contract allowed almost a full year for design and testing, including testing the brake under flight conditions.

At Goodrich, after much prototype testing during 1967, Lawson realized in December that Warren's brake was too small and could not withstand the demands of a normal aircraft landing. Lawson took his concerns to Warren, who had moved on to other projects. He assured Lawson there was no flaw in the brake design. Lawson continued testing the brake, but the results were always the same: the Goodrich brake could not meet the Air Force specification requirements.

Lawson then talked to his immediate supervisor, Sink, the project manager. Sink told Lawson that he had already assured LTV several times that Warren's brake design was a success. Sink put pressure on Lawson by pointing out to him that if the A7D design did prove faulty, it would be Sink who would be held accountable by Goodrich and LTV. Still, Lawson could not devise a way to make Warren's brake pass the required tests, even after using special cooling fans in the testing laboratory to help the brake design operate at lower temperatures.

In April 1968, after nearly a year of attempts to qualify the Warren brake, technical writer Kermit Vandivier was brought into the project to bring together Lawson's test data and write a report that would indicate that the tests were successful. Of course, Vandivier found many discrepancies between the military specifications and the qualification tests. Vandivier questioned whether he should compose a report that was so at odds with the military specifications. When he approached Lawson about the qualification tests, Lawson indicated that his superiors, Sink and Van Horn, had told him to do whatever it took to pass the brake design.

By the end of May 1968, Vandivier and Lawson were discussing between themselves the implications of what they were doing. In June, after flight tests at Edwards Air Force Base in California indicated the inadequacy of the brake design, Vandivier contacted an attorney. Vandivier told Lawson that his attorney had advised him that both he and Lawson were guilty of conspiracy to defraud. Fearing conspiracy charges, Lawson met with the attorney and then talked to the FBI. By this time, Vandivier, because of his earlier outspokenness, had been removed from the day-to-day operations of the project. He no longer openly expressed his concerns, but continued to relay information to the FBI.

As with many whistleblowers, Vandivier was frightened. He felt isolated, even from Lawson, who seemed not to want to be part of the whistleblowing. Most other employees were busy on other projects and, like Lawson, preferred to not get involved. Vandivier had a family of seven to feed. His situation looked grim.

The Dramatic Finale

In October, 1968, Vandivier submitted a letter of resignation, which included accusations against Goodrich. Sunderman, chief engineer for the Goodrich Plant, called him to his office and dismissed him immediately for disloyalty. Vandivier went to work for a local newspaper, and told his editor there about his experience at Goodrich. The editor passed his story to associates in Washington.

The story reached Senator William Proxmire. In May, Proxmire asked the Government Accounting Office (GAO) to review the brake qualification testing performed by the Goodrich plant in Troy. The GAO reported in July. On receiving the report, and without consulting Goodrich, in August, 1969, Senator Proxmire described Vandivier's allegations and the GAO investigation to the full Senate. Senator Proxmire then chaired a four-hour Congressional hearing before the Senate Subcommittee on Economy in Government.

When Proxmire questioned Lawson about the falsification of the report he prepared with Vandivier, Lawson admitted being naive about both government contracting and aircraft brake testing. He felt that had he known typical engineering practices better, he could have made the A7D brake work, although, for example, it might only have stopped the aircraft 300 times, instead of the 400 times specified. Vandivier admitted not having even a high school diploma, casting doubt on his qualifications to judge the adequacy of technical reports.

Evidence presented at the hearing indicated that brake-testing procedures were lax because, over time, faith in proven technologies, as well as a professional trust, had developed between Goodrich and its clients. Those in the industry, from government procurement officials through subcontractors, had become accustomed to and comfortable with modifying reports so that designs would appear to meet the rigorous requirements laid out by the military. It seemed that improper testing was common, even expected by government

(continued on next page)

officials, and was effectively standard industry practice.

Despite the GAO investigation and the Congressional hearing, no action was taken against Goodrich. No one at Goodrich lost his job. Warren and Van Horn continued working for Goodrich, their careers unaffected. Sink's career, however, seems to have ended here, although he was not fired. Although Goodrich wanted him to stay, Lawson left to go to work for LTV. Vandivier stayed with his writing career. Business in the aircraft brake industry went on as usual and little changed in terms of governmental contracting procedures.

References

Fielder, J. H. 1988, Give Goodrich A Break, *Business and Professional Ethics Journal*, Vol. 7, No. 1, Spring.

Texas A&M University, Ethics Web site, <http://ethics.tamu.edu/ethics/goodrich/goodric1.htm>

Thinking Like An Engineer: The Place of a Code of Ethics in the Practice of a Profession, *Philosophy & Public Affairs*, Vol. 20, No. 2, Spring 1991, pp. 150–167.

Vandivier, K. 1972, Why Should My Conscience Bother Me? In: Heilbroner, R. L., Ed., *In The Name Of Profit*, Doubleday, Garden City NY, pp. 3–31.

Index

Association of Consulting Engineers Council (ACEC), 586-87
AT&T, *see* American Telephone and Telegraph Company
Atanasoff, John V., 72, 74-75
Australia, 301, 356
 bridges, 330-32, 392-95
 heavy engineering industry, 270-73
 Institution of Engineers, Australia (IEAust), 562, 592
automobile
 development, 56-57
 electric self-starter, 52-53
 fuel efficiency, 60, 508-509
 social and economic impacts, 51
 see also highways
automobile industry *see* motor vehicle industry
aviation, 66-70
 early flight, 66-68
 sound barrier, 68-69, 315
 gas turbine, 69, 282
 see also aerospace industry; aircraft brake scandal; de Havilland Comet
Ayer, A.J., 522
Ayrton, Hertha, 148-49
Ayrton, William E., 42

B

Babbage, Charles, 71, 228
Bacon, Francis, 17, 517, 539
Bell Telephone Laboratories, 79, 307, 312, 315
benchmarking, 337
Benz, Karl, 52
Berkeley, George, 517
Bessemer, Henry, 36
BFGoodrich®, 598-600
Bhaskar, 519, 538
bicycle, 33, 48, 439

bioengineering/biomedical engineering, 87-89, 129
biomass, 474-76
biosphere, 454
 see also ecosphere
Bishop, Arthur, 367
Blacks, *see* African Americans; minorities
Boulton, Matthew, 25
Boyle, Robert, 18
Bradfield, J. J. C., 290, 330-32
bridges, 9, 28-29, 36, 392-95
 Sydney Harbour, 330-32
Brindley, James, 24,
Britain, *see* Great Britain
Brunel, Isambard Kingdom, 27, 29,
Brunel, Marc Isambard, 29
bubonic plague (Black Death), 14
Burroughs, William, 71
Bush, Vannevar, 72, 90, 285, 534
business enterprise, 221-25
 culture, 223-25
 horizontal and vertical integration, 222, 354-55

C

CAD, CADD, CAE, *see* computer-aided engineering
Cambridge University, 572
Camus, Albert, 519
Canada, 114, 301, 566-67, 577
 see also NAFTA
Canadian Council of Professional Engineers (CCPE), 592
canals, 24, 27-28, 39
 see also Panama Canal
capitalism, 165-66, 334, 410-413
 key concepts and issues, 184-85
 and scientific revolution, 17-22
 and social classes, 178-186
 see also chapters on economics, politics

Daimler, Gottlieb, 52
Darby, Abraham, 23, 98-101
de Forest, Lee, 71, 314
de Havilland Comet, 69, 84
de Lesseps, Ferdinand 36, 40
de Rochas, Beau, 52
defense *see* aerospace; military
Delphi Packard, *see* General
    Motors
demand pull, 276, 286, 304, 306, 309
    see also innovation, technology
democracy,
    and industry, 224-25, 235
    popular vote, 167
    public consultation, 200,
Denmark, 480
Department of Defense, 177
Department of Energy, 177
Derrida, Jacques, 524
Descartes, René, 17-18, 517
design, 276, 288-96, 310-11,
    330-32, 353
    key concepts and issues, 291,
    293
    central to engineering practice,
    289, 534, 543
    choice of criteria, 552
    contractor's responsibility for,
    331-32
    essential interactions and driving
    forces, 307-309
    functional specification, 85, 293,
    326
    heuristics, 291
    importance of good, 295-96
    models of, 292-93
    process, 289-93
    see also intellectual property, ISO
development, 397-452
    key concepts and issues, 399, 420
    aid, 405
    broader definition of, 415-21
    data for selected countries, 400-401

economic definition of, 398-402
    see also appropriate technology;
    colonization; green revolution;
    underdevelopment
    see also innovation; research and
    development; sustainable develop-
    ment
Diesel, Rudolph, 53
discourse, 524
    of engineering, 543, 574
    of science and business, 524, 543
diversity,
    in design teams, 295, 353
    in engineering management, 348-52
documentation, 346-47
    of quality, 288, 362-64
    see also ISO
Dostoevsky, Fyodor, 519
Dyer, Henry, 42

E

École Nationale des Ponts et
    Chaussées, 38, 226
ecologically sustainable development,
    455, 492-98
    see also development; sustainabil-
    ity; sustainable development;
    United Nations
ecology, 481-92
    key concepts and issues, 481
    global character, 592
    waste management, 476,
    491-92
    see also CFCs; ecologically
    sustainable development;
    greenhouse effect; ozone layer;
economic development, 19-22, 214
    see also development
economics, 211-73,
    key concepts and issues,
    212-13, 230, 240-41
    balance of payments, 253-58

Hong Kong Institution of
Engineers, 592
Hooke, Robert, 19, 552-53
households, 107-109
human-centered technology,
558-59
Hume, David, 517-18, 522
hydroelectric power,
475-76
see also Solomon Islands micro-
hydro-electricity plant; Tennessee
Valley Authority; Three Gorges
Project
hydrogen fuel, 506-509

I

IBM (International Business
Machines), 71-73, 307,
440, 442-44
idealism, 514, 518
ILO (International Labor Organization), 146
India, 9, 197, 199, 217, 412, 426
Arabic numerals, 10-11
see also colonization
indigenous populations,
411-12
Navajo technology 416-17
See also Alaska pipeline
Indonesia, 412-13, 488-89
industrial/manufacturing/production
engineering, 129, 580
industrial and manufacturing
industries *see* manufacturing;
production
industrial relations, 56-58,
145-47, 224-25, 394-95
Americans with Disabilities Act,
146
Civil Rights Act, 144, 146

see also harassment; labor; legal
issues; management;
production
Industrial Revolution, 22-28, 76, 143,
185, 525
industrialization, 31-35
industry economics, 222
industry policy, 268
local content criteria, 270-73
information technology, 338,
445
see also computers
infrastructure, 58-65
key concepts and issues, 64-65
see also bridges; electricity; highways; railroads; transport; etc.
innovation, 86-87, 213, 228,
275-332, 435
key concepts and issues, 276, 283
and creativity, 275
definition, 284
effects of trends and cycles,
302-307
government incentives for, 318
and legislation, 321-22
models of, 285-87
organizational context, 309-311
political context, 317-18
products, processes and systems
283, 288
some major innovations, 305
social context, 309-318
and social progress, 553
ten steps to, 285
see also demand pull; design;
invention; Schumpeter, Joseph;
technology push
Institute of Electrical and Electronics
Engineers (IEEE), 561-63, 585
Institution of Civil Engineers (ICivE)
26, 28-9
Institution of Electrical Engineers
(IEE), 36

Institution of Engineers, Australia
(IEAust), 562, 592
Institution of Engineers of Ireland, 592
Institution of Mechanical Engineers
(IMechE), 29, 36
Institution of Professional Engineers
New Zealand (IPENZ), 26, 533,
592
intellectual property, 283, 364-69
copyright, 367
copyright systems, 279-83, 285,
314-15
registered designs, 367
reverse engineering, 367
security procedures, 368-69
patent, trademark, design and
Intermediate Technology Development Group (ITDG), 422-23
internal combustion engine, 48, 52-53,
506-509
international engineering *see* globalization
Interstate Highway System, 60, 202-
204, 520
invention, inventors, 277-84
key concepts and issues, 283
see also innovation
Iowa State University, 74-75, 148
Iriri, *see* Solomon Islands
Iron Ring Ceremony 566-67
Islam, 10
ISO (International Standards Organisation), 322
ISO 9000, 341, 363, 421
ISO 14000, 363
Italy, 301

J

Japan, 76-77, 86, 114, 186, 218, 225,
230, 254, 260-61, 271, 301, 341,
383
consumer products, 73

employment, 58
government-business relationship, 317
industrialization, 35, 413
universities, 39, 42
see also colonization; Toyota
Jaspers, Karl, 519
Johns Hopkins University, 88
Joule, J. P., 461
Judah, Theodore D., 32

K

kaizen (continuous improvement), 383
Kant, Immanuel, 517-18
Kelvin, Lord, (William
Thomson), 461
Kettering, Charles F., 53
Keynes, John Maynard, 185, 233, 238-
41
Keynesian theory/ Keynesianism,
239-42, 244, 249-51
key concepts, 240-41
Kierkegaard, Søren, 519, 537
knowledge, 18, 516, 532, 539
workers, 199, 339, 387
Kondratieff cycle *see* cycles
Korea, 270-71, 301, 426
Kuhn, Thomas, 277, 523, 525-28, 530,
538
Kyoto conference, *see* United Nations
Organization

L

labor, 56-58, 146, 442
labor unions, 58, 362, 382
United Auto Workers (UAW), 57,
146, 384
see also industrial relations; social
power
Latin America, 373
see also Americas; Mexico;
Panama Canal

Toyota Production System, 380, 382-83
Transparency International, 373, 406-407, 436
transport industry, 580
    fuels, *see* energy
    investments, 62-63
    transport policy, 182
    transportation, 24, 319
    vehicle imports to Papua New Guinea, 418-19
    see also aerospace; canals; energy; highways; infrastructure; motor vehicle industry; railroads
Trevithick, Richard, 25

U

UAW, *see* United Auto Workers
UN, *see* United Nations
uncertainty, *see* risk
underdevelopment, 410-19
    historical origins, 410-14
    social implications, 402-10
    see also colonization; development
unemployment, 138-39
unions, *see* labor; labor unions
United Auto Workers, 57, 146, 384
United Kingdom, *see* Great Britain
United Nations Organization (UN), 194
    Conference on Environment and Development (UNCED, Rio de Janeiro, 1992), 195, 404, 493, 495
    Conference on Trade and Development (UNCTAD), 194, 196
    Development Program (UNDP), 398-403, 415-21
    Environment Program (UNEP), 194, 490
    Kyoto conference, 374, 406, 495
    Special Session of General Assembly (UNGASS), 404-406

United States of America (U.S.), 60, 76-77, 114, 121, 301-302
    American Association of Engineering Societies (AAES), 584
    economic tradition, 236-38
    energy use and reserves, 458-59, 464-73
    engineering profession, 579-80, 584-87
    Federal Reserve System, 241, 246
    and global economies, 253-60
    households and families, 107-109
    industrial challenge, 29-35
    Monroe Doctrine, 413-14
    National Bureau of Economic Research, 237, 242
    National Society of Professional Engineers (NSPE), 576-77, 584, 586
    political structure, 170
    political parties, 171
    salaries, 118
    see also ABET; individual engineering associations
universities, 38-42, 301, 572-74
    early western, 15, 572
    see also individual universities and colleges
University of Berlin, 39
University of California, 125, 148, 350
University of Kentucky, 208
University of Pennsylvania, 72, 75, 236
University of the Pacific (UOP), 125-26
University of Texas, 149
University of Toronto, 567
University of Wisconsin, 148, 150
upper class, 110-112, 183-84, 199
U. S. Army Corps of Engineers, 61, 178, 180-182, 262
U. S. Military Academy, West Point, 38, 180